Adobe® Creative Cloud®

ALL-IN-ONE

3rd Edition

by Jennifer Smith and Christopher Smith

Adobe® Creative Cloud® All-in-One For Dummies®, 3rd Edition

Published by: **John Wiley & Sons, Inc.,** 111 River Street, Hoboken, NJ 07030-5774, www.wiley.com

For general information on our other products and services, please contact our Customer Care Department within the U.S. at 877-762-2974, outside the U.S. at 317-572-3993, or fax 317-572-4002. For technical support, please visit https://hub.wiley.com/community/support/dummies.

Wiley publishes in a variety of print and electronic formats and by print-on-demand. Some material included with standard print versions of this book may not be included in e-books or in print-on-demand. If this book refers to media such as a CD or DVD that is not included in the version you purchased, you may download this material at http://booksupport.wiley.com. For more information about Wiley products, visit www.wiley.com.

Library of Congress Control Number: 2021932167

ISBN 978-1-119-72414-8 (pbk); ISBN 978-1-119-72416-2 (ebk); ISBN 978-1-119-72415-5 (ebk)

Manufactured in the United States of America

SKY10025518 03082021

Table of Contents

Introduction

Adobe software has always been highly respected by creative professionals. Adobe creates tools that allow you to produce designs for all things, from printed brochures and posters, to websites, applications, and advanced video productions. The Adobe Creative Cloud is the company's latest release of sophisticated and professional-level software that bundles many separate programs as a suite. Each program in the Creative Cloud works individually, or you can integrate the programs by using *Adobe Bridge,* an independent program that helps you control file management with thumbnails, metadata, and other organizational tools.

With Adobe Creative Cloud, a monthly membership offers you the entire collection of Adobe tools and more. Love print? Interested in websites and iPad apps? Ready to edit video? You can do it all. Plus, Creative Cloud members automatically get access to new products and exclusive updates as soon as they're released. And, with cloud storage and the ability to sync to any device, your files are always right where you need them. Creative Cloud is available for individuals or teams.

The first is Creative Services, which are hosted services that you can use in your production work, in the delivery of your content. This includes a font service like Typekit, which enables the use and delivery of a broad foundry of cloud fonts across all of your work. A second area is Digital Publishing, which enables publishing rich media to tablets via the cloud. And the third category is Business Catalyst, which supports designing and operating websites for small businesses, with prebuilt services for things like handling e-commerce, doing customer relationship management, and integrating with social networks.

The second pillar is Creative Community, which is the community of creative people (like you!) around the world and which enables you to connect more easily with other creative people — it's a place to share, to communicate, and to inspire each other with your work. The community is a critical part of our whole ecosystem, and it's a critical part of the cloud. At the center of this is the web presence of the Creative Cloud, which is `creative.adobe.com`. And one of the great things there is it will understand all the formats you're using in your creative work — PSD files, InDesign files, and Illustrator files. Where other cloud services might show you an icon describing the file type, this shows you the actual content, and you can interact with it in a context-sensitive way. It's a much deeper understanding of creative content.

And lastly, the third pillar is Creative Applications — and these are enabling you to create not only on personal computers, but also wherever you are with mobile devices, all connected through the Creative Cloud. This includes a whole new collection of Adobe touch apps to run on tablets and other mobile devices. In addition, membership also includes access to all Adobe creative desktop products you know and love, including Photoshop, Illustrator, Dreamweaver and Premiere and InDesign, Adobe XD, and more. You can download and install any of these applications you choose as part of your membership, and these are all connected to Creative Cloud via desktop sync. They also interact with the touch apps, and you can move files between desktop and touch as you're working.

Why Is It Important?

You can use the Adobe Creative Cloud programs to create a wide range of products, from illustrations, page layouts, websites, photographic compositions, video, and 3D images. Integrating the CC programs extends the possibilities for you as a designer. Don't worry about the programs being too difficult to figure out — just come up with your ideas and start creating!

About This Book

Adobe Creative Cloud All-in-One For Dummies is written in a thorough and fun way to show you the basic steps of how to use each program included in the Creative Cloud package. You find out how to use each program individually and how to work with the programs together, extending your projects even further. You find out just how easy it is to use the programs by following simple steps so that you can discover the power of the Adobe software. You'll be up and running in no time!

Here are some things you can do with this book:

- Create page layouts with text, drawings, and images in InDesign.
- Make illustrations such as logos, graphics, and data visualizations using drawing tools with Illustrator.
- Manipulate photographs by using filters and drawing or color correction tools with Photoshop.
- Create PDF (Portable Document Format) documents with Adobe Acrobat or other programs.

- Design web pages in Adobe XD that you can share with clients and team members as well as developers.
- Create animations such as GIFs on Photoshop, and User Experience interactions in Adobe XD.
- Create web images in various formats from Photoshop, Illustrator, and Adobe XD.

You discover the basics of how to create all these different kinds of things throughout the chapters in this book in fun, hands-on examples and clear explanations, getting you up to speed quickly!

Adobe Creative Cloud works for both Windows and the Macintosh. We cover both platforms in this book. When the keys you need to press or the menu choices you need to make differ between Windows and the Mac, we let you know by including instructions for both platforms. For example:

- Press the Alt (Windows) or Option (Mac) key.
- Choose Edit ⇨ Preferences ⇨ General (Windows) or InDesign ⇨ Preferences ⇨ General (Mac).

The programs in Creative Cloud often require you to press and hold down a key (or keys) on the keyboard and then click or drag with the mouse. For brevity's sake, we shorten this action by naming the key you need to hold down and adding a click or drag, like this:

- Shift-click to select multiple files.
- Move the object by Ctrl-dragging (Windows) or ⌘-dragging (Mac).

This book is pretty thick; you may wonder whether you have to read it from cover to cover. You don't have to read every page of this book to discover how to use the programs in the Creative Cloud. Luckily, you can choose bits and pieces that mean the most to you and will help you finish a project.

Icons supplement the material in each chapter with additional information that may interest or help you with your work. The Technical Stuff icons are helpful if you want to find out a bit more about technical aspects of using a program or your computer, but don't feel that you need to read these icon paragraphs if technicalities don't interest you.

Foolish Assumptions

You don't need to know much before picking up this book and getting started with Creative Cloud. All you have to know is how to use a computer in a very basic way. If you can turn on the computer and use a mouse, you're ready for this book. A bit of knowledge about basic computer operations and using software helps, but it isn't necessary. We show you how to open, save, create, and manipulate files using the CC programs so that you can start working with the programs quickly. The most important ingredient to have is your imagination and creativity — we show you how to get started with the rest.

Icons Used in This Book

What's a *For Dummies* book without icons pointing you in the direction of truly helpful information that's sure to speed you along your way? Here we briefly describe each icon we use in this book.

The Tip icon points out helpful information that's likely to make your job easier.

This icon marks a generally interesting and useful fact — something you may want to remember for later use.

The Warning icon highlights lurking danger. When we use this icon, we're telling you to pay attention and proceed with caution.

When you see this icon, you know that there's techie-type material nearby. If you're not feeling technical-minded, you can skip this information.

Beyond the Book

You can find a little more helpful information on `www.dummies.com`, where you can peruse this book's Cheat Sheet. To get this handy resource, go to the website and type *Adobe Creative Cloud All-in-One For Dummies Cheat Sheet* in the Search box.

The book you're holding right now contains six minibooks. A seventh, on the topic of Adobe Acrobat, is available in PDF form at www.dummies.com. This minibook describes Acrobat and explains how to create and edit PDF files; how to add text, images, and interactive elements to PDF files; how to use commenting and annotation tools; and how to secure PDF documents. You can access this minibook at www.dummies.com/go/adobeccaiofd3e.

Where to Go from Here

Adobe Creative Cloud All-in-One For Dummies is designed so that you can read a chapter or section out of order, depending on what subjects you're most interested in. Where you go from here is entirely up to you!

Book 1 is a great place to start reading if you've never used Adobe products or if you're new to design-based software. Discovering the common terminology, menus, and panels can be quite helpful for later chapters that use the terms and commands regularly!

You can find tips and tricks and more files for you to experiment with and investigate at www.agitraining.com/dummies.

1

Getting Started with the Creative Cloud Suite

Contents at a Glance

IN THIS CHAPTER

» What is included in the Creative Cloud?

» What applications are included in the Creative Cloud?

» What applications are covered in this book?

Chapter 1
The Creative Cloud

The Adobe Creative Cloud is a subscription-based service that includes a wide array of applications used in the communication, design, development, and marketing industries.

Over the last several versions, the Creative Cloud has increased in capabilities and application tools. Using the Creative Cloud you can build print, web, video, 3D, and application designs, in addition to building interactive designs that can be used for many purposes. In this book, the focus is mainly on design tools that are used for the creation of printed material, as well as on the designs for websites, mobile apps, and other interactive presentations.

To help you understand the breadth of capabilities, Table 1-1 lists the applications included in your Creative Cloud subscription at the time that this book was written. The bolded applications are covered in detail in this book.

TABLE 1-1 Creative Cloud Applications

Application	Description
Creative Cloud	Quick desktop access to Creative Cloud apps and services
Photoshop	Image editing, cropping, masking and more
Illustrator	Vector graphics for logos, data visualization and more
InDesign	Page design and layout for print and digital publishing
Acrobat Pro	Create, edit, and sign PDF documents and forms
Bridge	Centralize your creative assets; make it easy to find what you want
XD	Design and prototype user experiences for web and mobile
Lightroom	Digital photo processing and editing
Adobe Premiere Pro	Video production and editing
Premiere Rush	Create and share online videos anywhere
Fresco	Drawing and painting app built for the latest styllus and touch devices
After Effects	Cinematic visual effects and motion graphics
Dreamweaver	Design and develop modern, responsive websites
Animate	Flash Professional is now Adobe Animate CC. Interactive animations for multiple platforms.
Adobe Audition	Audio recording, mixing, and restoration
Character Animator	Animate your 2D characters in real time
Media Encoder	Quickly output video files for virtually any screen
InCopy	Collaborate with copywriters and editors
Prelude	Metadata ingest, logging, and rough cuts
Fuse CC	Make custom 3D characters for your Photoshop projects
Substance	3D painting and texturing
Spark	Easily create and share impactful visual stories — in minutes

Applications Covered in This Book

If you are a designer of print, web, or mobile content, you will need to know the core applications included in the Creative Cloud. This book is focused on the main design tools, Photoshop, InDesign, Illustrator, Acrobat DC, Adobe Bridge, and XD. You can read on for a brief introduction to each of these applications.

Crossing the Adobe Bridge

Adobe Bridge is truly an incredible application, especially within the Creative Cloud release, because the processing speed is greatly improved and new features are available.

Even though Adobe Bridge is part of the Creative Cloud, it does not install automatically with your other applications. The first time that you choose File ⇨ Browse in Bridge from your other Creative Cloud applications, you will be directed to the Creative Cloud app, where you can choose to download it on to your system. You can find out more about Adobe Bridge in Book 2.

Getting started with Photoshop CC

Photoshop is the industry standard software for web designers, video professionals, and photographers who need to manipulate bitmap images. Using Photoshop, you can manage and edit images by correcting color, editing photos by hand, and even combining several photos to create interesting and unique effects. Alternatively, you can use Photoshop as a painting program, where you can artistically create images and graphics.

Photoshop enables you to create complex text layouts by placing text along a path or within shapes. You can edit the text after it has been placed along a path; you can even edit the text in other programs, such as Illustrator CC, and join text and images into unique designs or page layouts.

Sharing images from Photoshop is easy to do. You can share multiple images in a PDF file, or upload images to an online photo service. You can even set up Photoshop to automatically export multiple assets for interactive apps or websites in one click.

It's hard to believe that Photoshop can be improved on, but Adobe has done it again in Adobe Photoshop CC. Book 3 shows you the diverse capabilities of Photoshop. From drawing and painting to image color correction, Photoshop has many uses for print and interactive design alike.

Introducing InDesign CC

InDesign is a diverse and feature-rich page layout program. With InDesign, you can create beautifully laid-out page designs. You can also execute complete control over your images and export them to interactive documents, such as Acrobat PDF. You can use InDesign to

- Use images, text, and even rich media to create unique layouts and designs.
- Import native files from Photoshop and Illustrator to help build rich layouts in InDesign that take advantage of transparency and blending modes.
- Export your work as an entire book, including chapters, sections, automatically numbered pages, and more.
- Create interactive PDF documents that can be used for website or application prototypes or wireframes.

InDesign caters to the layout professional, but it's easy enough for even beginners to use. You can import text from word processing programs (such as Microsoft Word, Notepad, or Adobe InCopy) as well as tables (say, from Microsoft Excel) into your documents and place them alongside existing artwork and images to create a layout. In a nutshell, importing, arranging, and exporting work are common processes when working with InDesign. Throughout the entire process, you have a large amount of control over your work, whether you're working on a simple one-page brochure or an entire book of 800-plus pages. Find out how you can take advantage of this feature-rich application in Book 4.

Using Illustrator CC

Adobe Illustrator is the industry's leading vector-based graphics software. Aimed at everyone from graphics professionals to interactive designers, Illustrator enables you to design layouts, logos for print, or vector-based images that can be imported into other programs, such as Photoshop, InDesign, and XD. Adobe also enables you to easily and quickly create files by saving Illustrator documents as templates (so that you can efficiently reuse designs) and using a predefined library and document size.

Illustrator also integrates with the other products in the Adobe Creative Cloud by enabling you to create PDF documents easily within Illustrator. In addition, you can use Illustrator files in Photoshop, InDesign, and the Adobe special effects program, After Effects.

Here are some of the things you can create and do in Illustrator:

- Create technical drawings (floor plans or architectural sketches, for example), logos, illustrations, posters, packaging, and web graphics.
- Create multiple screens for websites or mobile design.
- Align text along a path so that it bends in an interesting way.

- Lay out text into multicolumn brochures — text automatically flows from one column to the next.
- Create charts and graphs using graphing tools.
- Create gradients that can be imported and edited into other programs, such as Experience Design.
- Create documents quickly and easily using existing templates and included stock graphics in Illustrator.
- Save a drawing in almost any graphic format, including the Adobe PDF, PSD, EPS, TIFF, GIF, JPEG, and SVG formats.
- Save your Illustrator files for the web by using the Asset Export panel, or Export ⇨ Export As menu item.

Illustrator has new features for you to investigate, many of them integrated in the chapters in Book 5. Find out about new tools, including features to help you use patterns.

Working with Acrobat DC

Acrobat DC is aimed at both business and creative professionals and provides an incredibly useful way of sharing, securing, and reviewing the documents you create in your Creative Cloud applications.

Portable Document Format (PDF) is the file format used by Adobe Acrobat. It's used primarily as an independent method for sharing files. This format enables users who create files on either Macintosh or PC systems to share files with each other and with users of handheld devices or UNIX computers. PDF files generally start out as other documents — whether from a word processor or a sophisticated page layout and design program.

Although PDF files can be read on many different computer systems using the free Adobe Reader, users with the DC version of Adobe Acrobat can do much more with PDF files. With your version of Acrobat, you can create PDF documents, add security to them, use review and commenting tools, edit documents, and build PDF forms.

Use Acrobat to perform any of the following tasks:

- Create interactive forms that can be filled out online.
- Allow users to embed comments within the PDF files to provide feedback. Comments can then be compiled from multiple reviewers and viewed in a single summary.

- Create PDF files that can include MP3 audio, video, and even 3D files.
- Combine multiple files into a single PDF and include headers and footers as well as watermarks.
- Create secure documents with encryption.
- Combine multiple files into a searchable, sortable PDF package that maintains the individual security settings and digital signatures of each included PDF document.
- Use auto-recognize to automatically locate form fields in static PDF documents and convert them to interactive fields that can be filled electronically by anyone using Adobe Reader software.
- Manage shared reviews — without IT assistance — to allow review participants to see one another's comments and track the status of the review.
- Enable advanced features in Adobe Reader to enable anyone using free Adobe Reader software to participate in document reviews, fill and save electronic forms offline, and digitally sign documents.
- Permanently remove metadata, hidden layers, and other concealed information and use redaction tools to permanently delete sensitive text, illustrations, or other content.
- Save your PDF to Microsoft Word. You can take advantage of improved functionality for saving Adobe PDF files as Microsoft Word documents, retaining the layout, fonts, formatting, and tables.
- Enjoy improved performance and support for AutoCAD. Using AutoCAD, you can now more rapidly convert AutoCAD drawing files into compact, accurate PDF documents, without the need for the native desktop application.

Want to discover other great Acrobat improvements? Read Book 6 to find out all about Acrobat and PDF creation.

Prototyping your apps with Adobe XD

By designing or importing your art into Adobe XD artboards, you can go from idea to prototype quickly without building using code. XD lets you do what you do best . . . design! Take your touchable, interactive design and share it with others for feedback or user testing. Find out more about Adobe XD in Book 7.

Some of the things that you can do in Adobe XD include:

- Build clickable prototypes using multiple artboards as screens.
- Share your clickable application on mobile devices or on the web.
- Draw with vector tools.
- Copy and paste vector images and other assets from other Adobe applications.
- Import SVG and edit them directly in XD.
- Build lists of imagery and data quickly.
- Take advantage of layers and symbols.
- Create reusable components.
- Build shareable libraries.
- Add animation.
- Send your designs to development.

Integrating software

With so many great pieces of software in a single package, it's only natural that you'll want to start using the programs together to build exciting projects. You may want to design a book using InDesign (with photos edited in Photoshop and drawings created in Illustrator) and then create logos, buttons, and other art in Adobe Illustrator and import them into an interactive prototype using Adobe XD. Similarly, you may want to take a complex PDF file and make it into something that everyone can view online. All tools in the Adobe Creative Cloud are built to work together, and achieving these tasks suddenly becomes much easier to do because the products are integrated.

Integrating software is typically advantageous to anyone. Integration enables you to streamline the workflow among programs and sometimes team members. Tools exist that let you drop native images into Photoshop, InDesign, Illustrator, and XD. With Adobe Bridge, you can view files and investigate specific information about them, such as color mode and file size, before selecting them for placement.

Acquiring assets for this book

Many of the files that are referenced in this book are available right in the application sample folders that come with Creative Cloud. The path locations are defined when they are referenced, making it easy to find and use them in the provided step-by-step examples. In addition to these sample files, you can find tips and tricks and more files for you to experiment with and investigate at `www.agitraining.com/dummies`.

» Adding and updating additional apps

» Using CC libraries

» Creative Cloud extras

Chapter 2 Creative Cloud Application Management

Before we introduce the Creative Cloud, you should understand that the applications covered in this book do not run "in the cloud." Instead, they are rich desktop applications that reside on your computer and are downloaded from the cloud. Other online services such as registration validation, and updates continue online, and run typically in the background.

If you are looking for all of the apps included in the Creative Cloud go to `https://creativecloud.adobe.com/apps/all/desktop`. Here you will find apps built for mobile devices and the web. Keep in mind that the Creative Cloud requires a subscription, but Adobe has trial versions available for most of their applications.

Downloading Your First App

If you are new to the Creative Cloud and are just getting started, you can go to `www.adobe.com/creativecloud/catalog/desktop.html` and choose the app that you wish to download first. If you do not have an Adobe ID, you will need to set one up at this point.

Click Download to install the app you want. If you are not signed in, you are asked to sign in with your Adobe ID and password. Follow any onscreen instructions that are provided to you. Your app will then begin to download.

When you install your first application, the Creative Cloud desktop app is also installed, as shown in Figure 2-1. The Creative Cloud app manages the rest of the installation process; you can check your download progress in the status bar next to the app's name. If it does not automatically appear, look in the location where your applications normally are installed: for instance, in the Program Files folder (Windows) or the Applications folder (Mac OS).

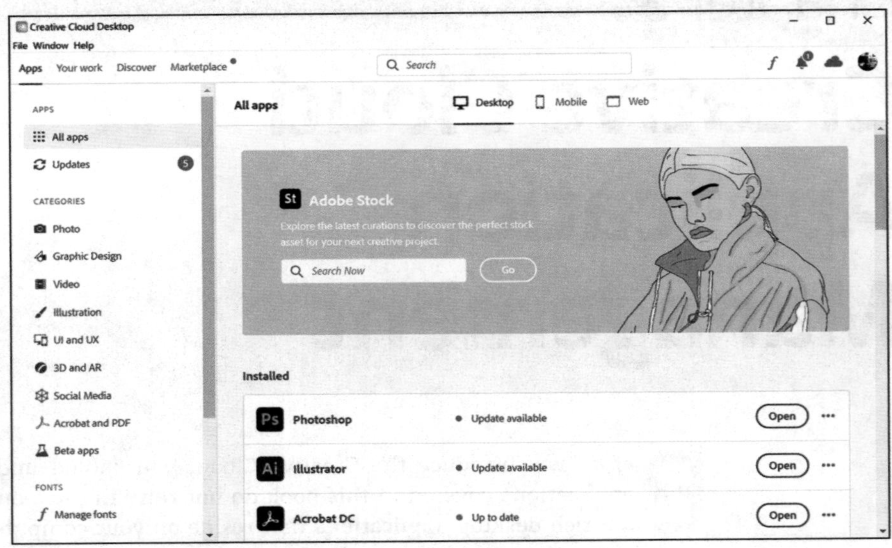

FIGURE 2-1: The Creative Cloud is not just the name of the suite of tools; it's also an application.

The Creative Cloud app is a desktop application that allows you to manage downloads and updates of your Creative Cloud applications. It lets you take advantage of other cloud-based services, such as storage and stock photos, and allows you to view work by you and others in the design community.

Note: If you are downloading a free trial subscription to Creative Cloud, you will be provided a different set of instructions that you should follow.

In the Creative Cloud app you have several choices: Install an App, Open an App, or Update an App. When you first launch the Creative Cloud app, the apps included in your subscription appear displayed. They may indicate that they can be installed, updated, or opened (if they were already installed), as shown in Figure 2-2. You can also click the dots to the right of the app to access additional services such as uninstall. Keep in mind that you do not have to install all of the apps; simply choose the ones you wish to use.

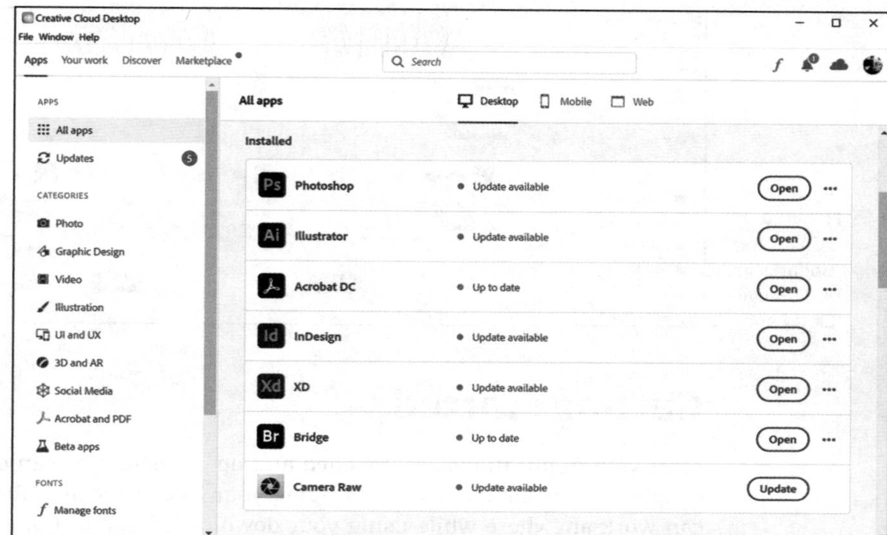

FIGURE 2-2: The status of your applications appears in the Creative Cloud app.

When you are done, launch your app by clicking the Open button from the list of apps that appear on the right column of the Creative Cloud app. You can also launch the app the same way you normally launch any other app on your computer. Your new app is installed in the same location where your applications are normally installed, such as the Program Files folder (Windows) or the Applications folder (Mac).

Checking for updates

The Creative Cloud desktop app indicates when updates are available for your installed apps. You can check for updates manually by following these steps:

1. **Open the Creative Cloud desktop app and look for Updates in the column on the left. A number appears if there are updates available. Click Updates to see the apps that should be updated.**
2. **Click the Gear icon in the upper-right to turn on the Auto-Update feature.**
3. **Click All Apps in the left column to see the status of your apps, as shown in Figure 2-3.**

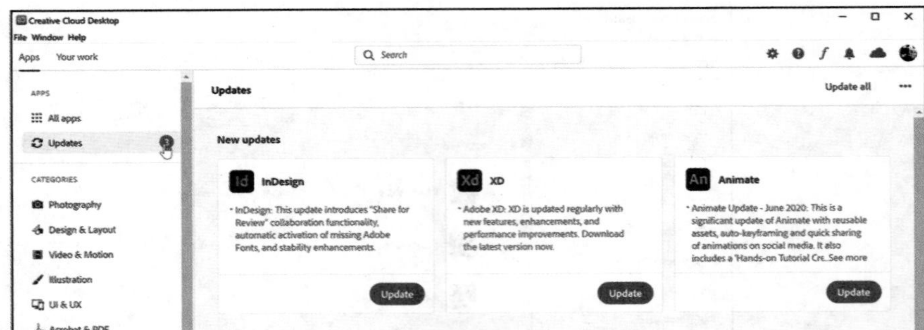

FIGURE 2-3: Check for updates in the Creative Cloud app.

Getting started

After your applications are installed and up-to-date, you can dive into the next chapter of this book and start bringing your ideas to reality. Remember that you can work anywhere while using your downloaded applications. Once in a while, and when connected to the Internet, Adobe checks your license to make sure it is valid and current. Count on this happening about once every 30 days, but don't expect it to cause any problems for you.

IN THIS CHAPTER

» Other hidden gems in the Creative Cloud

» Finding imagery and other assets through your Creative Cloud subscription

» Taking advantage of additional fonts and managing them

» Sharing and syncing your work using libraries

» Promoting yourself with Behance and Adobe Portfolio

Chapter 3
Creative Cloud Extras for You

Creative Cloud includes more than just applications. It includes extras that help you keep in touch with changes in your subscription and with recent downloads. It also includes assets for subscribers such as typefaces, cool images, and illustrations, as well as access to stock photos and prebuilt site management tools to help you connect with the creative community. You can also take advantage of the Creative Cloud to store, sync, and share important files, making collaboration easy.

Launching the Creative Cloud App

When you launch the Creative Cloud app, an application pane appears that includes a column on the left side, as shown in Figure 3-1. Launching the Creative Cloud app is one way to access many of the additional extras that come with your subscription, but there are other ways discussed in this chapter that you can access these extras right from the applications themselves.

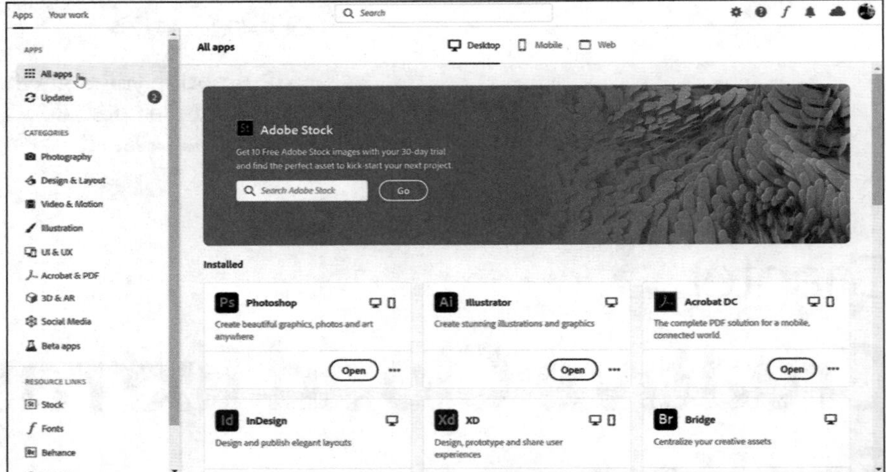

FIGURE 3-1: You can find many extras right in the Creative Cloud pane.

Apps

Find the status of the Creative Cloud applications in the Apps menu at the top of the workspace. Here you see the apps that are currently installed, represented by an Open button to the right. You can also find the apps that are not yet installed, represented by an Install button (or an Update button, if your installed version isn't current). In Figure 3-2, you see that you can click Updates to display a view that includes all the apps in need of updating.

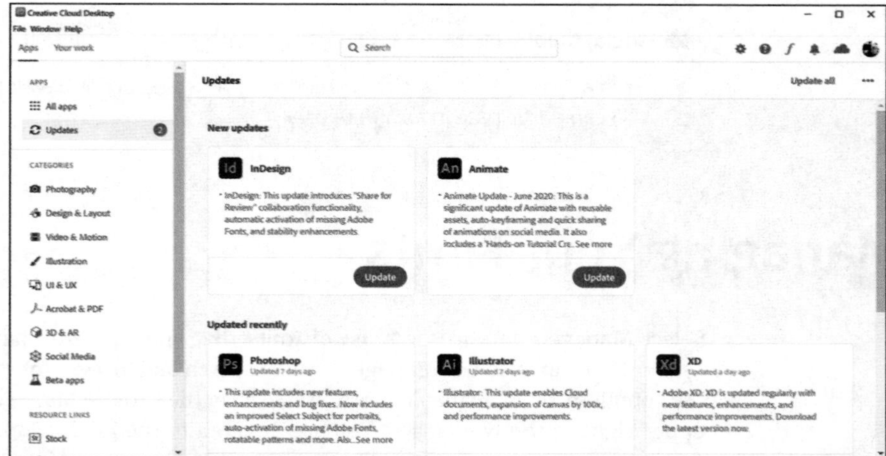

FIGURE 3-2: Keep track of the status of your applications and updates by selecting Updates.

Categories

Adobe Creative Cloud has done a nice job of breaking down the available apps and offering quick access by category. Many of the apps cross into multiple categories and even additional platforms such as mobile or web. In this book, we cover the desktop applications and discuss in detail Bridge, Photoshop, InDesign, Illustrator, Acrobat, and Adobe XD. We also discuss some of the web-based services available through Adobe Stock, Behance, and Portfolio.

The categories of applications included in the Creative Cloud are as follows:

- **Photography:** Quick access to Lightroom, Photoshop, Lightroom Classic, Bridge, and Camera Raw
- **Graphic Design:** Quick Access to Photoshop, Illustrator, InDesign, Acrobat DC, InCopy, Dimension, and Bridge
- **Video:** Quick Access to Premiere Pro, Premiere Rush, After Effects, Animate, Character Animator, Audition, Media Encoder, Prelude, Photoshop, and Bridge
- **Illustration:** Quick access to Photoshop, Fresco, Illustrator, Portfolio, and Behance
- **UI & UX:** Quick access to XD, Photoshop, Illustrator, Animate, After Effects, and Dreamweaver
- **Acrobat & PDF:** Quick access to one lonely app, Acrobat DC
- **3D & AR:** Quick access to Dimension and Photoshop

- **Social Media:** Quick access to another lonely app, Premiere Rush
- **Beta apps:** Quick access to current programs that are in development that you can participate in as a beta user

Managing Your Fonts

Select Manage Fonts to see a list of fonts that you currently have active on your system. You can also use Manage Fonts to search and browse for other fonts. Search the complete list of available fonts by selecting the Browse More Fonts button in the upper right of the workspace. You are directed to the Adobe Fonts website, where you can take advantage of advanced search filters in order to find the font you want.

Keep in mind that your Creative Cloud subscription includes the use of a limited number of typefaces that you can activate using Manage Fonts. When you are finished using the font, you can deactivate it. Fonts are discussed in more detail later in this chapter.

Resource Links

These are links to resources that are available through the web. Clicking on these topics redirects you to the appropriate website:

- **Stock:** Click on Stock to access Adobe Stock. This is a paid subscription in addition to your Creative Cloud subscription. You can use it to find reasonably priced royalty-free stock photos, illustrations, videos, and audio clips.
- **Tutorials:** Online tutorials for all experience levels are available for your Creative Cloud apps. Simply click this link to be taken to the Creative Cloud Tutorials page.
- **Portfolio:** Showcase your work by accessing your portfolio-building tools using Adobe Portfolio. Find templates and tools in order to create your own galleries of work and publish them to the web.
- **Behance:** Use Behance as a tool to network with others and to help you be found. Behance is a leading online platform that allows creative professionals to showcase and discover creative work. Behance members can Follow your profile and you can, in turn, follow other users. When you Follow someone, their updates such as projects appear in your For You feed.

» **Support Communities:** If you have a specific challenge or question you can get help from the Support community. Go to this site to post questions, or search other users' questions. This is a dynamic resource for those questions you just can't find answers to easily.

Your Work

The second menu item at the top is Your Work. Your Work essentially shows you what you have saved on the Adobe Cloud. Cloud storage is included with your subscription — in fact, 100 GB at the time of writing. On your desktop computer, Creative Cloud Files is a folder where you can store files that synchronize to Creative Cloud online storage. On your mobile device, the Creative Cloud app and many Adobe mobile apps also connect to that storage. You can also use a web browser to see, preview, and download those files. You can also access your files using the Creative Cloud app, as you are doing right now.

If you click on the Your Work header, three icons appear on the left, as shown in Figure 3-3: Show Libraries, Show Cloud, and Your Work links. It is possible that you have not yet stored any files to the cloud at this time, but you will have an opportunity to do that later in this chapter.

If you are collaborating with a team you can see files that are shared and followed in the Your Work section. In Figure 3-4, Libraries was selected from Your Work. You can see saved libraries, any libraries shared with you, and any libraries that you are following. When you follow a library, you are notified of any updates.

Libraries

As you start to work on projects, you may want to store or share items that you use on a regular basis. This is where the Libraries feature comes in handy. By using the Creative Cloud Libraries feature, you can save and share colors, character styles, imagery, and more. You can share them with just yourself, as you move from one computer to another, between applications, or with others on your team. Look for Libraries and the Libraries panels in your Creative Cloud applications, typically in the Window menu. You can add assets in this panel that you can then open in other applications.

If you select Your Work, you may not see any content at this time, but as you create libraries over time, you will see them here. Read on to see how you can create your own library. If you want to follow along, launch Photoshop by going to the Apps section of the Creative Cloud desktop app and selecting the Open button to the right of the Adobe Photoshop app.

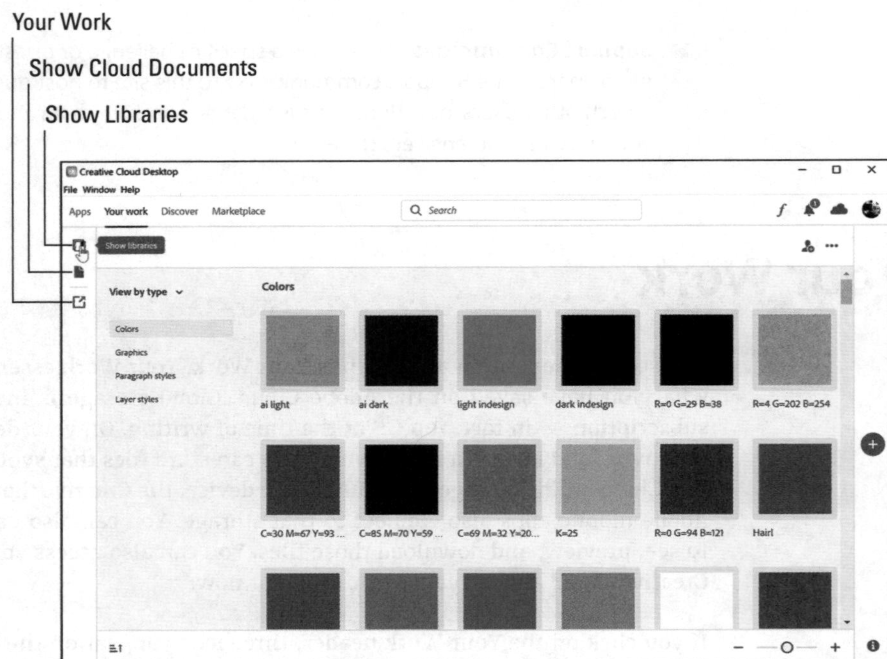

FIGURE 3-3: Click on Your Work to access files that are saved in the Adobe Cloud.

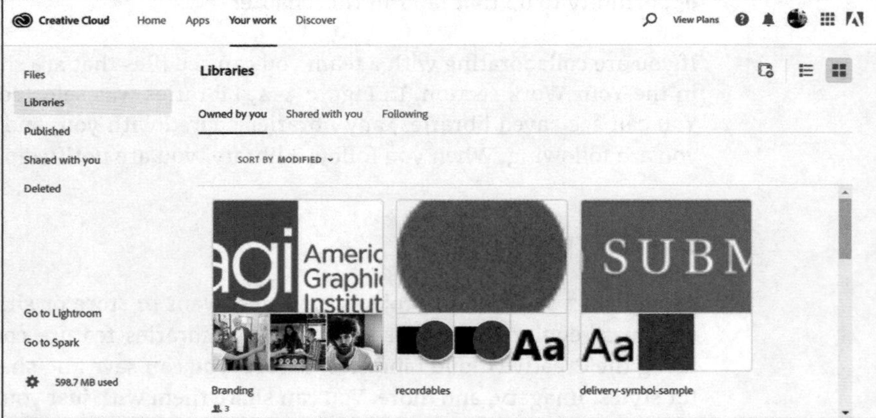

FIGURE 3-4: Your Work links contain your libraries and cloud documents. These are files that you have stored or are shared with you in the Creative Cloud.

Creating your own library

Creating a library can be completed directly in an application using the Libraries panel. Figure 3-5 shows the Libraries panel in Adobe Photoshop, which is like the Libraries panel in other Creative Cloud apps. Some applications access the library differently; those details are discussed in the relevant minibooks. Follow these steps to create your own library. As previously mentioned, this example uses

Adobe Photoshop, but these steps can also be used to create libraries in many of the other apps in the Creative Cloud.

1. **In Photoshop, choose Libraries from the Window menu.**

   The Libraries panel appears.

2. **Click the Create New Library button, add a name for the library, and press Create. The Libraries panel updates to offer the opportunity to add elements, such as colors, images, styles and more to the library, as shown in Figure 3-6.**

3. **Add items to the library by doing one of following:**
   - Drag and drop items to the Libraries panel
   - Select an element and use the Add Elements button at the bottom of the Libraries panel.
   - Add colors by selecting an object that has a fill or stroke color applied that you want to save and then click the Add Elements button at the bottom of the Libraries panel and select Fill Color, Stroke Color.
   - Add a type style by selecting text in your document that uses the type properties you wish to save and then clicking the Add Elements button at the bottom of the Libraries panel and selecting Character Style (for selected characters) or Paragraph Style (for entire paragraphs).

   After you have added elements to your library, you can access them easily from other applications by displaying the Libraries panel and then selecting your saved library from the list of libraries.

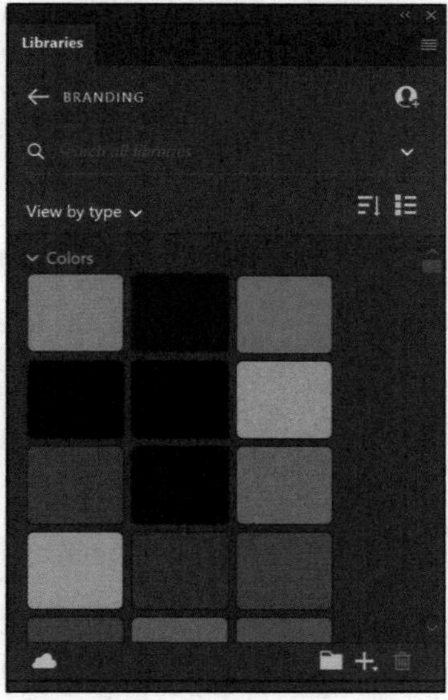

FIGURE 3-5: The Libraries panel is accessible from the Window menu in most Adobe Cloud apps.

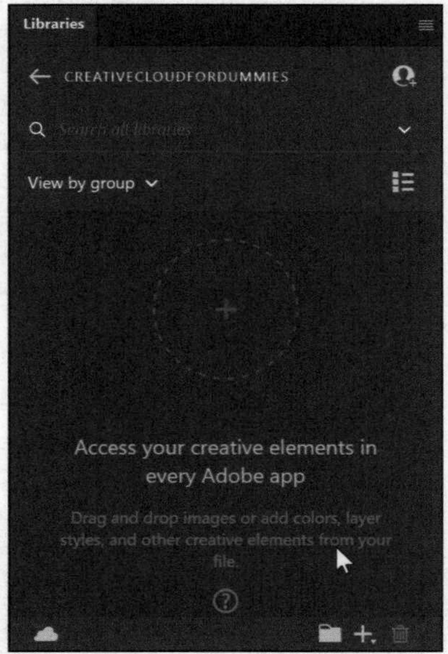

FIGURE 3-6: After your library is created, you can add elements such as images, fonts, colors, styles and more.

You can also share your library with others by clicking on the Share Library icon in the upper right of the Libraries panel. This take you to the Creative Cloud desktop app, where you can enter the email addresses of people with whom you would like to share this library.

Using the Libraries panel to add elements to your open documents

Access the elements that have been saved in libraries by simply opening the Libraries panel, Window⇨Libraries. In this example, we used the Photoshop app.

If you have not created a library, go back to the previous section and create a small library to use for this exercise. Then, follow these steps:

1. **Open an Adobe app. In this example, open Photoshop.**
2. **Select Window⇨Libraries to open the Libraries panel.**
3. **Click on View by Type to see categories of elements saved in your library, as shown in Figure 3-7.**

   Note: You can organize your library elements by using groups. Simply click the Create New Group folder icon at the bottom of the Libraries panel and then name your group. Drag and drop elements into this group to help you find items in your library:

   - *To use an image:* Click the image and drag it out to your page, and then click once on your page to release the image and place it. After it is placed, note that the image displays a little cloud icon in the upper-left corner, indicating that it is from the cloud.
   - *To use a color:* Keep in mind that you need to have saved a color to the library to apply it. Select an object to which you want to apply a saved library color, then open a saved library and choose a color.
   - *To apply a text style:* Select the text in your document to which you wish to apply a saved library style. Find your library in the Libraries panel and click on the saved library style.

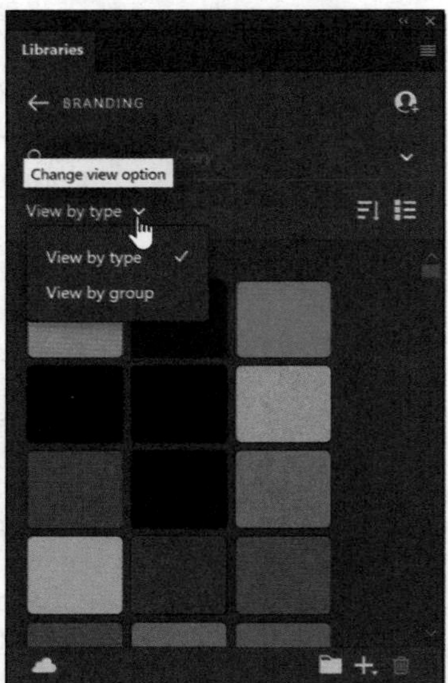

FIGURE 3-7: Organize library elements by type to find colors, images, styles, and other library elements more quickly.

Show Cloud Documents

When you select Show Cloud Documents, you see three options: Cloud Documents, Shared with You, and Deleted. Click on Cloud Documents to see the contents of the Creative Cloud files folder that exists on your computer. (See Figure 3-8.) The items in the Creative Cloud files folder on your computer are synced with the items in the cloud. You can use this available space in many ways: for storage of virtually any files, documents, images, branding elements and more.

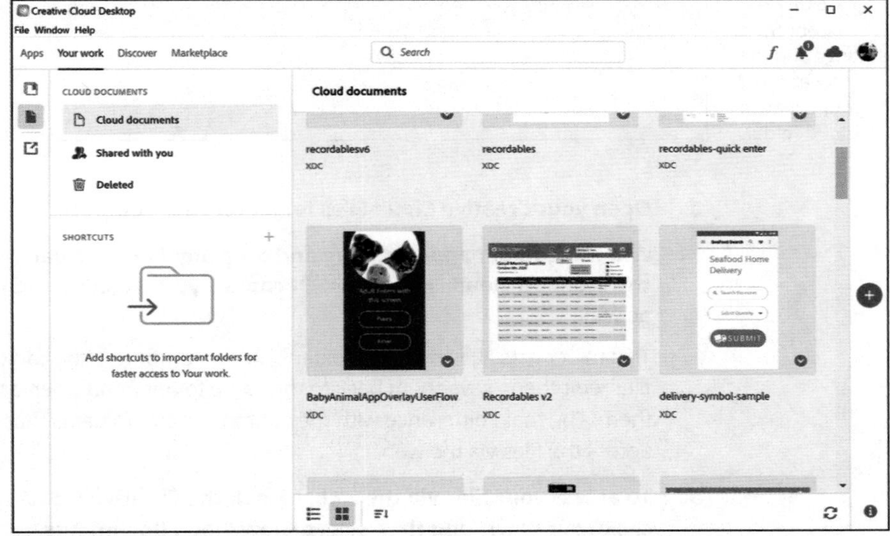

FIGURE 3-8: Click on Show Cloud Documents in order to see and manage the documents you saved to the Creative Cloud.

If you are looking for extra space to store or to easily share your files, look no further than your Creative Cloud files folder. Follow these steps in order to store files in the Creative Cloud:

1. **Open your Creative Cloud app.**
2. **Click the Cloud Activity icon (the cloud) in the upper-right of the Creative Cloud app and then click the gear icon.**

   The Sync settings appear here, as well as the location of the Creative Cloud files folder, as shown in Figure 3-9.

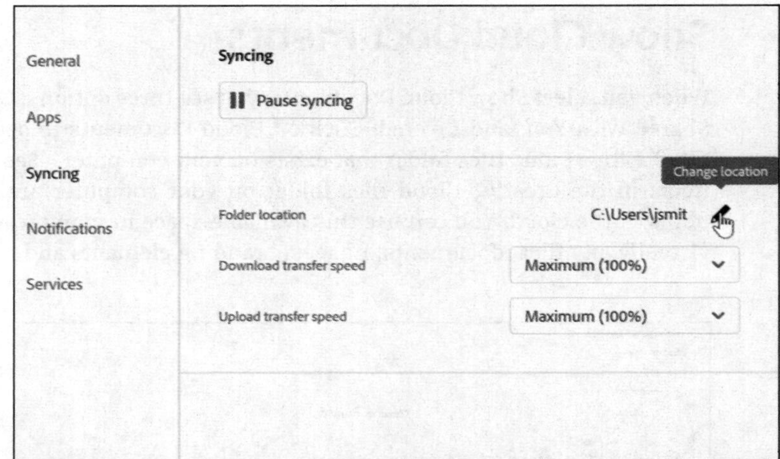

FIGURE 3-9: Your Sync settings and the location of your Creative Cloud folder.

3. **Open your Creative Cloud Files folder on your computer.**
4. **When the folder appears, drag and drop any files you want to save into the Creative Cloud Files folder. Of course, you can copy and paste files as well.**

   This folder acts like any other folder on your computer. You can open your files, edit them, save them back to the same folder or another, and delete them. The main difference with the contents in this folder is that you can also access the files via the web.
5. **To access your files via the web, launch the Creative Cloud application, click Your Work, and then select Show Cloud Documents.**

   Keep in mind that you can also manage your cloud documents right in the Creative Cloud app. Right-click on any document shown in the cloud documents to select functions such as Share, Delete, and more. See Figure 3-10.

Not only can you access your own files anywhere by signing into the Creative Cloud on any computer, but you can also share them with others. You can do this directly in the Creative Cloud Files folder or on the web. The following directions tell you how.

To share from the Creative Cloud Files folder:

1. **Click Your Work and then click Show Cloud Documents if they are not already visible.**
2. **Right-click on a file that you wish to share to see the contextual menu. (Ctrl-click if you do not have a right-clicking mouse.)**

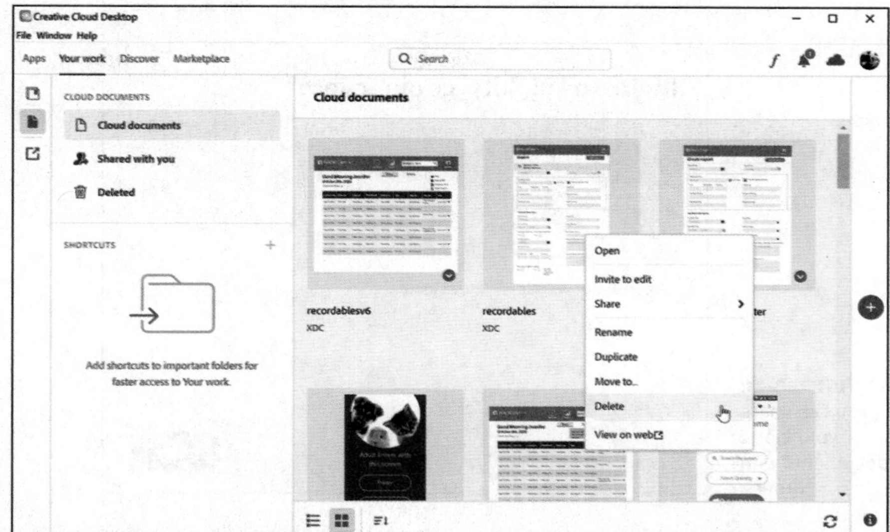

FIGURE 3-10: You can manage your cloud files in the Creative Cloud app.

3. **Choose the Creative Cloud menu item and Send Link from the submenu. On the Mac, choose Share Link from the contextual menu.**
4. **If you have not previously shared the selected file, you first need to make it a public link. When the dialog box appears indicating that your link is private, click the Create Public Link button.**
5. **Copy and paste the provided link into your correspondence with the people to whom you want to give access to your file or project.**

To share from the Creative Cloud app:

1. **Go to Your Work and select Show Cloud Documents.**
2. **Right-click (Ctrl-click if you do not have a right-clicking mouse) on a file's thumbnail.**
3. **Select Invite to Edit.**
4. **When the Share dialog box appears, enter the email of the person you wish to share the document with and any additional information in the Message text field, as shown in Figure 3-11.**

Keep in mind that you can also share directly from the application you are working in. You will find out how to do this later, when discussing the individual applications.

FIGURE 3-11: You can share your cloud documents with others.

Discovering Design Opportunities with New Fonts

Using fonts in the past was rather cumbersome. You had to purchase a license for a typeface or family, and then make sure that it was loaded correctly. The purchase of these fonts often limited your opportunity to try new typefaces. Then there was the issue with using typefaces for the web: Even though you have licensed a font, that doesn't mean the end user has access to it. Frequently this led to unexpected display results in various browsers. This is where Adobe Fonts comes to the rescue. By using Adobe Fonts, you can access an available font when it's needed, activating a license and adding it to your system so it is ready to use immediately.

Adobe Fonts allows you to easily access new and interesting fonts created by leading type foundries. By using Adobe Fonts you do not have to worry about licensing, and you can use fonts from Adobe Fonts on the web or in desktop applications. Look for new fonts to use in your web or print design by clicking on the Manage Fonts submenu in the Apps tab of the Creative Cloud App. You can sync your fonts with those available in Adobe Fonts or manage your existing font selections.

In the following steps, you have the opportunity to add a font using Adobe Fonts.

1. **In the Creative Cloud application, click Apps and then select Manage Fonts in the left column.**

   A window appears that shows your active fonts, which you can use in your Creative Cloud apps, as shown in Figure 3-12.

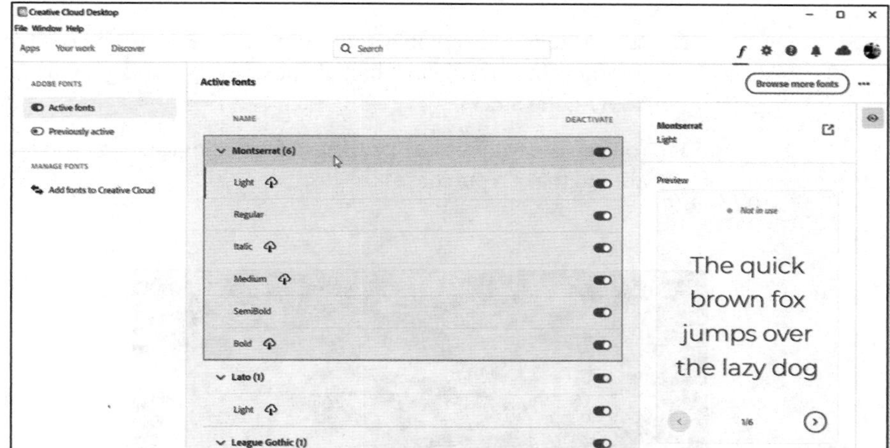

FIGURE 3-12: When you select Manage Fonts, you can activate, deactivate, and preview fonts.

2. **Click on Browse More Fonts and you will be directed to `fonts.adobe.com`, where you can filter fonts by serif or sans serif; choose different properties, such as x-height, width, or weight; and even select the language. You can also choose between two recommended categories: headings and paragraphs.**
3. **When you decide on a typeface, click it to enter the details page. There, press the Activate button to the right of the font family to add it to your list of available fonts.**

Finding Images with Adobe Stock

Adobe makes it easy for you to find a varied source of images, graphics, and videos. Simply click on Apps in your Creative Cloud app and then scroll down the left column until you see Stock, under Resource Links. Keep in mind that even though you can receive an introductory offer of ten images, this is a paid service that is not included with your subscription. However, pricing is reasonable and there are many benefits to using Adobe Stock:

- Adobe Stock is easy to locate. It is right there in your Creative Cloud app.
- You can choose File ➪ Search Adobe Stock from almost every application in Creative Cloud.
- You can search Adobe Stock right in the CC Libraries panel in InDesign, as shown in Figure 3-13. Just drag and drop the image that you want to use as a preview. If you decide to license it, just click the shopping cart icon in the upper-left of the image in the Libraries panel.

» Adobe Stock is searchable right from Adobe Bridge. You can type your search criteria right in the Search box of Adobe Bridge, and generate search results from Adobe Stock.

» You can easily download previews for your images until you are sure that you want to make a purchase.

FIGURE 3-13: Search and use Adobe Stock right in the Libraries panels.

Promoting Yourself with Behance

If you are promoting yourself, you should consider posting work on Behance. Behance is an online platform that allows users to showcase and discover creative work. Behance is used by creative professionals to find work and by those hiring in the creative field to discover creative professionals that fit their needs.

To create a portfolio, follow these simple steps:

1. **First of all, either click on Behance in the left column of the Creative Cloud app or go to** `www.behance.net` **and sign up. Look for the Log-in button in the top right and enter your Adobe ID and password.**

   This connects your Creative Cloud applications with Behance. Because you are signed in you can choose to save your projects to Behance right from your menu in many of the Creative Cloud apps. This is typically done by choosing File ➪ Share on Behance in your Adobe Creative Cloud applications.

If you have not used Behance before, take a look at the first screen, For You; it appears with projects that you can click on to investigate. These are projects from other creative professionals like you.

Projects are the primary way Behance members can showcase their work. A project is a group of image files, text files, videos, and other media that have a central theme or idea. For instance, it may be an example of all the visual and UX design components and assets for a mobile application, or a branding concept with print and web pieces.

The average project might have 10 to 15 images broken up with descriptive text titles. All public projects are published to your profile and can be viewed by anyone by default. By the way, each project has its own unique URL that you can use to share with others.

2. **Add your own project by clicking the Create a Project button at the top-right near your log-in information.**

   A window appears that offers you the ability to add assets, as shown in Figure 3-14.

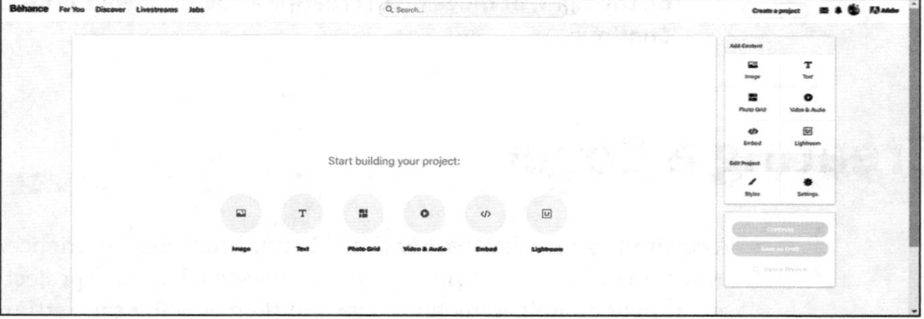

FIGURE 3-14: Create a new project and add files you wish to share with others.

3. **You can now follow the buttons in the middle of the screen. Start by clicking Files and then uploading files that you want to share with others. You can load PNG, JPG, or GIF files as content. Keep in mind that your selected content displays at a maximum of 1400 px unless a viewer clicks on it, in which case it will expand.**
4. **Click the Image button and then use the file browser window that appears to locate an image.**

   After you select an image, it is added to the screen.

5. **Click the Pencil icon, on the upper-left corner of the image, to reorder your project, add a caption, replace the image, or delete the image.**

   Note: To add an element, such as text, you must click the related icon in the Tools panel that appears on the right side of the window, as seen in Figure 3-15.

FIGURE 3-15: Add text and other elements to your project by using the menu items on the right.

6. **Continue adding more images by selecting to upload files using the menu on the right, or move on to creating a cover for your project by clicking Continue.**

Creating a Cover

A cover image is a key part of the Behance browsing experience. By creating a relevant cover, you allow viewers to quickly see what your project is about. You create a cover by uploading an image and then clicking the Settings icon at the right of the workspace, as shown in Figure 3-16.

In this window, you can add project information such as a title, tags, or list of the tools that were used to create the project. There are also options to help findability of your project by selecting categories. The information that you enter in this window allows you to be found by others. Be careful to select appropriate categories and enter all information about your project accurately.

By clicking on the Edit Cover Image link under the image you can scale and crop the image. This will be used as the thumbnail for your project that appears when viewers are browsing in Behance. Keep in mind that the cover image must be a minimum size of 808 x 632 px and GIF files do not animate as a cover image.

When you are done editing, click Save as Draft to continue working or click Publish to post your project in the Behance website.

FIGURE 3-16: Create a cover for your image.

Bonus! Adobe Portfolio

In addition to sharing your work on Behance, you can take advantage of Adobe Portfolio. Adobe Portfolio offers you the opportunity to easily build a beautiful visual resume. You can customize layouts and segment your own categories in extensive galleries of work. One of its greatest features is its compatibility with Behance. Build once in Behance, and your project automatically appears in your portfolio and vice versa.

Follow these steps to open and explore Adobe Portfolio:

1. **In your browser, go to** `https://portfolio.adobe.com/`**.**

   This opens a browser window with information about you. This is also the location in which you can launch Adobe Portfolio.

2. **Click View Adobe Portfolio in the upper-right of the screen.**
3. **If you have an Adobe ID, you have access to Adobe Portfolio and you can get started by clicking the Edit Your Sites or Create Your Site button in the middle of the screen. If you have not yet used the site, you may be directed right to the layout page.**
4. **In the next window, you are provided a selection of layouts, as seen in Figure 3-17. Click any layout you are interested in. The rest is easy: Simply follow the instructions provided directly to you on the screen.**
5. **As with Behance, you can choose to save your project when you are done or publish to the Adobe Portfolio site.**

The support community is one of the best places to go if you need help with a technical issue, or if you just want advice on how to better create your work. Each app has its own support forum that is supported by helpful users like you as well as staff from Adobe. Use this resource by selecting your app of interest in the left column and then using the Search This Community search box at the top, as shown in Figure 3-18.

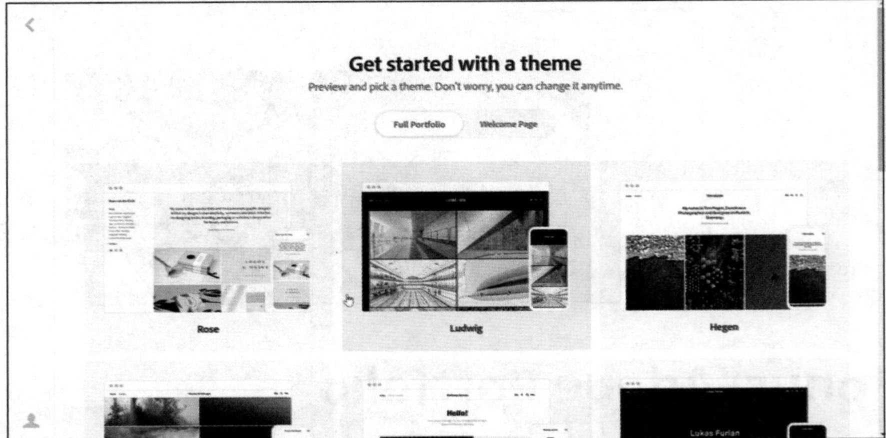

FIGURE 3-17: Choose from a selection of different layouts to start your portfolio.

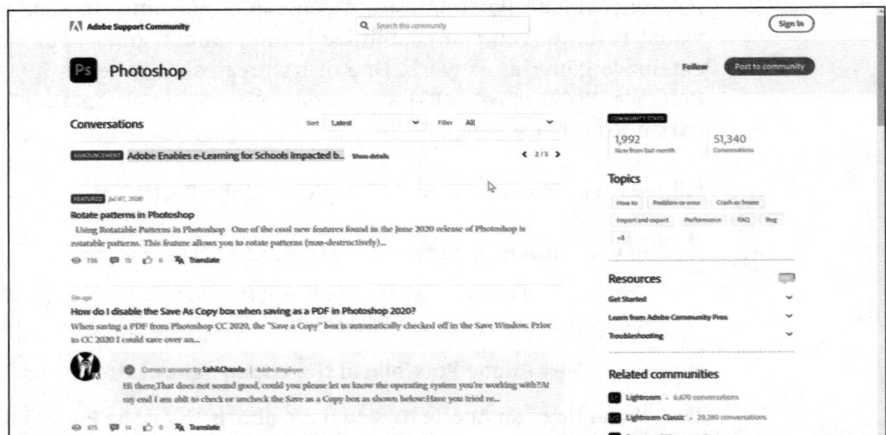

FIGURE 3-18: Get help from the community by searching existing issues, or post your own question.

If you can't find what you want, create your own post by selecting the Post to Community button in the upper right. Answers come quickly so you will want to check the Email Me checkbox at the bottom of your post message textbox.

It's Not Just Apps!

Wow . . . and you thought the Creative Cloud was just a collection of applications! As you can see, there is so much more to take advantage of when you have a subscription to the Creative Cloud. In the following minibooks, you find out how to use the applications to bring your designs to reality.

IN THIS CHAPTER

» Discovering common menus and dialog boxes

» Addressing Creative Cloud alerts

» Working with common menu options

» Understanding contextual menus

» Speeding up your workflow with shortcuts

» Changing preferences

Chapter 4 Using Common Menus and Commands

When you work in Adobe CC, you may notice that many menus, commands, and keyboard shortcuts are similar across the applications. These similarities can help you migrate more easily from one application to another. This chapter provides an overview of some of the common menus, dialog boxes, options, shortcuts, and preferences that exist in most or all of the applications in Adobe Creative Cloud.

Discovering Common Menus

When you work with applications in Adobe Creative Cloud, you may notice that many of the menus on the main menu bar are the same. Similar functionality makes finding important features easy, even when you're completely new to the software.

Menu items contain features that control much of the functionality in each application. A menu item may also contain features that are related to a particular task. For example, you might save from the File menu, or change your text in the Type menu. Some of the menu items that commonly appear in the Creative Cloud applications include the following:

- **File:** Contains many features that control the overall document, such as creating, opening, saving, printing, and setting general properties for the document. The File menu may also include options for importing or exporting data into or from the current document.
- **Edit:** Contains options and commands for editing the current document. Commands include copying, pasting, and selecting as well as options for opening preferences and setting dialog boxes that are used to control parts of the document. Commands for spell-checking are also common parts of the Edit menu.
- **Type:** Contains options related to type and typesetting, such as font selection, size, leading, and more.
- **View:** Contains options for changing the level of magnification of the document. The View menu also sometimes includes options for viewing the workspace in different ways; showing rules, grids, or guides; and turning alignment snapping on and off. *Snapping* helps with precise placement of selection edges, cropping marquees, slices, shapes, and paths.
- **Window:** Contains options primarily used to open or close whatever panels are available in the application. You can also choose how to view the workspace and save a favorite arrangement of the workspace.
- **Help:** Contains the option to open the Help documentation that's included with the application. This menu may also include information about updating the software, registration, and tutorials.

Figure 4-1 shows a list of items that appear under the File menu in Photoshop.

Each application has additional application-specific menus determined by the needs of the software. For example, you can use the Photoshop Image menu to resize the image or document, rotate the canvas, and duplicate the image, among other functions. InDesign has a Layout menu you can use to navigate the document, edit page numbering, and access controls for creating and editing the document's table of contents; we discuss these menus where appropriate throughout this book.

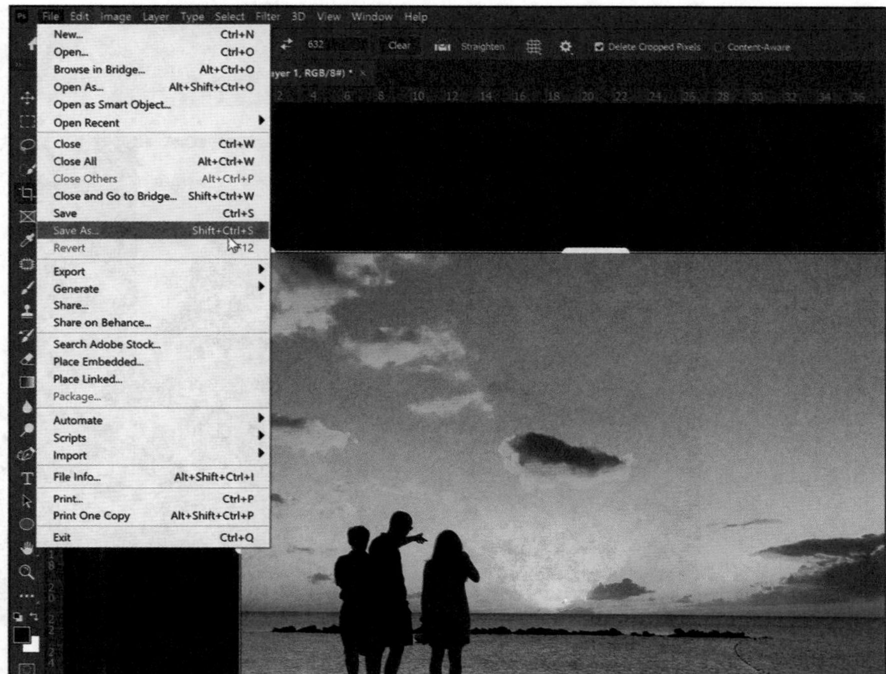

FIGURE 4-1: Menus in Photoshop let you choose and control different options.

Using Dialog Boxes

A *dialog box* is a window that appears when certain menu items are selected. It offers additional options in the form of drop-down lists, panes, text fields, checkboxes, and buttons that enable you to change settings and enter information or data as necessary. You use dialog boxes to control the software or your document in various ways. For example, when you open a new file, you typically use the Open dialog box to select a file to open. When you save a file, you use the Save As dialog box to select a location for saving the file, to name the file, and to execute the Save command.

Some dialog boxes also include tabs. These dialog boxes may contain many settings of different types that are organized into several sections by using tabs. A dialog box typically has a button that executes the particular command and one that cancels and closes the dialog box without doing anything. Figure 4-2 shows a common dialog box.

FIGURE 4-2: Using a dialog box to change filter settings.

A dialog box in Windows offers the same functionality as a dialog box on the Mac. Dialog boxes perform similar tasks and include the same elements to enter or select information. For example, here are some tasks you perform by using dialog boxes:

- Save a new version of a file
- Apply a filter to a selection
- Specify printing or page-setup options
- Set up preferences for the software you're using
- Check the spelling of text in a document
- Open a new document

REMEMBER

You cannot use the application you're working in until the dialog box is closed. Make sure to close the dialog box after you are finished making your changes. Close the dialog box by clicking a button (such as Save or OK) when you're finished or by clicking the Cancel button to close it without making any changes.

Encountering Alerts

Alerts, which are common on any operating system and in most applications, are similar to dialog boxes in that they're small windows that contain information. However, alerts are different from dialog boxes because you can't edit the information in them. Alerts are designed simply to tell you something and give you one or more options that you select by clicking a button. For example, an alert may indicate that you can't select a particular option. Usually you see an OK button to click to acknowledge and close the alert. You may see other buttons on the alert, such as a button to cancel what you were doing or one that opens a dialog box. Figure 4-3 shows a typical alert.

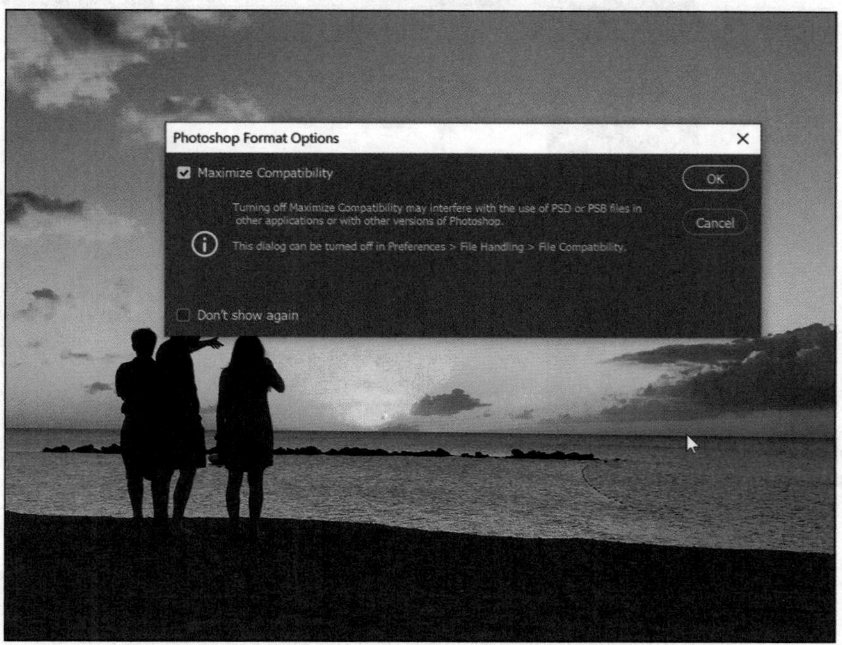

FIGURE 4-3: A simple choice: OK or cancel.

TIP

You can sometimes use an alert to confirm an action before executing it. Sometimes an alert window also offers the option (typically in the form of a checkbox) of not showing the alert or warning again. You may want to select this option if you repeatedly perform an action that shows the warning and you don't need to see the warning every time.

Discovering Common Menu Options

Various menu options are typically available in each of the Creative Cloud applications. However, within each of these menus, several other options are available. Some of them open dialog boxes — this type of option is typically indicated by an ellipsis that follows the menu option, as shown in Figure 4-4.

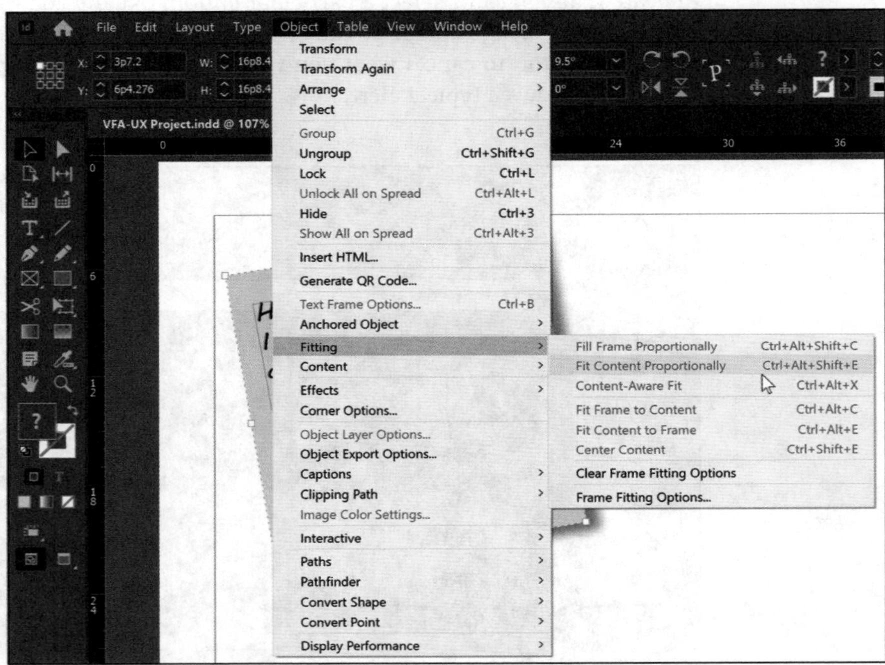

FIGURE 4-4: Choosing a menu option with an ellipsis opens a submenu.

The following menu options are found in several Creative Cloud applications, and these commands either perform similar (or the same) functions or they open similar dialog boxes:

- **New:** Creates a brand-new document in the native file format. For example, in InDesign, a new *INDD* (the extension for InDesign documents) file is created by choosing File ⇨ New ⇨ Document. You can sometimes choose the type of new file you want to create.
- **Open:** Opens a dialog box where you can choose a (supported) file to open on your hard drive or a disc.
- **Close:** Closes the current document. If it has unsaved changes, you're prompted to save those changes first.

- **Save:** Saves the changes you've made to the current document.
- **Save As:** Saves a new copy of the current document using a different name.
- **Import:** Imports a file, such as an image or sound file, into the current document.
- **Export:** Exports the current data to a specified file format. You can sometimes select several kinds of file formats to save the current data in.
- **Copy:** Copies the selected data to the computer's Clipboard.
- **Paste:** Pastes the data from the Clipboard into the current document.
- **Undo:** Undoes the most recent task you performed in the application. For example, if you just created a rectangle, the rectangle is removed from the document.
- **Redo:** Repeats the steps you applied the Undo command to. For example, if you removed that rectangle you created, the Redo command adds it back to the document.
- **Zoom In:** Magnifies the document so that you can view and edit its contents closely.
- **Zoom Out:** Scales the view smaller so that you can see more of the document at a time.
- **Help:** Opens the Help documentation for the current application.

About Contextual Menus

A *contextual menu* is similar to the menu types we describe in the previous sections; however, it's context-sensitive and opens when you right-click (Windows) or Control-click (Mac) on an object in your document. *Contextual* means that the menu dynamically changes depending upon what you have selected.

For example, if you right-click, when an image is selected, you see options referring to the image. However, if you right-click on a document's background, you typically see options that affect the entire document instead of just a particular element within it. This makes it easy to find the options you need quickly. Figure 4-5 shows a contextual menu that appears when you right-click (Windows) or Control-click (Mac) text in InDesign.

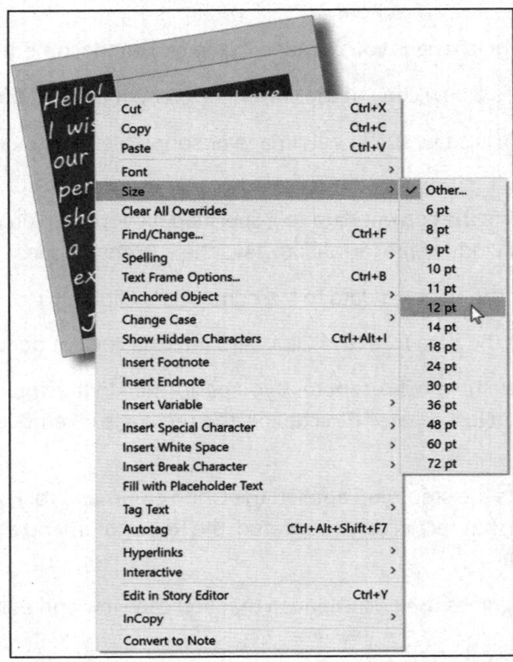

FIGURE 4-5: Open a contextual menu by right-clicking on a selected object.

REMEMBER

The tool you select in the Tools panel may affect which contextual menus you can access in a document. You may have to select the Selection tool first to access certain menus. If you want to open a contextual menu for a particular item in the document, make sure that the object is selected before you right-click (Windows) or Control-click (Mac).

TIP

If you're using a Mac, you can right-click to open a contextual menu if you have a two-button mouse. Otherwise, you press Control-click to open a contextual menu.

Using Common Keyboard Shortcuts

Shortcuts are key combinations that enable you to quickly and efficiently execute commands, such as save or open files or copy and paste objects. Many of these shortcuts are listed on the menus discussed in previous sections. If the menu option has a key combination listed next to it, you can press that combination to access the command rather than use the menu to select it. Figure 4-6 shows shortcuts associated with a menu item in InDesign.

For example, if you open the File menu, next to the Save option is Ctrl+S (Windows) or ⌘+S (Mac). Rather than choose File⇨Save, you can press the shortcut keys to save your file. It's a quick way to execute a particular command.

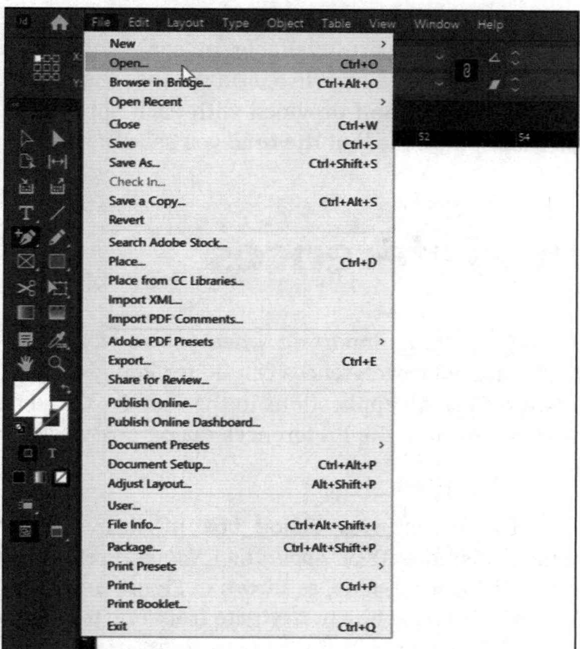

FIGURE 4-6: Shortcuts are shown next to their associated commands.

TIP

Some commonly used shortcuts in the Adobe Creative Cloud applications are listed in Table 4-1.

TABLE 4-1

Common Keyboard Shortcuts

Command	Windows Shortcut	Mac Shortcut
New	Ctrl+N	⌘ +N
Open	Ctrl+O	⌘ +O
Save	Ctrl+S	⌘ +S
Undo	Ctrl+Z	⌘ +Z
Redo	Shift+Ctrl+Z	Shift+⌘ +Z
Copy	Ctrl+C	⌘ +C
Paste	Ctrl+V	⌘ +V
Print	Ctrl+P	⌘ +P
Preferences (General)	Ctrl+K	⌘ +K
Help	F1 or sometimes Ctrl+?	F1 or sometimes ⌘ +?

Many additional shortcuts are available in each application in the Creative Cloud applications, and not all are listed on menus. You can find these shortcuts throughout the documentation provided with each application. Memorizing the shortcuts can take some time, but the time you save in the long run is worth it.

Changing Your Preferences

Setting your preferences is important when you're working with new software. Understanding what your preferences can do for you gives you a good idea about what the software does. All applications in the Creative Cloud have different preferences; however, the way the Preferences dialog box works in each application is the same.

You can open the Preferences dialog box in each application by choosing Edit ➪ Preferences (Windows) or *Application Name* ➪ Preferences ➪ General (Mac). The Preferences dialog box opens, as shown in Figure 4-7. Click an item in the list on the left side of the dialog box to navigate from one topic to the next.

Note: At the time of this writing, Adobe XD does not have a preference menu.

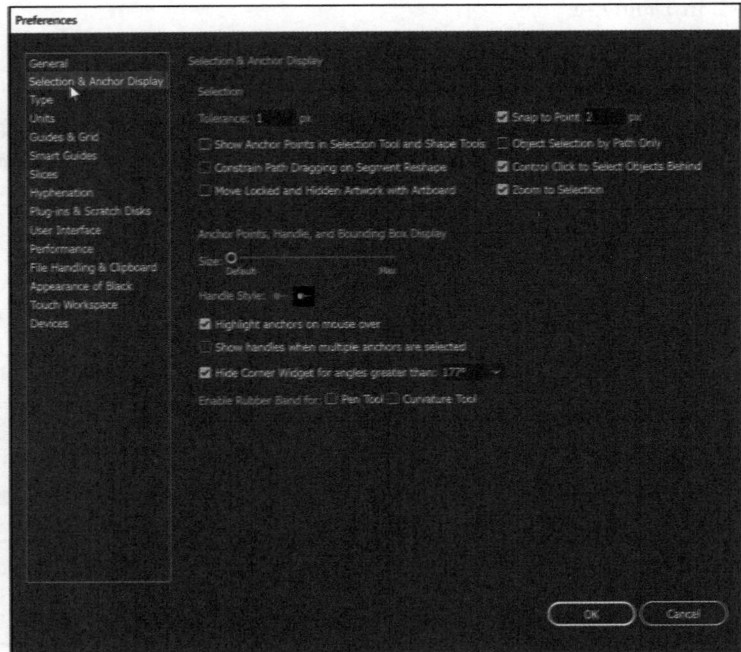

FIGURE 4-7: Adobe Illustrator preferences. Click an item on the left to navigate among topics.

The Preferences dialog box contains a great number of settings you can control by entering values into text fields using drop-down lists, buttons, checkboxes, sliders, and other, similar controls. Preferences can be quite detailed. However, you don't have to know what each preference does or even change any of them: Most dialog boxes containing preferences are quite detailed in outlining which features the preferences control and are therefore intuitive to use. Adobe also sometimes includes a Description area near the bottom of the dialog box. When you hover the mouse over a particular control, a description of that control appears in the Description area.

In some Preferences dialog boxes, a list box on the left side of the dialog box contains the categories of preferences you can change. When you finish changing the settings in that topic, select a new topic from the list and change the settings for another topic.

In some applications, not all settings you can modify are in the Preferences dialog box. For example, in Illustrator, you can change the color settings by choosing Edit⇨Color Settings to open the Color Settings dialog box. When you hover the mouse pointer over a particular drop-down list or button, a description of that control appears at the bottom of this extremely useful dialog box.

TIP

By launching Adobe Bridge (described in Book 2) and choosing Edit⇨Color Settings, you can change your color preferences across all Creative Cloud applications at one time, as shown in Figure 4-8.

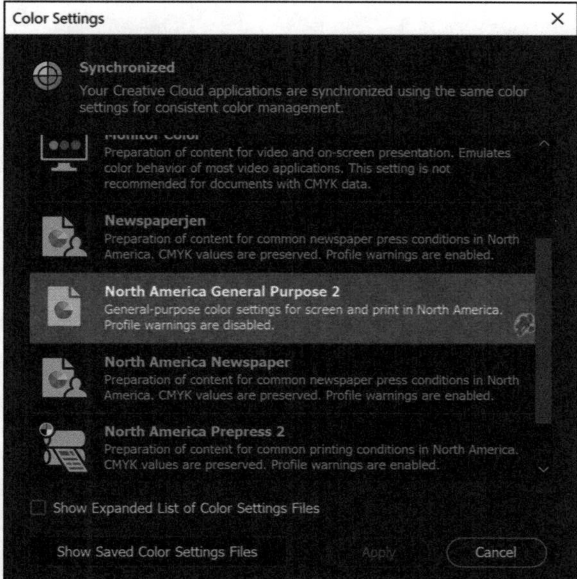

FIGURE 4-8: Change all color settings at one time using Adobe Bridge.

» Manipulating panels in the workspace

» Discovering different kinds of panels

» Getting to know the common panels in Adobe Creative Cloud

Chapter 5 Exploring Common Panels

When you first open a Creative Cloud app, you can find tools and options in three general places. As a default, the menu appears at the top, tools down the left, and then panels on the right. The panel is an integral part of working with most of the programs in the Creative Cloud because they contain many of the important options for features that you need in order to edit your documents.

The basic functionality of panels is quite similar across the programs in Adobe Creative Cloud, and the purpose of all panels is generally the same. Panels offer a great deal of flexibility in how you organize the workspace and the parts of it you use. The task you use each program for and the level of expertise you have may affect which panels you have open at a given moment. This chapter gives you an overview of how to work with the panels that are common in most of Creative Cloud applications.

Understanding the Common Workspace

One thing you immediately notice when opening applications in the Creative Cloud is the common workspace. Most of the applications look similar and have the same set of features to help you organize your workspace.

The tools in InDesign, Illustrator, and Photoshop appear on a space-saving, single-column toolbar, and panels (described in detail in the next section) are arranged in convenient, self-adjusting docks that you can widen to full size or narrow so that the panels are collapsed to icons.

Here are some pointers to help you navigate the workspace in the Creative Cloud applications:

- **To expand tools to two columns,** click the right-facing double arrows on the gray bar at the top of the Tools panel, as you see in Figure 5-1.
- **To collapse tools to a single column,** click the left-facing double arrows on the gray bar at the top of the Tools panel, as shown in Figure 5-2.
- **To expand a docked panel,** simply click the icon in the docking area. The panel you selected expands but goes away when you select a different panel.

  If you have difficulty identifying the panel, you can choose the panel you want from the Window menu.
- **To expand all docked panels,** click the left-facing double-arrow icon at the top of the docking area; put away the panels by clicking the right-facing double-arrow icon in the gray bar above them.
- **To undock a panel,** simply click the tab (where the panel name is located) and drag it out of the docking area. You can re-dock the panel by dragging the panel back into the docking area.

FIGURE 5-1: Click the arrows at the top of the Tools panel to show tools in two columns or one column.

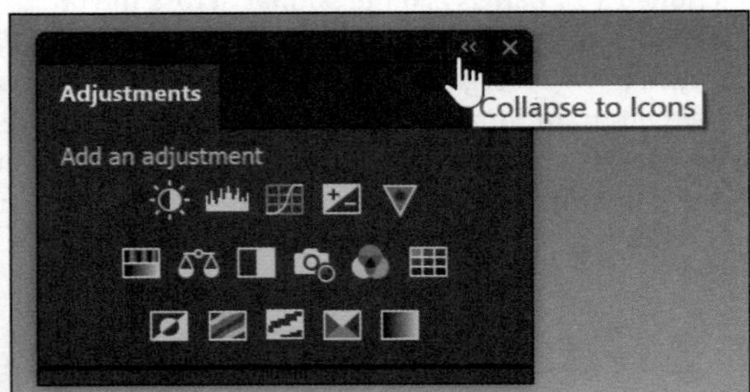

FIGURE 5-2: Click an icon to collapse or expand the panel.

Using Panels in the Workspace

Panels are small windows in a program that contain controls, such as sliders, menus, buttons, and text fields, that you can use to change the settings or attributes of a selection or an entire document. Panels may also include information about a section or about the document itself. You can use this information or change the settings in a panel to modify the selected object or the document you're working on.

Whether you're working on a Windows machine or on a Mac, panels are similar in the way they look and work. Here are the basic instructions for working with panels:

- **Open a panel:** Open a panel in a Creative Cloud program by using the Window menu: Choose Window and then select the name of a panel. For example, to open the Swatches panel (which is similar in many programs in the Creative Cloud), choose Window ⇨ Swatches.
- **Close a panel:** If you need to open or close a panel's tab or panel altogether, just choose Window ⇨ *Panel's Tab Name.* Sometimes a panel contains a close button (an X button in Windows or the red button on a Mac), which you can click to close the panel.
- **Organize the workspace:** Most Creative Cloud applications offer options for workspace organization. You can return to the default workspace, which restores panels to their original locations, by choosing Window ⇨ Workspace ⇨ Essentials (Default). You can also open frequently used panels, position them where you want, and save a customized workspace by choosing Window ⇨ Workspace ⇨ Save (or New) Workspace. Name the workspace and click OK; the saved workspace is now a menu item that you can open by choosing Window ⇨ Workspace ⇨ *Your Saved Workspace's Name.*

  TIP

  You can also choose from a wide range of included presets designed for a variety of specialized tasks.
- **Access the panel menu:** Panels have a panel menu, which opens when you click the menu icon in the upper-right corner of the panel, as shown in Figure 5-3. The panel menu contains a bunch of options you can select that relate to the active panel when you click the panel menu. When you select an option from the panel menu, it may execute an action or open a dialog box. Sometimes a panel menu has just a few options, but particular panels may have a bunch of related functionality and therefore have many options on the panel menu.
- **Minimize/maximize:** All you need to do to minimize a selected panel is click the Collapse to Icons double-arrow button on the title bar of the panel (if it's available). If the panel is undocked, you can also double-click the tab itself in

the panel. This action either partially or fully minimizes the panel. If it only partially minimizes, double-clicking the tab again fully minimizes it. Double-clicking the active tab when it's minimized maximizes the panel again.

Panels that partially minimize give you the opportunity to work with panels that have differing amounts of information, which simplifies the workspace while maximizing your screen real estate.

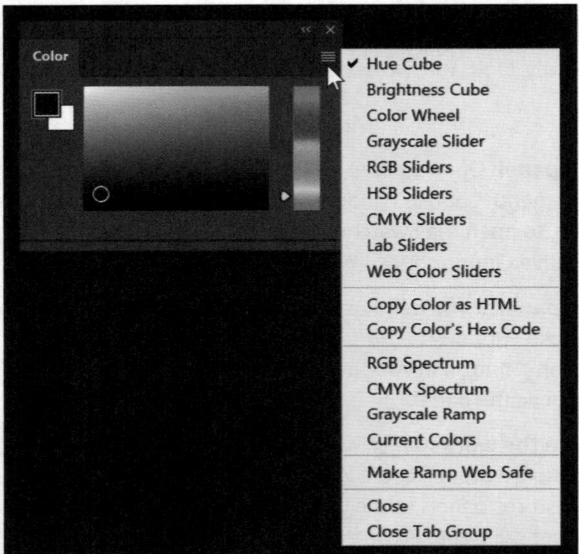

FIGURE 5-3: Displaying the panel menu.

Most panels contain tabs, which help you organize information and controls in a program into groupings. Panel tabs contain a particular kind of information about a part of the program; a single panel may contain several tabs. The name on the tab usually gives you a hint about the type of function it controls or displays information about, and it's located at the top of the panel. Inactive tabs are dimmed.

Moving panels

You can move panels all around the workspace, and you can add or remove single tabs from a panel. Each panel snaps to other panels, which makes it easier to arrange panels alongside each other. Panels can overlap each other as well. To snap panels to each other, drag the panel to a new location onscreen, as shown in Figure 5-4; you see the top bar of the panel become shaded, indicating that it's becoming part of another panel's group.

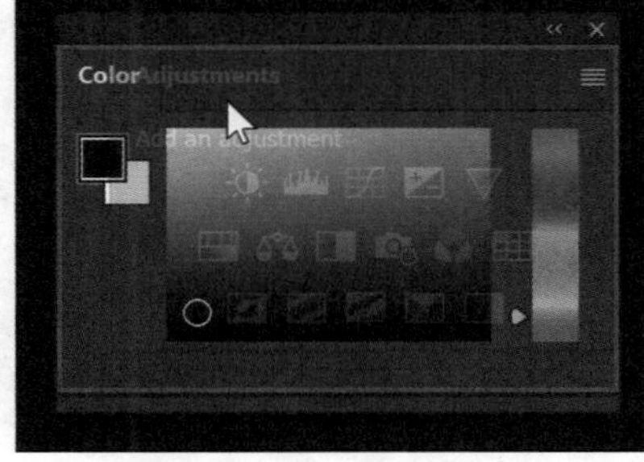

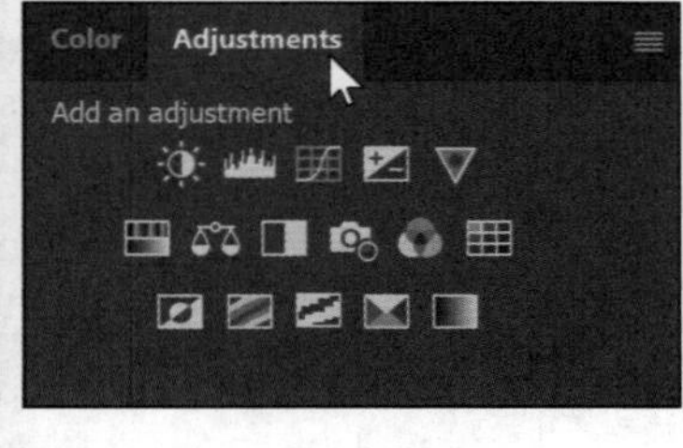

FIGURE 5-4: To move a panel, drag it by its tab to another panel.

TIP

Group similar tabs by moving them into a single grouped panel. Accessing different functions in your document becomes a lot easier because then you have less searching to do to find related functions for a task. If you want to return to the original workspace, you can choose Reset Workspace from the Window menu in the Workspace category.

TIP

You can hide all panels by pressing the Tab key. Press it again to reveal all panels you've hidden.

Looking at common panels

Many panels are similar across programs in the Creative Cloud. Although not every panel has exactly the same content in every program it's in, many have extremely similar content. You use these panels in similar ways, no matter which program or operating system you're using.

REMEMBER

Acrobat and XD do not contain numerous panels. Instead, they rely mainly (but not entirely) on a system of menus and toolbars.

The following panels are available in most, but not all, Creative Cloud programs. This list describes what you can do with each one:

- **Color:** Select or mix colors for use in the document you're working on. You can use different color modes and several ways of mixing or choosing colors in the Colors panel.
- **Info:** See information about the document itself or a particular selection you've made. The Info panel includes information on the size, positioning, and rotation of selected objects. You can't enter data into the Info panel: It only displays information without accepting it, so you have to use the Transform panel (described in this list) to make these modifications, if necessary.
- **Swatches:** Create a swatch library, which can be saved and imported into other documents or other programs. You can store on the Swatches panel any colors and gradients you use repeatedly.
- **Tools:** You use this important panel, sometimes called the *toolbox* (and not available in all Creative Cloud programs), to select tools — such as the Pencil, Brush, or Pen — to use in creating objects in a document.
- **Layers:** Display and select layers, change the layer order, and select items on a particular layer.
- **Align:** Align selected objects to each other or align them in relation to the document itself so that you can arrange objects precisely.
- **Stroke:** Select strokes and change their attributes, such as color, width/weight, style, and cap. The program you're using determines which attributes you can change.
- **Transform:** Display and change the *shear* (skew), rotation, position, and size of a selected object in the document. You can enter new values for each transformation.
- **Character:** Select fonts, font size, character spacing, and other settings related to using type in your documents.

2

Adobe Bridge

Contents at a Glance

- The Adobe Bridge workspace
- Changing the view
- Changing the workspace
- Using folders in Adobe Bridge
- Making a favorite
- Creating and locating metadata

Chapter 1
Organizing and Managing Your Files with Adobe Bridge

Adobe Bridge is an application included in the Creative Cloud collection of tools. Using Adobe Bridge, you can also organize and manage images, videos, and audio files, as well as preview, search, and sort your files without opening them in their native applications.

With Bridge, you can easily locate files using the Filters panel and import images from your digital camera right into a viewing area that allows you to quickly rename and preview your files. Adobe Bridge will save you time and frustration when looking for files you have worked on in the past; read this chapter to find out how to take advantage of its simple yet useful features.

Getting to Know the Adobe Bridge Workspace

Before you start, locate and launch Adobe Bridge. In the next few steps you will be taken through the steps to install Adobe Bridge, using the Creative Cloud app, and then shown how to launch the application.

If you open your Programs folder (PC) or Application folder (Mac), you will see a folder named Adobe. If you do not see this folder, you will want to review Book 1, "Getting Started with the Creative Cloud Suite," in order to find out how to download and install your Creative Cloud applications. This folder contains the applications that you downloaded and installed from the Creative Cloud app. If you do not see Adobe Bridge listed, follow these steps:

1. **Locate your Creative Cloud application in either your Programs (PC) or Applications (Mac) folder and double-click it to launch it.**
2. **In the Creative Cloud app, click the All Apps tab, scroll down until you see the Adobe Bridge app, and then click the Install button to the right. Follow the directions provided in order to install the application.**
3. **After Bridge is installed, you can press the Open button on the right to launch it, or you can find the application in your Programs or Applications folder.**

Now that you have launched Adobe Bridge, you might be wondering where to start. In this chapter, we introduce the file management features Adobe Bridge offers in the panels and other menu items shown in Figure 1-1.

Here are some additional details about these components:

- **Application bar:** The Application bar provides buttons for essential tasks, such as navigating the folder hierarchy, switching workspaces, and searching for files. The Application bar is discussed in detail later in this chapter.
- **Path bar:** The Path bar shows the path for the folder you're viewing and allows you to navigate the directory by clicking on the path locations.
- **Favorites panel:** The Favorites bar provides quick access to frequently browsed folders. Essentially these work the same as shortcuts.
- **Folders panel:** The Folders panel shows the folder hierarchy. You can use it to navigate folders.

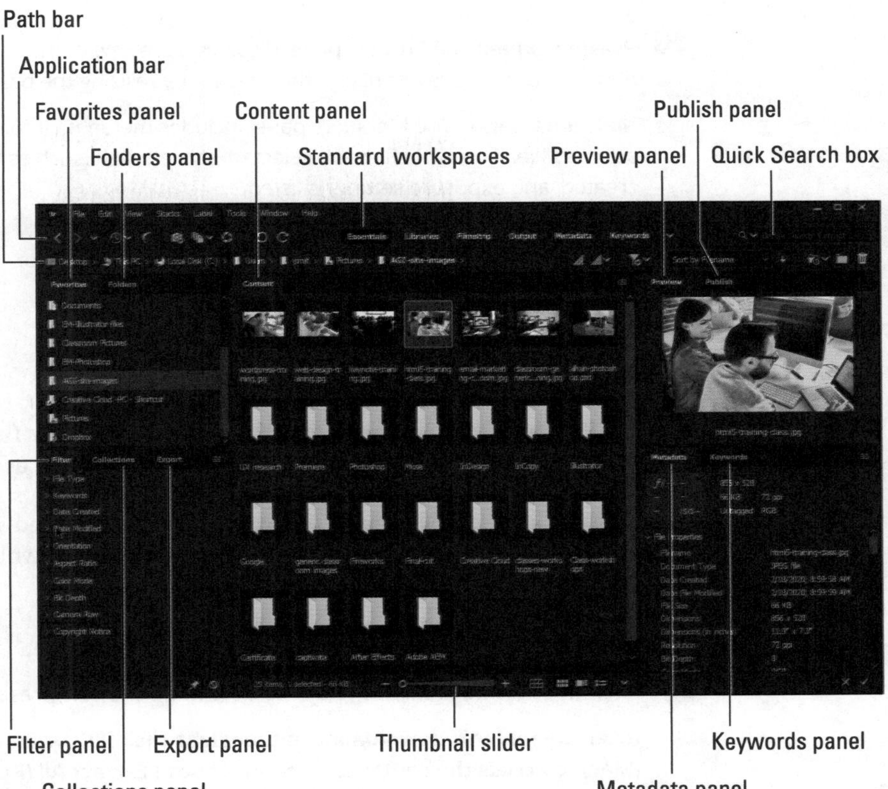

FIGURE 1-1: The Adobe Bridge workspace.

- **Filter panel:** The Filter panel lets you sort and filter files that appear in the Content panel. You can use this panel to easily find files of the same type, files created on the same date, and more.
- **Collections panel:** The Collections panel offers you the opportunity to use shortcuts to create collections of files that you can easily locate without moving in your directory.
- **Export panel:** The Export panel allows you to drag and drop images to export presets. This is helpful when you are creating batches of images that need to be saved in different sizes and formats.
- **Content panel:** The Content panel displays files specified by the navigational menu buttons, Path bar, Favorites panel, Folders panel, or Collections panel. Basically, these include thumbnails of your files within your selected folder.
- **Publish panel:** The Publish panel lets you upload photos to Adobe Stock from within the Bridge app. This might be used if you are a photographer who wants to submit images to Adobe Stock.

» **Preview panel:** The Preview panel displays a preview of the selected file or files. You can reduce or enlarge the preview by resizing the panel.

» **Metadata panel:** The Metadata panel includes metadata information for the selected file. If multiple files are selected, shared data (such as keywords, date created, and exposure setting) is listed.

» **Keywords panel:** The Keywords panel helps you organize your images by attaching keywords to them.

Practice makes perfect

Bridge offers an intuitive approach to using your system file directory. Keep in mind that Adobe Bridge is really just a visual method of accessing your files. It may help to dive right in and try out some of Adobe Bridge's features to understand its benefits. Some step-by-step lessons are included in this minibook; if you would like to follow along with them, you can download work files at `www.agitraining.com/dummies`.

1. **Go to** `https://www.agitraining.com/dummies`.
2. **Click the DummiesCCFiles folder to download the compressed folder.**
3. **After the folder is downloaded, either right-click (PC) or Control-click (Mac) to access the contextual menu. Choose Extract All (PC) or Unzip (Mac) and locate an easy-to-find location, perhaps the desktop.**

Finding and using your folders within Adobe Bridge

Let's say that you want to locate a picture in a folder of images that are named using numbers instead of descriptive names. By opening that folder in Adobe Bridge, you can see thumbnails for each image and even locate important information about each file's site, resolution, and more. To do so, follow the steps with your own files or download the DummiesCCfiles folder as instructed earlier:

1. **Select the Essentials workspace in Adobe Bridge.** This is the first button in the upper-right of the Bridge Workspace.
2. **Click the tab that says Folders, in the left column, to make sure it is brought forward.**

   If you have files located on your hard drive, you can navigate to them using the named folders icons in the Folders panel. Keep in mind that the folders and files that you see are exactly as you would see them if you were using your regular directory system.

3. **Click on Documents in the Folders panel to see saved documents that are stored in that folder. If you saved your DummiesCCFiles on the desktop, click on Desktop to see the folder in the Content pane, along with any other files that you might have on your desktop.**

4. **In the Content pane, double-click the DummiesCCFiles folder to open the contents.**

   You see additional folders that contain files that will be referenced throughout the book.

5. **Double-click the folder named Book02_Bridge to see the images within that folder, as well as a subfolder, as shown in Figure 1-2.**

6. **Now, double-click the Hockey folder. This folder contains images that were named automatically with a digital camera, as shown in Figure 1-3.**

   Finding the right file to open is much easier when you can see thumbnails of the images instead of just the name.

7. **Double-click the image to have it launch Adobe Photoshop.**

TIP

   Photoshop is typically the default destination for image files such as the JPGs in this folder, but you can change file associations by choosing Edit ➪ Preferences ➪ File Type Associations (PC) or Adobe Bridge ➪ Edit ➪ Preferences ➪ File Type Associations (Windows).

8. **Choose File ➪ Close in Adobe Photoshop and return to Adobe Bridge.**

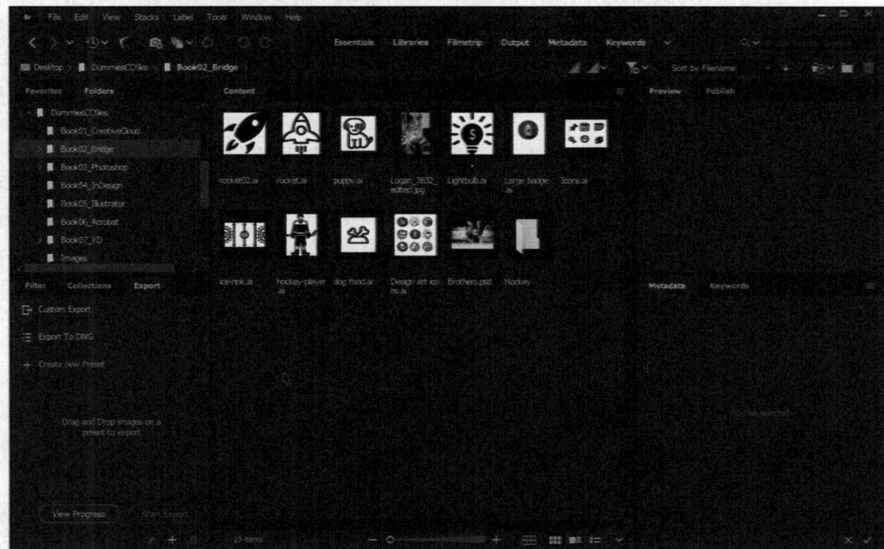

FIGURE 1-2: You can visually see the contents of a folder using Adobe Bridge.

FIGURE 1-3: A visual display of images in a folder.

TIP

From most Creative Cloud applications, you can choose File ⇨ Browse in Bridge to pop back into the Bridge application.

The default workspace

As a default, Bridge is opened in the default Essentials workspace. In this view, you have three columns that you can resize by clicking and dragging the borders separating the columns. This view also provides you with a Content pane, in the center, that displays thumbnails documents in your selected folder.

In addition to the Essentials workspace, there are several other preconfigured workspaces that you can access from the Application bar, as shown in Figure 1-4. Depending upon the information you want to showcase, you might want to switch from one workspace to another. Switching workspaces does not change the locations of your files and folders; it just changes the view.

Experiment with these to see which works best for you.

Here are explanations of the preconfigured workspaces:

- **Metadata:** Displays the Content panel in List view, along with the Favorites, Metadata, and Filter panels.
- **Essentials:** Displays the Favorites, Folders, Filter, Collections, Content, Preview, Metadata, and Keywords panels.

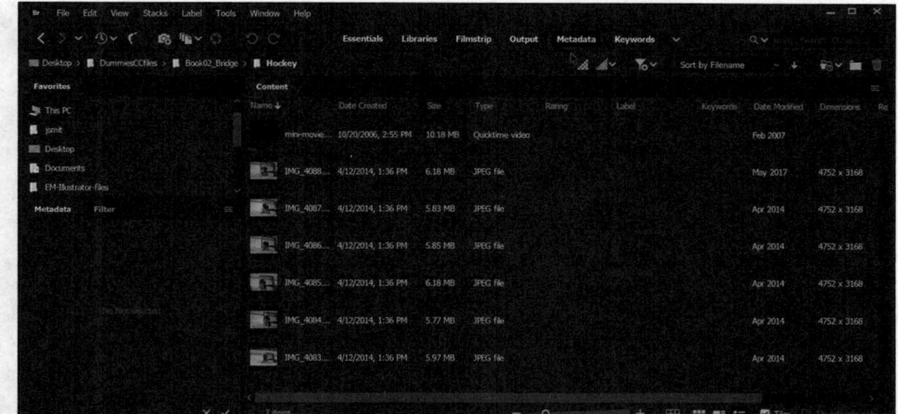

FIGURE 1-4: Change your workspace in the Application bar. The Metadata view was selected in this example.

- **Filmstrip:** Displays thumbnails in a scrolling horizontal row along with a preview of the selected item. This workspace also displays the Favorites, Folders, Filter, and Collections panels.
- **Keywords:** Displays the Content panel in Details view, as well as the Favorites, Keywords, and Filter panels.

Looking for a better view

You may want to resize your thumbnails in the Content pane, especially if you have many files. You can easily do this by using the Thumbnail slider in the lower portion of the Bridge workspace.

You can also change the presentation of the content by using the View options in the lower-right corner, as shown in Figure 1-5. Your choices are

- Click to lock to thumbnail grid
- View content as thumbnails
- View content as details
- View content as list

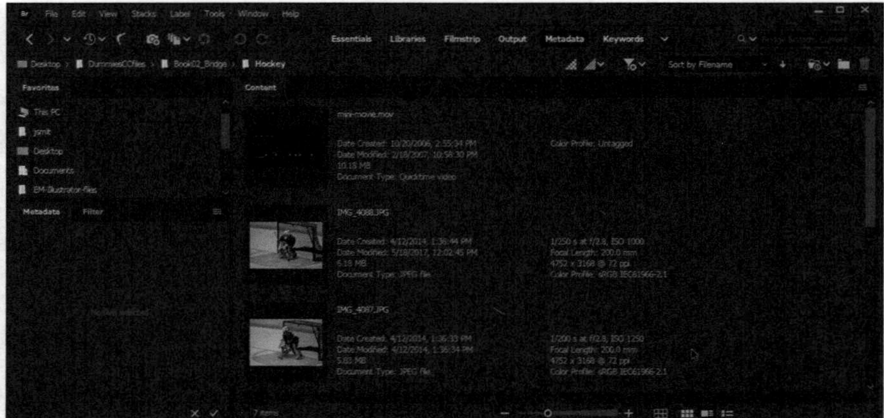

FIGURE 1-5: Change your view to bring out additional details. View content as details was selected for this example.

Creating a new folder

Perhaps you want to create a new folder. Whether you create a new folder in your system File directory or in the Finder for the Mac, it will be recognized by Bridge. In this next example, you navigate back using the Path bar, and create a new folder in the Book02_Bridge folder:

1. **Make sure you are still viewing the contents of the Hockey folder. Note that the location your are in is listed in the Path bar as shown in Figure 1-6.**

   This is the directory path to the folder that you have active and open at this time.

2. **Click on Book02_Bridge in the Path bar to navigate back to that folder.**
3. **To create a new folder inside this folder, click the Create New Folder button in the upper right, as shown in Figure 1-7, or right-click (PC)/Control-click (Mac) and select New Folder from the contextual menu that appears.**

   A new folder appears, ready to be renamed.

4. **Name the file** Rockets.

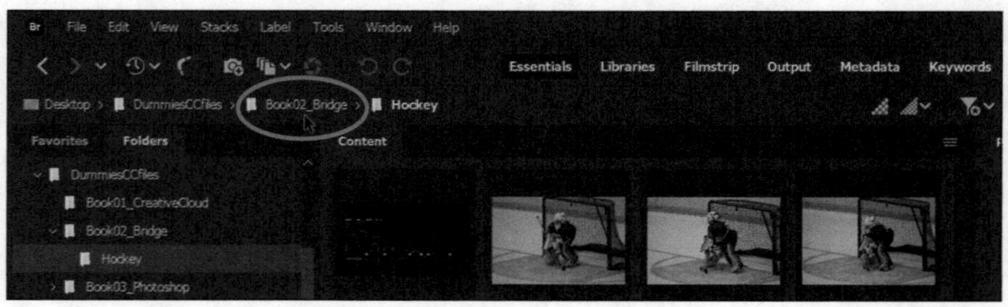

FIGURE 1-6: Use the Path bar to navigate back through your folders.

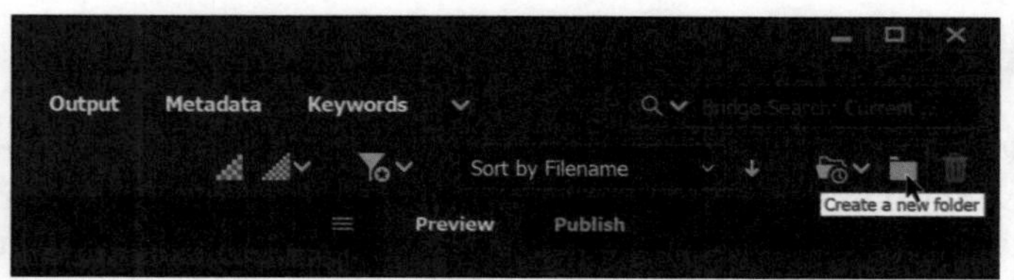

FIGURE 1-7: You can create new folders right in Adobe Bridge.

Moving a file to another folder

Next you select and move some files into your newly created folder:

1. **Click on one of the Rocket images and then, holding down the Ctrl key (PC) or Command key (Mac), click on the other rocket image.**

   Using the Ctrl or Command key allows you to select multiple files.

2. **With both rocket images selected, click and drag them over to your new Rockets folder, as shown in Figure 1-8. Release your mouse when the files are on top of the folder to move them into that folder.**

TIP

You can also move and copy files using the contextual menu. Right-click (PC) or Control-click (Mac) on any image file to see the contextual menu that appears with options that include moving and copying the selected file.

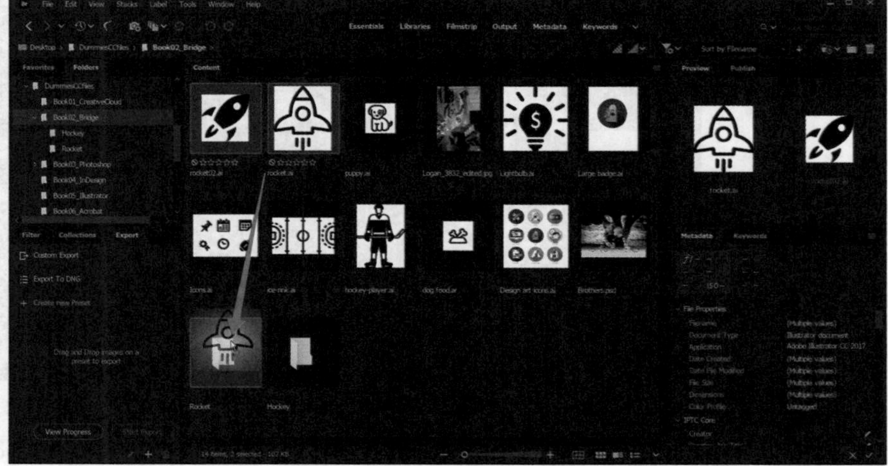

FIGURE 1-8: Move images from one folder to another in Adobe Bridge.

Making a favorite

When working in the Creative Cloud, you may return to certain folders over and over again. In that situation, it would help to have some Favorite folders established and saved in the Favorites panel. When your most popular folders are saved as favorites, accessing them requires only one click.

Follow these steps in order to save your own favorite. In the following steps, the files from the DummiesCCFile are used. You can follow along with your own files and folders if you like.

1. **Using the Folders panel, navigate your file directory system until you locate the DummiesCCFile folder. If you have followed instructions it should be easy to find on the Desktop.**

   In the example shown in Figure 1-9, the folder is located on the desktop.

2. **Make sure that the Favorites panel is forward. Check this by clicking on the tab that says Favorites.**
3. **Click on the DummiesCCFiles folder and drag it to the Favorites panel. Release the mouse when you see a blue line that indicates where the favorite will appear in the present list, as shown in Figure 1-9.**

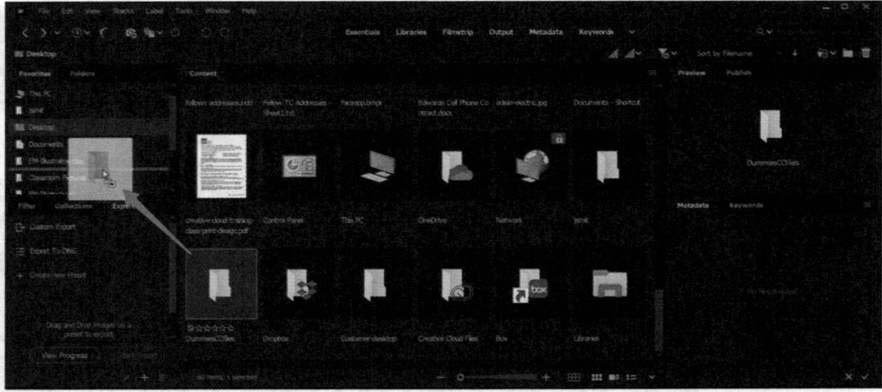

FIGURE 1-9: Click and drag a folder into the Favorites panel.

Now that you have saved the DummiesCCFiles as a favorite, you can access the folder anytime you like simply by clicking on the folder name in the Favorites panel.

Investigating the Application bar

Now that you have spent a little bit of time working in Bridge, the Application bar and its tools will make more sense to you. The Application bar runs along the top of the Bridge workspace. The tools and features are shown in Figure 1-10.

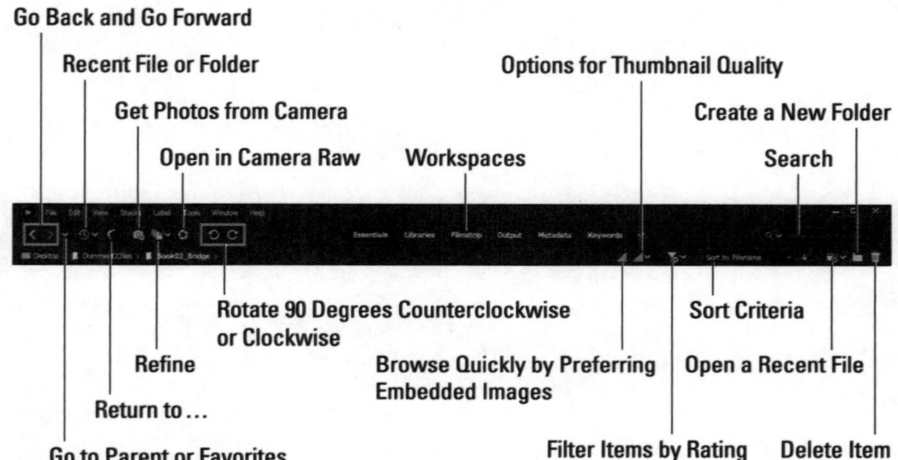

FIGURE 1-10: The Application bar.

There are many features included in the Application bar that help you with navigating, acquiring, and organizing your files. Some important features include Get Photos from Camera, which is covered in Chapter 3 of this minibook, and Sort Criteria, which is covered in Chapter 2.

Developing a Bridge habit

Now that you have experienced using Adobe Bridge, try making it a habit. Instead of selecting File⇨Open, choose File⇨Browse in Bridge and start recognizing the time savings.

» Creating a metadata template

» Searching for files using Adobe Bridge

» Searching by name or keyword

» Using the Filter panel

» Saving a collection

Chapter 2

Taking Advantage of Metadata in Adobe Bridge

Metadata is information that can be stored with images. This information travels with the file and makes it easy to search for and identify the file. Use metadata features and templates to help you save and find your files more easily.

You find out how to attach metadata to your images by following some step-by-step lessons included in this chapter. If you would like to follow along with them, you can download work files at `https://www.agitraining.com/dummies`.

Locating Your Files

You can follow these steps with your own files, or you can download the DummiesCCfiles folder, as instructed in Chapter 1 of this minibook.

1. **In Bridge choose Window ➪ Workspace ➪ Reset Standard Workspaces.**

   This ensures that you are in the Essentials view and that all the default panels for Adobe Bridge are visible. Alternatively, click Essentials in the Application bar at the top-right of the Bridge workspace. You might need to maximize your Bridge window after you reset the workspace.

2. **Click the Folders tab to make sure it is brought forward.**

   If you have files located on your hard drive, you can navigate to them using the named folders icons in the Folders panel.

3. **Click Documents in the Folders panel to see saved documents that are stored in that folder. If you saved your DummiesCCFiles on the desktop, click Desktop to see in the center Content pane the files you have sitting on the desktop.**

4. **In the Content pane, double-click the DummiesCCFiles folder to open the contents.**

   You see additional folders that contain files that will be referenced throughout the book.

5. **Double-click the folder named Book02_Bridge to open it and see the images within that folder, as well as a subfolder named Hockey.**

6. **Double-click on the folder named Hockey to see the images that are located inside the folder, as shown in Figure 2-1.**

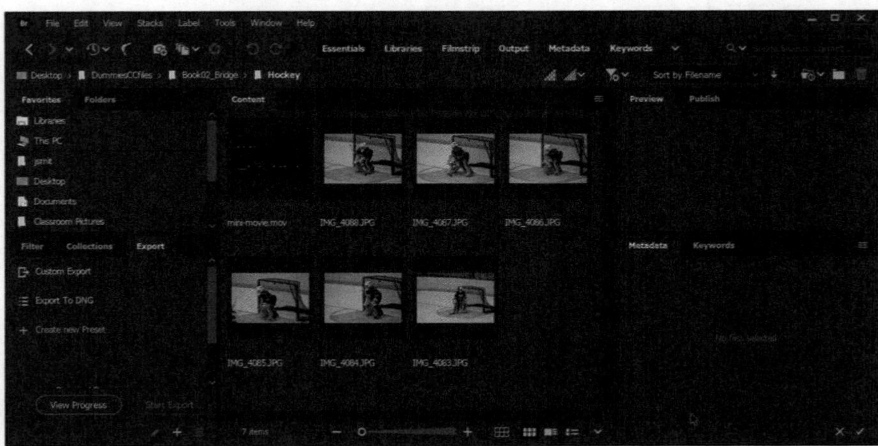

FIGURE 2-1: The contents of the Hockey folder which is inside the Book02_Bridge folder.

7. **Click once on IMG_4088.JPG, and look for the Metadata and Keywords panels in the lower-right area of the Adobe Bridge workspace.**

8. **If the Metadata panel is not visible, click the Metadata panel tab. In this panel, as shown in Figure 2-2, you see the image data that is stored**

with the file. Take a few moments to scroll through the data and view the information that was imported from the digital camera that was used to take the photo.

9. **If necessary, click the arrow to the left of IPTC Core to reveal its contents.**

   IPTC Core is the schema for XMP that provides a smooth and explicit transfer of metadata. Adobe's Extensible Metadata Platform (XMP) is a labeling technology that allows you to embed data about a file, known as metadata, into the file itself. With XMP, desktop applications and back-end publishing systems gain a common method for capturing and sharing valuable metadata.

10. **On the right side of this list, notice a series of pencil icons. These icons indicate that you can enter information in these fields.**

    Note: If you are not able to edit or add metadata information to a file, it could be locked. You can check by right-clicking on the file directly in Bridge and selecting Reveal in Explorer (Windows) or Reveal in Finder (Mac OS). In Windows, right-click the file, choose Properties, and uncheck Read-only; in Mac OS, right-click the file, choose Get Info, then change the Ownership (Sharing) and Permissions to Read and Write.

11. **Scroll down until you can see Description Writer and click the pencil next to it. All editable fields are highlighted, and a cursor appears in the Description Writer field.**

12. **Type your name, or type** Picture-by-me.

13. **Scroll up to locate the Description text field. Click the Pencil icon to the right and type** Goalie in net **to add a description for the image, as shown in Figure 2-3.**

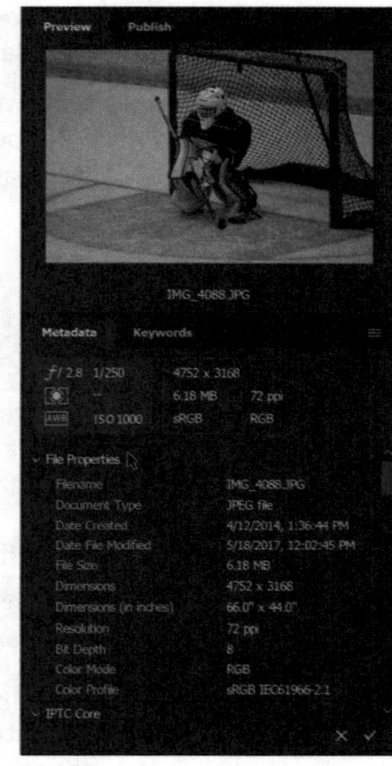

FIGURE 2-2: Scroll through the Metadata panel on the right side of the Bridge workspace.

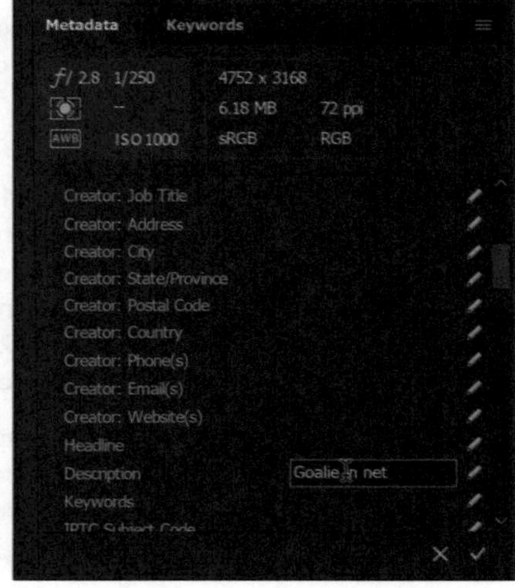

FIGURE 2-3: The description added to the image's metadata.

14. **Click the Apply button, located in the bottom-right corner of the Metadata panel, to apply your changes.**

    You have now edited metadata that is attached to the image; this information will appear whenever someone opens your image in Bridge or views the image information in Adobe Photoshop using File ➪ File Info.

Using Keywords

Using keywords — logical words to help you locate images more quickly — can reduce the amount of time it takes to find an image on a computer. Here's how you can add your own keywords:

1. **Keep the IMG_4088.JPG image selected, and then click the Keywords tab, which appears behind the Metadata panel.**

   A list of commonly used keywords appears.

2. **Click the New Keyword button at the bottom of the Keywords panel, as shown in Figure 2-4. Type** Goalie **into the active text field, and then press Enter (Windows) or Return (Mac).**

3. **Click to select the empty checkbox to the left of the Goalie keyword. This adds the Goalie keyword to the selected image.**

4. **With the Goalie keyword still selected, click the New Sub Keyword button. This is the button to the left of the New Keyword plus sign button. Type** Boy **into the active text field, then press Enter (Windows) or Return (Mac).**

5. **Click on the empty checkbox to the left of the Boy keyword to also apply it to this image.**

   You have now assigned a keyword and a sub keyword to the IMG_4088.JPG image. Now you will add a second keyword.

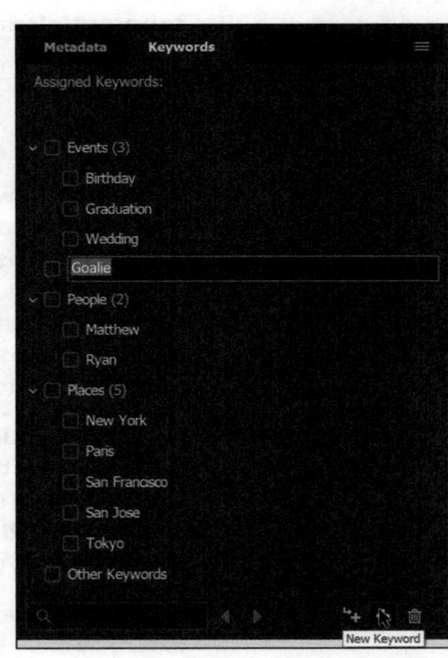

FIGURE 2-4: Click the New Keyword button to add your own unique keyword to the file.

6. **Select the Goalie keyword, and then click the New Keyword button at the bottom of the Keywords panel; a blank text field appears. Type** Peewee **and press Enter (Windows) or Return (Mac). Then, select the checkbox next to Peewee to assign the keyword to this image.**

7. **Right-click (Windows) or Control-click (Mac) on the Peewee keyword, and choose the option Rename. When the text field becomes highlighted, type** Youth, **press Enter (Windows) or Return (Mac). Make sure the Youth checkbox remains selected.**

TIP

You can also enter information directly into the image by opening the image in Adobe Photoshop, and then choosing File ➪ File Info. The categories that appear on the top include Description, Camera Data, IPTC, and IPTC Extension, among others. After it is entered in the File Info dialog box, the information is visible in Adobe Bridge.

Creating a Metadata Template

After you have added metadata to an image, you can easily apply it to more images by creating a metadata template. In this exercise, you apply the metadata template from the IMG_4088.JPG image to other images in the same folder:

1. **Make sure that IMG_4088.JPG is selected in Adobe Bridge.**

2. **Choose Tools ➪ Create Metadata Template.**

   The Create Metadata Template window appears. If you scroll down you will see the keywords that you applied to this file. In the next few steps you will apply those same keywords to many other files. Adding keywords to events, or photoshoots can help make finding the files you need easier.

3. **In the Template Name text field (at the top), type** Game photos.

   In the Create Metadata Template window, you can choose the information that you want to build into a template. In this exercise, we choose information that already exists in the selected file, but if you wanted to, you could add or edit information at this point.

4. **Select the checkboxes to the left of the following categories: Description, Description Writer, Keywords, and Date Created and then click Save, as shown in Figure 2-5.**

   You have just saved a template. Next, you apply it to the other hockey images in this folder.

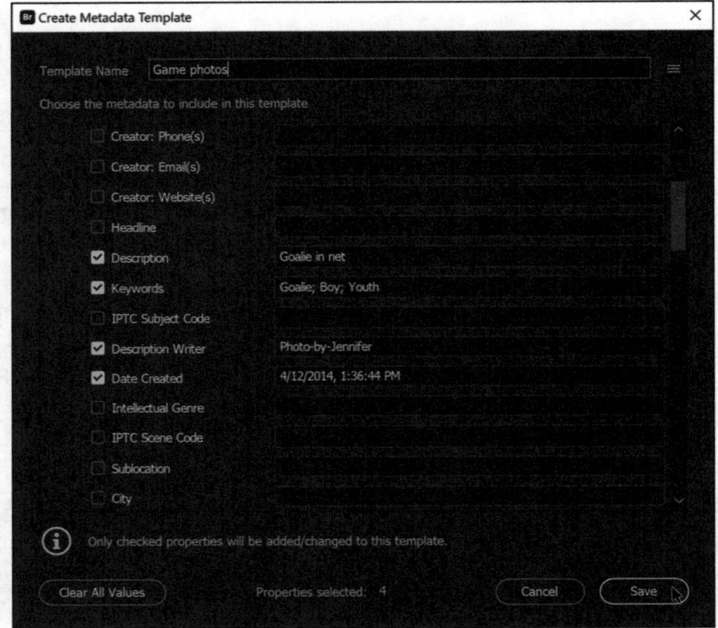

FIGURE 2-5: Select a file and check the information you want to save into a metadata template.

5. **Select the IMG_4087.JPG image, press and hold the Ctrl (Windows) or Command (Mac) key, and select the IMG_4083.JPG image.**

   All the images are selected.

6. **Choose Tools ⇨ Append Metadata and select Game Photos, as shown in Figure 2-6. Note that you can also choose Replace Metadata if you want to eliminate existing metadata.**

   The same metadata has now been applied to all the images at once.

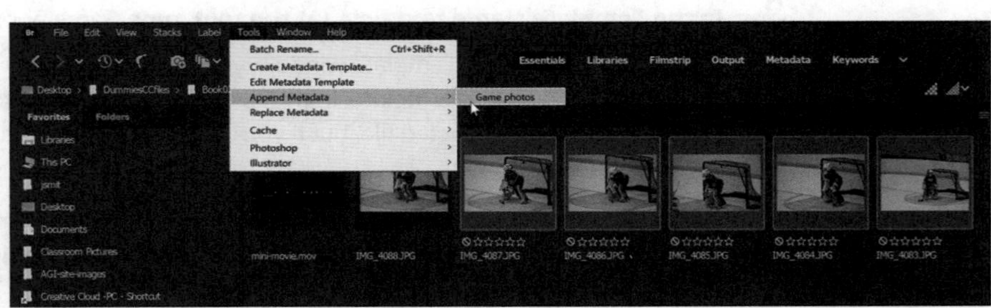

FIGURE 2-6: Choose the metadata template you want to use to add metadata to an image or images.

Searching for Files Using Adobe Bridge

Find the files that you want quickly and easily by using the Search tools built directly into Adobe Bridge and by taking advantage of the Filter panel.

In this example, you have a limited number of files to search within, but even so, you can still see how helpful these search features can be.

Searching by name or keyword

The benefit of adding metadata to your images is that you can use it to find your files later. Using the Find dialog box in Adobe Bridge, you can narrow your criteria down to make it easy to find your files when needed. Here's how:

1. **Make sure that you are still viewing the content in the Hockey folder.**
2. **Choose Edit ⇨ Find, or use the keyboard shortcut Ctrl+F (Windows) or Command+F (Mac OS).**

   The Find dialog box appears.
3. **Select Keywords from the Criteria drop-down menu, and type** Goalie **into the third text field (replacing Enter Text), as shown in Figure 2-7. Then press Enter (Windows) or Return (Mac).**

   Because you are looking within the active folder only, you get a result immediately. The image files that you tagged with the Goalie keyword are visible.

   Note: You can click the plus sign to the right of the criteria to add additional search types.

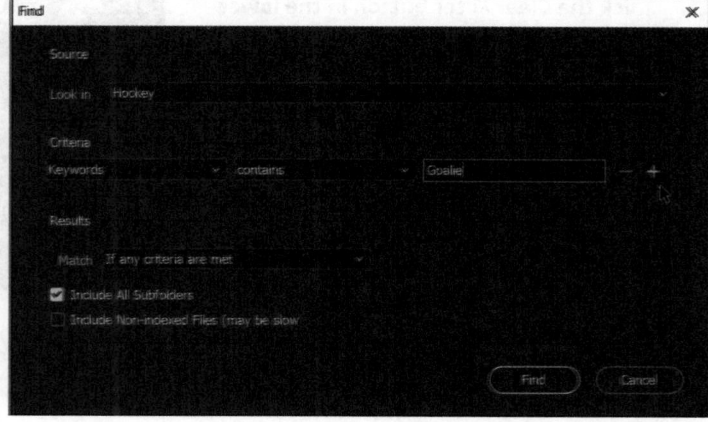

FIGURE 2-7: Search your folders using the tools built right into Adobe Bridge.

4. **Clear the search by clicking the X icon to the right of the New Search icon at the top of the results pane, as shown in Figure 2-8.**

FIGURE 2-8: Cancel the search results by clicking on the X in the Content pane.

Using the Filter panel

You can use the Filter panel to locate files whose location you can't remember. With the Filter panel, you can look at attributes such as file type, keywords, and date created or modified to narrow down the files that appear in the content window of Adobe Bridge. The following steps show you how:

1. **Make sure that you are still viewing the content of the Hockey folder. Notice that the Filter panel collects the information from the active folder, indicating the keywords that are being used, as well as modification dates and more.**
2. **Click to turn down the arrow next to Keywords in the Filter panel, and select Goalie from the list; notice that only images with the Goalie keyword applied are visible. Click Goalie again to deselect it and view all the images.**

   Find files quickly by selecting different criteria in the Filters panel.
3. **Click the Clear Filter button in the lower right of the Filter panel to turn off any filters.**
4. **Experiment with investigating file types as well. Only file types that exist in the selected folder appear in the list. If you are looking for an Adobe Illustrator file, you might see that there are none located in this folder, but you will see a QuickTime video file that you can select and preview right in Adobe Bridge, as shown in Figure 2-9.**

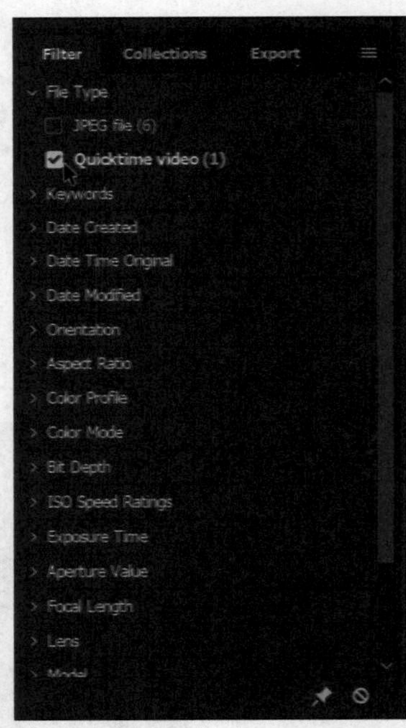

FIGURE 2-9: You can select File Types from the Filter panel to locate them easily.

5. **Again, click the Clear Filter button in the lower-right area of the Filter panel to turn off any filters. The Clear Filter button looks like a circle with a slash through it.**

Saving a Collection

If you like using favorites, you'll love using collections. A collection allows you to take images from multiple locations and access them in one central location. Understand that when you use a collection, Adobe Bridge essentially creates a shortcut, or alias, to your files and does not physically relocate them or copy them to a different location.

1. **If your Collections tab is not visible, choose Window ➪ Collections Panel or click the tab next to Filter.**

   The Collections panel comes forward.

2. **Click the gray area in the Content pane to make sure that nothing is selected, and then click the New Collection button in the lower-right area of the Collections panel. Type** Pictures of hockey **into the new collection text field, as shown in Figure 2-10. Press Return or Enter to confirm your new collection.**
3. **Navigate back to the Bridge02-Folder, and then take two random images and drag them to the Pictures of hockey collection.**
4. **Click the Pictures of hockey collection folder; notice that even though you can easily access the files you added to the collection, the files remain intact in their original location.**

FIGURE 2-10: Create a collection to keep selected Pictures easily accessible.

Taking Advantage of Metadata in Adobe Bridge

» Batch renaming your files

» Exporting your files

» Importing files as layers

Chapter 3
Using Automation Tools in Adobe Bridge

You have already discovered how Adobe Bridge can save you time by helping you organize and more easily find your images. In this chapter, you find out how to use automated features that cut out redundant tasks and save even more time.

Getting Files from Your Camera

With the Get Photos from Camera feature, you can connect your camera, card reader, or mobile device to the computer using a supported cable. If you want to give this feature a try, follow these steps:

1. **Simply attach your phone via USB, insert a card into your card reader, or connect your camera via cable.**

   Note: At the time of writing this feature was not working correctly with iPhones on the Mac platform.

2. **Launch Bridge.**
3. **Click the Get Photos from Camera button. The Adobe Bridge CC – Photo Downloader dialog box appears, as shown in Figure 3-1.**

4. **Click open the Get Photos From drop-down menu and select your device.**

   Note: If Bridge does not recognize your device, make sure that you have the latest version installed by launching your Creative Cloud App. Look in the Apps section to see if an update is available. You can also try connecting your device when Bridge is not open and launching it after the device is connected.

   In the Create Subfolders drop-down menu, you can use a standard format for naming your folders or choose Custom to create your own naming convention.

5. **If you choose to have the Adobe Downloader rename your files, you can choose a standard naming convention using the date, or choose Custom name and create your own custom name.**

6. **The options at the bottom offer you the opportunity to see the files in Bridge, convert the files to DNG (Digital Negative Files), delete the files off the device, and save them to another location.**

7. **Click the triangle to the left of the Advanced Dialog button to see a more detailed preview of your files. In this more detailed view, you can individually select and deselect files to be downloaded, as shown in Figure 3-2.**

8. **In Save Options, select a location to which you wish the files to be transferred. And press Get Media.**

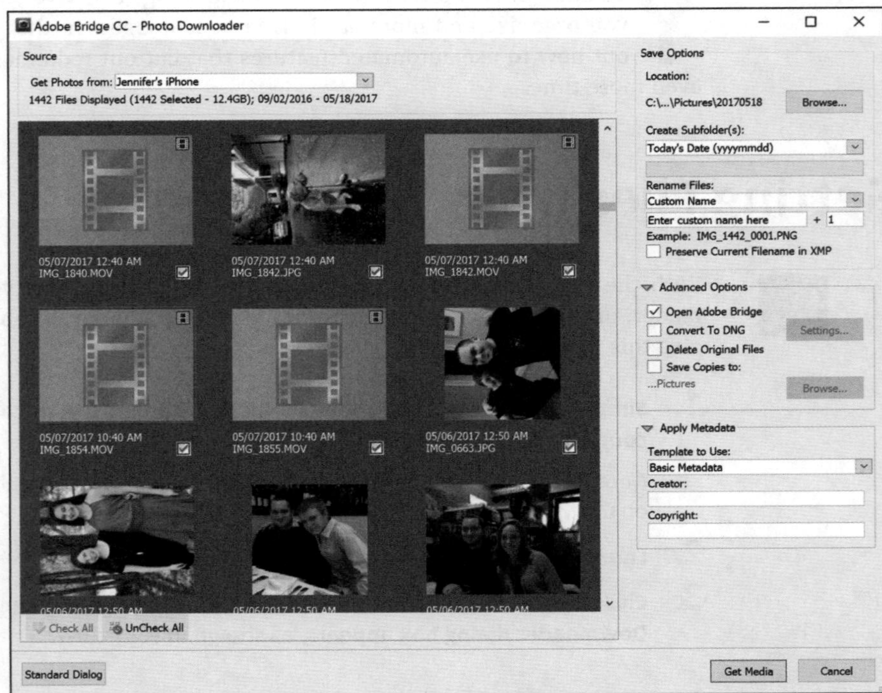

FIGURE 3-1: Download your photos quickly using the Photo Downloader available in Adobe Bridge.

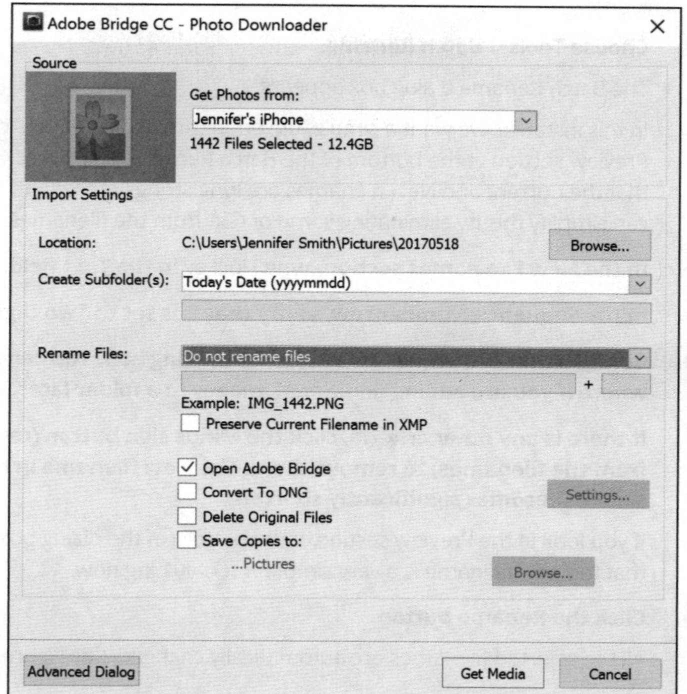

FIGURE 3-2: The Advanced view allows you to select your images.

Automation Tools in Adobe Bridge

Adobe Bridge provides many tools to help you automate tasks. In this section, you discover how to use a few helpful features that will save you from repeating redundant tasks.

Batch renaming your files

You might have noticed that in the Book02Bridge ⇨ Hockey folder, many files contain generic filenames. These images were downloaded from *a digital camera*, and instead of changing the names immediately, we opted to change them simultaneously using the batch rename feature in Adobe Bridge. To do so, follow these steps:

1. **In Bridge, navigate to the Book02Bridge folder and then open the Hockey folder that is inside.**
2. **Choose Edit ⇨ Select All, or press Ctrl+A (Windows) or ⌘ +A (Mac).**

3. **Choose Tools ⇨ Batch Rename.**

   The Batch Rename dialog box appears.

   In this instance, we want a simple, uncomplicated name. If you look in the Preview section at the bottom of the Batch Rename dialog box, you can see that the Current and New filenames are long strings of text and numbers. You can simplify this by eliminating some of text from the filenames.

4. **In the New File names section, type** Hockey **in the Text field.**
5. **In the Sequence Number row, verify that it is set to Two Digits.**
6. **Confirm that the sequence number is starting at 1. You can start it anywhere if you are adding additional images to a folder later.**
7. **If there is any other criteria, click the Minus sign button (remove this text from the filenames) to remove them. The New filename in the Preview section becomes significantly shorter.**

   If you look in the Preview section at the bottom of the dialog box, you can see that the new filename is a very simple Hockey01.jpg now.

8. **Click the Rename button.**

   All the selected filenames are automatically changed. See Figure 3-3.

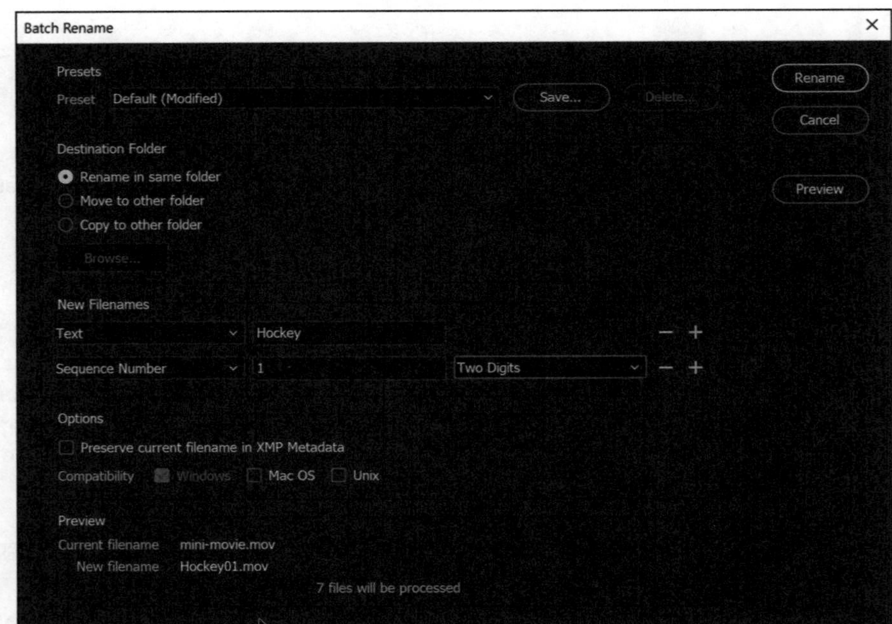

FIGURE 3-3: You can change multiple files names simultaneously in Adobe Bridge.

Exporting your files from Bridge

If you are looking for a method in which to save batches of files in a specific size and format, you might be interested in the Export tab. Using the Export feature in Photoshop allows you to prepare assets for projects like web pages and applications. Using this feature, you can easily select all your images and drag them on top of a preset in the Export tab.

If you want to choose custom settings, you can click and drag your images on top of Custom Export. You will see the dialog box shown in Figure 3-4.

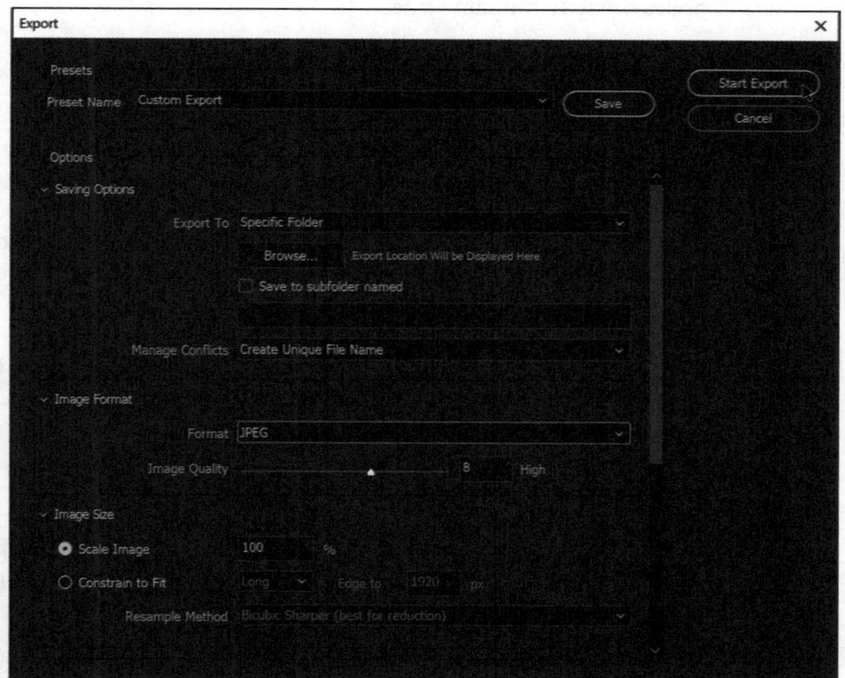

FIGURE 3-4: Resize and export files in other formats quickly using the Export feature in Bridge.

Loading files as layers

Adobe Bridge comes with a variety of Photoshop tools that you can use in Bridge as well. In this example, you select three images that you want to incorporate into one composited image. Instead of opening all three images and cutting and pasting or dragging them into one file, you use the Load Files into Photoshop Layers feature.

Make sure that you are still in the Hockey folder. Hold down your Ctrl (Windows) or ⌘ (Mac) key and click on any three images. All three images are selected.

Select Tools⇨Photoshop. Note that there are many tools that you can use in this menu item; for this example, select the Load Files into Photoshop Layers option. A script immediately launches Photoshop (if it is not already open) and a new layered file is created from the selected images.

You should ensure that your selected images are approximately the same pixel dimensions before running this script; otherwise, you might have to make some transformation adjustments in Photoshop. In this example, the images are approximately the same size.

Building a contact sheet

If you come from the traditional photography world, you may be familiar with contact sheets. Before digital photography, contact sheets were thumbnail images created from film negatives and were used to help identify which images should be processed. Even though a film negative environment isn't common anymore, contact sheets can be very useful when trying to identify which images you choose to save or edit.

Fortunately there is a contact feature built right into the Adobe Bridge tools. Follow these steps to create your own using your own images, or any of our sample images:

1. **Within Bridge, open a folder of images and either Ctrl/⌘-click on the images you want to include in a contact sheet or press Ctrl+A (Windows)/⌘+A (Mac) to select all the images. If you would like to use our sample image, navigate to the folder named Flowers in the Images folder.**
2. **Choose Tools⇨Photoshop⇨Contact Sheet II.**

   Photoshop is launched and the Contact Sheet II dialog box appears, as shown in Figure 3-5.
3. **Choose your paper size and resolution. The default is 8 inches x 10 inches and 300 dpi, which works fine for reliable results.**

   Choose how many thumbnails you want across and down using the rows and columns. The fewer rows and columns, the larger the thumbnail images will be.

   Check Use Filename as Caption if you want to see the filename listed underneath the thumbnail.
4. **Press OK to see the final Contact Sheet created, as shown in Figure 3-6.**

Contact Sheet II
Source Images
Use: Bridge
6 files selected
OK
Cancel
Load...
Save...
Reset...
Press the ESC key to Cancel processing images
Document
Units: inches
Width: 8
Height: 10
Resolution: 300 pixels/inch
Color Profile: sRGB IEC61966-2.1
Flatten All Layers
Mode: RGB Color
Bit Depth: 8-bit
Thumbnails
Place: across first
Columns: 5
Rows: 6
Rotate For Best Fit
Use Auto-Spacing
Vertical:
Horizontal:
Use Filename as Caption
Font: Arial Regular 12 pt

FIGURE 3-5: The Contact Sheet II dialog box offers size and resolution options.

FIGURE 3-6: The resulting contact sheet.

3

Photoshop CC

Contents at a Glance

IN THIS CHAPTER

» Discovering the Photoshop tools

» Navigating your work area

» Changing screen modes in Adobe Photoshop CC

Chapter 1 Getting into Photoshop CC Basics

In this chapter, you are introduced to the Photoshop CC work area and tools. You'll also find out how to neatly organize and hide panels. You will also find out how to do basic tasks, such as open an image, crop it to a different size, and then resave it.

The Start Screen

When you launch Photoshop, unless you have changed your preferences, you see a start screen, as shown in Figure 1-1. This start screen can making it easier for you to find files that you have recently worked on or jump right into a new document.

If you would rather not have this screen every time you launch Photoshop, you can choose Edit ⇨ Preferences ⇨ General and uncheck Auto show the Home Screen. On the Mac you would choose Photoshop ⇨ Preferences ⇨ General and uncheck Auto show the Home Screen.

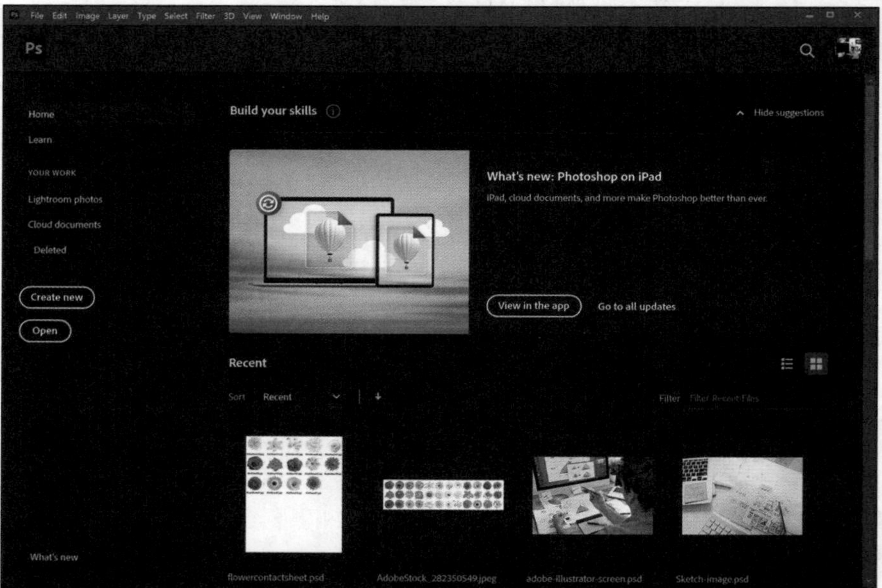

FIGURE 1-1: The Start screen in Adobe Photoshop appears as a default.

The New Document Window

If you are still using Photoshop at its default settings, you can click the New button in the start screen, or choose File⇨New. The initial New Document dialog window appears. Just like the start screen, the default New Document window may be more than you have asked for, but it is meant to offer you more control and options when creating new files. Note in Figure 1-2 that you can choose the type of document that you are working on, such as Photo, Print, Art & Illustration, Web, Mobile, Film, and Video.

Each one of these selections offers standard size and resolution defaults for the medium selected in addition to templates that you are able to open and customize for your own needs.

If this is a little much for you and you want the standard, smaller, more compact New Document dialog box, choose the Close button in this window and then select Edit⇨Preferences⇨General and check Use Legacy "New Document" Interface. (On the Mac, choose Photoshop⇨Preferences⇨General and check Use Legacy "New Document" Interface.)

FIGURE 1-2: The updated New Document dialog box has many options from which to choose. In this example the Photo category was selected.

Sample Images Can Help You Investigate More Features

Of course it helps to have an image open while exploring Adobe Photoshop. You are free to use your own or use the samples image files located here: `https://www.agitraining.com/dummies`. This downloadable file includes files that can be used throughout the entire *Adobe Creative Cloud All-in-One For Dummies* book.

1. **Go to** `https://www.agitraining.com/dummies`**.**
2. **Click on the DummiesCCFiles folder to download the compressed folder.**
3. **After the folder is downloaded, right-click (PC) to access the contextual menu and choose Extract All and locate an easy-to-find location or Control-click (Mac) and choose Open.**

   Inside the DummiesCCFiles folder, you find separate folders for each minibook. Images are available for you to use in the Book03_Photoshop folder. There are also additional miscellaneous images in the Images folder at the top level of this folder.

Opening an Image

You can open an existing Photoshop image in one of several ways:

- » If you see the start screen, you can click on the Open button and navigate to the folder that contains the image you would like to open.
- » Choose File⇨Open, select the file in the Open dialog box, and then click the Open button.
- » Choose File⇨Browse in Bridge. By selecting Browse in Bridge instead of Open, you launch the Adobe Bridge application. Read more about Adobe Bridge later in this section and also in Book 2.

REMEMBER

Photoshop can open a multitude of file formats, even if the image was created in another application, such as Illustrator or another image-editing program. However, you have to open the image in Photoshop by choosing File⇨Open or, using Adobe Bridge, by selecting an image and dragging it to the Photoshop icon on the taskbar (Windows) or Dock (Mac). If you double-click an image file (one that wasn't originally created in Photoshop, or from different versions) in a folder, the image may open only in a preview application.

If you're opening a folder of images that you want to investigate first, choose File⇨Browse in Bridge to open Adobe Bridge, the control center for Adobe Creative Cloud. You can use Adobe Bridge to organize, browse, and locate the assets you need to create your content. Adobe Bridge keeps available, for easy access, native AI, INDD, PSD, and Adobe PDF files as well as other Adobe and non-Adobe application files. Adobe Bridge does not install by default with Adobe Photoshop. If you do not have it loaded you can read through the directions and highlights of using Adobe Bridge in Book 2, "Adobe Bridge."

Getting to Know the Tools

You use tools to create, select, and manipulate objects in Photoshop CC. When you have Photoshop open, the Tools panel appears along the left side of the workspace (see Figure 1-3), and panels appear on the right side of the screen. (We discuss panels in the later section "Navigating the Work Area.")

FIGURE 1-3: The Photoshop CC workspace includes the Tools panel.

TIP

In the Tools panel, look for a tooltip when you hover the cursor over any one of the tools. Following the tool name is a letter in parentheses, which is the keyboard shortcut you can use to access that tool. Simply press the Shift key along with the key command you see to access any hidden tools. In other words, pressing P activates the Pen tool, and pressing Shift+P rotates through the hidden tools under the Pen tool. When you see a small triangle in the lower-right corner of the tool icon, you know that the tool contains hidden tools.

Table 1-1 lists the Photoshop tools, describes what each one is used for, and specifies in which chapter you can find more information about each one.

TABLE 1-1 Photoshop CC Tools

Button	Tool	What You Can Do with It	See This Chapter in Book 3
	Move (V)	Move selections or layers	3
	Marquee (M)	Select the image area	3
	Lasso (L)	Make freehand selections	3

(continued)

TABLE 1-1 *(continued)*

Button	Tool	What You Can Do with It	See This Chapter in Book 3
	Object Selection (W)	Auto selects objects	3
	Crop (C)	Crop an image	1
	Frame (K)	Creates placeholder frames for images	1
	Eyedropper (I)	Eyedropper	7
	Spot Healing Brush (J)	Retouch flaws	7
	Brush (B)	Paint the foreground color	7
	Clone Stamp (S)	Copy pixel data	7
	History Brush (Y)	Paint from the selected state	7
	Eraser (E)	Erase pixels	7
	Gradient (G)	Create a gradient	7
	Blur	Blur pixels	7
	Toning (O)	Dodge, burn, saturate	N/A
	Pen (P)	Create paths	4
	Type (T)	Create text	8

Button	Tool	What You Can Do with It	See This Chapter in Book 3
	Path Selection (A)	Select paths	4
	Hidden Direct Selection (A)	Select individual anchor points on paths	4
	Vector Shape (U)	Create vector shapes	8
	Hand (H)	Navigate page	8
	Zoom (Z)	Increase or decrease the view	1

TIP

Looking for the Magic Wand tool, shown in Figure 1-4? Click and hold the Quick Selection tool in the Tools panel to access it.

FIGURE 1-4: Some popular tools may be hidden; click and hold on the arrow in the lower right of the tool to see additional tools.

Navigating the Work Area

Navigating in Photoshop isn't much different than getting around in other Adobe applications. Most Creative Cloud applications make extensive use of panels. In the following sections, you have an opportunity to experience the Photoshop workspace and configure those panels.

As a default, you are in the Essentials workspace, as shown in Figure 1-5. If you are happy with the panels that appear while in that workspace there is no reason to change your workspace at this time.

Changing your workspace

As you improve your skills, you may want to try some of the other workspaces that Photoshop offers. Workspaces simply show or hide different panels and tools based upon the tasks you are hoping to complete. For instance, if you choose

Window➪Workspace, as shown in Figure 1-6, you can choose between workspaces that are specifically set up to help you with 3D editing, graphics and web, motion graphics, painting, and photography. We show you how you can create your own custom workspace later in this chapter.

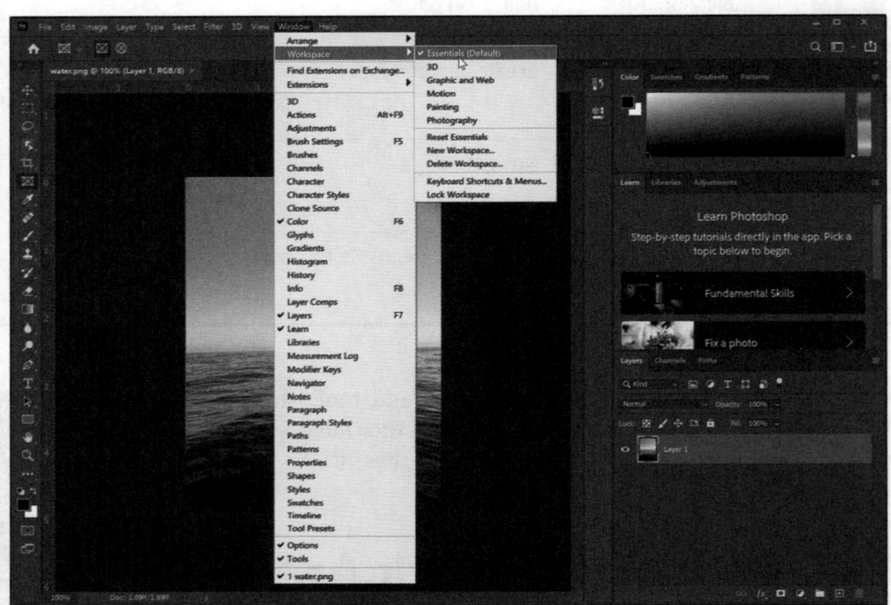

FIGURE 1-5: As a default, you are in the Essentials workspace.

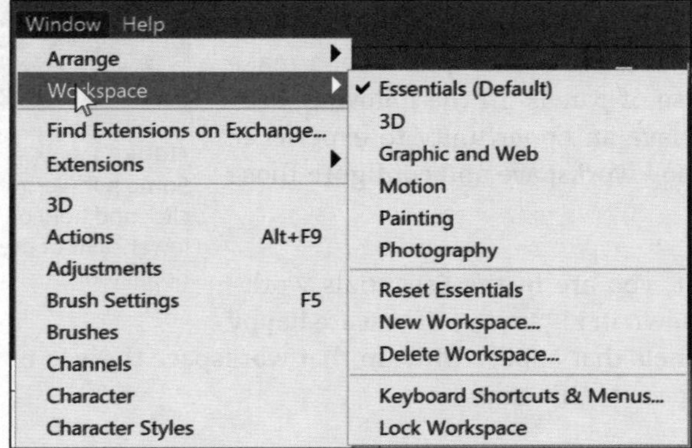

FIGURE 1-6: Choose among a variety of workspaces.

Resetting your workspace

Even if you're happy with the default panels that appear, you may end up moving, closing, or inadvertently collapsing important panels. When working in any workspace, including Essentials, you can easily reset the panel configuration by selecting Windows ⇨ Workspace ⇨ Reset Essentials (or the workspace name). Read the following section to discover how to create and save your own workspace in Photoshop.

Docking and saving panels

Panels, panels everywhere — do you really need them all? Maybe not just yet, but as you increase your skill level, you'll start to take advantage of more of those Photoshop panels. Panels give you easy access to important functions. In this chapter we touch on panel organization, but if you want more details about using panels there is a fast read in Book 1, Chapter 5, "Exploring Common Panels."

REMEMBER

When you work in Photoshop, keep in mind these two keyboard shortcuts can help temporarily clear up distractive tools and panels while you are working:

- Press Tab to switch between hiding and showing the tools and panels.
- Press Shift+Tab to hide the panels, leaving only the Tools panel visible. Press Shift+Tab to show them again.

If you find that you are frequently using the same panels, close the less used panels that you don't need and arrange the others onscreen where you want them.

Follow one of these three methods to close a panel:

- **Choose Window ⇨ *Panel's Tab Name*.** If the panel name is checked, it is open; selecting the panel name in the list closes the panel.
- **Click on the panel menu in the upper right of the panel and choose Close, as you see in Figure 1-7.**
- **Click and drag on a panel name to undock it.** When undocked, the panel displays a close button (an X button in Windows or the red button on a Mac) that you can click to close the panel.

To open and relocate panels that you frequently use, choose Window ⇨ *Panel's Tab Name*.

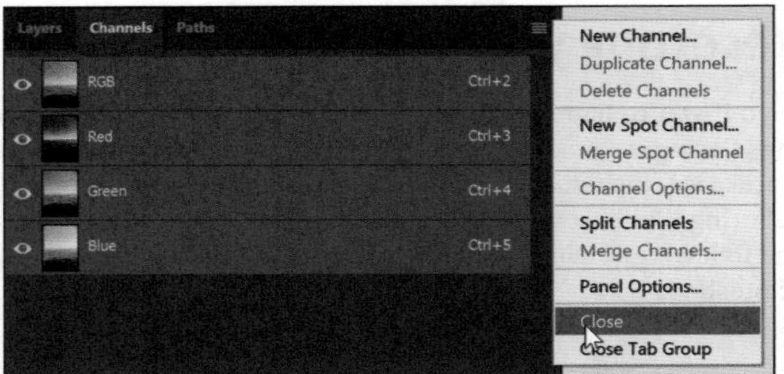

FIGURE 1-7: Closing a panel that you do not use frequently.

Saving your workspace

After you have displayed and arranged the panels you frequently use, follow these steps to save the configuration:

1. **Choose Window ➪ Workspace ➪ New Workspace.**

   The New Workspace dialog box appears.

2. **Name your workspace and click Save.**
3. **Any time you want the panels to return to your saved locations, choose Window ➪ Workspace ➪ *Name of Workspace*.**

   Name of Workspace is the name you supplied in Step 2.

Taking advantage of workspace features

In addition to the Essentials workspace, Photoshop has workspaces you can take advantage of to open the panels you need for specific tasks. You can use some of these workspaces for photography, painting, or design, for example.

Increase your work area by turning panels into icons, as shown in Figure 1-8. Do so either by right-clicking the tab of a panel and selecting Collapse to Icons or by clicking the double arrows.

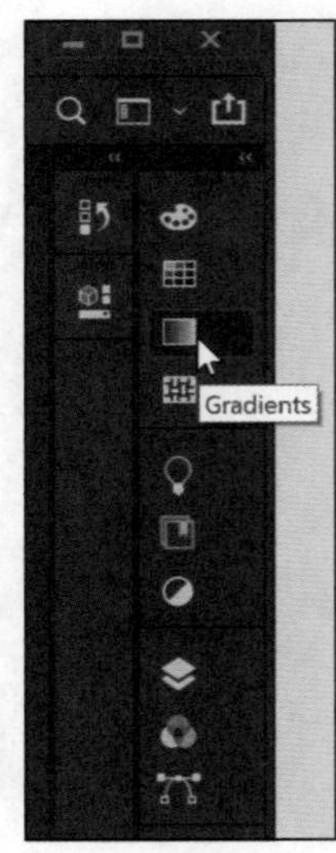

FIGURE 1-8: Turn panels into icons.

Zooming in to get a better look

Images that look fine at one zoom level may look extremely bad at another. You'll zoom in and out quite often while working on images in Photoshop. You can find menu choices for zooming on the View menu, but a quicker way to zoom is to use the keyboard commands listed in Table 1-2.

TABLE 1-2 Zooming and Navigation Keyboard Shortcuts

Command	Windows Shortcut	Mac Shortcut
Actual size	Ctrl+1	⌘ +1
Fit in window	Ctrl+0 (zero)	⌘ +0 (zero)
Zoom in	Ctrl++ (plus sign) or Ctrl+spacebar	⌘ ++ (plus sign) or ⌘ +spacebar
Zoom out	Ctrl+– (minus) or Alt+spacebar	⌘ +– (minus) or Option+spacebar
Hand tool	Spacebar	Spacebar

This list describes a few advantages of working with the Zoom tool to get a better look at your work:

- **100 percent view:** Double-clicking the Zoom tool in the Tools panel gives you a 100 percent view. Do it before using filters to see a more realistic result of making changes.
- **Zoom marquee:** While the Zoom marquee tool is selected, uncheck Scrubby Zoom from the options at the top to add additional control to the Zoom marquee tool. Drag from the upper-left corner to the lower-right corner of the area you want to zoom to. While you drag, a marquee appears; when you release the mouse button, the marqueed area zooms to fill the image window. The Zoom marquee gives you much more control than just clicking the image with the Zoom tool. Zoom out again to see the entire image by pressing Ctrl+0 (Windows) or ⌘ +0 (Mac). Doing so fits the entire image in the viewing area.

TIP

- **Keyboard shortcuts:** If a dialog box is open and you need to reposition or zoom to a new location on an image, you can use the keyboard commands without closing the dialog box.
- **A new window for a different look:** Choose Window ⇨ Arrange ⇨ New Window to create an additional window for the frontmost image. This technique is helpful when you want to view the entire image (say, at actual size) to see the results as a whole yet zoom in to focus on a small area of the image to do some fine-tuning. The new window is linked dynamically to the original window so that when you make changes, the original and any other new windows created from the original are immediately updated.
- **Cycle through images:** Press Ctrl+Tab (Windows) or ⌘ +~ (tilde) (Mac) to cycle through open images.

Choosing Your Screen Mode

You have a choice of three screen modes in which to work. Most users start and stay in the default (standard screen) mode until they inadvertently end up in another. The following modes are accessible by clicking and holding Screen Mode, located at the bottom of the Tools panel, as shown in Figure 1-9:

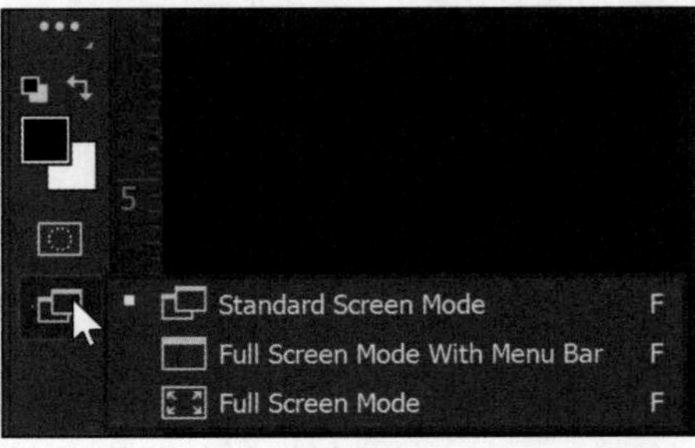

FIGURE 1-9: Change your screen mode.

- **Standard Screen Mode:** In this typical view, an image window is open, but you can see your desktop and other images open behind it.
- **Full Screen Mode with Menu Bar:** In this view, the image is surrounded to the edge of the work area with neutral gray. Working in this mode prevents you not only from accidentally clicking out of an image and leaving Photoshop, but also from seeing other images behind the working image.
- **Full Screen Mode:** A maximized document window fills all available space between docks and resizes when dock widths change.

Getting Started with Basic Tasks in Photoshop CC

Unless you use Photoshop as a blank canvas for painting, you may rarely create a new file in Photoshop. This is because you typically have a source image you start with that may have been generated by a digital camera, stock image library, or scanner.

The following sections show you how to open an existing image file in Photoshop, create a new image (if you want to use Photoshop to paint, for example), crop an image, and save an edited image.

Cropping an image

A simple but essential task is to crop an image. *Cropping* means to eliminate all parts of the image that aren't relevant to create a dynamic composition.

DISCOVER CAMERA RAW

If you haven't discovered the Camera Raw capabilities in Adobe Photoshop, you'll want to give them a try. The Camera Raw format is available for image capture in many cameras. Simply choose the format in your camera's settings as Raw instead of JPEG or TIFF. These Raw files are a bit larger than standard JPEG files, but you capture an enormous amount of data with the image that you can retrieve after opening. (See `www.adobe.com` for a complete list of cameras that support Camera Raw.)

A Camera Raw file contains unprocessed picture data from a digital camera's image sensor, along with information about how the image was captured, such as camera and lens type, exposure settings, and white balance setting. When you open the file in Adobe Photoshop CC, the built-in Camera Raw plug-in interprets the Raw file on your computer, making adjustments for image color and tonal scale.

When you shoot JPEG images with your camera, you're locked into the processing done by your camera, but working with Camera Raw files gives you maximum control over images, such as controlling their white balance, tonal range, contrast, color saturation, and image sharpening. Cameras that can shoot in Raw format have a setting on the camera that changes its capture mode to Raw. Rather than write a final JPEG file, a Raw data file is written, which consists of black-and-white brightness levels from each of the several million pixel sites on the imaging sensor. The actual image hasn't yet been produced, and unless you have specific software, such as the plug-in built in to Adobe Photoshop, opening the file can be difficult, if not impossible.

To open a Camera Raw file, simply choose File ⇨ Browse in Bridge. Adobe Bridge opens, and you see several panels, including the Folders, Content, Preview, and Metadata panels. In the Folders panel, navigate to the location on your computer where you've saved Camera Raw images; thumbnail previews appear in the Content panel. Think of Camera Raw files as photo negatives. You can reprocess them at any time to achieve the results you want.

Right-click (Windows) or Control-click (Mac) a JPEG or TIFF file and choose Open in Camera Raw from the contextual menu. This is a great way to experiment with all the cool features available with this plug-in, but your results aren't as good as if you used an actual Raw file.

If Adobe Photoshop CC doesn't open your Raw file, you may need to update the Raw plug-in. (See `www.adobe.com` for the latest plug-in.) The plug-in should be downloaded and placed in this location in Windows: C:\Program Files\Common Files\Adobe\Plug-Ins\CC\File Formats. On the Macintosh, place the plug-in here: Library\Application Support\Adobe\Plug-Ins\CC\File Formats.

Cropping is especially important in Photoshop. Each pixel, no matter what color, takes up the same amount of information, so cropping eliminates unneeded pixels and reduces file size and processing time. For that reason, you should crop images before you start working on them.

You can crop an image in Photoshop CC in two ways:

- Use the Crop tool.
- Select an area with the Marquee tool and choose Image ➪ Crop.

To crop an image by using the Crop tool, follow these steps:

1. **Press C to access the Crop tool.**

   Make sure that the preset (in the upper-left corner of the control panel) is set to Ratio, or, if you wish, to a specific preset size in the Presets drop-down list, as shown in Figure 1-10.

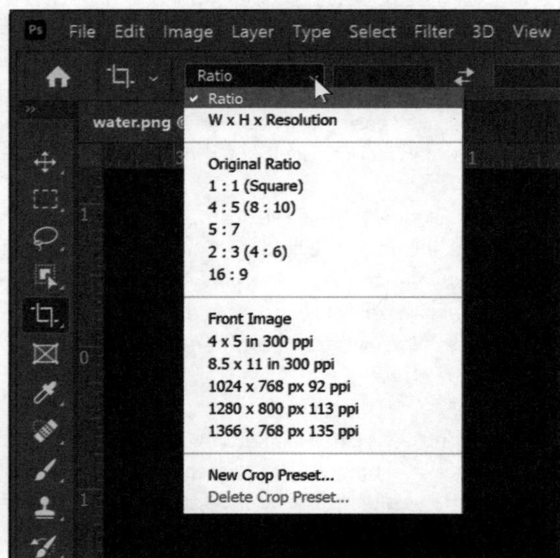

FIGURE 1-10: Select the Crop tool preset before you click and drag.

2. **Click and drag over the area of the image you want to keep.**
3. **If you need to adjust the crop area, drag the handles in the crop-bounding area.**

4. **When you're satisfied with the crop-bounding area, double-click in the center of the crop area or press the Return or Enter key to crop the image.**
5. **If you want to cancel the crop, press the Esc key.**

TIP

Ever scan in an image that ends up crooked? When using the Crop tool, if you position the cursor outside any handle, a rotate symbol appears. Drag the crop-bounding area to rotate it and line it up the way you want it cropped. When you press Return or Enter, the image straightens out.

Using grids as you crop

As a default, a three-row and three-column grid appears after you click and drag over an area with your Crop tool. This grid can be used as a guide to help you create more dynamically cropped images. Images whose points of interest appear on the intersections of this grid, or that can be divided into the sections of the grid, are generally considered to be more interesting. (This is the called the *Rule of Thirds.*) See how the image in Figure 1-11 fills the bottom one-third of the image with water, and the upper two-thirds with sky.

FIGURE 1-11: Use of the Rule of Thirds overlay in the Crop tool.

If you prefer to change or turn off the overlay, select the Never Show Overlay option from the Set the Overlay Options drop-down list, as shown in Figure 1-12. Keep in mind that you must have already clicked and dragged out your crop area before you can make changes to the overlay.

FIGURE 1-12: Select from a multitude of overlays to use as guides, or turn the overlay off.

Using the Frame tool to create cropped images

Frames allow you to build quick and easy placeholders for images. By defining the size and position of a frame, you essentially allow yourself the ability to hide or show parts of your image by simply pushing the image around within the frame you created.

There are several ways you can create a frame; in this chapter, the first two methods are covered.

- Use the Frame Tool to draw an empty rectangular or elliptical frame in your image, and then add an image inside the frame.
- Click and drag a frame over an existing image to define the masked area.
- Convert any existing shape or text to a frame. This method is discussed in Chapter 8 of this minibook.

Creating an empty frame with the Frame tool

Creating an empty frame offers the functionality of an image placeholder. The empty frame essentially determines the position and size of the image before it has been selected and placed.

1. **Open an image.**
2. **Choose Window⇨Layers to show the Layers panel if it is not visible.**
3. **Click on the Create new layer icon at the bottom of the Layers panel, as shown in Figure 1-13.**

4. **Select the Frame tool.**
5. **In the Tool Options bar, choose a rectangular frame or an elliptical frame, as shown in Figure 1-14.**
6. **Click and drag to create a new frame on the canvas.**
7. **With the frame selected, choose File⇨Place Linked or Place Embedded. In the dialog box that appears, select an image that you want to place in the selected frame.**

   The image is placed as Linked or Embedded Smart Object, as shown in Figure 1-14.

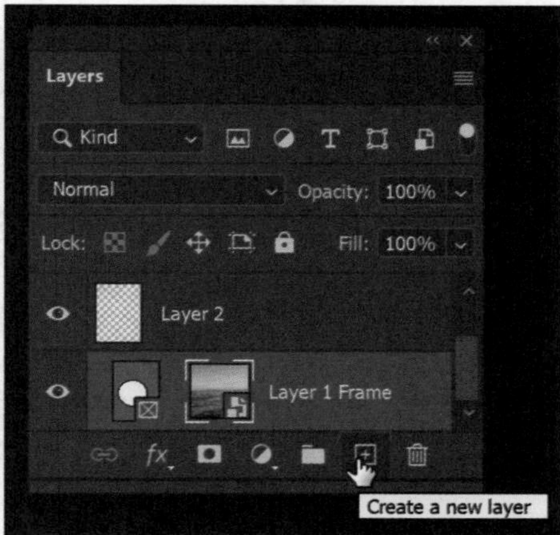

FIGURE 1-13: Add a new empty layer to your image.

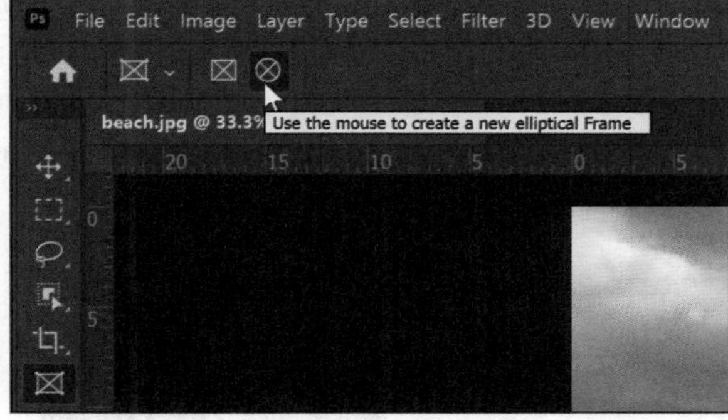

FIGURE 1-14: Select the Frame tool, and the select the shape you want to use from the Options bar.

TIP

You can drag an image from your local disk, Library, or Adobe Stock asset right into a frame.

Creating a frame by clicking and dragging over an existing image

Open an image in Photoshop. Clicking and dragging a frame works with any layer in Photoshop. Here's how:

1. **The first step is to verify that you have a pixel layer to work with and not a Background layer. Choose Window⇨Layers and check to see whether your image has a Background layer. If it shows Background layer, as shown in Figure 1-15, double-click on the layer to open the New Layer dialog box and press OK.**

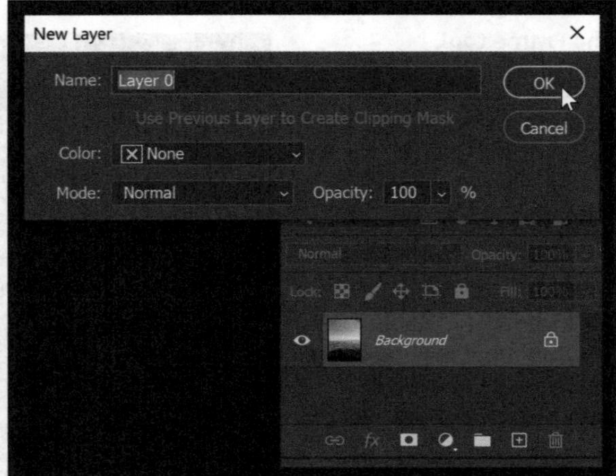

FIGURE 1-15: Make sure to convert a Background layer to a regular layer by double-clicking on it.

2. **Select the Frame tool. In the Tool Options bar, choose a rectangular frame or an elliptical frame.**
3. **Click and drag the frame over the image to create a mask with your frame, as you see in Figure 1-16.**
4. **Using the Layers panel, reposition the image within the frame.**

   For more about moving frames and images around using the Layers panel, see Chapter 8 of this minibook.

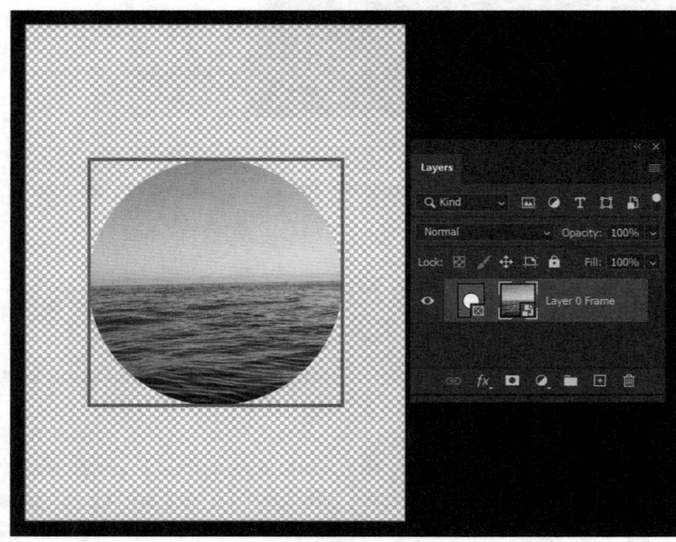

FIGURE 1-16: A frame allows you to mask an image into a shape.

Saving images

Save an image file by choosing File⇨Save. If you're saving the file for the first time, the Save As dialog box appears. Notice in the Format drop-down list that you have plenty of choices for file formats. (File formats are discussed in more detail in Chapter 9 of this minibook.) You can always play it safe by choosing the native Photoshop (PSD) file format, which supports all Photoshop features, such as channels, layers, text, and other vector objects that are discussed throughout this minibook. Choosing certain other formats may eliminate layers, channels, and other special features.

Many users choose to save a native Photoshop file as a backup to any other file format. Be sure to have a backup or an original file saved as a native Photoshop (PSD) file when you start taking advantage of layers and other outstanding Photoshop elements. As a Creative Cloud user, keep in mind that you can use the native file format for Photoshop in all other Creative Cloud applications.

» Understanding Photoshop image modes

» Working in black and white, RGB, or CMYK

Chapter 2
Messing with Mode Matters

Before taking on working with any imagery in Photoshop, you should understand color modes and the importance of setting up the most optimum color settings. A strong foundation in the use of color plays into the production and editing of all imagery, grayscale, color image, content for websites, movies, and more. Read through this chapter to help you to feel confident in the choices you make.

Working with Bitmap Images

You may wonder how images in Photoshop show subtle changes in color values. To create those smooth gradations from one color to the next, Photoshop takes advantage of pixels and anti-aliasing. *Bitmap images* (or *raster images*) are based on a grid of pixels, as shown in Figure 2-1. The grid is smaller or larger depending on the resolution of your image. The number of pixels along the height and width of a bitmap image are the pixel dimensions of an image, measured in pixels per inch (ppi). The more pixels per inch, the more detail in the image.

FIGURE 2-1: You can see the grid that makes up the pixel resolution of an image when you zoom in.

Unlike *vector graphics* (mathematically created paths), bitmap images can't be scaled without losing detail. (See Figure 2-2 for an example of a bitmap image and a vector graphic.) Generally, you should use bitmap images at or close to the size you need. If you resize a bitmap image, it can become jagged on the edges of sharp objects. On the other hand, you can scale vector graphics and edit them without degrading sharp edges.

FIGURE 2-2: Bitmap versus vector.

Photoshop can work on both bitmap and vector art. (In the path line around the vector shape layer, notice that the path isn't pixelated or broken down into a step pattern created by the pixels.) Combining the two technologies gives you, as a designer, incredible opportunities.

TIP

For information on changing and adjusting image resolution, see Chapter 6 of this minibook.

Choosing the Correct Photoshop Mode

Choose Image⇨Mode to view the available image mode choices. Selecting the right mode for an image is important because each one offers different capabilities and results. For example, if you choose Bitmap mode, you can work only in black and white. That's it — no shades of color, not even gray. Most features are disabled in Bitmap mode, which is fine if you're working on art for a black-and-white logo, but not for most images. If, instead, you work in RGB (Red, Green, Blue) mode, you have full access to Photoshop's capabilities.

Read on to see which image mode is best for your needs. When you're ready to make your mode selection, open a file and choose Image⇨Mode to make a selection. You can read descriptions of each image mode in the following sections.

Keep in mind, if you are not sure which mode will work best, stay in the RGB mode. The RGB mode offers access to most all of Photoshop's features and can easily be changed to a different mode later.

REMEMBER

A *channel* simply contains the color information in an image. The number of default color channels in an image depends on its color mode. For example, a CMYK image has at least four channels — one each for cyan, magenta, yellow, and black information. Grayscale has one channel. If you understand the printing process, think of each channel representing a plate (color) that, when combined, creates the final image.

Bitmap

Bitmap mode offers little more than the capability to work in black and white. Many tools are unusable, and most menu options are grayed out in this mode. If you're converting an image to bitmap, you must convert it to grayscale first.

Grayscale

Use Grayscale mode, shown in Figure 2-3, if you're creating black-and-white images with tonal values specifically for printing to one color. Grayscale mode supports 256 shades of gray in 8-bit color mode. Photoshop can work with grayscale in 16-bit mode, which provides more information but may limit your capabilities when working in Photoshop.

FIGURE 2-3: Grayscale supports 256 shades of gray.

When you choose Image ⇨ Mode ⇨ Grayscale to convert to Grayscale mode, a warning message asks you to confirm that you want to discard all color information. If you don't want to see this warning every time you convert an image to grayscale, select the option not to show the dialog box again before you click Discard.

TIP

Using the Black & White adjustment is the best way to create a good grayscale image. Simply click and hold the Create New Fill or Adjustment Layer button at the bottom of the Layers panel and choose Black & White. Set the sliders to achieve the best black-and-white image. After you have adjusted the colors using the Black & White adjustment, choose Image ⇨ Mode ⇨ Grayscale.

Duotone

Use Duotone mode when you're creating a one to four-color image created from spot colors (solid ink, such as Pantone colors). You can also use Duotone mode to create monotones, tritones, and quadtones. If you're producing a two-color job, duotones create a beautiful alternative to full color.

REMEMBER

The Pantone Matching System (PMS) helps keep printing inks consistent from one job to the next. By assigning a numbered Pantone color, such as 485 for red, you eliminate the risk of one vendor (printer) using fire engine red and the next using orange-red for your company logo.

To create a duotone, follow these steps:

1. **Choose Image ➪ Mode ➪ Grayscale.**
2. **Choose Image ➪ Mode ➪ Duotone.**
3. **In the Duotone dialog box, select Duotone from the Type drop-down list.**

   Your choices range from *monotone* (one-color) up to *quadtone* (four-color). Black is assigned automatically as the first ink, but you can change it, if you like.
4. **To assign a second ink color, click the white swatch immediately under the black swatch.**

   The Color Picker appears, as shown in Figure 2-4.

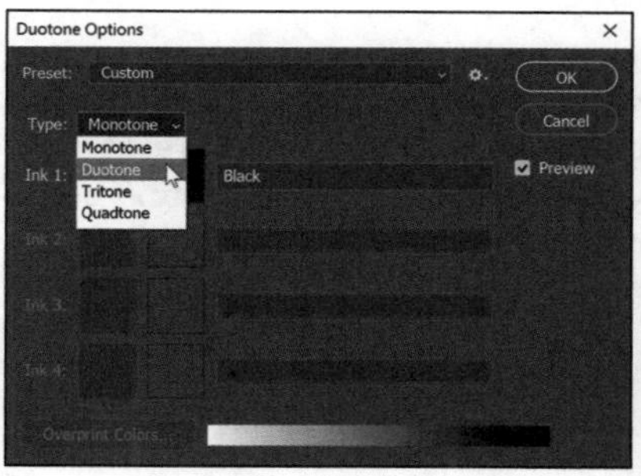

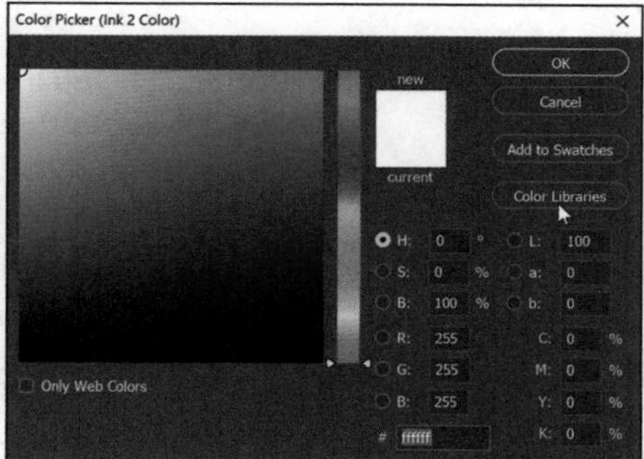

FIGURE 2-4: Click the white swatch to open the Color Picker dialog box and then click on Color Libraries.

5. **Click the Color Libraries button, and then select Pantone Solid Coated as the library, it should be set by default.**
6. **Now comes the fun part: Type (quickly!) the Pantone or PMS number you want to access, and then click OK. (See Figure 2-5.) If you type too slowly, the incorrect number will appear.**

   There's no text field for you to enter the number, so don't look for one. Just type the number while the Color Libraries dialog box is open.

   Try entering **300** to select PMS 300. You can already see that you've created a tone curve.
7. **Click the Curve button to the left of the ink color to further tweak the colors.**
8. **Click and drag the curve to adjust the black in the shadow areas, perhaps to bring down the color overall. Then experiment with the results.**

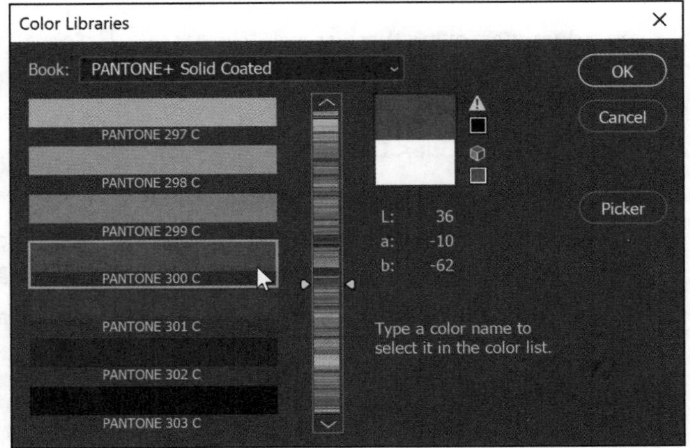

FIGURE 2-5: In the Color Library dialog box, type a Pantone number in quickly, for instance 3-0-0 for PMS 300; there is no text field.

9. **(Optional) If you like your duotone settings, store them by clicking the small Preset Options button to the right of the Preset drop-down list, as shown in Figure 2-6. Type a name into the Name text box, browse to a location on your computer, and then click Save.**

TIP

You can also use one of the presets that Adobe provides. Do this by selecting an option from the Presets drop-down menu at the top of the Duotone dialog box.

Click the Preset Options button to find your saved presets.

Duotone images must be saved in the Photoshop Encapsulated PostScript (EPS) format in order to support the spot colors. If you choose another format, you risk the possibility of converting colors into a build of CMYK (Cyan, Magenta, Yellow, and Black).

10. **Click OK when you're finished.**

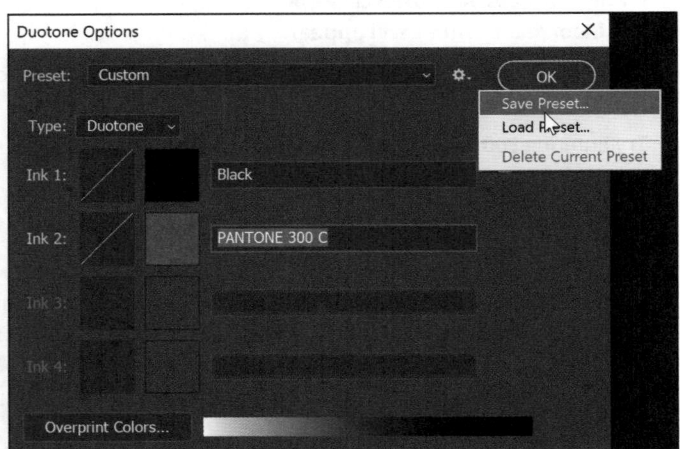

FIGURE 2-6: Save your duotone by clicking the Preset Options button.

Index color

Typically you don't edit your images in the Index color mode, but you probably have saved a file in this mode. If you have ever saved a GIF file out of Photoshop, you have experienced working with Indexed color firsthand. Indexed Color mode (see Figure 2-7) uses a color lookup table (CLUT) to create the image from a limited palette of colors. In this example, the flower was set to reduce to 10 colors.

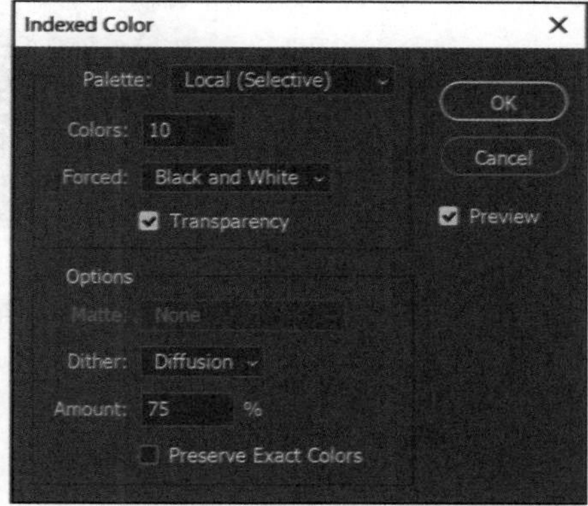

FIGURE 2-7: Index color uses a limited number of colors to create an image.

A CLUT contains all colors that make up an image, like a box of crayons used to create artwork. If you have a box of only eight crayons that are used to color an image, you have a CLUT of only eight colors. Of course, your image would look much better if you used the 64-count box of crayons with the sharpener on the back, but those additional colors also increase the size of the CLUT and the file size.

The highest number of colors that can be in Indexed Color mode is 256. When saving web images, you often have to define a color table in order to keep a file size small. We discuss saving files for the web in Chapter 10 of this minibook.

TIP

If your image is not in Indexed color, choose Image⇨Mode⇨Indexed Color, then, choose Image⇨Mode⇨Color Table to see the color table making up an image.

RGB

RGB (Red, Green, Blue) mode, shown in Figure 2-8, is the standard format you work in if you import images from a digital camera or import images from a copier or scanner. For complete access to Photoshop's features, RGB is the best color mode to work in. If you're working on images for use on the web, color copiers, desktop color printers, or onscreen presentations, stay in RGB mode.

REMEMBER

If you're having an image printed on a press (for example, if you're having it professionally printed), it must be separated. Don't convert images to CMYK mode until you've finished the color correction and you know that your color settings are accurate. A good print service may want the RGB file so that it can complete an accurate conversion.

FIGURE 2-8:
RGB creates the image from red, green, and blue.

CMYK

CMYK (Cyan, Magenta, Yellow, Black) mode is used for final separations for the press. Use a good magnifying glass to look closely at anything printed in color, and you may see the CMYK colors that created it. A typical four-color printing press has a plate for each color and runs the colors in the order of cyan, magenta, yellow, and then black.

WARNING

Don't take lightly the task of converting an image into this mode. You need to make decisions when you convert an image to CMYK, such as where to print the file and on which paper stock, so that the resulting image is the best it can be. Talk to your print provider for specifications that are important when converting to CMYK mode.

LAB color

The LAB (Lightness, A channel, and B channel) color mode is used by many high-end color professionals because of its wide color range. Using LAB, you can make adjustments to *luminosity* (lightness) without affecting color. In this mode, you can select and change an *L* (lightness or luminosity) channel without affecting the *A* channel (green and red) and the *B* channel (blue and yellow).

LAB mode is also good to use if you're in a color-managed environment and want to easily move from one color system to another with no loss of color.

TIP

Some professionals prefer to sharpen images in LAB mode because they can select just the Lightness channel and choose Filter ⇨ Sharpen ⇨ Unsharp Mask to sharpen only the gray matter of the image, leaving the color noise-free.

Multichannel

Multichannel is used for many things; you can end up in this mode and not know how you got there. Deleting a channel from an RGB, a CMYK, or a LAB image automatically converts the image to Multichannel mode. This mode supports multiple spot colors.

Bit depth

You have more functionality in 16-bit and even 32-bit mode. Depending on your needs, however, you may spend most of your time in 8-bit mode, which is more than likely all you need.

Bit depth, or *pixel depth* or *color depth,* measures how much color information is available to display or print each pixel in an image. Greater bit depth means more available colors and more accurate color representation in the digital image. In Photoshop, this increase in accuracy also limits some available features, so don't use it unless you have a specific request or need for it.

To use 16-bit or 32-bit color mode, you also must have a source to provide you with that information, such as a scanner or camera that offers a choice to scan in either mode.

» Painting selections the easy way

» Refining your selections

» Keeping selections for later use

Chapter 3 Making Selective Changes

Photoshop is well known for its capability to create compositions and for its retouching capabilities. How you create these masterpieces is by perfecting the many tools that can be used to make selective changes. As you practice the tools and techniques covered in this chapter, you will become better at making your selective changes less obvious. A professional photo-retoucher does not want to produce images that look contrived, unless that is the look they are going for. For instance, if they intend to create a humorous image, such as putting swapping heads, or putting someone in an unrealistic location.

In this chapter, you try out several selection methods and see how to use the available tools to make images look as though you *haven't* retouched them. This chapter provides a plethora of tips and tricks that can save you time and help make your images look professional and convincing.

Getting to Know the Selection Tools

Think of *selections* as windows in which you can make changes to pixels. Areas not selected are *masked,* which means that they're unaffected by changes, much like the windows and door frames you cover with tape before painting the walls. In this section, we briefly describe the selection tools and show you how to use them. You must be familiar with these tools in order to do *anything* in Photoshop.

REMEMBER

As with all Photoshop tools, the Options bar (viewed across the top of the Photoshop window) changes when you choose different selection tools. The keyboard commands you read about in this section exist on the tool Options bar and appear as buttons across the top.

If you move a selection with the Move tool, pixels move as you drag, leaving a blank spot in the image. To *clone* a selection (to copy and move the selection at the same time), Alt+drag (Windows) or Option+drag (Mac) the selection with the Move tool.

The Marquee tool

The Marquee tool is a common selection tool. You can use it to make a rectangular or elliptical selections. Throughout this section, we describe creating (and then deselecting) an active selection area; we also provide you with tips for working with selections.

The Marquee tool includes the Rectangular Marquee (for creating rectangular selections), Elliptical Marquee (for creating round or elliptical selections), and Single Row Marquee or Single Column Marquee tools (for creating a selection of a single row or column of pixels). You can access these other Marquee tools by selecting the arrow in the lower-left of the Marquee tool on the Tools panel, as shown in Figure 3-1.

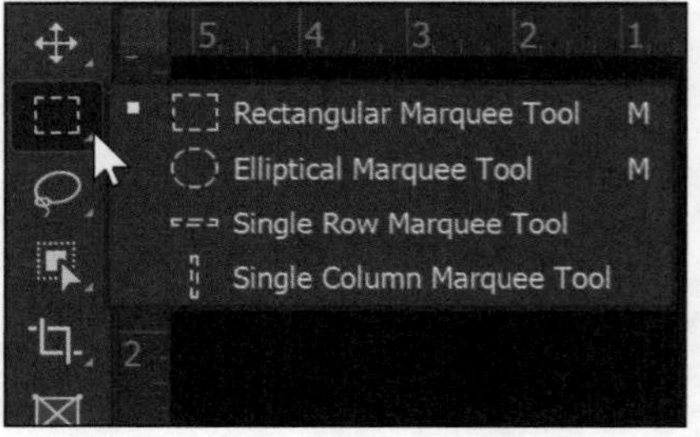

FIGURE 3-1: Change to other Marquee tools by clicking on the arrow in the Tools panel.

To create a selection, select one of the Marquee tools (remember that you can press M to do this), and then drag anywhere on your image. When you release the mouse button, you create an active selection area. When you're working on an active selection area, the effects you choose are applied to the whole selection.

To deselect an area, you have three choices:

- Choose Select ⇨ Deselect.
- Press Ctrl+D (Windows) or ⌘ +D (Mac).
- While using a selection tool, click outside the selection area.

Creating rectangular and elliptical selections

How you make a selection is important because it determines how realistic your edits appear on the image. You can use the following tips and tricks when creating both rectangular and elliptical selections:

- » Add to a selection by holding down the Shift key; drag to create a second selection that intersects the original selection (see the left image in Figure 3-2). The two selections become one big selection.

FIGURE 3-2: You can add to a selection by holding down the Shift key.

- » Delete from an existing selection by holding down the Alt (Windows) or Option (Mac) key, and then dragging to create a second selection that intersects the original selection where you want to take away from the original selection (shown on the right in Figure 3-2).
- » Constrain a rectangle or an ellipse to a square or circle by Shift-dragging; make sure that you release the mouse button before you release the Shift key. Holding down the Shift key makes a square or circle only when there are no other selections. (Otherwise, it adds to the selection.)

- Make the selection from the center by Alt-dragging (Windows) or Option-dragging (Mac); make sure that you release the mouse button before releasing the Alt (Windows) or Option (Mac) key.
- Create a square or circle from the center outward by Alt+Shif+-dragging (Windows) or Option+Shift+dragging (Mac). Again, make sure that you always release the mouse button before releasing the modifier keys.
- When making a selection, hold down the spacebar before releasing the mouse button to drag the selection to another location.

Setting a fixed size

If you've created an effect that you particularly like — say, changing a block of color in your image — and you want to apply it multiple times throughout, you can do so. To make the same selection multiple times, follow these steps:

1. **With the Marquee tool selected, select Fixed Size from the Style drop-down list on the Options bar.**

   You can also select Fixed Ratio from the Style drop-down list to create a proportionally correct selection not fixed to an exact size.

2. **On the Options bar, type the Width and Height values into the appropriate text fields.**

   You can change ruler increments by choosing Edit ➪ Preferences ➪ Units & Rulers (Windows) or Photoshop ➪ Preferences ➪ Units & Rulers (Mac).

3. **Click the image.**

   A selection sized to your values appears.

4. **With the selection tool, drag the selection to the location you want selected.**

TIP

Shift-drag a selection to keep it aligned to a 45-degree or 90-degree angle.

Making floating and nonfloating selections

When you're using a selection tool, such as the Marquee tool, your selections are *floating* by default, which means that you can drag them to another location without affecting the underlying pixels. You know that your selection is floating by the little rectangle that appears on the cursor. (See the left image in Figure 3-3.)

FIGURE 3-3: The Float icon is used on the left, and the Move icon is used on the right.

If you want to, however, you can move the underlying pixels. Using the selection tool of your choice, just hold down the Ctrl (Windows) or ⌘ (Mac) key to temporarily access the Move tool; the cursor changes to a pointer with scissors, denoting that your selection is nonfloating. Now, when you drag, the pixel data comes with the selection (as shown on the right in Figure 3-3).

TIP

Hold down Alt+Ctrl (Windows) or Option+⌘ (Mac) while using a selection tool and drag to clone (copy) pixels from one location to another. Add the Shift key, and the cloned copy is constrained to a straight, 45-degree, or 90-degree angle.

The Lasso tool

Use the Lasso tool for *freeform selections* (selections of an irregular shape). To use the Lasso tool, click and drag to create a path surrounding the area that you want to select. Be careful! If you don't return to the starting point to close the selection before you release the mouse button, Photoshop completes the path by finding the most direct route back to your starting point.

TIP

As with the Marquee tool, you can press the Shift key to add to a lasso selection and press the Alt (Windows) or Option (Mac) key to delete from a lasso selection.

Hold down the Lasso tool to show the hidden Lasso tools:

- **Polygonal Lasso tool:** Click a starting point, and then click and release from point to point until you come back to close the selection.

» **Magnetic Lasso tool:** Click to create a starting point, and then hover the cursor near an edge in your image. The Magnetic Lasso tool is magnetically attracted to edges; as you move the cursor near an edge, the Magnetic Lasso tool creates a selection along that edge. Click to manually set points in the selection; when you return to the starting point, click to close the selection.

You may find that the Polygonal Lasso and the Magnetic Lasso tools make less-than-ideal selections. Take a look at the later section "Painting a selection with the Quick Mask tool" for tips on making finer selections.

The Object Selection tool

With the right image the Object Selection tool can help you select like a pro in only a matter of seconds. This tool recognizes the edges of objects that have a fair amount of contrast to them. It doesn't work well with every image, but for many it is a huge timesaver.

To use the Object Selection tool, simply select the tool and then click and drag to draw a rectangular region around the object that you want to select, as shown in Figure 3-4. The Object Selection tool automatically selects the object inside the region.

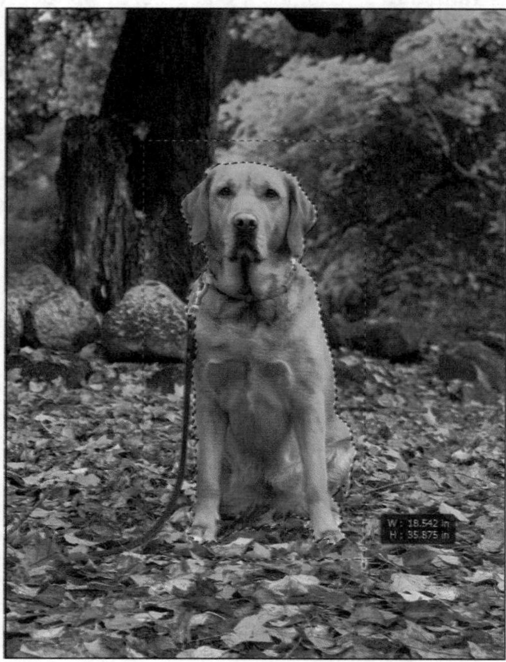

FIGURE 3-4: Click and drag a rectangular selection with the Object Selection tool in order to automatically select a portion of your image.

The Quick Selection tool

The Quick Selection tool lets you quickly "paint" a selection with a round brush tip of adjustable size. You can find the Quick Selection tool by holding down on the arrow in the lower-right portion of the Object Selection tool. This reveals the hidden tools. One the Quick Selection tool is selected, click and drag over a part of an image that you want to select. Almost like you are painting a selection. As you click and drag you can watch as the selection expands outward and automatically follows defined edges in the image. A Refine Edge command lets you improve the quality of the selection edges and visualize the selection in different ways for easy editing.

Follow these steps to find out how you can take advantage of this tool:

1. **Open a file that requires a selection.**

   If you downloaded the DummiesCCFiles, as noted in Chapter 1 of this minibook, you can use the file named `Model01` in the Book03_Photoshop folder.

2. **Select the Quick Selection tool.**
3. **Position the cursor over the area you want to select. Notice the brush size displayed with the cursor; click and drag to start painting the selection.**

   TIP

   You can adjust the size of the painting selection by pressing the left bracket key ([) to make the brush size smaller or the right bracket key (]) to make the brush size larger.

   By using the Add to Selection or Subtract from Selection buttons on the Options bar, you can paint more of the selection or deselect active areas. (See Figure 3-5.) Keep in mind that the Add to Selection option is on by default.

If you would rather use keyboard shortcuts to add or delete to the selection, you can hold down the Alt (Windows) or Option (Mac) key to delete from a selection. As a default, you do not have to hold down any key to add to a selection. If adding to the selection is not working, your default Add to Selection tool might have been switched, as pointed out in Figure 3-6. Click on the Center selection button in order to turn Add to Selection back on, or simply hold down the Shift key and paint to add additional content to your selection.

The Quick Selection tool does not make a perfect selection, but it provides a great start for even the most difficult selections. Keep reading to see how you can take your selection even further with the Select and Mask feature in Photoshop.

FIGURE 3-5: Use the Quick Selection tool to paint your selection.

FIGURE 3-6: When the Quick Selection tool is active, you have additional selection options available in the control panel.

The Magic Wand tool

The Magic Wand tool is also hidden within the Object Selection and Quick Selection tool. To access it, click on the arrow in either the Object Selection, or Quick Selection tool, whichever one you used last, and then select the Magic Wand tool. The Magic Wand tool is particularly helpful when you're working on an image of high contrast or with a limited number of colors. This tool selects individual pixels of similar shades and colors. Select the Magic Wand tool, click anywhere on an image, and hope for the best — the Magic Wand tool isn't magic. You decide how successful this tool is. What we mean is that *you* control how closely matched each pixel must be in order for the Magic Wand tool to include it in the selection. You do so by setting the tolerance level on the Options bar.

When you select the Magic Wand tool, a Tolerance text field appears on the Options bar. As a default, the tolerance is set to 32. When you click with a setting of 32, the Magic Wand tool selects all pixels within 32 shades (steps) of the color you clicked. If it didn't select as much as you want, increase the value in the Tolerance text field (all the way up to 255). The number you enter varies with each individual selection. If you're selecting white napkins on an off-white tablecloth, for example, you can set the number as low as 5 so that the selection doesn't leak into other areas. For colored fabric with lots of tonal values, on the other hand, you might increase the tolerance to 150.

TIP

Don't fret if you miss the entire selection when using the Magic Wand tool. Hold down the Shift key and click in the missed areas. If the tool selects too much, choose Edit⇨Undo (step backward) or press Ctrl+Z (Windows) or ⌘ +Z (Mac), reduce the value in the Tolerance text field, and try again.

Painting a selection with the Quick Mask tool

If you have fuzzy selections (fur, hair, or leaves, for example) or you're having difficulty using the selection tools, the Quick Mask tool can be a huge help because it enables you to paint your selection uniformly in one fell swoop. If you want to follow along in this section, you can open the image named duck that is located in the Book03Photoshop folder in the sample files that have been provided for you.

To enter Quick Mask mode, create a selection, and then press Q. (Pressing Q again exits you from Quick Mask mode.) You can also click the Quick Mask button at the bottom of the Tools panel. If you have a printing background, you'll notice that the Quick Mask mode, set at its default color (red), resembles something that you may want to forget: rubylith and amberlith. (Remember slicing up those lovely films with X-ACTO blades before computer masking came along?) In Quick Mask mode, Photoshop shows your image as it appears through the mask. The clear part is selected; what's covered in the mask isn't selected.

TIP

Change the default red color of the mask by double-clicking the Edit in Quick Mask Mode button (at the bottom of the Tools panel). The tooltip says Edit in Standard Mode if you're already in Quick Mask mode. This opens the Quick Mask Options dialog box, shown in Figure 3-7.

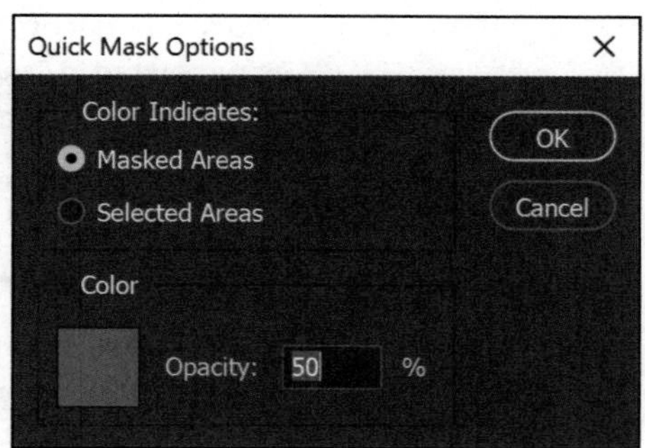

FIGURE 3-7: Change the color of the Quick Mask.

To create and implement a quick mask, follow these steps:

1. **Use the Quick Selection tool or another selection tool to create a selection before you enter the Quick Mask mode. This gives you a kickstart into making a much better mask.**

   In the example of the Duck image, the Quick Selection tool was selected and the duck was painted with the tool. The result is shown in Figure 3-8. You can find the Duck image in the Dummies files that you downloaded.

FIGURE 3-8: Start with a selection and then press Q.

2. **Press Q to enter Quick Mask mode.**
3. **Press D to change the foreground and background color boxes to the default colors of black and white.**

4. **Select the Brush tool and start painting with black in the clear area of the image in Quick Mask mode, as shown in Figure 3-8.**

   It doesn't have to be pretty; just get a stroke or two in there.

5. **Press Q to return to Selection mode.**

   You're now out of Quick Mask mode. Notice that where you painted with black (it turned red in Quick Mask mode), the pixels are no longer selected, as shown in Figure 3-9.

6. **Press Q again to reenter Quick Mask mode, and then press X.**

   Pressing X switches the foreground and background colors, giving you white in the foreground and black in the background.

7. **With the Brush tool, paint several white strokes in the red mask area. If you are working on the duck image, paint over your red lines that you created, and paint over the red eye to make sure that it is selected.**

   The white strokes turn clear in Quick Mask mode.

8. **Press Q to return to Selection mode.**

   The area you painted white in Quick Mask mode is now selected.

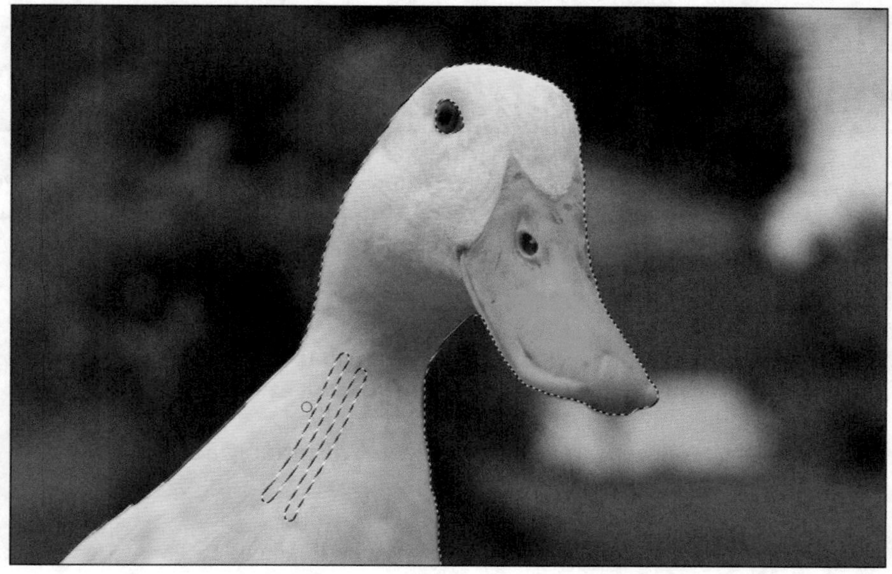

FIGURE 3-9: Painting with black in the Quick Mask mode masks out the area and deselects it.

When you're in Quick Mask mode, you can paint white over areas you want selected and black over areas you don't want selected. When painting in Quick Mask mode, increase the brush size by pressing the] key. Decrease the brush size by pressing the [key. You might need to paint the duck's eye with white if it is not selected.

In Selection mode, your selection seems to have a hard edge; you can soften those hard edges by using a softer brush in Quick Mask mode. To make a brush softer, press Shift+[. To make a brush harder, press Shift+].

Because Quick Mask mode makes selections based on the mask's values, you can create a mask by selecting the Gradient tool and dragging it across the image in Quick Mask mode. When you exit Quick Mask mode, it looks as though there's a straight-line selection, but the selection transitions the same as your gradient did. Choose any filter from the Filters menu and notice how the filter transitions into the untouched part of the image to which you applied the gradient.

If you're working in Quick Mask mode, choose Window⇨Channels to see that what you're working on is a temporary alpha channel. See the later section "Saving Selections" for more about alpha channels.

Manipulating Selections with Refine Selection

After you practice creating selections, you'll find that working with the selections — painting, transforming, and feathering them — can be easy and fun.

Transforming selections

Don't deselect and start over again if you can just nudge or resize a selection. You can scale, rotate, and even distort an existing selection.

Follow these steps to practice transforming a selection:

1. **Open any image.**

   You may wish to open the image named `FlatPanel.jpg` that is in the Book03_Photoshop folder.

2. **Practice by creating a rectangular selection of any size using the rectangular Marquee tool, and then choose Select⇨Transform Selection.**

   You can use the bounding box to resize and rotate your selection:

   - Drag the handles to make the selection larger or smaller. Drag a corner handle to adjust width and height simultaneously.
   - Position the cursor outside the bounding box to see the Rotate icon; drag when it appears to rotate the selection. Shift+drag to constrain to straight, 45-degree, or 90-degree angles.
   - Ctrl+drag (Windows) or ⌘ +drag (Mac) a corner point to distort the selection.

3. **Press the Esc key in the upper-left of your keyboard in order to escape the transformation tool without making any changes. Now you will make a selective change using the transform feature.**

4. **Choose Select⇨Deselect or press Ctrl+D (PC) or Command+ D (Mac) to deselect any pixels in your image.**

Transforming selections

In this section, you use the Transform Selection feature to make a selective change in the image.

1. **Using the Rectangular Marquee tool, click and drag out another rectangle that spans the inside of the monitor, as shown in Figure 3-10. Size does not have to be exact.**

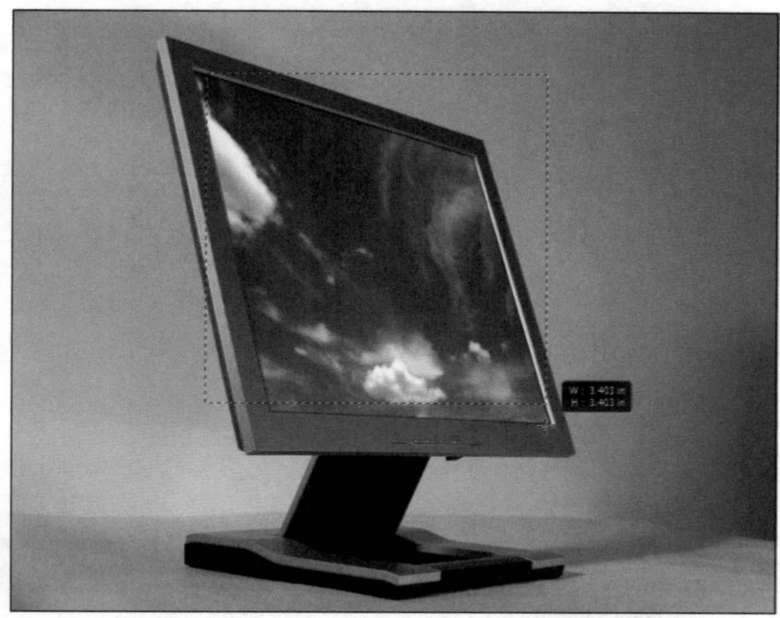

FIGURE 3-10: Click and drag a rectangular marquee selection over the monitor.

2. **Choose Select⇨Transform Selection.**
3. **Hold down the Ctrl (PC) or Command (Mac) key and click on the lower-right corner of your selection. Pull that corner to match the lower-right corner of the monitor.**
4. **Ctrl- or Command-click on each corner, dragging the corner to match the monitor, as shown in Figure 3-11. When finished, click on the checkmark in the upper-right corner of the Options to confirm the transformation.**
5. **Choose Image⇨Adjustments⇨Hue/Saturation and drag the Hue slider to the left or the right to see that the color inside your selection changes, but the rest of your image is not affected, as shown in Figure 3-12.**

   When you change the Hue of an image, you keep the underlying gray shading but change the color. You will discover more about working with hues in Chapter 7 later in this minibook.
6. **Click OK when you've chosen the color you are happy with. Choose Select⇨Deselect to turn off your selection. You can either save or close your image.**

FIGURE 3-11: Ctrl- or Command-click and drag each corner to match the inside of the monitor.

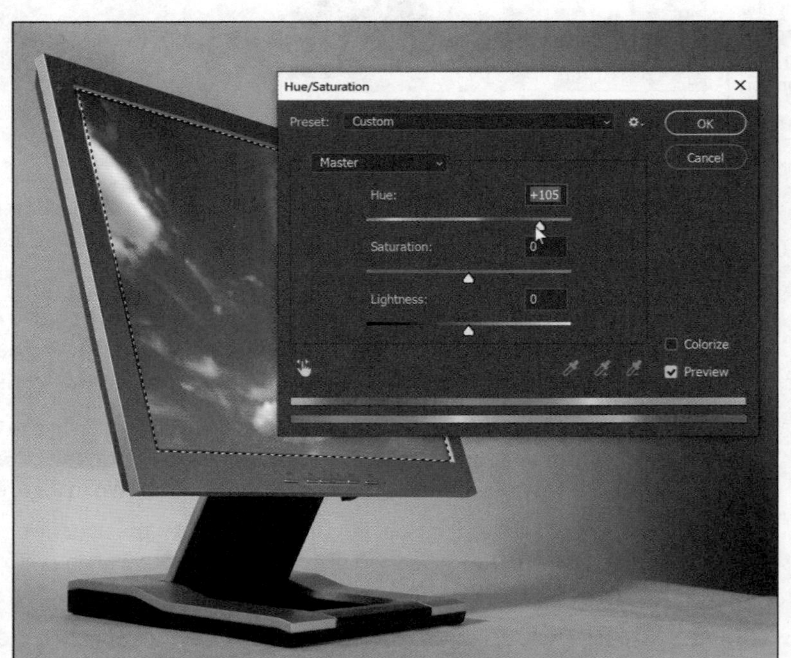

FIGURE 3-12: The hue change affects only the inside of the selection.

Feathering

There are many techniques to make your retouches discreet. By *feathering* a selection (fading the selection edges), you create a natural-looking transition between the selection and the rest of the image, as you see in Figure 3-13.

FIGURE 3-13: No feathering applied to the selection (left); feathering applied to the selection (right).

To feather a selection, follow these steps:

1. **Create a selection.**

   Use any selection technique that you like. You could make a quick selection with the Rectangular or Elliptical Marquee tools, or create a custom selection with the Lasso or Quick Selection tool.

2. **When you have any selection tool active, the Select and Mask button becomes available in the Options across the top of the Photoshop work area. Click the Select and Mask button.**

   You are then entered into the Select and Mask workspace. This dialog box on the right provides you with a selection preview option and with many settings that can improve your selection; for right now, however, just focus on feathering.

3. **In the Properties panel, click on the View option and change it to View on Layers, as shown in Figure 3-14.**

   This allows you to see the image and any transparency that will be applied when you confirm the mask that you are refining.

4. **In the Global Refinements section of the Properties panel, click and drag the slider for Feather to the right until you see the right amount of feathering, or *vignetting*, for your image, as in Figure 3-15.**

   TECHNICAL STUFF

   The results of the feathering depend on the resolution of the image. A feather of 20 pixels in a 72 ppi (pixels per inch) image is a much larger area than a feather of 20 pixels in a 300 ppi image. Typical amounts for a nice vignette on an edge of an image are 20 to 50 pixels. Experiment with images to find what works best for you.

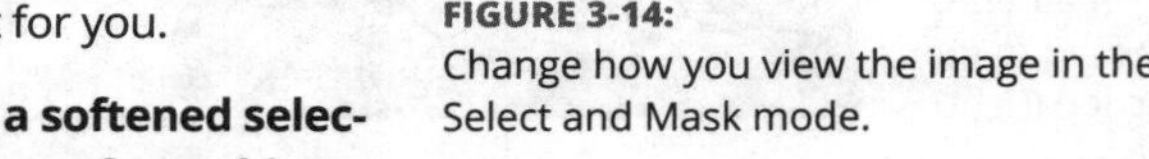

FIGURE 3-14: Change how you view the image in the Select and Mask mode.

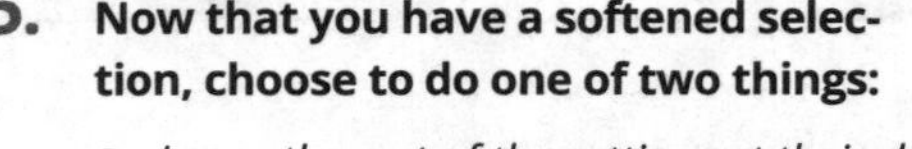

5. **Now that you have a softened selection, choose to do one of two things:**

   - *Leave the rest of the settings at their default and click OK.* This enables you to use the feathered selection to apply subtle changes in the image, such as a curve change to lighten or darken an image, as you see in Figure 3-16.

     This feathering effect creates a nice, soft edge to your image, but it's also useful when retouching images. If you want to try the change shown in Figure 3-13, follow these steps: a.) Make sure that your feathered selection is still active; if it isn't, try Select ➪ Reselect; b.) choose Image ➪ Adjustments ➪ Curves; c.) click in the center of the curve to add an anchor point and drag up to lighten the image; this lightens the midtones of the image.

     Notice how the lightening fades out so that the correction has no definite edge. You can have more fun like this in Chapter 6 of this minibook, in which we cover color correction.

   - *Choose Layer Mask in the Output To drop-down list of the Select and Mask workspace, and then click OK.* You can do this if you are still in the Select and Mask mode. This option keeps only the feathered selection, making the area outside your selection transparent, as you see in Figure 3-17.

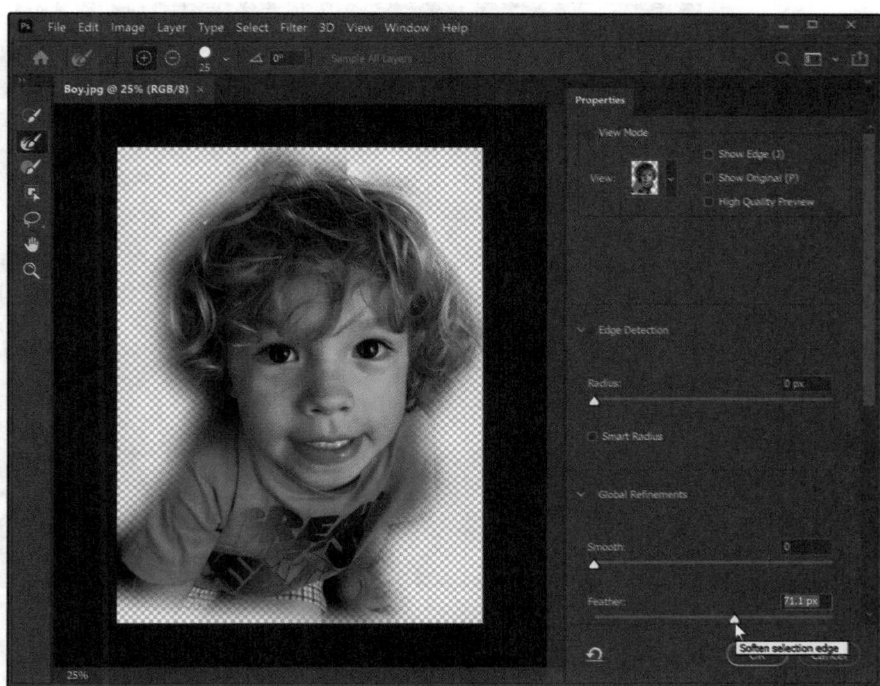

FIGURE 3-15: Click and drag the Feather slide to soften the edges of your selection.

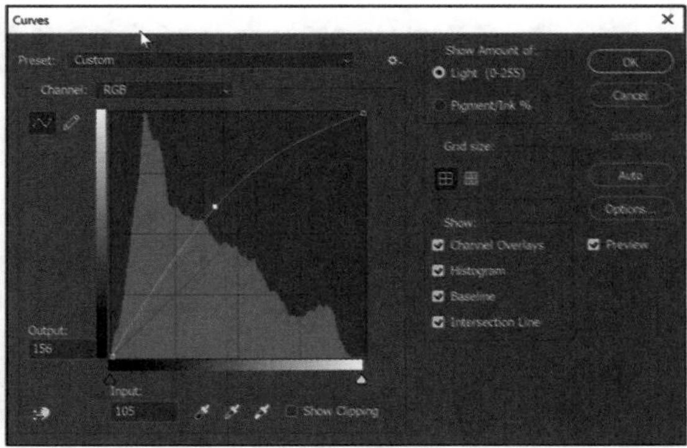

FIGURE 3-16: A curve correction was applied to make this section lighter.

If you are no longer in the Select and Mask mode, choose Window⇨Layers and click on the Add a mask button at the bottom of the panel as shown in Figure 3-17.

When you choose either method, a layer mask is created, which you can see in Window⇨Layers. Hold the Shift key and click the thumbnail of that mask in the Layers panel to turn it off and on.

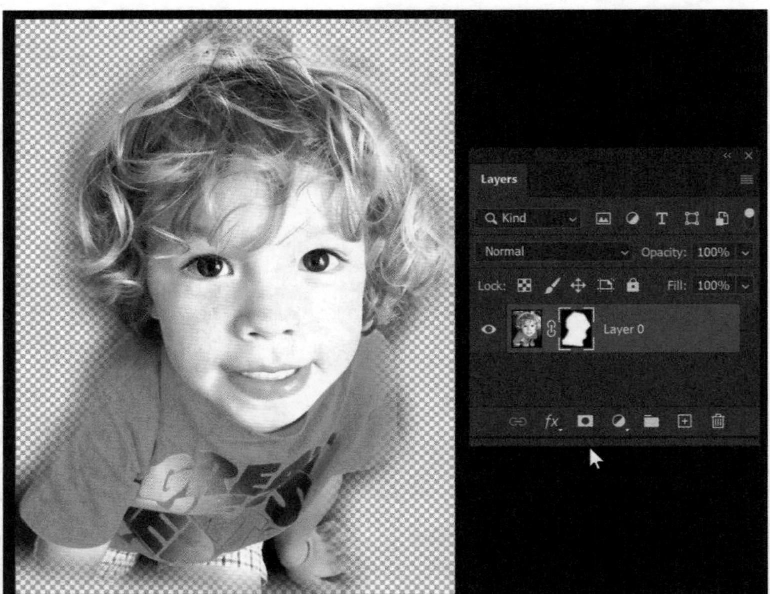

FIGURE 3-17: The selection used for a mask.

Tweaking the edges of a selection in the Select and Mask workspace

You can also use the Select and Mask feature to create more difficult selections. In this section, you review the capabilities of the Select and Mask feature. If you like, you can use the sample image named `YellowPuffball.jpg` for this example.

Follow these steps:

1. **If using our lesson files from Book 1, Chapter 1, choose File⇨Open and locate the image named `YellowPuffball.jpg` in the Book03_Photoshop folder.**
2. **Locate the Magic Wand tool; depending upon the last tool you used, it could be hidden in either the Object Selection or Quick Selection tool in the Tools panel.**
3. **Click on the white background of the Puffball image to automatically select it, as shown in Figure 3-18.**
4. **Inverse the selection so the Yellow Puffball is selected by choosing Select⇨Inverse.**
5. **With any other selection tool active, click the Select and Mask button on the Options bar at the top of the Photoshop window.**

   You are now entered into the Select and Mask mode.

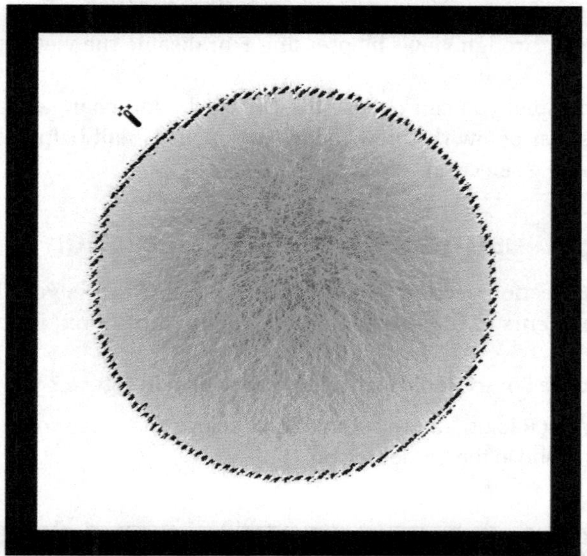

FIGURE 3-18: Use the Magic Wand tool to select the white background in the image.

Select an appropriate preview method

You can choose a preview method by selecting the View drop-down list at the top of the Properties panel. This preview option helps you to better see the effects of the changes you are making in this dialog box.

This list describes the other selections you can choose from the View drop-down list of the Refine Edge dialog box:

- **Onion Skin:** Produces a preview similar to what you might see if you were looking through onion skin paper and similar to an animation-style onion skin scheme.
- **Marching Ants:** Shows the marquee selection with no image masked.
- **Overlay:** Shows the default, slightly transparent mask. It's typically red unless you have changed the mask color.
- **On Black:** Reveals black around the masked area.
- **On White:** Reveals white around the masked area.
- **Black and White:** Reveals the masked area as black and the nonmasked area as white. This option is helpful if you want to see only the mask and not the image area.
- **On Layers:** Bound to be the most popular view if you're creating composites. From this view, you can see the layers and how they're affected by your selection.

TIP

You can cycle through views by pressing F or disable the view by pressing X.

For this example, you can choose the On Black, and change the Opacity to 100% using the slider below the view mode section. This will help you to see the effect of changing the selection.

Making adjustments with Edge Detection

The Edge Detection section of the Properties panel offers you the opportunity to make refinements to the selection edge. To see it in action, follow these steps:

1. **Select the Smart Radius checkbox, as shown in Figure 3-19.**

   The Smart Radius feature automatically adjusts the radius for hard and soft edges found in the border region.

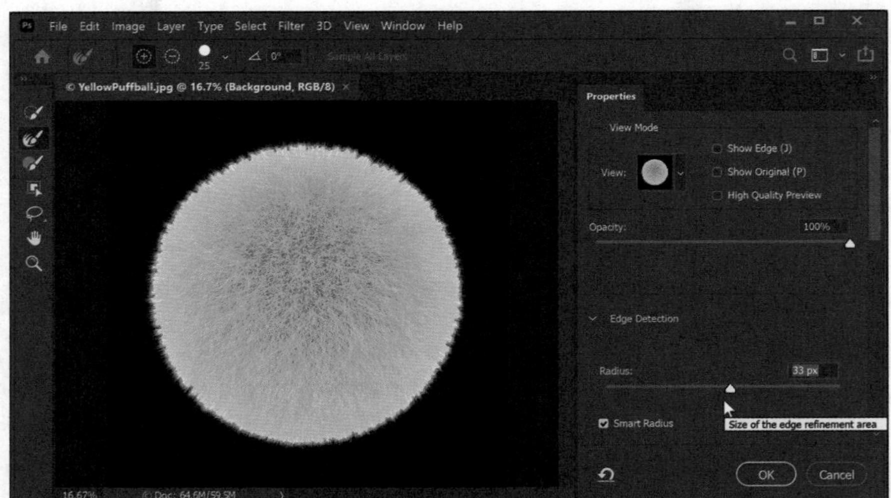

FIGURE 3-19: Refining the edge of a selection.

2. **Increase the radius by dragging the slider to the right.**

   The Radius slider lets you precisely adjust the border area in which edge refinement occurs. Depending on the content, you will increase or decrease the radius.

3. **If you are interested in seeing the radius, select the Show Edge checkbox in the upper-right corner of the Select and Mask workspace.**

Using the Refine Radius tool

To further improve the selection, click the Refine Radius tool, shown in Figure 3-20. By pressing either the right (]) or left bracket ([) key, you can increase or decrease the size of the Refine Radius tool and edit your edge selection by painting the edge of it. If your results are a little too drastic, hold down the Alt or Option key while painting over the area with the Refine Radius tool again. This technique is especially helpful for creating selections of hair or fur.

FIGURE 3-20: Use the Refine Radius tool to make difficult selections, like fur, easier to create.

Making additional refinements

But wait — there's more! You can find additional refinements in the Adjust Edge section of the Properties panel in the Select and Mask workspace:

- **Smooth:** Reduces irregular areas in the selection boundary
- **Feather:** Creates a soft-edged transition between the selection and its surrounding pixels
- **Contrast:** Sharpens selection edges and removes fuzzy artifacts
- **Shift Edge:** Contracts or expands to shrink or enlarge the selection boundary

Selecting color decontamination

Color decontamination can create unwanted artifacts, and when this option is selected, the output option is set to create a new layer, which helps avoid over-writing the original color pixels.

Choosing output settings

As discussed earlier in this chapter, you can choose how you want selections implemented: only selection edges, a new layer, or a layer mask, for example. If you're like most professionals, you want to make sure not to affect the original imagery unless you're certain that you have the selection nailed down.

Here are your output options on the Output To drop-down list:

- **Selection:** The typical selection of a dashed selection, referred to as *marching ants*
- **Layer Mask:** Creates a mask on the active layer
- **New Layer:** Duplicates the active layer and applies the refined selection to the transparency of the layer
- **New Layer with Mask:** Duplicates the active layer and uses the refined selection or mask as the layer mask
- **New Document:** The same as the New Layer option but creates the new layer in a new document
- **New Document with Layer Mask:** The same as the New Layer with Mask option but creates the new layer in a new document

If you want to play it safe, select Layer Mask from the Output To drop-down list. This option offers you the opportunity to edit the layer mask, using the painting tools just as you did with the Quick Mask tool, or to turn off the mask (by Shift-clicking the mask in the Layers panel). Read more about how to work with layer masks in Chapter 8 of this minibook.

If you worked hard to make the perfect selection, don't forget to select the Remember Settings checkbox so that your settings apply the next time you open the Refine Edge dialog box.

Selecting the Subject

In this chapter, you discover multiple methods for making selections. Depending on your image's composition and color, you may have to use a combination of these selection techniques. If you are fortunate, you might be able to get away the Select Subject feature, a one-click selection solution. The following steps show you how. You can follow along using the `LadywithPuppies.jpg` image in the Book03_Photoshop folder:

1. **Open any image, or locate the image named `pinkpuppies.jpg` in our Book03_Photoshop folder.**
2. **Select either the object Selection, Quick Selection, or Magic Wand tool.**
3. **Click the Select Subject button in the Options bar at the top of the image, as shown in Figure 3-21.**

   After a moment, the woman and puppies are automatically selected.

FIGURE 3-21: Select either the Object, Quick or Magic Wand selection tool and then click on Select Subject.

4. **Clean up the selection further by choosing the Select and Mask button. Change the view to be On Layers, and use the Refine Edge Brush tool to clean up the selection around her hair.**
5. **In the Output To settings section, at the bottom of the panel, choose Output To: Layer Mask and click OK.**

   Your selection and mask are now complete.

Saving Selections

The term *alpha channel* sounds complicated, but it's simply a saved selection. Depending on the mode you're in, you already have several channels to contend with. A selection is just an extra channel that you can call on at any time.

To create an alpha channel, follow these steps:

1. **Create a selection that you want to save.**
2. **Choose Select ⇨ Save Selection.**
3. **Name the selection and click OK.**

   An additional named channel that contains your selection appears in the Channels panel.

To load a saved selection, follow these steps:

1. **Choose Select ⇨ Load Selection.**

   The Load Selection dialog box appears.

2. **Select a named channel from the Channel drop-down list.**

   If you have an active selection and then choose to load a selection, you have additional options. With an active selection, you can select one of the following options:

   - *New Selection:* Eliminate the existing selection and create a new selection based on the channel you select.
   - *Add to Selection:* Add the channel to the existing selection.
   - *Subtract from Selection:* Subtract the channel from the existing selection.
   - *Intersect with Selection:* Intersect the channel with the existing selection.

3. **Click OK.**

TECHNICAL STUFF

Other Adobe applications, such as InDesign, Illustrator, Premiere, and After Effects, also recognize alpha channels.

IN THIS CHAPTER

» Selecting the Pen tool for the task

» Using a path as a selection

» Creating curved paths

» Creating clipping paths

» Putting shape layers to work

Chapter 4
Using the Photoshop Pen Tool

The Pen tool is the ultimate method to make precise selections. You can also use it to create vector shapes and clipping paths (silhouettes). In this chapter, you see how to take advantage of this super multitasking tool. This chapter also shows you how to apply paths made with the Pen tool to create selections, shapes, and silhouettes using clipping paths.

If you're interested in fundamental practices when using the Pen tool, check out Book 5, Chapter 4, where we cover the Pen tool in more detail. Using that lesson helps you in Photoshop, too, because the Pen tool works essentially the same way in both Illustrator and Photoshop.

We recommend that you use the Pen tool as much as you can to truly master its capabilities. At first it may seem awkward, but it gets easier! Knowing how to effectively use the Pen tool puts you a grade above the average Photoshop user, and the quality of your selections will show it. Read Book 5 on Adobe Illustrator, Chapter 4 for tips and tricks on how to use the Pen tool. It works very much the same in all the Adobe programs.

Selecting the Right Pen Tool for the Task

Select the Pen tool from the Tools panel, and note the default mode setting of the Pen tool on the left side of the Options bar. You can select the following options from the Pick Tool Mode drop-down list, as you see in Figure 4-1:

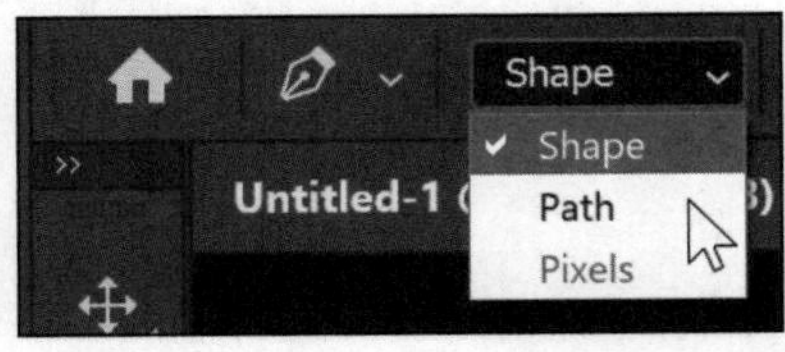

FIGURE 4-1: Select the mode of the Pen tool from the Options bar.

- **Shape:** Creates a new shape layer, a filled layer that contains the vector path.
- **Path:** Creates a path only; no layer is created.
- **Pixels:** Creates pixels directly on the image. No editable path or layer is created. This option may not be useful to new users, but some seasoned users prefer it because it's the only way to access the Line tool from earlier versions.

Using a Path as a Selection

You can use the Pen tool to create precise selections that would be difficult to create using other selection methods. The Pen tool produces clean edges that print well and can be edited using the Direct Selection tool. Before using the Pen tool, make sure that you select Path from the Pick Tool Mode drop-down list on the Options bar.

To create a path as a selection (which is extremely helpful when you're trying to make a precise selection), follow these steps to create a simple straight line path:

1. **Open the Triangle.jpg file located in the Book03_Photoshop folder inside the DummiesCCFiles downloaded in Book 1, Chapter 1, or create a new blank file.**

   It is best to practice on an image that has a straight or angular item in it.

2. **Select the Pen tool.**

   REMEMBER

   Make sure that Path is selected from the Pick Tool Mode drop-down list on the Options bar or else you'll create a shape layer.

3. **Click on one of the corners of the triangle to place your first anchor point, and then click on the next corner. Notice that the anchor points are connected with a path.**

4. **Continue to click on the corner, when you reach the first point you made you will see a circle, indicating that you are closing this path.**

   When you return to your original starting point, a closed circle appears, indicating that you are closing this path, as you see in Figure 4-2.

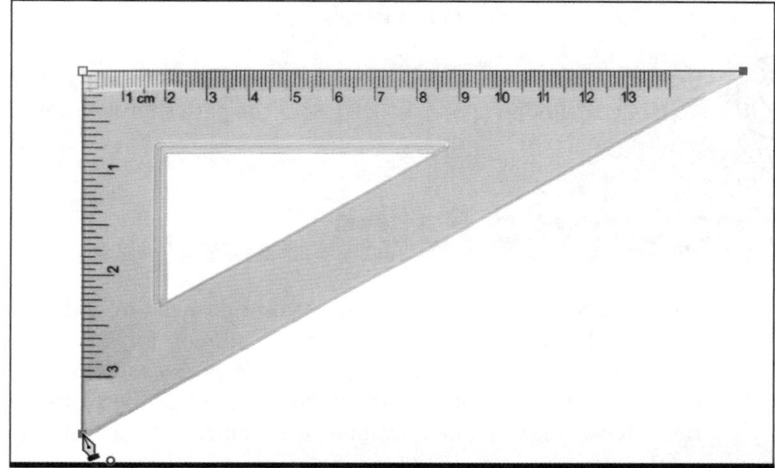

FIGURE 4-2: Using the Pen tool, click from one segment to another; close it at the end.

5. **Choose Window ➪ Paths.**

   In the Paths panel, you can create new paths and activate existing paths, apply a stroke, or turn paths into selections by clicking the icons at the bottom of the panel. (See Figure 4-3.)

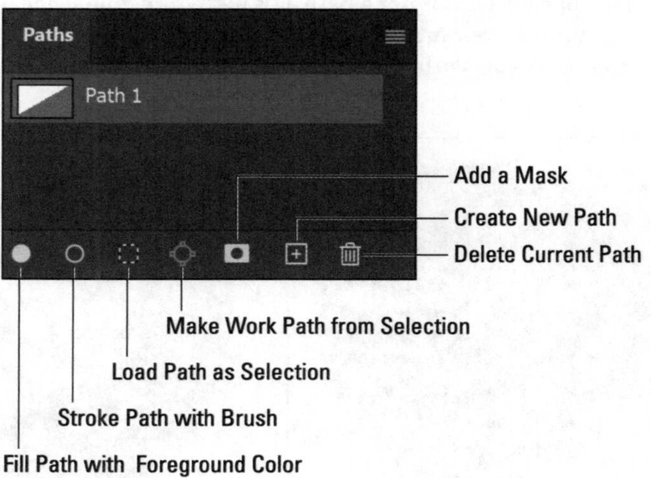

FIGURE 4-3: The Paths panel and its options.

Using the Photoshop Pen Tool

6. **Click and drag the Work Path down to the Create New Path icon at the bottom of the Paths panel.**

   The path is now named Path 1 and is saved. You can also double-click to rename the path if you like.

7. **Click the Load Path as Selection icon.**

   The path is converted into a selection.

TIP

Use this quick and easy method for turning an existing path into a selection: Ctrl-click (Windows) or ⌘-click (Mac) the path thumbnail in the Paths panel.

Creating Curved Paths

To create paths that are on a curve, you first need to understand how the Bézier curve works. A Bézier curve in its most common form is a simple cubic equation that can be used in any number of useful ways, but especially when creating curves in a path. Originally developed by Pierre Bézier in the 1970s for CAD/CAM operations, it became the underpinnings of the entire Adobe PostScript drawing model.

To get a better understanding of Bézier curves, work through this simple exercise:

1. **Open an image. You can use our example in the Book03_Photoshop folder called `Apple.jpg`.**

   In this example, an image of an apple is used.

2. **Click and drag the edge of the object that you are selecting. A directional line appears. (This is not a path; it is just a directional line determining the curve of your future path.) Drag along the edge of the object and then release, as you see in Figure 4-4.**

FIGURE 4-4: Click and drag to make curved paths.

3. **Move the cursor along the edge of the object that you want to select and then click and drag again.**

   Pay attention to the directional lines and have them flow, either on the shape or along the side of the shape. The path you want should fall into place with just some minor tweaking.

4. **Continue clicking and dragging to make the path, and then close the shape.**

   Don't worry if the path is not perfect.

5. **Choose the Direct Selection tool by holding down the Path Selection tool, as you see in Figure 4-5.**

   Use this tool to adjust your path by clicking anchor points and directional lines and repositioning or shortening them to provide you with a more accurate selection, as you see in Figure 4-6.

FIGURE 4-5: Select the hidden Direct Selection tool.

FIGURE 4-6: Edit your path with the Direct Selection tool.

Using the Photoshop Pen Tool

Clipping Paths

If you want to create a beautiful silhouette that transfers well to other applications as you see Figure 4-7, create a clipping path. A *clipping path* is essentially a vector path that defines what part of the image is visible and what part is transparent. Typically a path is created around an object that you want to silhouette.

Keep in mind that you do not have to define a clipping path if you are using your image in other Adobe applications, such as InDesign and Illustrator. A clipping path is helpful if you are trying to keep an image transparent in apps such as PowerPoint, Word, or various video-editing applications.

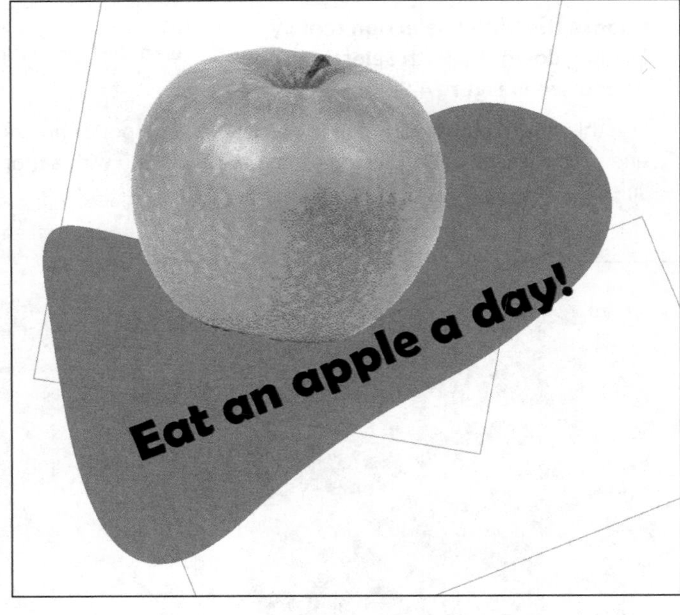

FIGURE 4-7: Clipping paths enable you to create silhouettes in other applications. This is an example of a clipping path of the apple placed into InDesign.

To create a clipping path, follow these steps:

1. **Use the Pen tool to create a path around the image area that will become the silhouette. You can use the same apple image if you like.**
2. **In the Paths panel, choose Save Path from the panel menu (click the menu icon in the upper-right corner of the panel), as shown in Figure 4-8, and then name the path.**

   If Save Path is not visible, your path has already been saved; skip to Step 3.

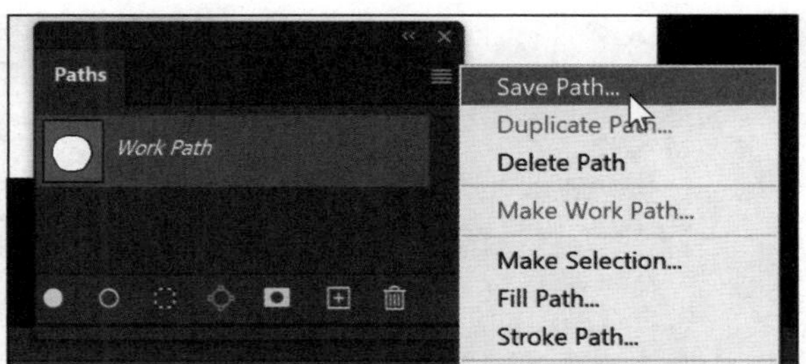

FIGURE 4-8: Convert your work path to a saved path.

3. **From the same panel menu, choose Clipping Path.**
4. **In the Clipping Path dialog box, select your path from the drop-down list, if it's not already selected; click OK.**

   Leave the Flatness Device Pixels text field blank unless you need to change it. The flatness value determines how many device pixels are used to create your silhouette. The higher the amount, the fewer points are created, thereby allowing for faster processing time. This speed comes at a cost, though: If you set the flatness value too high, you may see (if you look close) straight edges instead of curved edges.
5. **Choose File ➪ Save As and, from the Format drop-down list, select Photoshop EPS; in the EPS Options dialog box that appears, accept the defaults and click OK.**

   If you see PostScript errors when printing, choose Clipping Path from the panel menu and increase the value to 2 pixels in the Flatness Device Pixels text field. Keep returning to this text field and increasing the value until the file prints, or give up and try printing your document on another printer.

   TIP: If you're placing this file in other Adobe applications, such as InDesign, you don't need to save the file as EPS; you can leave it as a Photoshop (.psd) file.

Here's an even faster method you can use to create a clipping path that you can use in other Adobe applications, such as InDesign and Illustrator:

1. **Create a path around the item you want to keep when you create the clipping path.**
2. **In the Layers panel, click the Add Layer Mask button, and then click the Add Layer Mask button again.**

   A layer vector mask is created, and everything outside the path becomes transparent, as shown in Figure 4-9.

   You can still edit the path by using the Direct Selection tool.
3. **Save the file in the `.psd` format.**
4. **Choose File ➪ Place to put the image, with its clipping path included, into other Adobe applications.**

FIGURE 4-9: Creating a saved path that supports transparency the easy way, with layers.

Using Shape Layers

When you start creating a path with the Pen tool, Photoshop creates a *path.* You can also choose to make a vector shape or a fill (which essentially creates pixeled, nonvector shapes and paths).

Shape layers can be very useful when the goal of your design is to seamlessly integrate vector shapes and pixel data. A shape layer can contain vector shapes that you can then modify with the same features of any other layer. You can adjust the opacity of the shape layer, change the blending mode, and even apply layer effects to add drop shadows and dimension. Find out how to do this in Chapter 8 of this minibook.

Create a shape layer with any of these methods:

- **Create a shape with the Pen tool.** With the Pen tool, you can create interesting custom shapes and even store them for future use. We show you how in the following section.
- **Use a Vector Shape tool, as shown in Figure 4-10.** *Vector shapes* are premade shapes (you can even create your own) that you can create by dragging the image area with a shape tool.

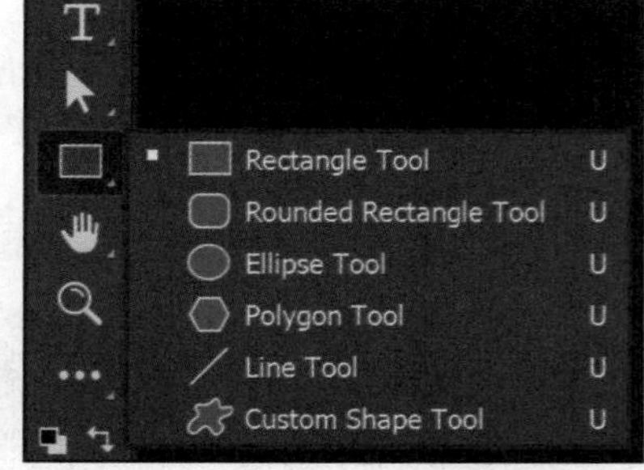

FIGURE 4-10: The Vector Shape tools.

- **Import a shape from Illustrator.** Choose File ➪ Place and choose an `.ai` file; when the Options window appears, choose to place the file as a shape layer or a path. This action imports an Illustrator file as a shape layer or path into Photoshop.

In Photoshop CC, you can access the Vector shape and path tool options directly from the Options bar, as shown in Figure 4-11, when either the Pen tool or a vector shape tool is selected.

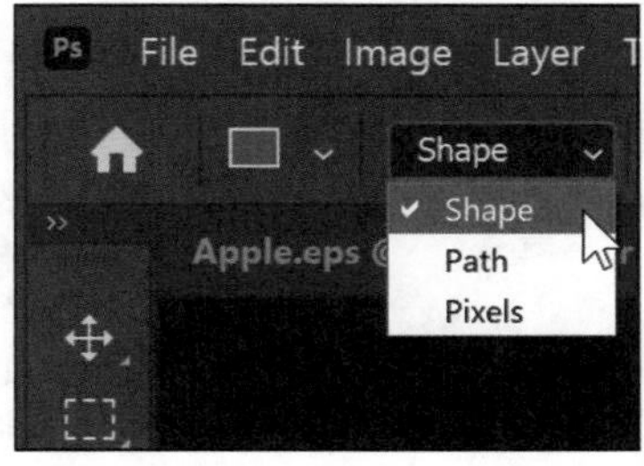

FIGURE 4-11: You can switch to the Vector shape tool with the Options bar.

Creating and using a custom shape

Perhaps you like the wave shape (see Figure 4-12) that's been cropping up in design pieces all over the place.

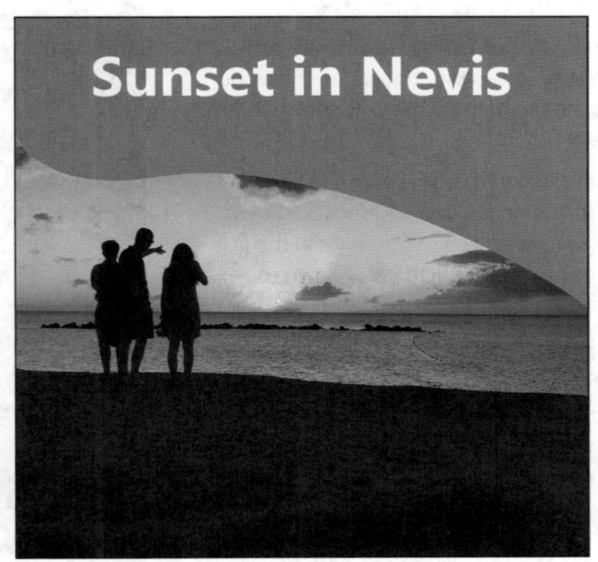

FIGURE 4-12: A custom wave shape integrated with an image in Photoshop.

You can create a wavy shape like that, too. With any image or blank document open, just follow these steps:

1. **Select the Pen tool, and then select Shape from the Pick Tool Mode drop-down list on the Options bar.**

   Refer to Figure 4-11.

2. **Click and drag with the Pen tool to create a wavy shape.**

   Don't worry about the size of the shape. The shape is a vector, so you can scale it up or down to whatever size you need without worrying about making jagged edges. Just make sure to *close* the shape (that is, return to the original point with the end point).

   When you create the shape, it fills in with your foreground color. Try to ignore it if you can; the next section shows you how to change the fill color.

3. **With the shape still selected, choose Edit ➪ Define Custom Shape, name the shape wave, and click OK.**

   After you save your custom shape, you can re-create it at any time. If you don't like the shape, choose Window ➪ Layers to open the Layers panel, and then drag the shape layer you just created to the Trash Can in the lower-right corner of the panel. This deletes the layer, but not the saved shape. If you want to experiment with your custom shape now, continue with these steps.

4. **Click and hold the Rectangle tool to access the other hidden vector tools; select the last tool, the Custom Shape tool.**

   When the Custom Shape tool is selected, a Shape drop-down list appears on the Options bar at the top of the screen, as shown in Figure 4-13.

   You have lots of custom shapes to choose from, including the one you've just created. If you just saved a shape, yours is in the last square; you may have to scroll down to select it.

5. **Click to select your custom shape; click and drag in the image area to create your shape.**

   You can make the shape any size you want. Hold the Shift key while resizing if you want to maintain the original poroportions.

6. **To resize the shape, choose Edit ➪ Free Transform Path, press Ctrl+T (Windows) or ⌘+T (Mac), grab a bounding box handle, and drag.**

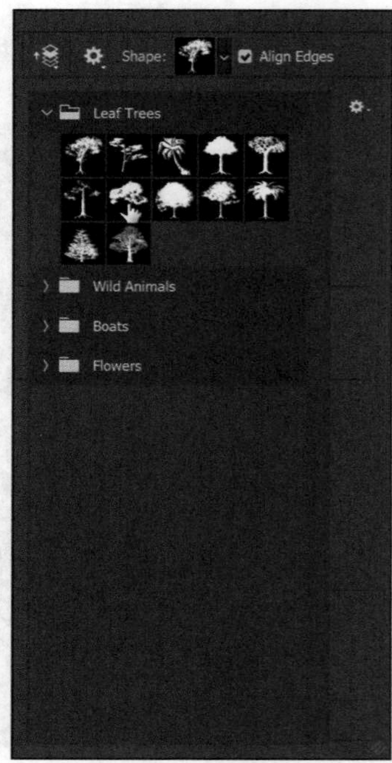

FIGURE 4-13: A Shape drop-down list appears on the Options bar.

TIP

Because a shape is created on its own layer, you can experiment with different levels of transparency and blending modes in the Layers panel. Figure 4-14 shows shapes that are partially transparent.

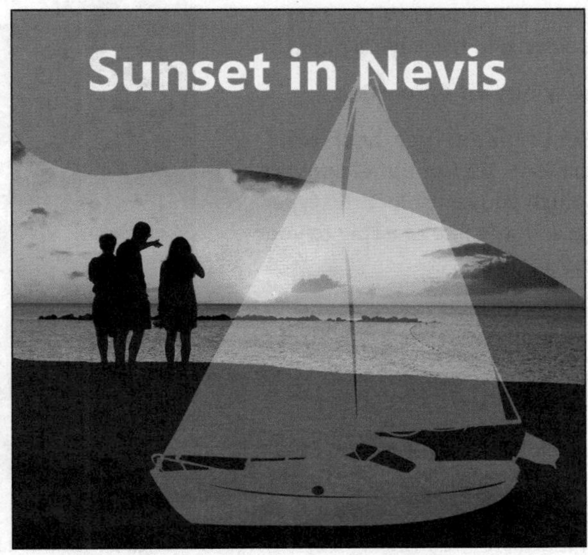

FIGURE 4-14: Experiment with blending modes and opacity changes on shape layers.

Changing the color of the shape

When you create a shape with a shape tool, the shape takes the color of the current foreground color. To change the color of an existing shape, choose Window⇨Properties. If you have a basic Shape layer selected, not a custom shape, you see that the Properties panel offers you options for changing the fill and stroke of your shape, as shown in Figure 4-15. The shape layer is considered a live shape because you can make nondestructive changes on this shape, as long as it remains a shape layer and is not flattened into the other parts of the image.

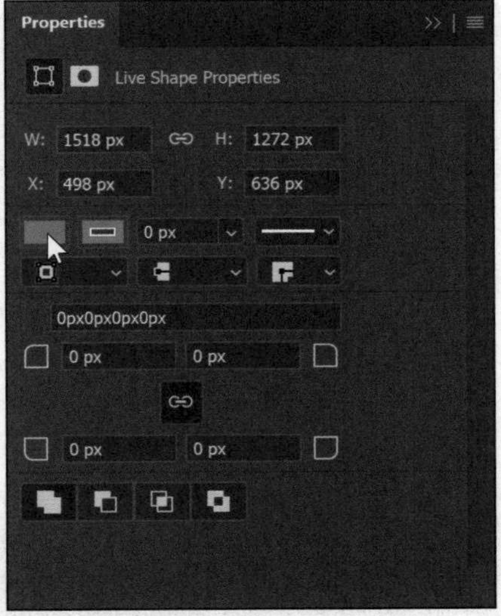

FIGURE 4-15: You can apply non-destructive changes to the fill, stroke, and more using the Properties pane.

To change the fill, follow these steps:

1. **Create a standard vector shape, such as with the Rectangle tool.**
2. **Make sure that you have the shape layer selected in the Layers panel.**
3. **Choose Window ⇨ Properties.**
4. **Click Set Shape Fill Type, as shown in Figure 4-16. A color panel appears from which you can select the type of fill you would like to apply to your selected shape.**
5. **Choose whether you want No fill (clear), a Solid Color fill, a Gradient fill, or a Pattern fill. For this example, use a Solid Color fill.**
6. **From the swatches, choose the color you wish to use. If you prefer to use the Color Picker, click the icon to the right of Pattern fill in the Properties panel.**

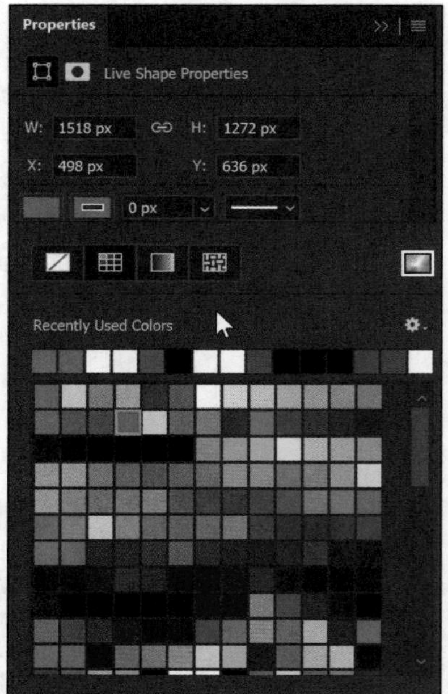

FIGURE 4-16:
Changing the fill in a shape.

Changing the color fill of a custom shape

If you create a custom shape, you may see only some Mask features in the Properties panel. If that is the case, follow these steps in order to change the color fill:

1. **Make sure you are on the Shape tool and then select the custom shape layer in the Layers panel.**
2. **Double-click the thumbnail of the shape layer in the Layers panel, as shown in Figure 4-17, or click the Fill color box on the Options bar across the top of the Document window.**

   The Color Picker appears.
3. **To select a new color, drag the Hue slider up or down or click in the large color pane to select a color with the saturation and lightness you want to use. Click OK when you're done.**

TIP

With the Color Picker open, you can also move outside the Picker dialog box and sample colors from other open images and objects.

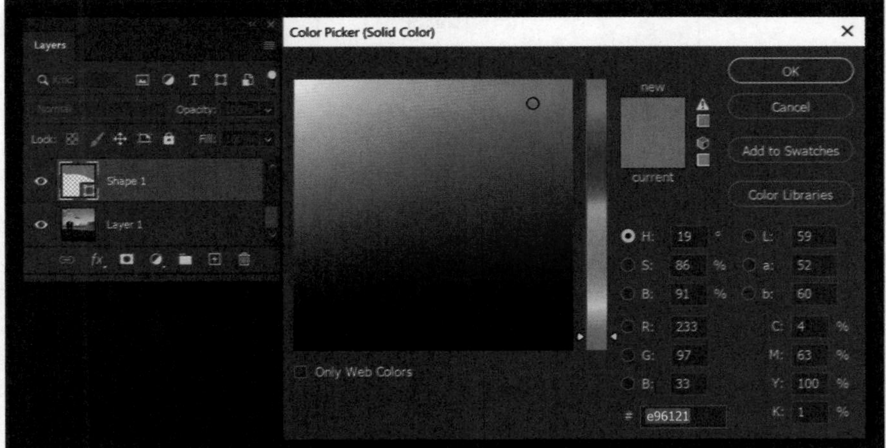

FIGURE 4-17: Double-click on the layer thumbnail to choose a fill color from the Picker or from a source outside of the Picker.

Changing the stroke on a live vector shape

In addition to changing the fill, you can also change the color, width, and even type of stroke or border on your shape. As with the fill, the Properties panel is used to make these changes, but if you have a custom shape you may not have access to these same controls. The next set of steps take advantage of effects in order to apply a stroke to a custom shape.

To apply a stroke to a live shape, follow these steps:

1. **Create a simple rectangular shape.**
2. **Choose Window ⇨ Properties.**
3. **Select Set Shape Stroke Type and choose the type of stroke you wish to apply, as shown in Figure 4-18.**
4. **Note that you can also change the line width, cap, stroke type, and more. See Figure 4-19 to see some of the other stroke- and shape-related properties you can set in the Properties panel.**

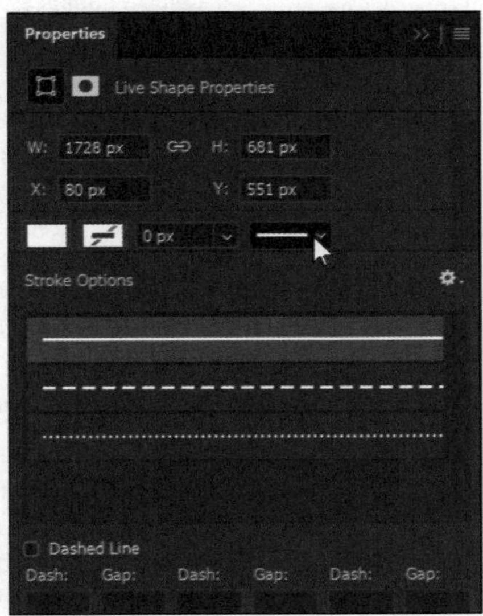

FIGURE 4-18: Choose the type of stroke that you would like to apply.

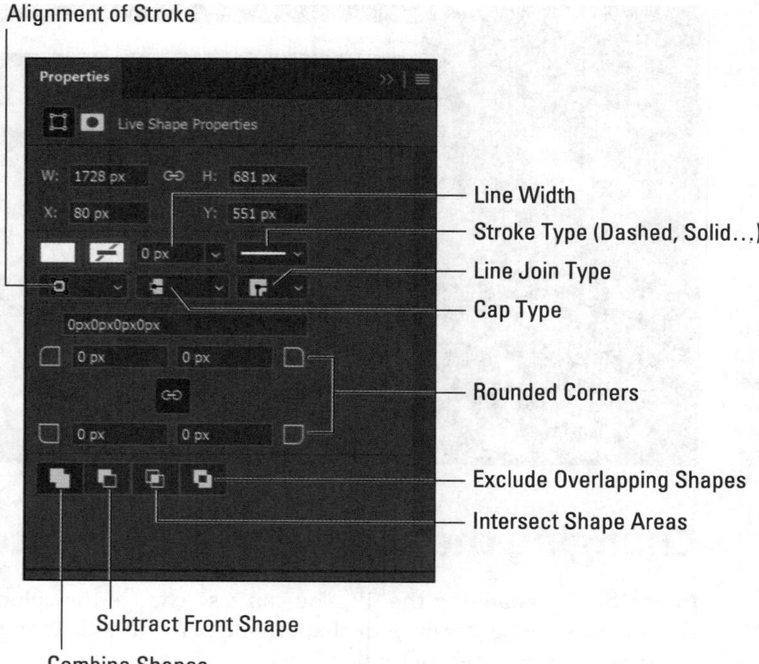

FIGURE 4-19: Stroke- and shape-related properties in the Properties panel.

Creating a stroke on a custom shape

If you want to create a stroke on a custom shape, you may not have the same properties available to you. In order to apply a stroke to a custom shape, follow these steps:

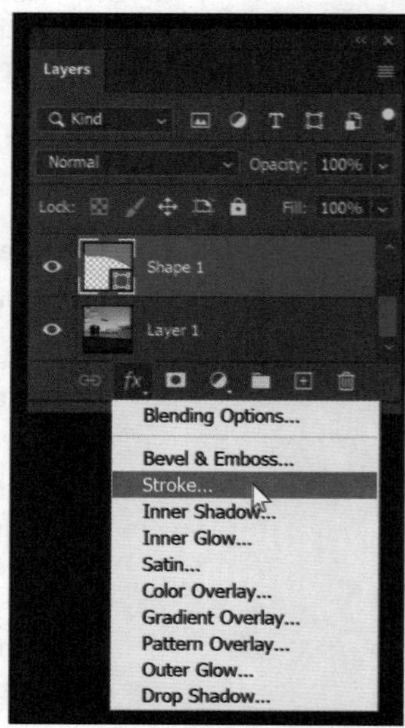

FIGURE 4-20: You can use Layer styles in order to add a stroke to a custom shape.

1. **Select the custom shape layer in the Options bar.**
2. **Choose Stroke from the menu that opens when you click the Add a Layer Style button at the bottom of the Layers panel, as shown in Figure 4-20.**
3. **When the Layer Style dialog box appears, click the Color swatch to choose a color, as shown in Figure 4-21, and then change any other properties such as Size (width), Position (alignment), Fill Type, and more. Click OK when you are finished making changes.**

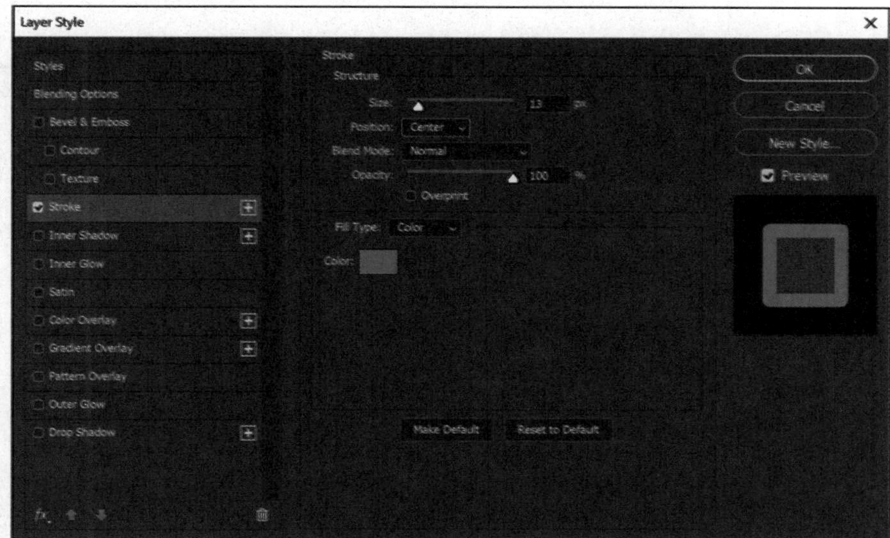

FIGURE 4-21: Use the Layer Styles feature in order to apply properties for the Stroke.

Removing a shape layer

Because the Pen tool now has multiple options, you may unexpectedly create a shape layer. You can delete a shape layer by dragging the layer thumbnail to the Trash Can in the lower-right corner of the Layers panel.

If you want to keep your path but throw away the shape layer, choose Window⇨Paths. Then drag the shape vector mask to the Create New Path icon, shown in Figure 4-22, which creates a saved path. Now you can throw away the shape layer.

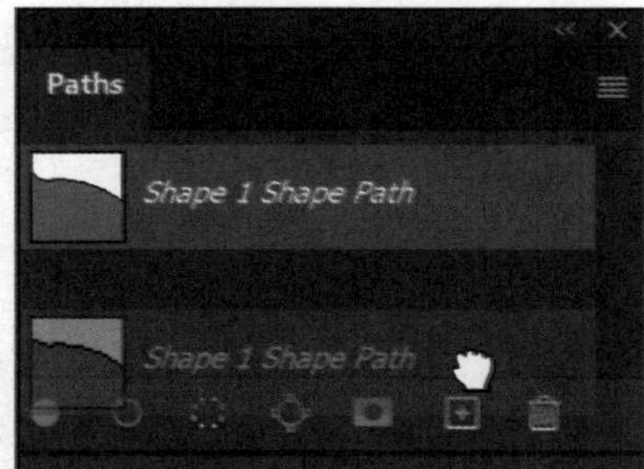

FIGURE 4-22: Drag the shape path to the Create New Path icon.

» **Adjusting file size**

» **Applying the Unsharp Mask filter to your image**

Chapter 5 Creating Images in the Right Resolution for Print and the Web

Something as important as setting the right resolution for your images deserves its own chapter. In this chapter, you discover the recommended resolution for your images, which is dependent upon where the image is to be used. Are you going to use the image in a high-quality annual report or are you going to email it to your family? This is important information to know right from the start. Combine the information about resolution along with the steps to use the color and retouching tools we show you in Chapter 7 of this minibook, and you should be ready to roll with great imagery.

Creating Images for Print

To begin, you need to know the existing resolution of your image. To see an open image's present size and resolution in Photoshop, choose Image⇨Image Size. The Image Size dialog box appears, as shown in Figure 5-1.

At the top of the Image Size dialog box, you see the pixel dimensions noted.

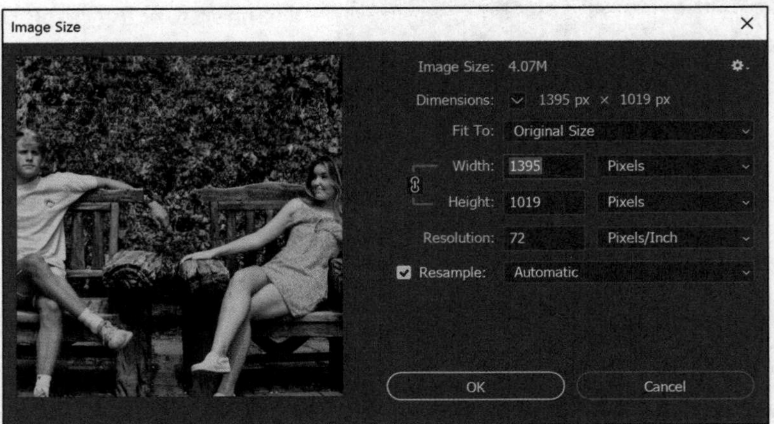

FIGURE 5-1: The Image Size dialog box.

The Width and Height text fields are in inches in the English default; you can change the increments to pixels by selecting Pixels from the measurement unit drop-down menu. The Width and Height text fields can be used for onscreen resizing. By switching the measurement unit back to inches, you can determine the size at which the image will print. The Resolution text field determines the resolution of the printed image; a higher value means a smaller (more pixels crammed in each inch), which typically produces a more finely detailed printed image.

Before you decide on a resolution, you should understand what some of the resolution jargon means:

- **ppi (pixels per inch):** The resolution of an image — how Photoshop measures resolution of images.
- **dpi (dots per inch):** The resolution of an image when printed.
- **lpi (lines per inch):** The varying dot pattern that printers and presses use to create images. (See Figure 5-2.) This dot pattern is referred to as the *lines* per inch, even though it represents rows of dots. The higher the lpi, the finer the detail and the less dot pattern or line screen you see.
- **Dot gain:** The spread of ink as it's applied to paper. Certain types of paper spread a dot of ink farther than others. For example, newsprint has a high dot gain and typically prints at 85 lpi; a coated stock paper has a lower dot gain and can be printed at 133–150 lpi and even higher.

Human eyes typically can't detect a dot pattern in a printed image at 133 dpi or higher.

FIGURE 5-2: The dot pattern used to print images is referred to as lpi (lines per inch).

TIP

Deciding the resolution or dpi of an image requires backward planning. If you want to create the best possible image, you should know where it'll print *before* deciding its resolution. Communicate with your printer service if the image is going to press. If you're sending an image to a high-speed copier, you can estimate that it will handle 100 lpi; a desktop printer handles 85 lpi to 100 lpi.

The resolution formula

When creating an image for print, keep this formula in mind:

2 × lpi = ppi (pixels per inch)

This formula means that if your image is going to press using 150 lpi, set your image at 300 ppi. To save space, many people in production use 1.5 × lpi because it reduces the file size significantly and you get similar results; you can decide which one works best for you.

Changing the resolution

Using the Image Size dialog box is only one way that you can control the resolution in Photoshop. Even though you can increase the resolution, do so sparingly and avoid it, if you can. The exception is when you have an image that's large in dimension size but low in resolution, like those you typically get from a digital camera. You may have a top-of-the-line digital camera that produces 72 ppi images, but at that resolution, the pictures are 28 x 21 inches (or larger)!

To increase the resolution of an image without sacrificing quality, follow these steps:

1. **Open any image.** If you want an unretouched digital image to work with you can use the image named LookingOut.jpg that is located in your Book03_Photoshop folder that you downloaded in Book 1, Chapter 1.
2. **Choose Image ⇨ Image Size.**

   The Image Size dialog box appears, as shown in Figure 5-3.

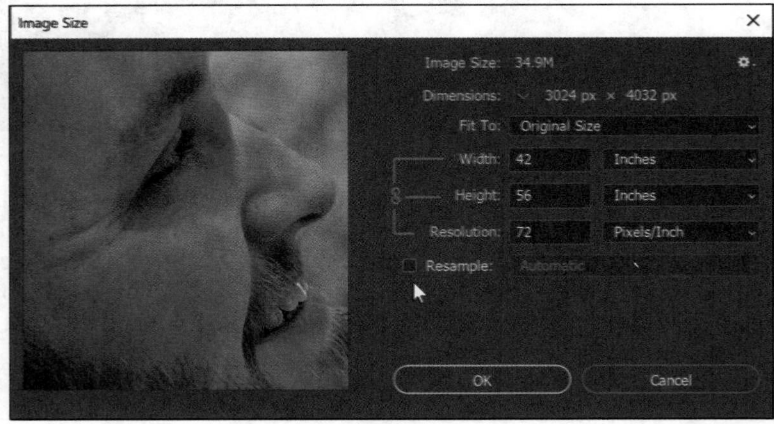

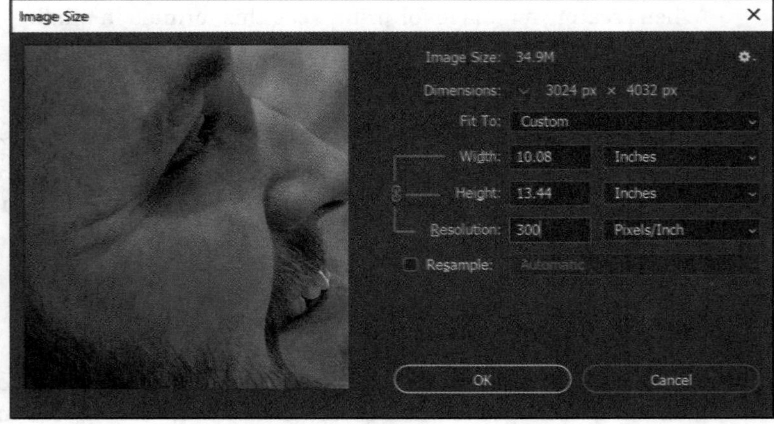

FIGURE 5-3: The Image Size dialog box, before and after deselecting Resample and changing the resolution.

3. **Deselect the Resample Image checkbox.**

   This way, Photoshop doesn't add any pixels. You will also notice that when you uncheck Resample that the image is now measured in Inches instead of pixels. The measurement default can vary depending upon your preferences and location.

4. **Enter a resolution in the Resolution text field.**

   Photoshop keeps the *pixel size* (the size of the image onscreen) the same, but the *document size* (the size of the image when printed) decreases when you enter a higher resolution, as you see in Figure 5-3.

5. **If the image isn't the size you need, select the Resample Image checkbox and type the size in the Width and Height text fields.**

   It's best to reduce the size of a bitmap image, such as a digital photo, rather than increase it.

REMEMBER

   You can also deselect the Resample Image checkbox and essentially play a game of give-and-take to see what the resolution will be when you enter the intended size of your printed image in the Width and Height text fields.

   Images can typically be scaled from 50 to 120 percent before looking jagged. (To scale by a percentage, select Percent from the drop-down lists beside the Width and Height text fields.) Keep these numbers in mind when placing and resizing images in a page layout application such as InDesign.

6. **Click OK when you're finished; double-click the Zoom tool in the Tools panel to see the image at its onscreen size.**

TIP

To increase the resolution *without* changing the image size, follow these steps. (This situation isn't perfect because pixels that don't presently exist are created by Photoshop and may not be totally accurate. Photoshop tries to give you the best image, but you may see some loss of detail.)

1. **Choose Image ➪ Image Size.**
2. **When the Image Size dialog box appears, make sure that the Resample Image checkbox is selected.**
3. **Enter the resolution you need in the Resolution text field, click OK, and then double-click the Zoom tool to see the image at its actual size.**

TIP

Reset your Image Size dialog, or any dialog box in Photoshop by holding down the Alt or Option button. The Cancel button changes to Reset!

Determining the Resolution for Web Images

Have you ever received an image via email and, after spending 10 minutes downloading it, discovered that the image is so huge that all you can see on the monitor is your nephew's left eye? Many people are under the misconception that if an image is 72 dpi, it's ready for onscreen presentation. Actually, pixel dimension is all that matters for viewing images on the web; this section helps you make sense of this concept.

Many people view web pages in their browser windows in an area of about 1024 x 768 pixels. You can use this figure as a basis for any images you create for the web, whether the viewer is using a 14-inch or a 21-inch monitor. (Remember that people who have large monitors set to high screen resolutions don't necessarily want a web page taking up the whole screen.) If you're creating images for a web page or to attach to an email message, you may want to pick a standard size to design by, such as 600 x 400 pixels at 72 ppi.

To use the Image Size dialog box to determine the resolution and size for onscreen images, follow these steps:

1. **Have an image open and choose Image ⇨ Image Size.**

   The Image Size dialog box appears.

2. **To make the image occupy half the width of a typical browser window, choose Pixels for the unit type in the Width drop-down list and type** 300 **(half of 600) in the Width text field.**

   If a little chain link is visible to the right, the Constrain Proportions checkbox is selected, and Photoshop automatically determines the height from the width you entered.

3. **Click OK and double-click the Zoom tool to see the image at its actual onscreen size.**

   That's it! Whether your image is 3,000 or 30 pixels wide doesn't matter; as long as you enter the correct pixel dimension, the image works beautifully.

Applying the Unsharp Mask Filter to an Image

When you resample an image in Photoshop, it can become blurry. A good practice is to apply the Unsharp Mask filter. You can see the difference in detail in the images shown in Figure 5-4. This feature sharpens the image based on levels of contrast while keeping smooth the areas that have no contrasting pixels. You have to set up this feature correctly to get good results.

FIGURE 5-4: The image with (left) and without (right) unsharp masking applied.

Here's the down-and-dirty method of using the Unsharp Mask filter:

1. **Open any image, or use the `SillyDog.jpg` image located in the Book03_Photoshop folder.**
2. **Choose View⇨100% or double-click the Zoom tool.**

   REMEMBER

   When you're using a filter, view your image at its actual size to best see the effect.

3. **Choose Filter⇨Sharpen⇨Unsharp Mask.**

   In the Unsharp Mask dialog box that appears, set these three options:

   - *Amount:* The Amount value ranges from 0 to 500. The amount you choose has a lot to do with the subject matter. Sharpening a car or appliance at 300 to 400 is fine, but if you do this to the CEO's 75-year-old wife, you may suffer an untimely death because every wrinkle, mole, or hair will magically become more defined. If you're not sure which amount to use, start with 150 and play around until you find an Amount value that looks good. In Figure 5-4 we used a higher amount because this is an image of a dog, and we wanted to see more texture in the fur.

- *Radius:* The Unsharp Mask filter creates a halo around the areas that have enough contrast to be considered an edge. Typically, leaving the amount between 1 and 2 is fine for print, but if you're creating a billboard or poster, increase the size.
- *Threshold:* This option is the most important one in the Unsharp Mask dialog box. The Threshold setting determines what should be sharpened. If you leave it at zero, you see noise throughout the image, much like the grain you see in high-speed film. Increase the setting to 10, and it triggers the Unsharp Mask filter to apply the sharpening only when the pixels are ten shades or more away from each other. The amount of tolerance ranges from 1 to 255. Apply too much and no sharpening appears; apply too little and the image becomes grainy. A good number to start with is 10.

To compare the original state of the image with the preview of the Unsharp Mask filter's effect in the Preview pane of the Unsharp Mask dialog box, click and hold the image in the Preview pane; this shows the original state of the image. When you release the mouse button, the Unsharp Mask filter is previewed again.

4. **When you've made your choice, click OK.**

   The image appears to have more detail.

Once in a while, stray colored pixels may appear after you apply the Unsharp Mask filter. If you feel this is a problem with your image, choose Edit⇨Fade Unsharp Mask immediately after applying the Unsharp Mask filter. In the Fade dialog box, select the Luminosity blend mode from the Mode drop-down list and then click OK. This step applies the Unsharp Mask filter to the grays in the image only, thereby eliminating the sharpening of colored pixels.

You can also choose Filter⇨Convert for Smart Filters before you apply the Unsharp Mask filter. Smart filters let you undo all or some of any filter, including sharpening filters you apply to a layer. Find out how by reading Chapter 8 in this minibook.

IN THIS CHAPTER

» Understanding the histogram

» Getting ready to correct an image

» Making a good tone curve

» Editing adjustment layers

» Testing your printer

Chapter 6 Creating a Good Image

When discovering all the incredible things you can do in Photoshop, you can easily forget the basics. Yes, you can create incredible compositions with special effects, but if the humans in your image look green, their alien skin tone will detract from the final result. Make it part of your image-editing process to create clean images before adding the artsy filters and complex compositions. Color correction isn't complicated, and by following the steps in this chapter you can produce almost magical results. In this chapter, you see how to use the values you read in the Info panel and use the Curves panel to produce quality image corrections.

Reading a Histogram

If you like, open our example image to reference as you progress through this chapter. Find `ThreeTeens` among the Dummies CCFiles inside the Book03_Photoshop folder you downloaded in Book 1, Chapter 1.

Before making adjustments, look at the image's *histogram*, which displays an image's tonal values, to evaluate whether the image has sufficient detail to produce a high-quality image. Once you have an image open in Photoshop CC, choose Window⇨Histogram to display the Histogram panel.

Note: A Warning (yield sign icon) in the upper-right of the Histogram means that the histrogram needs to be refreshed. Simply click the Warning icon to make sure the histogram is current.

REMEMBER

The greater the range of values in the histogram, the greater the detail. Poor images without much information can be difficult, if not impossible, to correct. The Histogram panel also displays the overall distribution of shadows, midtones, and highlights to help you determine which tonal corrections are needed.

Figure 6-1 shows a good, full histogram that indicates a smooth transition from one shade to another in the image. Figure 6-2 shows that when a histogram is spread out and has gaps in it, the image is jumping too quickly from one shade to another, producing a posterized effect. *Posterization* is an effect that reduces tonal values to a limited amount, creating a more defined range of values from one shade to another. It's great if you are looking for that effect, but if you want a smooth tonal change from one shadow to another, it's not so great.

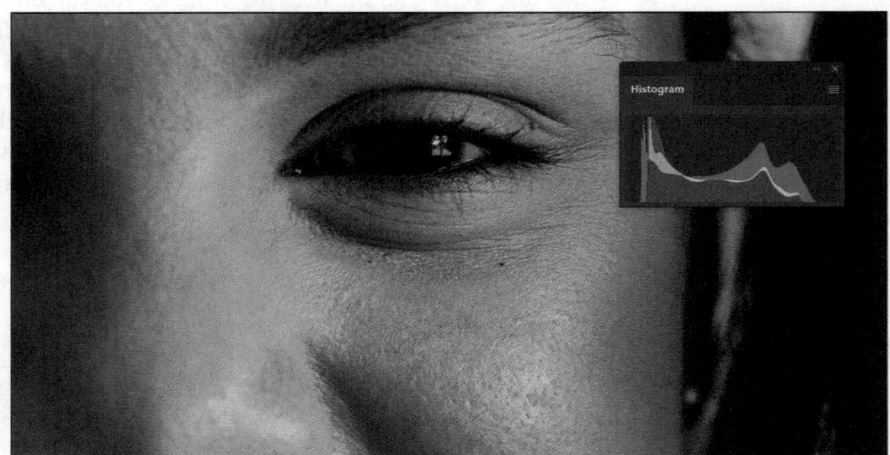

FIGURE 6-1: A histogram showing smooth transitions from one color to another.

You can produce this effect yourself by simply selecting Image⇨Adjustments⇨Posterize. Even though you are intentionally posterizing the image yourself, posterization can happen when you overwork an image. In this chapter, we teach you how to avoid taking photo corrections too far.

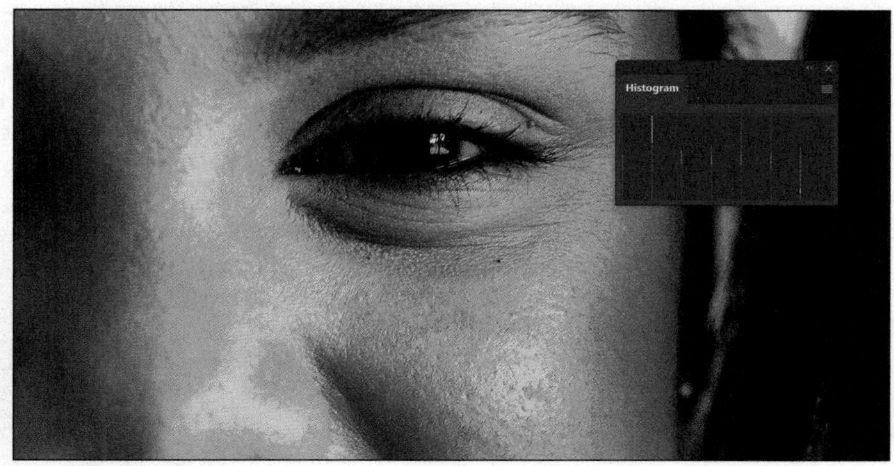

FIGURE 6-2: A histogram showing a lack of smoothness in the gradation of color.

So how do you create a good histogram? If you're scanning, make sure that your scanner is set for the maximum number of colors. Scanning at 16 shades of gray gives you 16 lines in your histogram — not good.

TIP

If you open an image and it displays a weak histogram, you should rescan or reshoot the image if you can. If you have a good histogram to start with, keep it that way by not messing around with multiple tone correction tools. Most professionals use the Curves feature — and that's it. Curves (choose Image⇨Adjustments⇨Curves), if used properly, do all the adjusting of levels (brightness and contrast) and color balance, all in one step. You can read more about curves in the section "Creating a Good Tone Curve," later in this chapter.

Figure 6-3 shows what happens to a perfectly good histogram when someone gets too zealous and uses the entire plethora of color correction controls in Photoshop. Keep in mind that even with the best efforts, your Histogram will break up a bit when adjusting your image. You just don't want it to be dramatic.

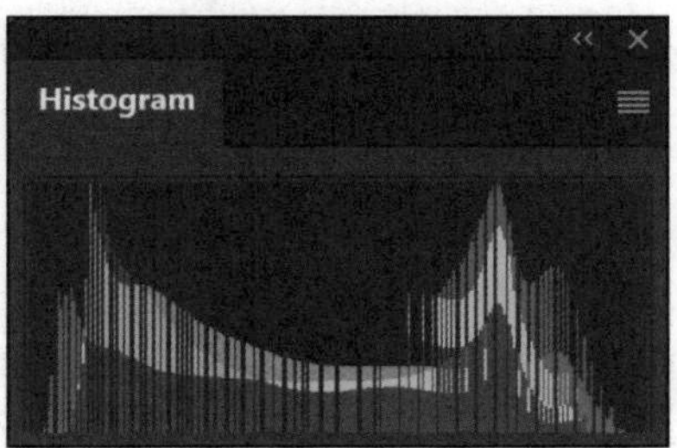

FIGURE 6-3: Tonal information is broken up.

REMEMBER

Remember, if the Warning icon appears while you're making adjustments, click on the icon or double-click anywhere on the histogram to refresh the display.

Breaking into key types

Don't panic if your histogram is smashed all the way to the left or right. The bars of the histogram represent tonal values. You can break down the types of images, based on their values, into three key types that are defined below:

- **High key:** An extremely light-colored image, such as the image shown in Figure 6-4. Information is pushed toward the right in the histogram. Color correction has to be handled a little differently for these images to keep the light appearance to them.

FIGURE 6-4: A high key image is a light image.

- **Low key:** An extremely dark image, such as the one shown in Figure 6-5. Information is pushed to the left in the histogram. This type of image is difficult to scan on low-end scanners because the dark areas tend to blend together with little definition.
- **Mid key:** A typical image with a full range of shades, such as the image shown in Figure 6-6. These images are the most common and easiest to work with. In this chapter, we deal with images that are considered mid key.

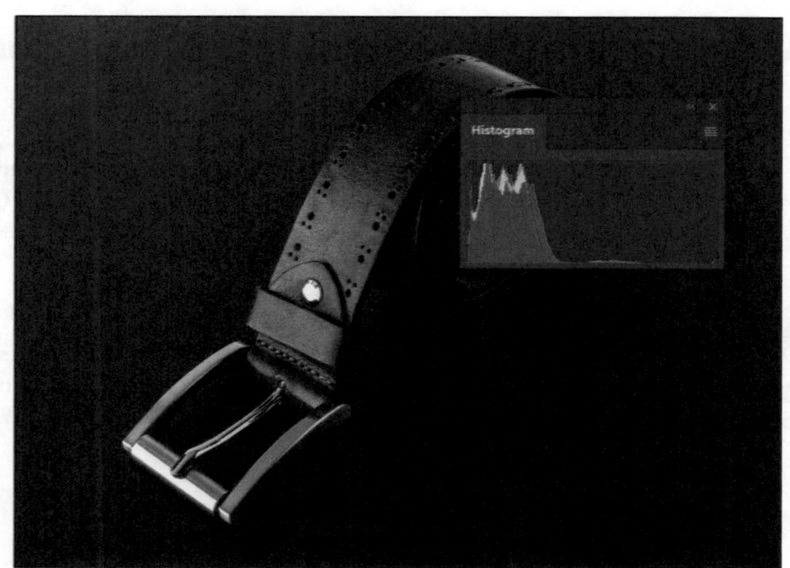

FIGURE 6-5: A low key image is a dark image.

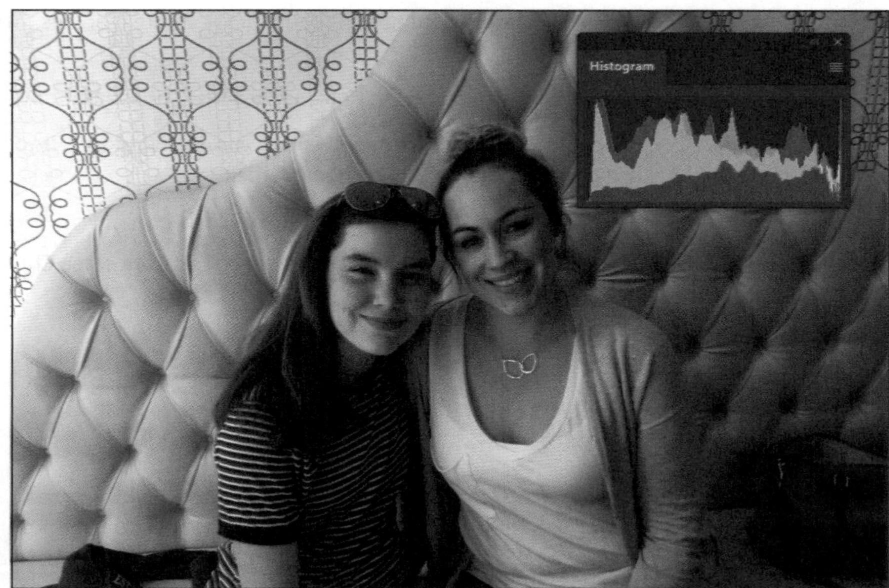

FIGURE 6-6: A typical image with a full range of values is a mid key image.

Starting the process to improve your image

To produce the best possible image, avoid correcting in the CMYK (Cyan, Magenta, Yellow, Black) mode. If your images are typically in RGB (Red, Green, Blue) or LAB mode (L for lightness, and A and B for the color-opponent dimensions), keep them in that mode throughout the process. Convert them to CMYK only when you're finished manipulating the images and only if someone requests they be in that mode.

REMEMBER

Don't forget! If your end result will be a printed image, you can press Ctrl+Y (Windows) or ⌘ +Y (Mac) to toggle on and off the CMYK preview so that you can see what your image will look like in CMYK mode without converting it. You will know that you are in this mode when you see *RGB/8#/CMYK* in the tab that contains your file name.

Set up these items before starting any color correction:

1. **Select the Eyedropper tool (the keyboard shortcut is I); on the Options bar, select the 3 by 3 Average setting from the Sample Size drop-down list.**

   The 3 by 3 Average setting gives you more accurate readings than the Point Sample setting.

2. **If the Histogram panel isn't already visible, choose Window ➪ Histogram.**
3. **If the Info panel isn't already visible, choose Window ➪ Info to show the Info panel so that you can check values.**
4. **Make sure that your color settings are correct.**

   If you're not sure how to check or set up color settings, see Chapter 2 of this minibook.

Creating a Good Tone Curve

A *tone curve* represents the density of an image. To produce the best image, you should find the highlight (lightest) and shadow (darkest) points in it. An image created in less-than-perfect lighting conditions may be washed out or have odd color casts. See Figure 6-7 for an example of an image with no set highlight and shadow. Check out Figure 6-8 to see an image that went through the process of setting a highlight and shadow.

FIGURE 6-7: The image is murky before defining a highlight and shadow.

FIGURE 6-8: The tonal values are opened after highlight and shadow have been set.

To make the process of creating a good tone curve more manageable, we've broken the process into four parts:

- Find the highlight and shadow.
- Set the highlight and shadow values.
- Adjust the midtone.
- Find a neutral.

Even though each part has its own set of steps, you must complete all four parts to improve your image's tone curve (unless you're working with grayscale images, in which case you can skip the neutral part). In this example, an adjustment layer is used for the curve adjustments. The benefit is that you can turn off the visibility of the adjustment later or double-click the adjustment layer thumbnail to make ongoing edits without destroying your image. If you want to follow along, open the image named Goalie in the Book03_Photoshop images folder.

Finding the highlight and the shadow

In the noncomputer world, you'd spend a fair amount of time trying to locate the lightest and darkest parts of an image. Fortunately, you can cheat in Photoshop by using some features in the Curves panel. Here's how to access the panel:

1. **With an image worthy of adjustment — one that isn't perfect already — choose Window ⇨ Layer (if the Layers panel isn't already open).**
2. **Click and hold the Create New Fill or Adjustment Layer button at the bottom of the Layers panel and select Curves, as shown in Figure 6-9.**

   The Adjustments panel appears with the Curves panel active.

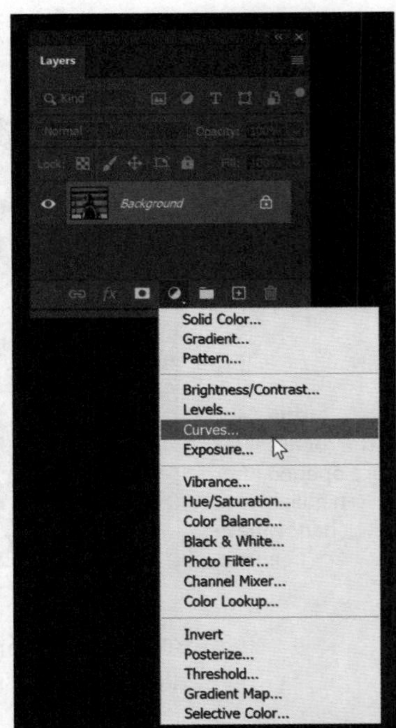

FIGURE 6-9: Access the Curves panel with the Create New Fill or Adjustment Layer button.

Notice the grayed-out histogram behind the image in the Curves panel. The histogram helps you determine where you need to adjust the image's curve.

TECHNICAL STUFF

If you're correcting in RGB (as you should be), the tone curve may be the opposite of what you expect. Instead of light to dark displaying as you expect, RGB displays dark to light. Think about it: RGB is generated with light, and no RGB means that there's no light and you therefore have black. If you turn on all RGB full force, you create white. Try pointing three filtered lights — one red, one green, and one blue. The three lights pointed in one direction create white.

If working with RGB confuses you, simply select Curves Display Options from the panel menu in the upper-right corner of the Adjustments panel. When the Curves Display Options dialog box appears, as shown in Figure 6-10, select the Pigment/Ink % radio button and click OK. If you are planning to follow this example, leave the setting at Light (2-255) so you don't get confused.

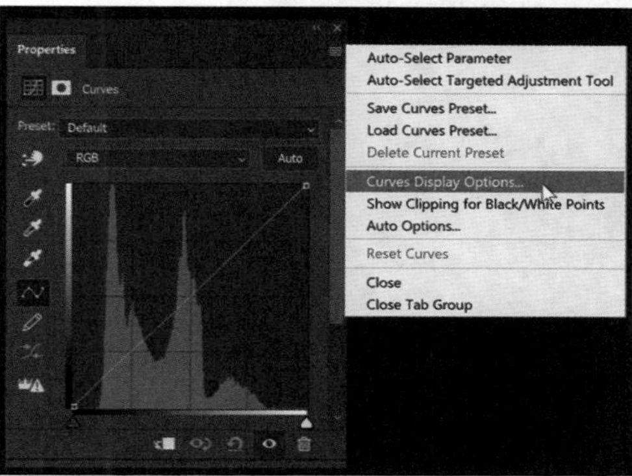

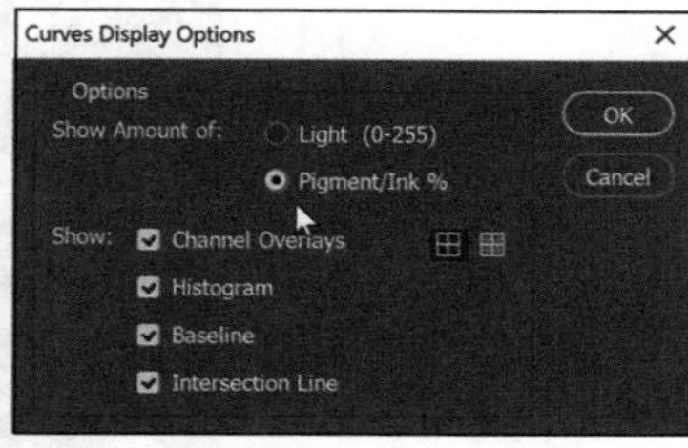

FIGURE 6-10: View the curve using light or pigment. If following the example you want to leave this at Light (0–255).

Note that in the Curves panel, you see a Preset drop-down list that offers quick fixes using standard curves for certain corrections. These settings are great for quick fixes, but for the best image you will want to create a custom curve.

The first thing you need to do in the Curves panel is determine the lightest and darkest parts of the image — referred to as *locating the highlight and shadow:*

In order for you to make adjustments to the curve, click the Click and Drag button to turn on click-and-drag for the image to modify the curve icon. It looks like a finger pointing to the left, as you see in Figure 6-11.

1. **Before starting the correction, click the Set Black Point eyedropper once (labeled in Figure 6-11).**
2. **Hold down the Alt (Windows) or Option (Mac) key and click the Shadow input slider (labeled in Figure 6-11).**

   When you Alt-click (Windows) or Option-click (Mac), the clipping preview turns on, revealing the darkest area of the image, as shown in Figure 6-12.

   If you don't immediately see a dark area in the clipping preview, you can drag the shadow input slider to the right while holding down the Alt key or Option key. The entire screen will turn white, and as you start dragging you will see the darkest areas of the image appear first. Note where the darkest area of the image appears because you will mark it in the next step.

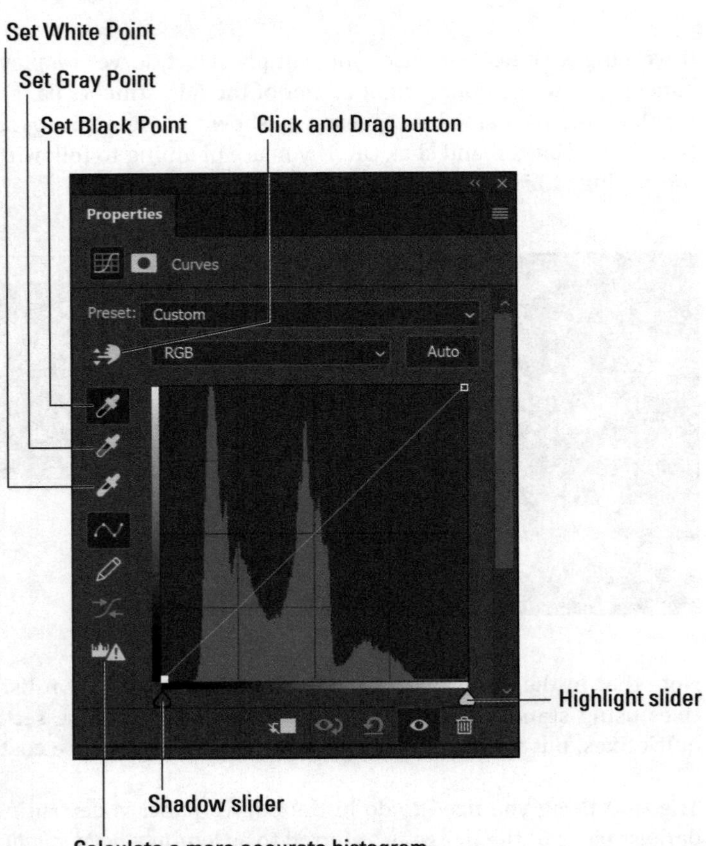

FIGURE 6-11: The critical tools on the Curves panel.

3. **Hold down the Shift key and click directly on the image where you saw the darkest area of the image appear.**

   This step adds a color sampler, essentially marking the spot, on the image that helps you reference that point later, as shown in Figure 6-14.

4. **Repeat Steps 1 through 3 with the Highlight input slider. Select the Set White Point eyedropper (labeled in Figure 6-11) in the Curves panel.**

5. **Hold down the Alt key or Option key and click the highlight input slider.**

   Again, you can drag the slider toward the left if the lightest point doesn't immediately show up. Initially your preview will be black and as you drag to the left you will see white appear, as shown in Figure 6-13.

   When you locate the lightest point, as indicated by the lightest point in the clipping preview, you can Shift+click to drop a second color sampler, as shown in Figure 6-13.

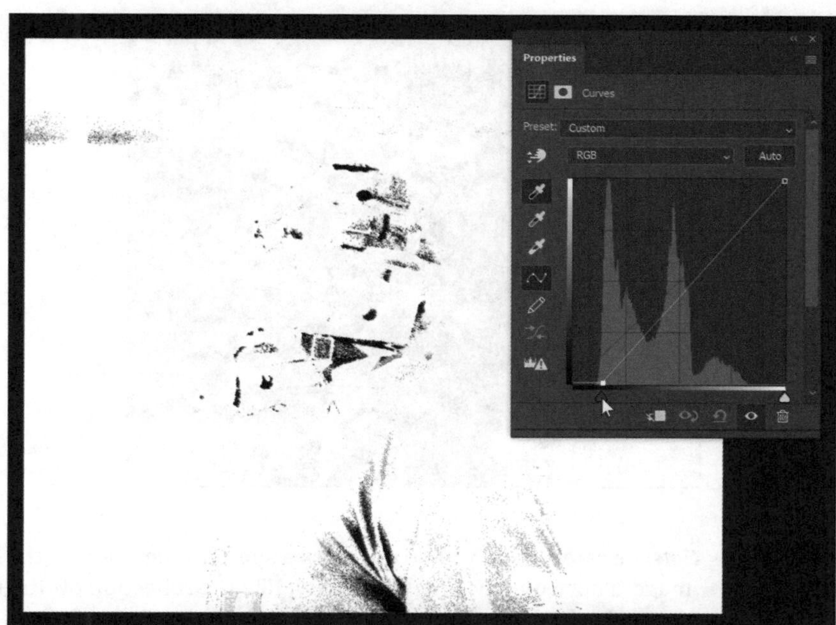

FIGURE 6-12: Alt- or Option-click on the Shadow slider and then drag slightly to the right to see the darkest part of the image.

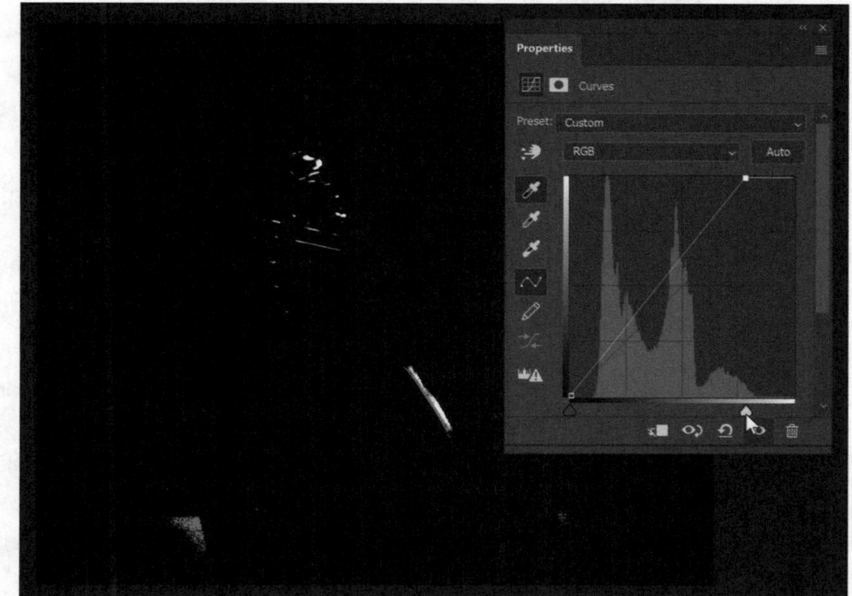

FIGURE 6-13: Alt- or Option-click on the Highlight slider and drag towards the left to find the lightest part of the image.

Creating a Good Image

FIGURE 6-14: When you Shift-click on the image you create a color sampler that essentially marks a spot for reference later.

Note: Be careful that you do not mark a specular highlight as the lightest part of an image. This would be anything shiny, like the reflection off the top of the helmet. A specular highlight can include light streaming in a window, a reflection off a bumper, or the light from a candle. Instead, slide in a little more and pick a light element that is not shining. A specular highlight can throw off your tone curve and make your image look washed out.

Setting the highlight and shadow values

After you determine the lightest and darkest points in an image, you can set their values. Follow these steps:

Note: If the Properties panel automatically switches to the Info panel while following these steps, Choose Window⇨Properties to display the Properties panel again.

1. **Double-click the Set White Point eyedropper, the white eyedropper on the left side of the Curves panel.**

   When you double-click the Set White Point eyedropper, the Color Picker dialog box appears.

2. **Enter a generic value for the lightest point in your image: Type** 5 **in the Cyan text box, type** 3 **in the Magenta text box, type** 3 **in the Yellow text box, leave the Black text box at 0 (zero), and then click OK.**

   The Black value helps to correct most images for print and online.

3. **When you receive an alert message asking whether you want to save the new target colors as defaults, click Yes.**
4. **With the Set White Point eyedropper still selected, click the color sampler for the lightest part of the image that you dropped on the image.**

   Now, set the shadow point.
5. **Double-click the Set Black Point eyedropper.**

   The Color Picker dialog box appears.
6. **Type** 65 **in the Cyan text box, type** 53 **in the Magenta text box, type** 51 **in the Yellow text box, type** 95 **into the Black text box, and then click OK.**

   As with the highlight value, the Black value is a generic value that works for most print and online images. Keep in mind that if you set your color settings up correctly, you don't have to change the Set Black Point value.
7. **With the Set Black Point eyedropper still selected, click the color sampler you dropped on the image, indicating the darkest point in the image.**

At this point, you should see a difference in your image already. Keep going, however, because there are just a few more steps.

Adjusting the midtone

You may have heard the phrase, "open up the midtones," which essentially means to lighten the midtonal values of an image. In many cases, opening up the midtones is necessary to add contrast and bring out detail in an image.

To adjust the midtones, follow these steps:

1. **In the Curves panel, click the middle of the curve ramp to create an anchor point; drag up slightly.**

   The image lightens. Move only a small amount and be careful to observe what's happening in the Histogram panel (which you should always have open when making color corrections). You don't want to make too drastic of a change that will introduce posterization into your image. See Figure 6-15.

   Because you set highlight and shadow (see the preceding section) and are now making a midtone correction, you see the bars in the histogram spreading out.

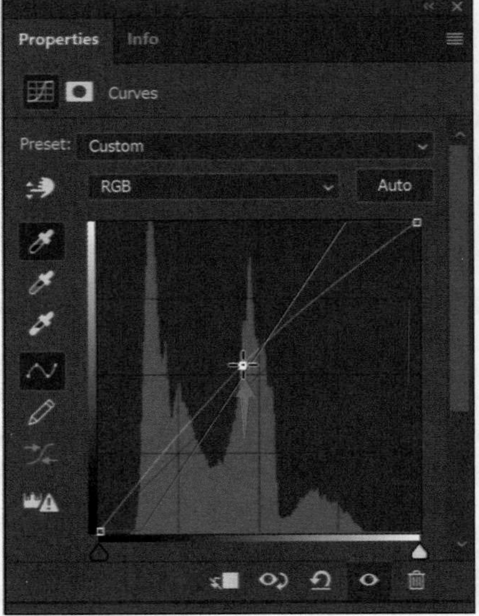

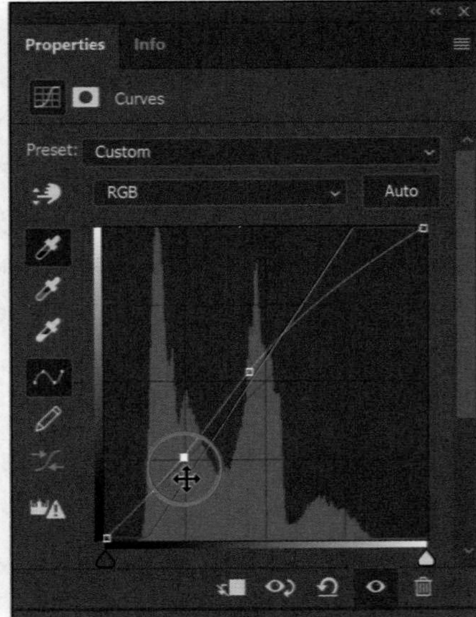

FIGURE 6-15: Click and drag up on the tone curve to adjust the midtones and three-quarter tones in your image.

2. **To adjust the three-quarter tones (the shades around 75 percent), click halfway between the bottom of the curve ramp and the midpoint to set an anchor point.**

   Use the grid in the Curves panel to find it easily. Adjust the three-quarter area of the tone curve up or down slightly to create contrast in the image. Again, keep an eye on your histogram! In the color-correction world, you just created an S-curve. You adjusted the midtones, and also the three-quarter tones which essentially turns your straight curve into more of an S.

   If you're working on a grayscale image, the tonal correction is done.

   If you're working on a color image, keep the Curves panel open for the final steps, which are outlined in the next section.

Finding a neutral

The last steps in creating a tone curve apply only if you're working on a color image. The key to understanding color is knowing that equal amounts of color create gray. By positioning the mouse cursor over gray areas in an image and reading the values in the Info panel, you can determine which colors you need to adjust.

1. **With the Curves panel open, position it so that you can see the Info panel.**

   If the Info panel is buried under another panel or a dialog box, choose Window ➪ Info to bring it to the front or undock it from its current location.

2. **Position the cursor over an image and, in the Info panel, look for the RGB values in the upper-left section.**

   You see color values and then forward slashes and more color values. The numbers before the slash indicate the values in the image before you opened the Curves panel; the numbers after the slash show the values now that you've made changes in the Curves panel. Pay attention to the values after the slashes.

3. **Position the cursor over something gray in your image.**

   It can be a shadow on a white shirt, a countertop, a road — anything that's a shade of gray. Look at the Info panel. If your image is perfectly color-balanced, the RGB values following the forward slashes should all be the same.

4. **If the color isn't balanced, click the Set Gray Point eyedropper in the Curves panel and click the neutral or gray area of the image.**

   The middle eyedropper (Set Gray Point) is a handy way of bringing the location you click closer together in RGB values, thereby balancing the colors.

   Note in Figure 6-16 the difference in the images after the Set Gray Point was clicked on the gray bleachers in the image.

FIGURE 6-16: Using the Set Gray Point tool, you can click on a gray in your image to balance the colors.

Curves can be as complex or as simple as you make them. As you gain more confidence in using them, you can check neutrals throughout an image to ensure that all unwanted color casts are eliminated. You can even individually adjust each color's curve by selecting it from the Channel drop-down list in the Curves panel.

When you're finished with color correction, using the Unsharp Mask filter on your image is a good idea. Chapter 5 of this minibook shows you how to use this filter.

Editing an Adjustment Layer

You may make a curve adjustment only to discover that some areas of the image are still too dark or too light. Because you used an adjustment layer, you can turn off the correction or change it repeatedly with no degradation to the quality of the image.

Here are the steps to take if you still have additional adjustments to make to an image, such as lightening or darkening other parts of the image. You can use the `ChevronRoad.jpg` file in your Book03_Photoshop folder if you would like to follow along:

1. **Select the area of the image that needs adjustments.**

   See Chapter 4 of this minibook for a refresher on how to make selections in Photoshop. In this example we just made an elliptical selection around a section of the image that we want to lighten.

2. **Choose Select ⇨ Modify ⇨ Feather to soften the selection.**

   The Feather dialog box appears.

3. **Enter a value into the Feather dialog box.**

   If you're not sure what value will work best, enter **25** in the Feather Radius text field and click OK. In this example we feathered 50 pixels.

   TIP

   You can also click the Refine Edge button in the Options panel (when you have a selection tool active) to preview the feather amount.

4. **If the Layers panel isn't visible, choose Windows ⇨ Layers; click and hold the Create New Fill or Adjustment Layer icon and select Curves.**

5. **In the Curves panel, click the middle of the curve ramp to create an anchor point; drag up or down to lighten or darken the selected area.**

   Notice in the Layers panel (see Figure 6-17) that the adjustment layer, Curves 1 by default, has a mask to the right of it. This mask was automatically created from your selection. The selected area is white; unselected areas are black.

6. **With the adjustment layer selected in the Layers panel, use the Brush tool to paint white to apply the correction to other areas of the image; paint with black to exclude areas from the correction.**

   You can even change the opacity with the Brush tool in the Options bar at the top to apply only some of the correction!

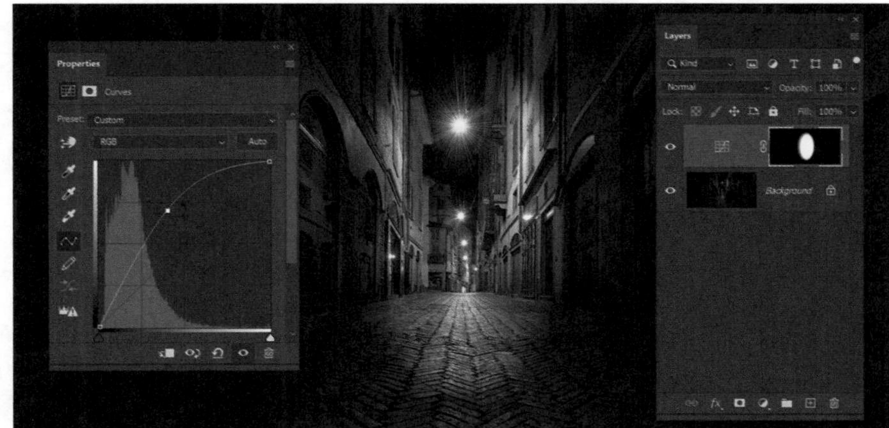

FIGURE 6-17: Paint on the adjustment layer mask.

Testing a Printer

If you go to all the trouble of making color corrections to images and you still see printed images that look hot pink, it may not be your fault. Test your printer by following these steps:

1. **Create a neutral gray out of equal RGB values. (Double-click the Fill Color swatch in the Tools panel.)**
2. **Create a shape, using neutral gray as the fill color.**

   For example, you can use the Ellipse tool to create a circle or oval.
3. **Choose File ➪ Print and click OK to print the image from a color printer.**

If you're seeing heavy color casts, adjust the printer; cleaning or replacing the ink cartridge may fix the problem. Check out Chapter 10 of this minibook for more about printing Photoshop files.

» Getting to know foreground and background colors

» Introducing painting and retouching tools

» Using the Content-Aware feature

» Discovering blending modes

» Saving presets for tools

Chapter 7
Working with Painting and Retouching Tools

This chapter explains how to use the painting and retouching tools in Photoshop. Have fun and be creative! Because Photoshop is pixel-based, you can create incredible imagery with the painting tools. Smooth gradations from one color to the next, integrated with blending modes and transparency, can lead from super-artsy to super-realistic effects. In this chapter, you discover fundamental painting concepts and also find out how to use tools to eliminate wrinkles, blemishes, and scratches. Don't you wish you could do that in real life?

Using the Swatches Panel

Use the Swatches panel to store and retrieve frequently used colors. The Swatches panel enables you to quickly select colors and gives you access to many other color options. By using the panel menu, shown in Figure 7-1, you can select from a multitude of color presets, as shown in Figure 7-1.

To sample and store a color for later use, follow these steps. You can open any image, or use one of our examples on the Book03_Photoshop folder:

1. **To sample a color from an image, select the Eyedropper tool in the Tools panel and click a color in the image.**

   Alternatively, you can use any of the paint tools (the Brush tool, for example) and Alt-click (Windows) or Option-click (Mac).

   The color you click becomes the foreground color.

2. **If the Swatches panel isn't already open, choose Window ➪ Swatches.**
3. **Store the color in the Swatches panel by clicking the New Swatch button at the bottom of the Swatches panel.**

Any time you want to use that color again, simply click it in the Swatches panel to make it the foreground color.

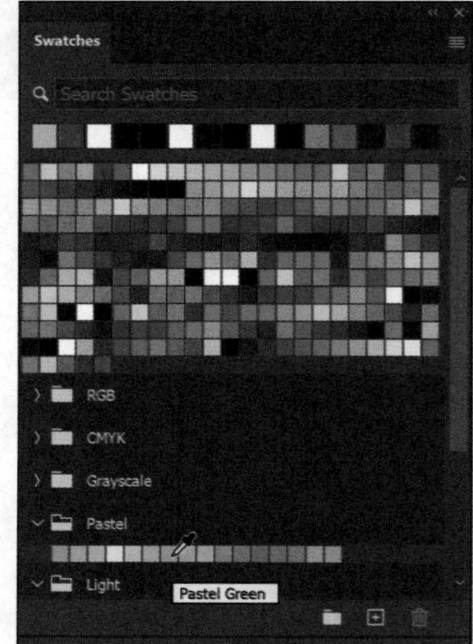

FIGURE 7-1:
Choose Window ➪ Swatches, and then click the panel menu to access additional color options.

Choosing Foreground and Background Colors

At the bottom of the Tools panel, you can find the foreground and background color swatches. The *foreground color* is the color you apply when using any of the painting tools. The *background color* is the color you see if you erase or delete pixels from the Background layer.

Choose a foreground or background color by clicking the swatch, which then opens the Color Picker dialog box, shown in Figure 7-2. To use the Color Picker, you can either enter values in the text fields on the right or slide the hue slider.

Pick the *hue* (color) you want to start with and then click in the color panel to the left to choose the amount of light and saturation (grayness or brightness) you want in the color. Select the Only Web Colors checkbox to choose one of the 216 colors in the web-safe color palette. The hexadecimal value used in HTML documents appears in the text field in the lower-right corner of the Color Picker.

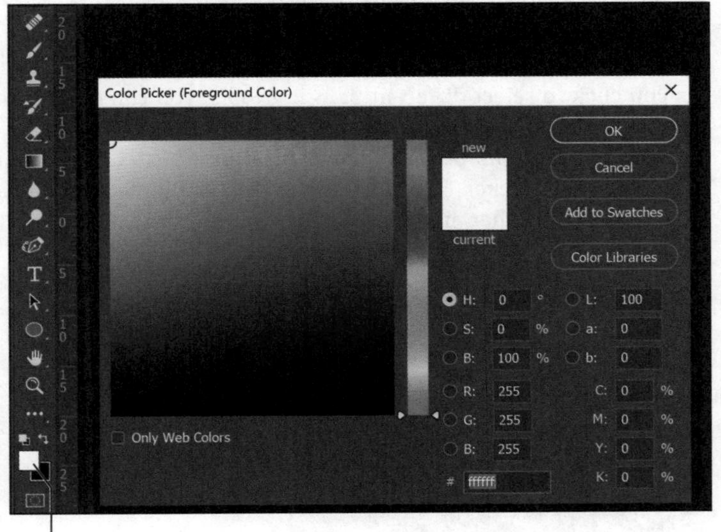

FIGURE 7-2: Use the Color Picker or enter numeric values to select a color in the color panel.

To quickly save a color, click the Add to Swatches button directly in the Color Picker.

The Painting and Retouching Tools

Grouped together in the Tools panel are the tools used for painting and retouching. The arrow in the lower-right area of a tool icon indicates that the tool has more related hidden tools; simply click and hold down on a tool icon to see additional painting and retouching tools, as shown in Figure 7-3. In this chapter, we show you how to use the Brush, Clone Stamp, Eraser, Gradient, Healing Brush, History Brush, Patch, Red Eye, and Spot Healing Brush tools. You also discover ways to fill shapes with colors and patterns.

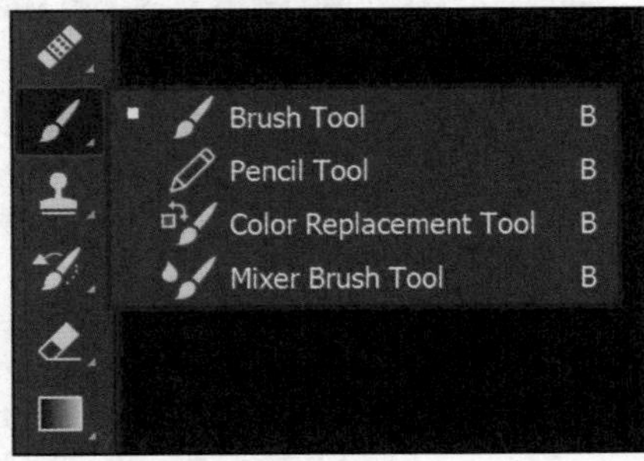

FIGURE 7-3: The painting and retouching tools have hidden tools available as well.

Changing the brush

As you click to select different painting tools, you can change the size and diameter of the tool. You can do this by accessing the Brush menu (second from the left) on the Options bar, as shown in Figure 7-4. Click the arrow to open the Brushes Preset picker. You can use the Master Diameter slider to make the brush size larger or smaller and to change the hardness of the brush.

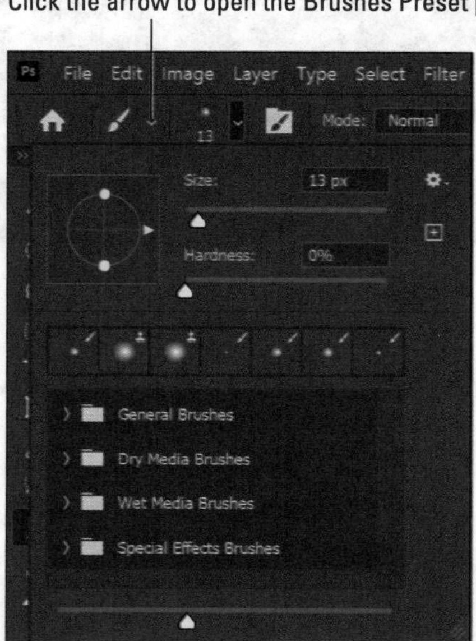

FIGURE 7-4: The Brushes Preset picker.

TIP

The hardness refers to the fuzziness of the edges; a softer brush is more feathered and soft around the edges, whereas a harder edge is more definite. (See Figure 7-5.)

TIP

If you don't feel like accessing the Brushes Preset picker every time you want to make a change, press the right bracket (]) several times to make the brush diameter larger or press the left bracket ([) to make the brush diameter smaller. Press Shift+] to make the brush harder or Shift+[to make the brush softer.

Choose Window ➪ Brushes to see a list of brush presets, plus more brush options you can use to create custom brushes. You can also choose other brush libraries from the list of brush preset folders that are listed in the lower section of the Brushes panel.

FIGURE 7-5: The inside border was created with a brush set to a hardness of 100%. The outside brush was set to a hardness of 0%.

TIP

Access the Brushes Preset picker while you're painting by right-clicking (Windows) or Control-clicking (Mac) anywhere in the image area. Double-click a brush to select it; press Esc to hide the Brushes Preset picker.

The Spot Healing Brush tool and the Content-Aware feature

The Spot Healing Brush tool was a great tool to begin with, but now, with the Content-Aware feature, it's even better. No matter what level of Photoshop user you are, you'll appreciate the magic in this tool option.

In its default settings, the Spot Healing Brush tool quickly removes blemishes and other imperfections in images. Click a blemish and watch it paint matching texture, lighting, transparency, and shading to the pixels being healed. The Spot Healing Brush tool doesn't require you to specify a sample spot — it automatically samples from around the retouched area.

Now take this concept a step further by selecting the Content-Aware option on the Options bar. The concept is the same, but if you look at Figure 7-6, you can see that painting with the Spot Healing Brush sets into action an incredible number of calculations that attempt to render pixels similar in detail to its surroundings. If you want to try this feature on the same image you can use the image named Couples in the Book03_Photoshop folder.

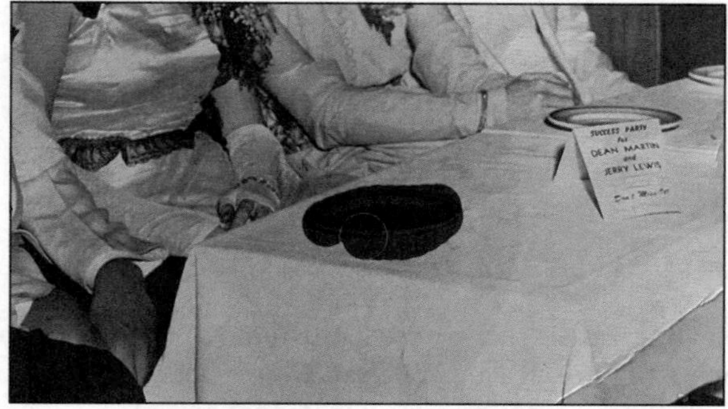

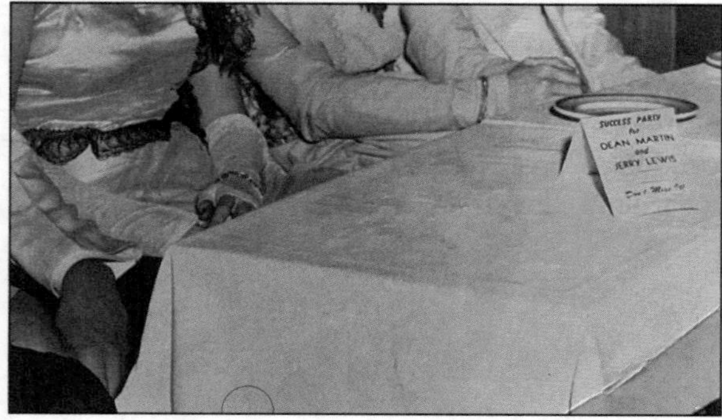

FIGURE 7-6: Using the Spot Healing Brush tool with the Content-Aware option enabled.

Does this work every time? Of course not. Does it work often enough to save you hours of work? Yes! If you want to follow along, open the image named `beach` that is located in the Book03_Photoshop folder located in the DummiesCCFiles here: `www.agitraining.com/dummies`.

You can also take advantage of the Content-Aware feature by using the Fill feature. Simply follow these steps:

1. **Select the area to be replaced by the new content.**

   It is best to make a tight selection when using this feature.

2. **Choose Edit ➪ Fill and, in the Fill dialog box, select Content-Aware from the Use drop-down list, shown in Figure 7-7.**

   The content is replaced.

   Note that if you are working on a Background layer, you can just press Delete and see the same options for replacing with the Content-Aware feature.

FIGURE 7-7: Use the Content-Aware feature in the Fill dialog box.

The Healing Brush tool

You can use the Healing Brush tool for repairs, such as eliminating scratches and dust from scanned images. The difference between the Spot Healing Brush tool and the Healing Brush tool is that a sample spot is required before applying the Healing Brush. Follow these steps to use this tool:

1. **Open an image that has a tear or other need of repair. You can use the Couples image again if you like.**
2. **Select the Healing Brush tool in the Tools panel.**

   It's a hidden tool of the Spot Healing Brush tool.

3. **Find an area in the image that looks good and then Alt-click (Windows) or Option-click (Mac) to sample that area.**

   For example, if you want to eliminate a tear in an image, such as the one in Figure 7-8, choose a tear-free area of image near the tear.

4. **Position the mouse cursor over the area to be repaired and start painting.**

   The Healing Brush tool goes into action, blending and softening to create a realistic repair of the area.

5. **Repeat Steps 2 through 4 as necessary to repair the additional tears, scratches, blemishes, and wrinkles.**

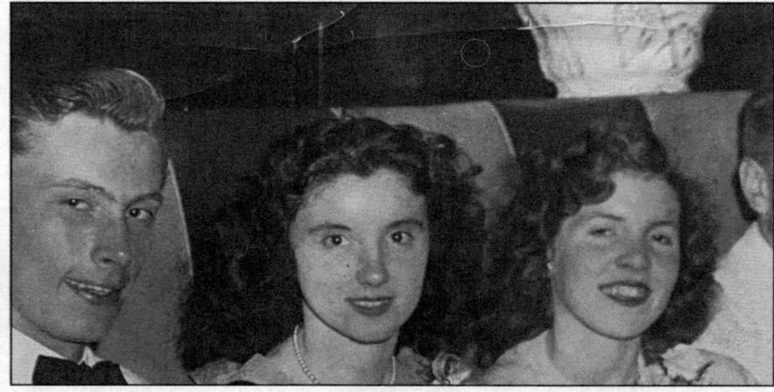

FIGURE 7-8: Position the cursor over the area that you want to use as a source for the Healing Brush tool and Alt-click or Option-click.

The Patch tool

Hidden behind the Healing Brush tool in the Tools panel is the Patch tool. Use it to repair larger areas, such as a big scratch or a large area of skin, by following these steps:

1. **Click and hold the Healing Brush tool to select the Patch tool; on the Options bar, select Normal from the Patch drop-down menu and then select the Destination tab.**

   You can patch either the source area or the destination — it's up to you. We recommend dragging a good source over the area that needs to be repaired.

2. **With the Patch tool still selected, drag to create a marquee around the source you want to use as the patch.**

   The source is an unscratched or wrinkle-free area.

3. **After you create the marquee, drag the selected source area to the destination to be repaired.**

   The Patch tool clones the selected source area while you drag it to the destination (the scratched area); when you release the mouse button, the tool blends in the source selection and repairs the scratched area!

Make the patch look better by choosing Edit ➪ Fade Patch Selection immediately after you apply the patch. Adjust the opacity until no telltale signs show that you made a change.

The Red Eye tool

So you finally got the group together and shot the perfect image, but red eye took over! *Red eye* is caused by a reflection of the camera's flash in the retina of your photo's subject or subjects. You see this effect more often when taking pictures in a dark room, because the subject's pupils are wide open. If you can, use your camera's red-eye-reduction feature. Or, use a separate flash unit that you can mount on the camera farther from the camera's lens.

You'll love the fact that red eye is extremely easy to fix in Photoshop. Just follow these steps:

1. **Select the Red Eye tool (hidden behind the Spot Healing Brush tool).**
2. **Click and drag to surround the red-eye area.**

   You should see a change immediately, but if you need to make adjustments to the size or the darkness amount, you can change options on the Options bar.

The Brush tool

Painting with the Brush tool in Photoshop is much like painting in the real world. However, unlike a real brush, the Brush tool offers you nifty keyboard shortcuts that can be much more productive when painting. These shortcuts are truly outstanding, so make sure that you try them while you read about them. By the way, the keyboard commands you see in Table 7-1 work on all the painting tools.

TABLE 7-1 Brush Keyboard Shortcuts

Task	Windows	Mac
Choose the Brush tool	B	B
Increase the brush size	]	]
Decrease the brush size	[	[
Harden the brush	Shift+]	Shift+]
Soften the brush	Shift+[	Shift+[
Sample the color	Alt-click	Option-click
Switch the foreground and background colors	X	X
Change the opacity by a given percentage	Type a number between 1 and 100	Type a number between 1 and 100

If you're really into brushes, don't miss out on the many useful options available in the Brushes settings panel (choose Window ➪ Brush Settings to open it), as shown in Figure 7-9.

You have several attribute choices, most of which have dynamic controls in the menu option. These options enable you to vary brush characteristics by tilting or applying more pressure to a stylus pen (if you're using a pressure-sensitive drawing tablet), among other things.

A warning sign indicates that you don't have the appropriate device attached to use the selected feature, such as a pressure-sensitive drawing tablet.

The following options are available in the Brush Settings panel:

- **Brush Tip Shape:** Select from these standard controls for determining brush dimensions and spacing.
- **Shape Dynamics:** Change the size of the brush as you paint.
- **Scattering:** Scatter the brush strokes and control the brush tip count.

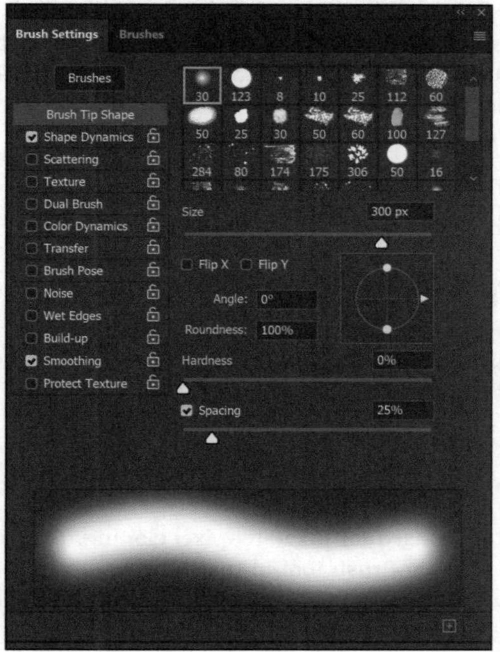

FIGURE 7-9: Additional options are available for painting in the Brush Settings panel.

TIP

- **Texture:** Choose from preexisting patterns or your own.

  Create a pattern by selecting an image area with the Rectangular Marquee tool. Choose Edit ➪ Define Pattern, name the pattern, and then click OK. The pattern is now available in the Brush panel's Texture choices.

- **Dual Brush:** Use two brushes at the same time.
- **Color Dynamics:** Change the color as you paint.
- **Transfer:** Adjusts the dynamics for the build-up of the paint.
- **Brush Pose:** Adjusts the tilt of the brush. You won't notice a difference when you have a regular rounded brush selected. Activate Pose when you use a beveled brush and change the X and Y tilt using the sliders to see how the angle of the brush is changed.
- **Noise:** Adds a grainy texture to the brush stroke.
- **Wet Edges:** Makes the brush stroke appear to be wet by creating a heavier amount of color on the edges of the brush strokes.
- **Build-up:** Gives airbrush features to the Brush tools and builds up paint based on time instead of movement. Simply hold down the mouse in one area to see the paint expand. You can also enable the effects of an airbrush by selecting the Enable Airbrush-Style Build-Up Effects button in the Options bar.

TIP

If you click and hold the Brush tool on the image area, the paint stops spreading. Turn on the Airbrush feature and notice that when you click and hold, the paint keeps spreading, just like using a can of spray paint. You can use the Flow slider on the Options bar to control the pressure.

- **Smoothing:** Smooths the path created with the mouse.
- **Protect Texture:** Preserves the texture pattern when applying brush presets.

After reviewing all the available brush options, you may want to start thinking about how you'll apply the same attributes later. Saving the Brush tool attributes is important as you increase your skill level. See the "Saving Presets" section later in this chapter to see how.

The Clone Stamp tool

The Clone Stamp tool is used for pixel-to-pixel cloning. The Clone Stamp tool is different from the Healing Brush tool in that it does no automatic blending into the target area. You can use the Clone Stamp tool for removing a product name from an image, replacing a telephone wire that's crossing in front of a building, or duplicating an item.

Here's how to use the Clone Stamp tool:

1. **With the Clone Stamp tool selected, position the cursor over the area you want to clone and then Alt-click (Windows) or Option-click (Mac) to define the clone source.**
2. **Position the cursor over the area where you want to paint the cloned pixels and then start painting.**

   Note the crosshair at the original sampled area, as shown in Figure 7-10. While you're painting, the crosshair follows the pixels you're cloning.

TIP

When using the Clone Stamp tool for touching up images, you should resample many times so as to not leave a seam where you replaced pixels. A good clone stamper Alt-clicks (Windows) or Option-clicks (Mac) and paints many times over until the retouching is complete.

Choose Window ➪ Clone Source to open the Clone Source panel, shown in Figure 7-11. With this handy little panel, you can save multiple clone sources to refer to while working. Even better, you can scale, preview, and rotate your clone source — before you start cloning.

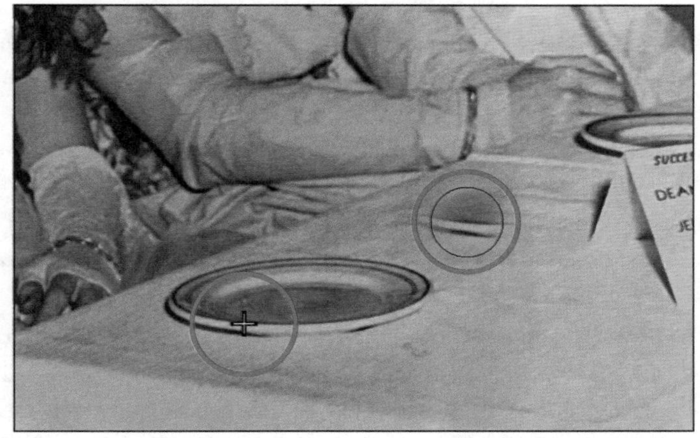

FIGURE 7-10: A crosshair over the source shows your clone source.

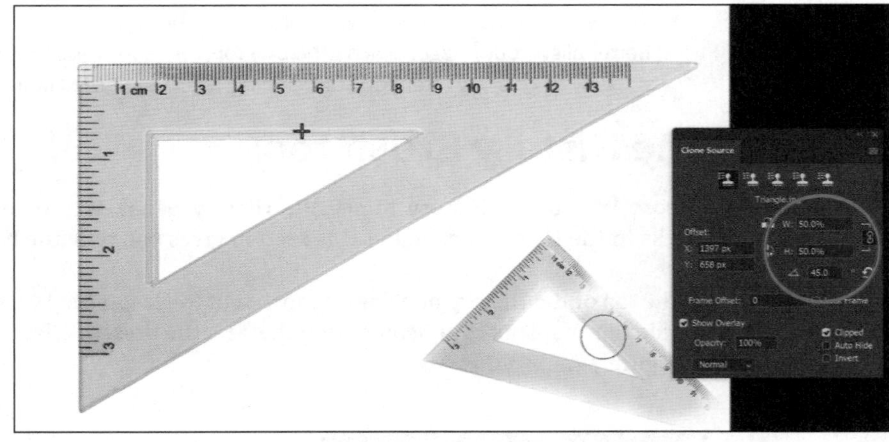

FIGURE 7-11: Additional options in the Clone Source panel. In this example, the triangle is rotated 45 degrees and scaled to 50% while being cloned.

TIP

The Clone Source panel can be extremely helpful with difficult retouching projects that involve a little more precision.

Follow these steps to experiment with this fun and interactive panel:

1. **If the Clone Source panel isn't visible, choose Window ➪ Clone Source.**

   The Clone Source icons across the top have yet to be defined. The first stamp is selected as a default.

2. **Alt-click (Windows) or Option-click (Mac) in the image area to record the first clone source.**

3. **Click the second Clone Source icon at the top of the Clone Source panel and then Alt-click (Windows) or Option-click (Mac) somewhere else on the page to define a second clone source.**

   Repeat as needed to define more clone sources. You can click the Clone Source icons at any time to retrieve the clone source and start cloning.

4. **Enter any numbers you want in the Offset X and Y, W and H, and Angle text boxes in the center section of the Clone Source panel to set up transformations before you clone.**

5. **Select the Show Overlay checkbox to see a preview of your clone source.**

   Whatever you plan to do, it's much easier to see a preview *before* you start cloning. If you don't use the Clone Source panel for anything else, use it to see a preview of your clone source before you start painting. If it helps to see the clone source better, select the Invert checkbox.

   You see an *overlay* (or preview) before cloning begins. This overlay helps you better align your image, which is helpful for precision work. If you want the preview to go away after you start cloning, select the Auto Hide checkbox.

The History Brush tool

Choose Window ➪ History to see the History panel. We could play around for weeks in the History panel, but this section gives you only the basic concepts.

At the top of the History panel is a snapshot of the last-saved version of the image. Beside the snapshot is an icon noting that it's the present History state, as shown in Figure 7-12.

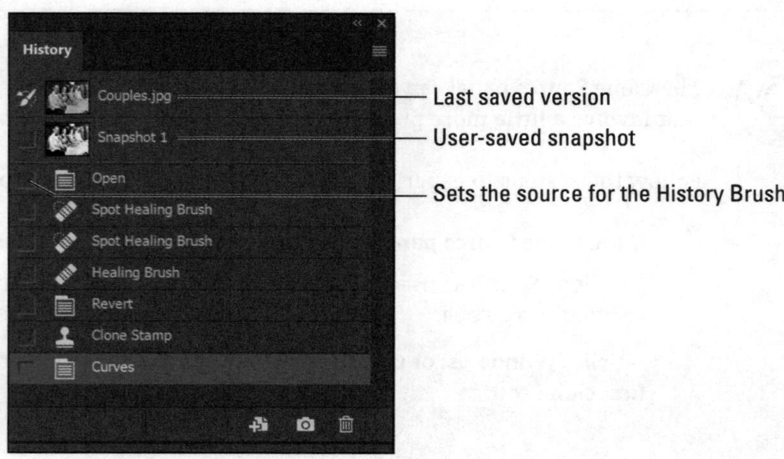

FIGURE 7-12: The History panel helps you see and select different states in your workflow to revert to.

When you paint with the History Brush tool, it reverts by default back to the way the image looked in the last saved version. You can click the empty square to the left of any state in the History panel (refer to Figure 7-12) to make it the source for the History Brush tool. Use the History Brush tool to fix errors and add pizzazz to images.

The Eraser tool

You may not think of the Eraser tool as a painting tool, but it can be! When you drag the image with the Eraser tool, it rubs out pixels to the background color. (Basically, it paints with the background color.) If you're dragging with the Eraser tool on a layer, it rubs out pixels to reveal the layer's transparent background. (You can also think of using the Eraser tool as painting with transparency.)

The Eraser tool uses all the same commands as the Brush tool. You can make an eraser larger or softer or more or less opaque. Even better, follow these steps to use the Eraser tool creatively:

1. **Open any color image and apply a filter.**

   For example, we chose Filter ⇨ Blur ⇨ Gaussian Blur. In the Gaussian Blur dialog box that appears, we changed the blur to 5 and then clicked OK to apply the Gaussian Blur filter.

2. **Select the Eraser tool and press 5 to change it to 50 percent opacity.**

   You can also use the Opacity slider on the Options bar.

3. **Hold down the Alt (Windows) or Option (Mac) key to repaint 50 percent of the original image's state before applying the filter.**

4. **Continue painting in the same area to bring the image back to its original state.**

   The original sharpness of the image returns where you painted.

Holding down the Alt (Windows) or Option (Mac) key is the key to erasing the last saved version (or history state). This tool is incredible for fixing mistakes or removing applied filters.

The Gradient tool

If you want to practice with this tool you can open any image from the Book03_Photoshop folder. In this example, we use the image named `Watersky.jpg`.

1. **So as to not affect your image, Choose Window>Layers and click on the Create a new Layer icon (Plus sign) at the bottom of the panel.**
2. **Press "D" to return your foreground and background colors to the default of black and white. As a default, the Gradient tool uses your foreground and background colors to create the gradient.**

   You can also click in the Gradient slider in the Options bar and from the basic folder choose foreground to background if your gradient is not set to the default black to white.
3. **Choose the Gradient tool and click and drag across an image area to create a gradient in the direction and length of the mouse motion. A short drag creates a short gradient; a long drag produces a smoother, longer gradient. See Figure 7-13.**

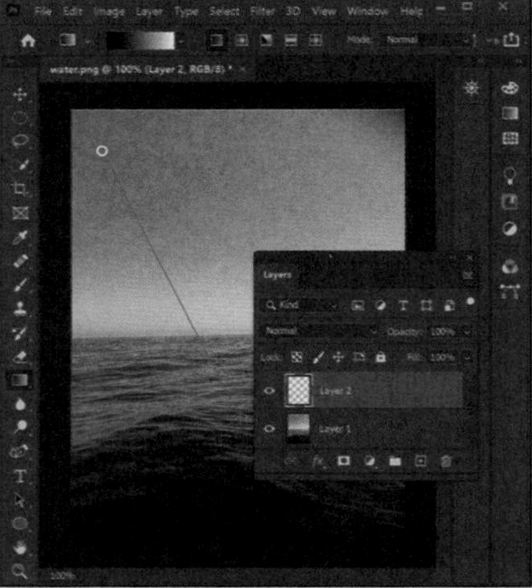
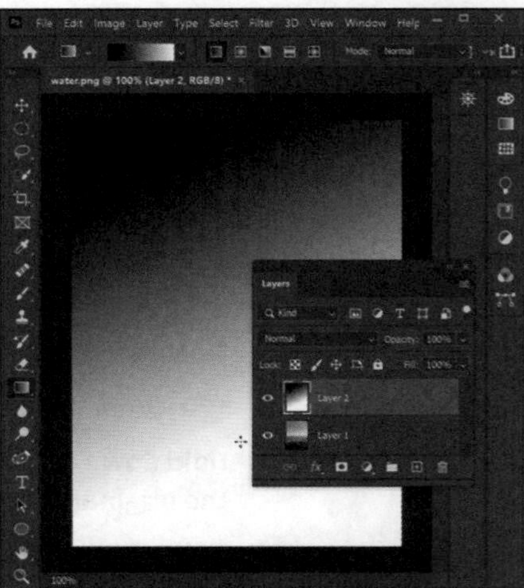

FIGURE 7-13: Create a new layer and then, using the Gradient tool, click and drag to create a custom gradient in the length and direction that you drag.

4. **To see how you can combine gradients with the rest of your imagery, click on the Blending Mode drop-down menu in the Layers panel and choose different blending options, as shown in Figure 7-14. Blending options are discussed in more detail in the next section.**

As previously mentioned, gradients are created using the current foreground and background colors. Click the arrow on the Gradient button on the Options bar to assign a different preset gradient.

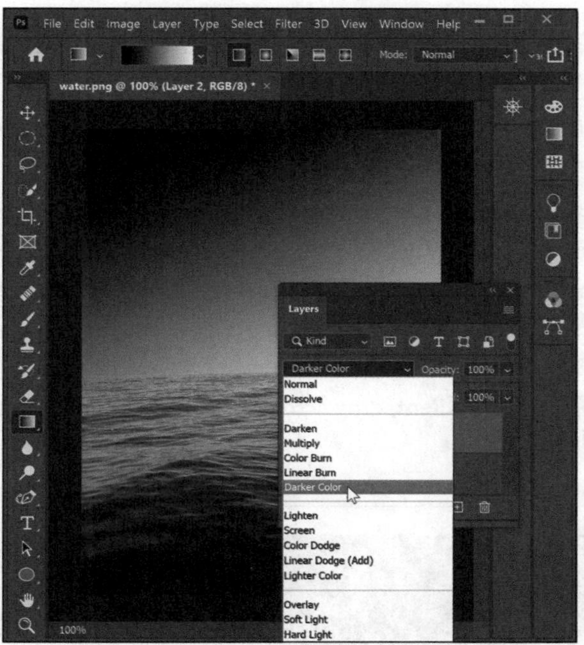

FIGURE 7-14: Change the blending mode of the new layer you created and added a gradient to.

To customize your gradient even further follow these steps:

1. **Choose the Gradient tool and then from the Options bar, choose the type of gradient you want: Linear, Radial, Angle, Reflected, or Diamond.**
2. **Click the Gradient Editor button (second from the left) on the Options bar, as shown in Figure 7-15.**

   The Gradient Editor dialog box appears. In the folders you will see some preset gradients, which you select to start with.

   At the bottom of the gradient preview, you see a gradient bar with two or more *stops,* which is where new colors are inserted into the gradient. They look like little house icons. Use the stops on top of the gradient slider to determine the opacity.

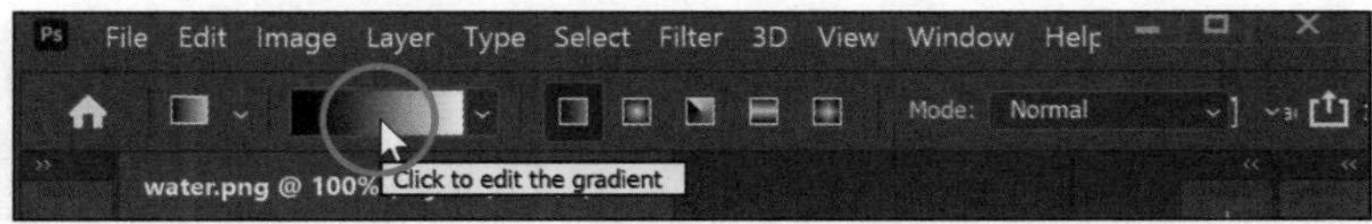

FIGURE 7-15: Click on the Gradient in the Options bar in order to open up the Gradient Editor.

3. **Click a stop and click the color swatch to the right of the word *Color* to open the Color Picker and assign a different color to the stop.**
4. **Click anywhere below the gradient preview to add more color stops.**

   Drag a color stop off the Gradient Editor dialog box to delete it.
5. **Click the top of the gradient preview to assign different stops with varying opacity, as shown in Figure 7-16.**
6. **When you are finished editing the gradient, name it and then click the New button.**

   The new gradient is added to the preset gradient choices.
7. **To apply your gradient, drag across a selection or image with the Gradient tool.**

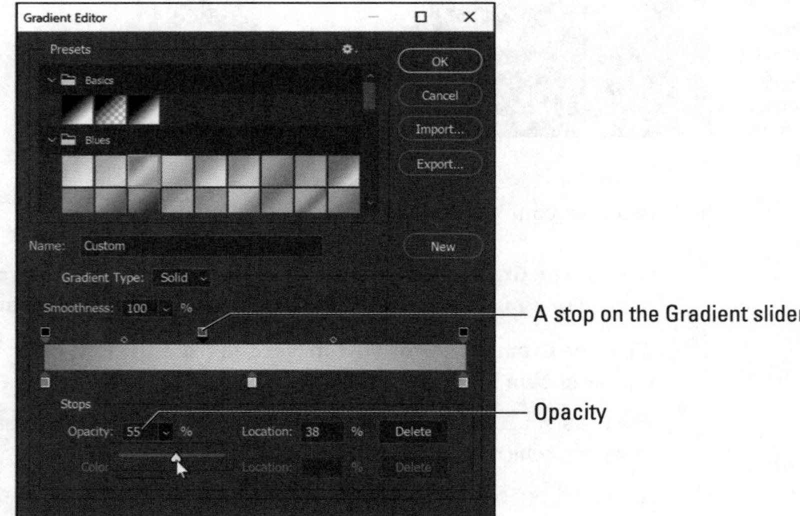

FIGURE 7-16: Assigning varying amounts of opacity using the stops on top of the gradient slider.

Blending Modes

You can use blending modes to add flair to the traditional opaque paint. Use blending modes to paint highlights or shadows that allow details to show through from the underlying image or to colorize a desaturated image. You access the blending modes for paint tools from the Options bar.

You can't get an accurate idea of how the blending mode works with the paint color and the underlying color until you experiment. (That's what multiple undos are for!) Alternatively, you can copy to a new layer the image you want to experiment with and hide the original layer.

This list describes the available blending modes:

- **Normal:** Paints normally, with no interaction with underlying colors.
- **Dissolve:** Gives a random replacement of the pixels, depending on the opacity at any pixel location.
- **Behind:** Edits or paints only on the transparent part of a layer.
- **Darken:** Replaces only the areas that are lighter than the blend color. Areas darker than the blend color don't change.
- **Multiply:** Creates an effect similar to drawing on the page with magic markers. Also looks like colored film that you see on theater lights.
- **Color Burn:** Darkens the base color to reflect the blend color. If you're using white, no change occurs.
- **Linear Burn:** Looks at the color information in each channel and darkens the base color to reflect the blending color by decreasing the brightness.
- **Darker Color:** Compares the total of all channel values for the blend and base color and displays the lower value color.
- **Lighten:** Replaces only the areas darker than the blend color. Areas lighter than the blend color don't change.
- **Screen:** Multiplies the inverse of the underlying colors. The resulting color is always a lighter color.
- **Color Dodge:** Brightens the underlying color to reflect the blend color. If you're using black, there's no change.
- **Linear Dodge:** Looks at the color information in each channel and brightens the base color to reflect the blending color by increasing the brightness.
- **Lighter Color:** Compares the total of all channel values for the blend and base color and displays the higher-value color.
- **Overlay:** Multiplies or screens the colors, depending on the base color.
- **Soft Light:** Darkens or lightens colors, depending on the blend color. The effect is similar to shining a diffused spotlight on the artwork.
- **Hard Light:** Multiplies or screens colors, depending on the blend color. The effect is similar to shining a harsh spotlight on the artwork.
- **Vivid Light:** Burns or dodges colors by increasing or decreasing the contrast.

- **Linear Light:** Burns or dodges colors by decreasing or increasing the brightness.
- **Pin Light:** Replaces colors, depending on the blend color.
- **Hard Mix:** Paints strokes that have no effect with other Hard Mix paint strokes. Use this mode when you want no interaction between the colors.
- **Difference:** Subtracts either the blend color from the base color or the base color from the blend color, depending on which one has the greater brightness value. The effect is similar to a color negative.
- **Exclusion:** Creates an effect similar to, but with less contrast than, Difference mode.
- **Subtract:** Subtracts pixel values of one layer with the other. In case of negative values, black is displayed.
- **Divide:** Divides pixel values of one layer with the other.
- **Hue:** Applies the hue (color) of the blend object to underlying objects but keeps the underlying shading or luminosity intact.
- **Saturation:** Applies the saturation of the blend color but uses the luminance and hue of the base color.
- **Color:** Applies the blend object's color to underlying objects but preserves the gray levels in the artwork. This mode is helpful for tinting objects or changing their colors.
- **Luminosity:** Creates a resulting color with the hue and saturation of the base color and the luminance of the blend color. This mode is the opposite of Color mode.

Painting with color

This section provides an example of using blending modes to change and add color to an image. An example of using a blending mode is tinting a black-and-white (grayscale) image with color. You can't paint color in Grayscale mode, so follow these steps to add color to a black-and-white image:

1. **Open an image in any color mode and choose Image⇨Mode⇨RGB, this is to make sure that you will see the effect.**
2. **If the image isn't already a grayscale image, choose Image⇨Adjustments⇨Desaturate.**

   This feature makes it appear as though the image is black and white, but you're still in a color mode and can apply color.

3. **Choose a painting tool (the Brush tool, for example) and, from the Swatches panel, choose the first color you want to paint with.**
4. **On the Options bar, select Color from the Mode drop-down list and then use the Opacity slider to change the opacity to 50 percent.**

   You can also just type **5**.
5. **Start painting!**

   Color Blending mode is used to change the color of pixels while keeping intact the underlying grayscale (shading).

TIP

Another way to bring attention to a certain item in an RGB image (such as those cute greeting cards that have the single rose in color and everything else in black and white) is to select the item you want to bring attention to. Choose Select ⇨ Modify ⇨ Feather to soften the selection a bit — 5 pixels is a good number to enter in the Feather Radius text field — then click OK. Then choose Select ⇨ Inverse. With everything else selected, choose Image ⇨ Adjustments ⇨ Desaturate. Everything in the image looks black and white, except for the original item you selected.

Filling selections

If you have a definite shape that doesn't lend itself to being painted, you can fill it with color instead. Make a selection and choose Edit ⇨ Fill to access the Fill dialog box. From the Use drop-down list, you can select from the following options to fill the selection: Foreground Color, Background Color, Color (to open the Color Picker while in the Fill dialog box), Content-Aware, Pattern, History, Black, 50% Gray, or White.

If you want to use an existing or saved pattern from the Brushes panel, you can retrieve a pattern by selecting Pattern in the Fill dialog box as well. Select History from the Use drop-down list to fill with the last version saved or the history state.

If you'd rather use the Paint Bucket tool, which fills based on the tolerance specified on the Options bar, it's hidden in the Gradient tool. To use the Paint Bucket tool to fill with the foreground color, simply click the item you want to fill.

Saving Presets

All Photoshop tools enable you to save presets so that you can retrieve them from a list of presets. The following steps show you an example of saving a Tool preset for a brush, but you can use the same method for all other tools as well:

1. **Choose a brush size, color, softness, or any other characteristic.**
2. **Select Window⇨Brushes.**
3. **Click on the Brushes panel menu and choose New Brush Preset. (See Figure 7-17.)**
4. **Type a descriptive name in the Name text field (leave the Include Color checkbox selected if you want the preset to also remember the present color) and then click OK.**

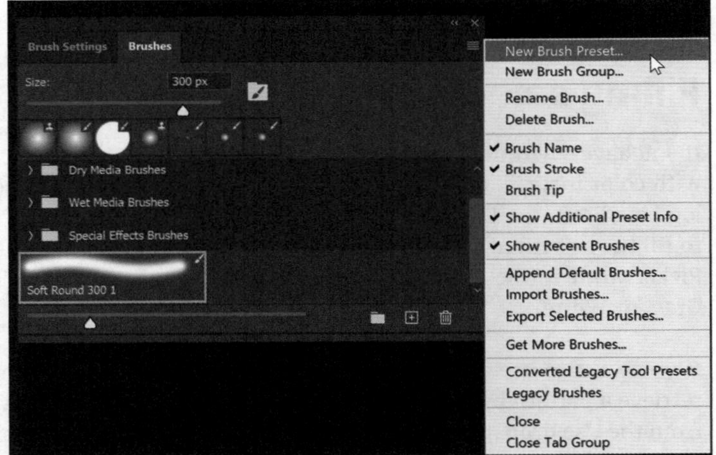

FIGURE 7-17: Save your brush properties as a new brush preset.

» Using text as a layer

» Implementing layer masks

» Organizing layers

Chapter 8 Using Layers

By using layers, you can make realistic additions to an image that can later be removed, edited, and changed with blending modes and transparency settings. This chapter shows you how to create composite images using helpful layer features. Even if you're an experienced Photoshop user, read this chapter to discover great keyboard shortcuts that can help improve your workflow.

Creating and Working with Layers

Layers make creating *composite images* (images pieced together from other, individual images) easy because you can separate individual elements in an image onto their own layer. Much like creating collages by cutting pictures from magazines, you can mask out selections on one image and place them on a layer in another image. When pixel information is on its own layer, you can move it, transform it, correct its color, or apply filters only to that layer, without disturbing pixel information on other layers.

The best way to understand how to create and use layers is to, well, create and use layers. The following steps show you how to create a new, layered image:

1. **Choose File ⇨ New to create a new document.**

   Depending upon your preferences one of two dialog boxes appear.

2. **Choose Photo ⇨ Default Photoshop file, or Choose Default Photoshop File from the Document Type dropdown menu and then select the Transparent option from the Background Contents area, and click OK.**

   Because you selected the Transparent option, the layer appears as a checkerboard pattern, which signifies that it is transparent, versus a white background layer.

TIP

   If you don't like to see the default checkerboard pattern where there's transparency, choose Edit ⇨ Preferences ⇨ Transparency and Gamut (Windows) or Photoshop ⇨ Preferences ⇨ Transparency and Gamut (Mac). In the Preferences dialog box that appears, you can change the Grid Size drop-down list to None to remove the checkerboard pattern entirely. If you don't want to remove the transparency grid, you can change the size of the checkerboard pattern or change the color of the checkerboard.

REMEMBER

   When you open an existing document (say, a photograph), the image is typically the background layer, this image is transparent because we indicated that we wanted it that way when you created the new document.

3. **Create a square with the Rectangular Marquee tool on Layer 1.**
4. **Choose to fill the rectangular Marquee with green by selecting Edit ⇨ Fill.**
   - Choose Color from the Contents drop-down menu; the Color Picker appears.
   - When the Color Picker appears, click and drag up on the Hue slider until you see green colors in the Picked Color pane. See Figure 8-1.
   - In the large square click on any green color in the large Color Picker. Click OK, and then click OK again. The rectangle is filled with Green.

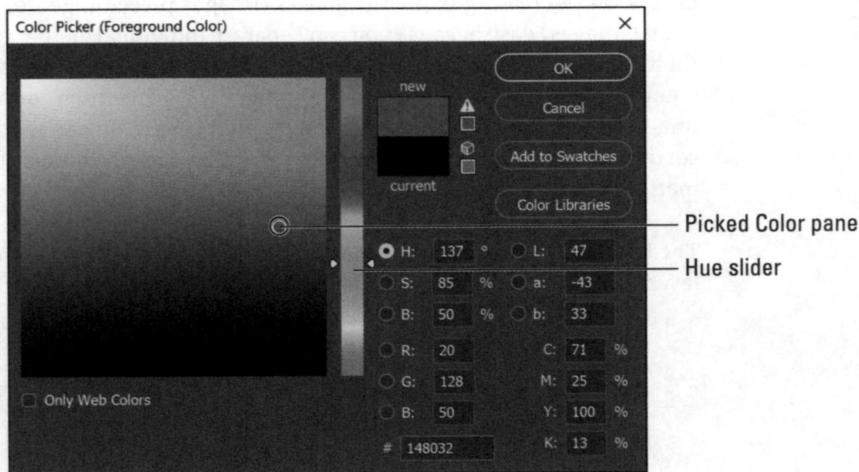

FIGURE 8-1: Click and drag the double-arrows up on the hue slider until you see green colors in the Color picker.

TIP

If you have a foreground color that you want to use as a fill, you can also use the key command Alt+Delete (Windows) or Option+Delete (Mac) to fill the selected area with your foreground color.

5. **Rename the layer by double-clicking the layer name (Layer 1) in the Layers panel and typing a short, descriptive name; in this example,** green rectangle**.**

   A good practice is to name layers based on what they contain; for this example, we named Layer 1 the catchy name *green rectangle.*

6. **Deselect any pixels by selecting Select⇨Deselect or pressing Ctrl+D (Windows) or Cmd+D (Mac).**

7. **Create a new layer by Alt-clicking (Windows) or Option-clicking (Mac) on the New Layer button at the bottom of the Layers panel.**

   The New Layer dialog box appears. By holding down Alt (Windows) or Option (Mac), you can name the layer while you create it.

8. **In this example, you add a circle, so name the layer** Red circle**. Click OK.**

9. **Create a circle using the Elliptical Marquee tool on the new layer.**

   In this example, we created a red circle by using the Elliptical Marquee tool and filled it with red.

TIP

   The keyboard shortcut for Edit⇨Fill is Shift+F5 (Windows) Shift+Delete (Mac and Windows). After the dialog box opens, you choose Contents⇨Color and select a red color from the Color Picker.

   The new shape can overlap the shape on the other layer, as shown in Figure 8-2.

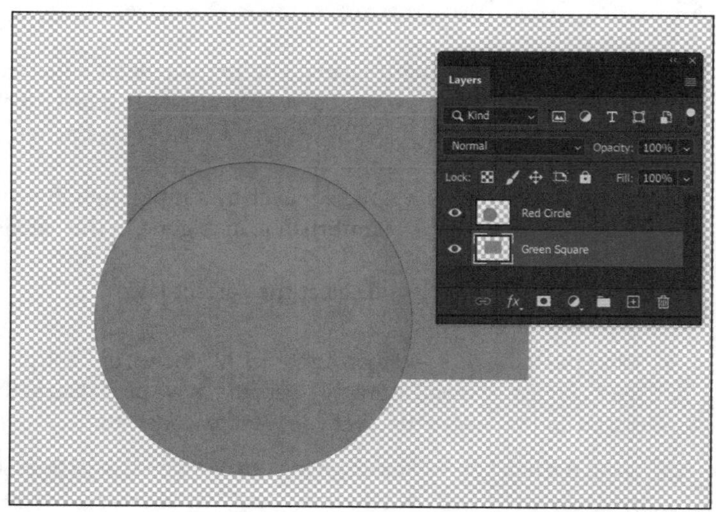

FIGURE 8-2: The circle overlaps the square.

10. **Press Ctrl + D or Cmd+D to deselect your elliptical pixels.**
11. **Choose File⇨Save and save this file as Layer Exercise. This is just in case you have an error while working and have to start again. Continue following the steps in this exercise.**

Duplicating a Layer

Perhaps you want to create a duplicate of a layer for your composite. This technique can be helpful for creating do-it-yourself drop shadows and for adding elements to an image, such as more apples in a bowl of fruit.

1. **In the Layers panel, select the layer that you wish to duplicate or clone. In this example, the Red Circle is used.**
2. **Press and hold on Alt (Windows) or Option (Mac) and drag the layer to the New Layer button at the bottom of the Layers panel to duplicate it. Again, by holding down Alt (Windows) or Option (Mac), you can name the layer while you create it. Name the file** Another Red Circle **and then click OK.**

   You now have three layers.

You can choose to keep this file open for practice as you go through the rest of this chapter or choose File⇨Close at this time.

Selecting a Layer

When you start working with layers, you may move or adjust pixels only to discover that you accidentally edited pixels on the wrong layer. Select the layer you plan to work on by clicking the layer name in the Layers panel.

REMEMBER

Photoshop CC represents a selected layer by simply highlighting the layer in the Layers panel. The indicator paintbrush icon is gone in this version.

Here are some tips to help you select the correct layer:

- Select the Move tool and then right-click (Windows) or Control-click (Mac) to see a contextual menu listing all layers that have pixel data at the point you clicked and to choose the layer you want to work with.
- Get in the habit of holding down the Ctrl (Windows) or ⌘ (Mac) key while using the Move tool and when selecting layers. This technique temporarily turns on

the Auto Select feature, which automatically selects the topmost visible layer that contains the pixel data you clicked.

» Press Alt+[(Windows) or Option+[(Mac) to select the next layer down from the selected layer in the stacking order.

» Press Alt+] (Windows) or Option+] (Mac) to select the next layer up from the selected layer in the stacking order.

Controlling the visibility of a layer

Hide layers that you don't immediately need by clicking the eye icon in the Layers panel. To see only one layer, Alt-click (Windows) or Option-click (Mac) the eye icon of the layer you want to keep visible. Alt-click (Windows) or Option-click (Mac) the eye icon again to show all layers.

Rearranging the stacking order

Layers are like clear pieces of film lying on top of each other. Change the stacking order of the layers in the Layers panel by dragging a layer until you see a black separator line appear, indicating that you're dragging the layer to that location. You can also use these helpful commands to move a layer:

Command	Windows Shortcut	Mac Shortcut
Move selected layer up	Ctrl+]	⌘ +]
Move selected layer down	Ctrl+[	⌘ +[

Creating a Text Layer

When you create text in Photoshop, the text is created on its own layer. By having the text separate from the rest of the image, applying different styles and blending modes to customize the type, as well as repositioning the text, is simplified.

To create text in Photoshop follow these steps.

1. **Choose the Type tool and then do one of the following:**
   - Click the image area. A curser appears and then you can start typing.
   - Click and drag to create a text area. When you create a text area the text will stay within the confines of that area.

Using Layers

2. **Use the Options bar, shown in Figure 8-3, to change the font, size, alignment, and more.**

3. **If you want to change the color, click on the Color Picker, choose a color by sliding the Hue slider on the right until you reach the color you wish to pick from, and then click on a color in the large Picked Color pane on the left. You can also position your cursor outside the color picker and click on your image to sample a color.**

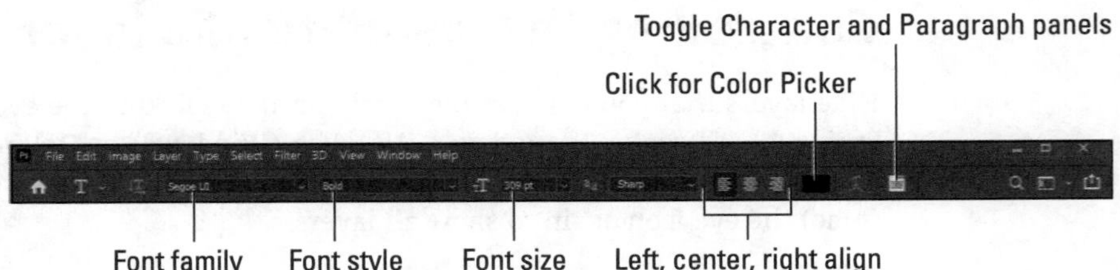

FIGURE 8-3: The Text tool options.

REMEMBER

When you're finished typing, you must confirm your text entry by selecting the checkbox (on the right of the Options bar) or pressing Ctrl+Enter (Windows) or ⌘ +Return (Mac).

Warping text

When you click the Create Warped Text button on the Options bar, the Warp Text dialog box appears. By clicking on the Style drop-down menu, you can choose from a variety of warp settings. After you select a setting, you can customize it using the options in the Warp Text dialog box.

Keep in mind that you can still edit text that's been warped. To remove a warp, click the Create Warp Text button again and select None from the Style drop-down list.

Fine-tuning text

For controls such as leading, baseline shift, and paragraph controls, click the Toggle Character and Paragraph Panels icon near the right end of the Options bar, as shown in Figure 8-3.

Use the keyboard commands in Table 8-1 to fine-tune text in Photoshop. Make sure that you have text selected when you use these shortcuts.

TABLE 8-1 Helpful Typesetting Key Commands

Task	Windows	Mac
Increase the font size	Shift+Ctrl+>	Shift+⌘ +>
Decrease the font size	Shift+Ctrl+<	Shift+⌘ +<
Increase the kerning (the cursor must be between two letters)	Alt+→	Option+→
Decrease the kerning (the cursor must be between two letters)	Alt+←	Option+←
Increase the tracking (on several selected letters)	Alt+→	Option+→
Decrease the tracking (on several selected letters)	Alt+←	Option+←
Increase or decrease the leading (on several selected lines)	Alt+↑ or Alt+↓	Option+↑ or Option+↓

TIP

To change the font, drag over the font family name on the Options bar and then press the up-arrow key (↑) to move up in the font list or the down-arrow key (↓) to move down in the font list.

After you're finished editing text, confirm or delete the changes by clicking the buttons on the right end of the Options bar.

TIP

If you'd rather use key commands to confirm or delete changes, press the Esc key to cancel text changes; press Ctrl+Enter (Windows) or ⌘ +Return (Mac) to commit text changes (or use the Enter key on the numeric keypad).

Check that spelling!

We can't be perfect all the time! Check your spelling easily using the Edit⇨Check Spelling menu item. Photoshop works with a standard spell checker, you can ignore words that are unique, and add them to your own dictionary if you like.

Note: The spell checker in Photoshop defaults to your language defaults. If you use multiple languages, you can change the default language by selecting Edit⇨Preferences⇨Interface (Windows) or Photoshop⇨Preferences (Mac) and changing the selection in the UI Language menu.

Using Layer Masks

In the following sections, we show you how to create a layer mask from a selection or a pen path. A *layer mask* covers areas of the image that you want to make transparent and exposes pixels that you want visible. Masks can be based on a selection you've created by painting on the mask itself. You can also take advantage of the Pen tool to create a path around the object you want to keep visible.

Creating a layer mask from a selection

You need to have two images open to follow these steps where we show you how to create layer masks from a selection. If you wish to use the same files used in tour example you can open the images named `Diver.jpg` and `Goldfish.jpg` from the Book03_Photoshop folder that you downloaded in Book 1, Chapter 1.

Note: If you plan on combining images, choose Image⇨Image Size to make sure that the images are approximately the same resolution and pixel dimensions. Otherwise, you may be astonished by the disproportionate images in your composite.

1. **Open two images that you wish to combine. In this example the Diver and Goldfish images are used.**
2. **Choose Window⇨Arrange⇨Tile Vertically or Horizontally to position the images in separate windows.**
3. **Select the Move tool, and then click one image and drag it to the other image window. In this example the Goldfish is dragged over to the diver image.**

   A black border appears around the image area when you drop an image into another image window. By dragging and dropping an image, you automatically create a new layer on top of the active layer.

   Hold down the Shift key when dragging one image to another to perfectly center the new image layer in the document window.

   You can also drag a layer (background or regular layer) to another image file. First make sure that no selection is active. Then use the Move tool to drag the layer from the artboard up to the tab of the image you want it to copy the image to.
4. **Using any selection method, select a part of the image that you want to keep. If using the Goldfish image you can choose to use the Object Selection tool and click and drag over the Goldfish to easily select it. If you are working in an older version of Photoshop you can also use the Quick Selection tool and paint the selection of the Goldfish.**
5. **Choose Select⇨Modify⇨Feather to soften the selection.**

   Five pixels should be enough.
6. **Click the Add Layer Mask button at the bottom of the Layers panel.**

   A mask is created to the right of the layer, leaving only your selection visible, as shown in Figure 8-4.

FIGURE 8-4: Use the Object Selection tool to surround the Fish and make a selection. Click the Layer Mask button while you have an active selection to create a custom layer mask from your selection.

7. **If you click the Layer thumbnail in the Layers panel, the mask thumbnail shows corner edges, indicating that it's activated.**

   While the layer mask is active, you can paint on the mask.

8. **Select the Brush tool and paint black while the mask thumbnail is selected to cover areas of the image that you don't want to see; press X to switch to white and paint to expose areas on the image that you do want to see.**

   TIP

   If you are not seeing satisfactory results, check to make sure you have a default brush selected. You can reset your brush by selecting the Brush tool and then right-clicking on the Brush icon in the Options at the top and selecting Reset Tool.

   Experiment with this feature by changing the opacity of your brush to make the image slightly transparent where you paint. The keyboard shortcut to change opacity is to just press a number value. For instance, press 2 for a 20% opacity brush, or 57 for a 57% opacity brush.

9. **If you feel your fish is too large or if you want to change its angle, choose Edit⇨Free Transform, or press Ctrl-T (Windows) or Cmd-T (Mac). When the handles appear, you can grab a corner handle and drag to scale the fish, or position your cursor outside any corner until you see the curved arrow icon. Click and drag to change the angle, as shown in Figure 8-5.**

10. **If you are happy with your transformation, click on the Commit checkmark up in the Options bar or press the Enter or Return key. Press the Esc key if you want to cancel the transformation.**

If you are following along with our exercise file, keep it open for the next section.

TIP

You can create a smooth transition from one image to another by dragging the Gradient tool across the image while the layer mask is selected in the Layers panel. This essentially fades one image into another.

FIGURE 8-5: Use the Free Transform feature in order to scale and rotate your new layer.

Creating a vector mask from a pen path

A *vector mask* masks a selection, but it does so with the precision you can get only from using a path. The following steps show you another, slightly more precise, way to create a layer mask by using a pen path. Feel free to use our example file named Starfish when following these steps:

1. **If you are using our example file, click on the topmost layer in your Layers panel and then choose File⇨Place Embedded. Locate your file by navigating to the Book03_Photoshop folder and then select the image named `Starfish.jpg`.**

   By placing an embedded image into your document, your file will always travel with the image. If you selected Place Linked, you would need to share the Starfish file with your existing file when sending this image file to others. Press Enter to confirm the placement of the image.

2. **Select the Pen tool and make sure that Path is selected in the Pick tool mode drop-down menu in the Options bar. Using the Pen tool, click from point to point to make a closed pen path.**

   Do your best to click from one point to another around one of the starfish. The Pen path does not have to be perfect because you can clean it up later.

   If you already have an image with a path, choose Window⇨Paths and click a path to select it.

   See Chapter 4 of this minibook for more about details about how to use the Pen tool.

3. **Click once on the Add Layer Mask button at the bottom of the Layers panel. Nothing happens at this point because you do not have a selection. Click again on the same icon; it now states Add Vector Mask. See Figure 8-6.**

   The mask is built from your path.

FIGURE 8-6: Create a pen path around the starfish and then click on the Add layer mask button twice.

4. **You can now clean up your mask using the Direct Selection tool. It is hidden in the Path Selection tool. Using the Direct Selection tool, you can click and drag your anchor points to fix the mask in places you want to improve it, as shown in Figure 8-7.**

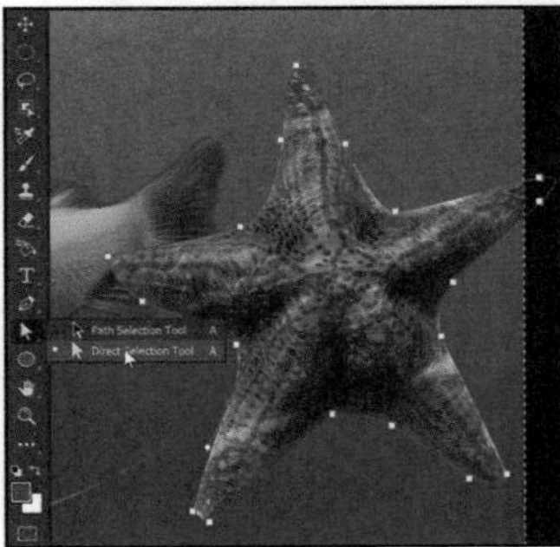

FIGURE 8-7: Clean up your selection using the Direct Selection tool and by painting into your layer mask that is in between the image and the vector mask.

5. **For extra credit, click on the blank layer mask that is sandwiched between the image of the starfish and the vector mask and paint varying opacity levels of black using the Brush tool. Notice that you have a bonus layer mask when working with vector masks.**
6. **If you no longer want a vector mask, drag the thumbnail to the Trash icon in the Layers panel. An Alert dialog box appears, asking whether you want to discard the mask or apply it. Click the Discard button to revert your image to the way it appeared before applying the mask, or click the Apply button to apply the masked area.**

Organizing Your Layers

As you advance in layer skills, you'll want to keep layers named, neat, and in order. In this section, we show you some tips to help you organize multiple layers.

Activating multiple layers simultaneously

Select multiple layers simultaneously by selecting one layer and then Shift-clicking to select additional layers. The selected layers are highlighted. Selected layers move and transform together, making repositioning and resizing easier than activating each layer independently.

Select multiple layers to keep their relative positions to each other and take advantage of alignment features. When you select two or more layers and choose the Move tool, you can take advantage of alignment features on the Options bar (see Figure 8-8). Select three or more layers for distribution options.

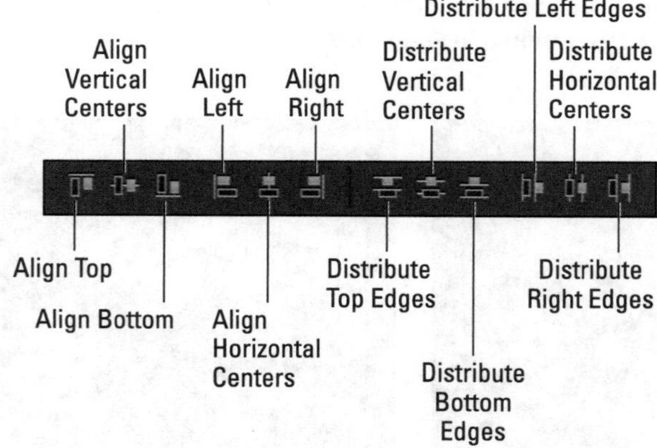

FIGURE 8-8: Align multiple layers simultaneously with the Move tool.

Layer groups

After you start using layers, you'll likely use lots of them, and your Layers panel will become huge. If you often scroll to navigate from one layer to another, take advantage of *layer groups,* which essentially act as folders that hold layers you choose, as shown in Figure 8-9. Just as with a folder you use for paper, you can add, remove, and shuffle around the layers within a layer group. Use layer groups to organize layers and make the job of duplicating multiple layers easier.

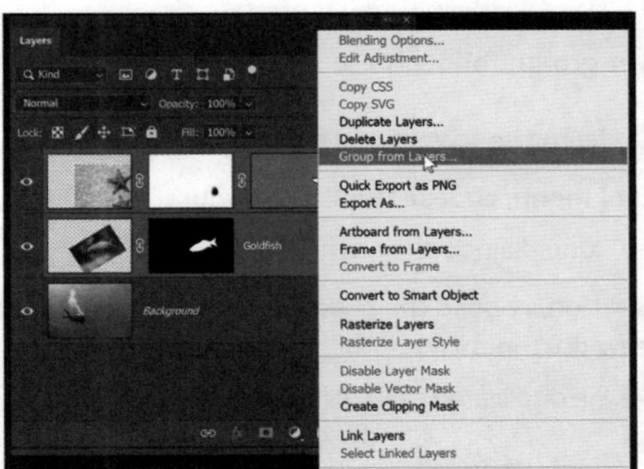

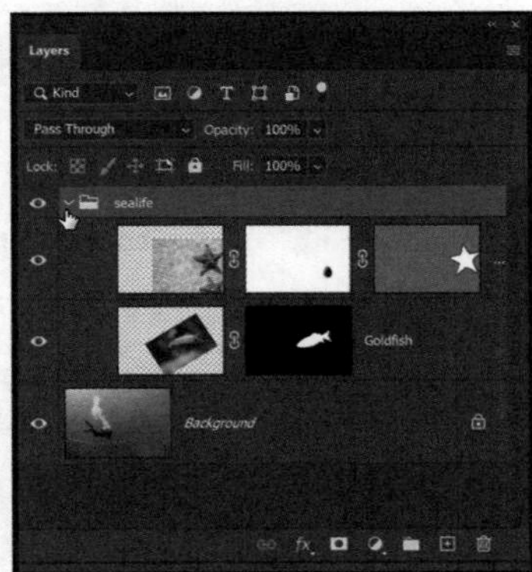

FIGURE 8-9: Select the layers that you want to group together and then choose New Group from Layers.

To create a layer group, follow these steps:

1. **After creating several layers, Shift-click to select the layers you want to group together in a set.**
2. **Choose New Group from Layers from the Layers panel menu, name the group, and then click OK.**

   That's it. You've created a layer group from your selected layers.

TIP

With the Blending Mode drop-down list in the Layers panel, you can change all layers within a group to a specific blending mode, or you can use the Opacity slider to change the opacity of all layers in a group at one time. *Pass through* in the blending mode indicates that no individual blending modes are changed.

After you create a layer group, you can still reorganize layers within the group or even drag additional layers in or out. You can open and close a layer group with the arrow to the left of the group name.

Duplicating a layer group

After you create a layer group, you may want to copy it. For example, you may want to copy an image, such as a button created from several layers topped off with a text layer. The most efficient way to make a copy of that button is to create a layer group and copy the entire group. To copy an image made up of several layers that aren't in a layer group would require you to individually duplicate each layer — how time-consuming!

To duplicate a layer group, follow these steps:

1. **Select a group from the Layers panel.**
2. **From the panel menu, choose Duplicate Group.**

   The Duplicate Group dialog box appears.
3. **For the destination, choose the present document or any open document or create a new document.**

   Be sure to give the duplicated set a distinctive name!
4. **Click OK.**

In this chapter, you were introduced to the basics of using layers. As you might have discovered, using layers in Photoshop is how you unlock many of the possibilities when creating composites and editing existing images. Read the next chapter to take layers to the next level.

Filter Capabilities in the Layers Panel

To make it easier to identify layers you can filter the display in the Layers Panel. Filter categories include Kind, Name, Effect, Mode, Attribute, and Color.

If you would rather not use the filters, click the Turn Layer Filtering On/Off toggle switch to the right of the filters.

Merging and Flattening Images

Merging layers combines several selected layers into one layer. *Flattening* occurs when you reduce all layers to one Background layer. Layers can increase file size, thereby also tying up valuable processing resources. To keep down file size, you may choose to merge some layers or even flatten the entire image to one Background layer.

Merging

Merging layers is helpful when you no longer need every layer to be independent, such as when you have a separate shadow layer aligned to another layer and don't plan to move it again, or when you combine many layers to create a composite and want to consolidate it to one layer.

To merge layers (in a visual and easy way), follow these steps:

1. **Turn on the visibility of only the layers you want merged.**
2. **Choose Merge Visible from the Layers panel menu.**

   That's it. The entire image isn't flattened, but the visible layers are now reduced to one layer.

TIP

To merge visible layers on a *target* (selected) layer that you create while keeping the visible layers independent, create a blank layer and select it. Then hold down Alt (Windows) or Option (Mac) when choosing Merge Visible from the panel menu.

Flattening

WARNING

If you don't have to flatten your image, don't! Flattening an image reduces all layers to one Background layer, which is necessary for certain file formats. After you flatten an image, you can't take advantage of blending options or reposition layered items. (Read more about saving files in Chapter 10 of this minibook.)

If you absolutely must flatten layers, keep a copy of the original, unflattened document for additional edits later.

To flatten all layers in an image, choose Layer➪Flatten Image or choose Flatten Image from the panel menu on the Layers panel.

» Using and saving styles

» Taking advantage of Smart Objects

» Building designs with artboards

Chapter 9
Going Beyond the Basics with Layers

In this chapter, you take the basics that you just learned and apply them to more advanced layer options. Keep in mind that non-destructive editing is the goal in Photoshop, so these tips will help you create incredible compositions that can be returned to their original source if necessary.

In this chapter, you find out how to apply styles, such as embossing, drop shadows, and more. You also discover how to use Smart Objects in order to avoid overwriting your original image with filters and size changes.

Using Layer Styles

Layer styles are wonderful little extra effects that you can apply to layers to create drop shadows and bevel and emboss effects and to apply color overlays, gradients, patterns and strokes, and more.

Applying a style

To apply a layer style (for example, the drop shadow style, one of the most popular effects) to an image, follow these steps. If you would like to follow along open the image named SunsetBeach from the Book03_Photoshop folder that you downloaded in Book 1, Chapter 1:

1. **Create a layer on any image. In this example, a text layer is created by selecting the Type tool and clicking once on the text layer. You can also press Ctrl+Enter (Windows) or Cmd+Return to commit a text layer.**

   For example, you can create a text layer to see the effects of the layer styles.

2. **With the layer selected, click and hold the Add a Layer Style button at the bottom of the Layers panel; from the menu options, choose Drop Shadow, as shown in Figure 9-1.**

   In the Layer Style dialog box that appears, you can choose to change the blending mode, color, distance spread, and size of a drop shadow. You should see that the style has already applied to your text. Position the cursor on the image area and drag to visually adjust the position of the drop shadow.

FIGURE 9-1: Click on the *fx* button at the bottom of the Layers panel to add a style.

3. **When you are happy with the drop shadow, click OK to apply it.**

TIP

To apply another effect and change its options, click on the Layer Style button in the Layers panel and choose the name of the layer style from the menu that appears — Bevel and Emboss, for example. In the dialog box that appears, change the settings to customize the layer style and click OK to apply it to your image. For example, if you choose Bevel and Emboss from the Layer Styles menu, you can choose from several emboss styles and adjust the depth, size, and softness. In Figure 9-2, you see two styles applied to the SunsetBeach image.

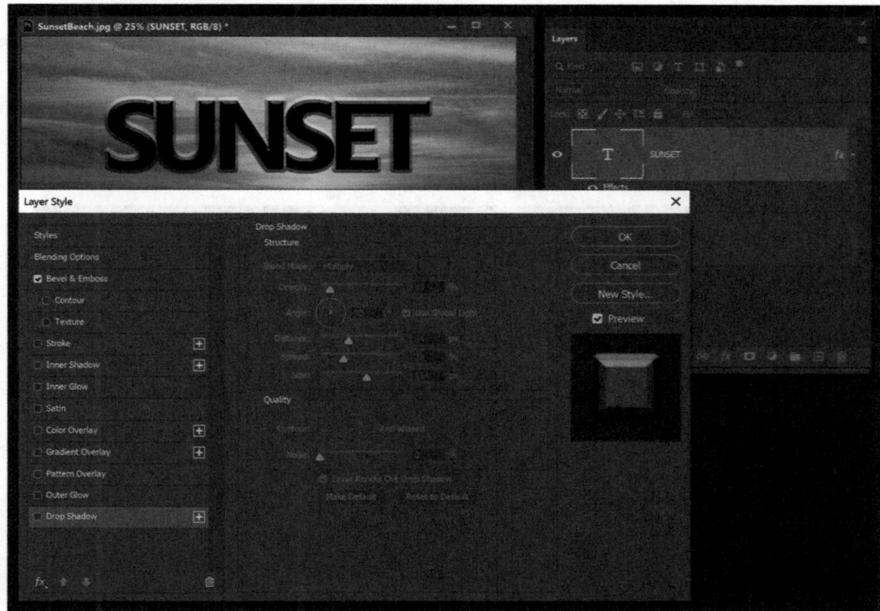

FIGURE 9-2: Two styles applied to the same layer.

Here are some consistent items you see in the Layer Style dialog box:

- **Contour:** Use contours to control the shape and appearance of an effect. Click the arrow to open the Contour fly-out menu to choose a contour preset or click the preview of the contour to open the Contour Editor and create your own edge.
- **Angle:** Drag the crosshair in the angle circle or enter a value in the Angle text field to control where the light source comes from.
- **Global light:** If you aren't smart about lighting effects on multiple objects, global light makes it seem as though you are. Select the Use Global Light checkbox to keep the angle consistent from one layer style to another.
- **Color:** Whenever you see a color box, you can click it to select a color. This color can be for the drop shadow, a highlight, the shadow of an emboss, or a color overlay.

Creating and saving a style

If you come up with a combination of attributes you like, Choose Window➪Styles and click the Create New Style button at the bottom of the Styles panel. By clicking this button while you have the layer with the styles applied to it active, you can name the style and store it in the Styles panel as shown in Figure 9-3. You can later retrieve the style at any time by choosing Window➪Styles. If it helps, click the panel menu button and choose either Small or Large List to change the Styles panel to show only the names of the styles. Notice that Adobe provides organized sets of styles that you can try if you like.

FIGURE 9-3: You can easily save combinations of styles so that you can apply them to other layers.

After you apply a layer style to a layer, the style is listed in the Layers panel. You can turn off the visibility of the style by turning off the eye icon or even throw away the layer style by dragging it to the Layers panel's Trash icon.

Thinking about opacity versus fill

In the Layers panel, you have two transparency options: one for opacity and one for fill. Opacity affects the opacity of the entire layer, including effects. Fill, on the other hand, affects only the layer itself, but not layer styles. Figure 9-4 shows what happens when the Drop Shadow and Emboss effects are applied to text and the fill is reduced to 0 percent. This technique gives you the opportunity to make it look like text is chiseled into rock, wood, or even plastic for the look of a personalized credit card.

FIGURE 9-4: A text layer with styles applied and the fill reduced to 0 percent.

Smart, Really Smart! Smart Objects

Choose File➪Place (embedded or linked) and place an image, an illustration, or even a movie into a Photoshop document. A new layer is created and — even better — a Smart Object is created. The double-square icon in the lower-right corner of the layer thumbnail indicates that a layer is a Smart Object. This means that you have much more flexibility in the placement of your images because the original pixel information has essentially been embedded into your image.

Have you ever placed a logo only to find out later that you need it to be three times its size? Resizing is no longer an issue, because the Photoshop Smart Object is linked to an embedded original. If the original is vector, you can freely resize the image repeatedly without worrying about poor resolution. Want to change the spelling of the Illustrator logo you placed? Just double-click the Smart Object, and the embedded original is opened directly in Adobe Illustrator. Make your changes, save the file, and — *voilà* — the file is automatically updated in the Photoshop file.

What could be better? Smart Filters, of course. You can apply Smart Filters to any Smart Object layer, or even convert a layer to use Smart Filters, by selecting Filters➪Convert for Smart Filters. After a layer has been converted to a Smart Object, you can choose filters, any filters, and apply them to the layer. If you want to paint out the effects of the filter on the layer, simply paint with black on the Filter Effects mask thumbnail. Paint with different opacities of black and white to give an artistic feel to the filter effect, as shown in Figure 9-5. You can even turn off filters by turning off the visibility on the Filter Effects thumbnail by clicking the eye icon to the left of the thumbnail.

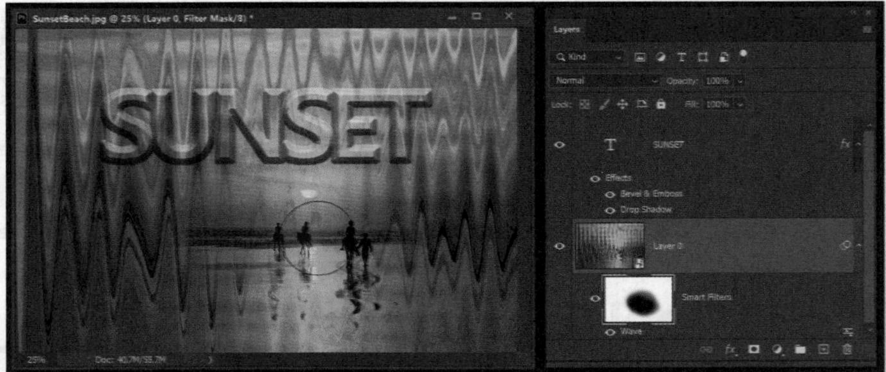

FIGURE 9-5: Cover the filter effects by painting on the Filter Effects thumbnail.

Finding Tools and Features That Are Hidden

Keep in mind that Photoshop has some hidden UI. As we navigate through additional layer features you may discover that the tool being referenced is nowhere to be found. Don't sweat, the tool is there if you are using the latest version, which is, for this printing, Photoshop 2021. Find any hidden tools and features by simply clicking the Search icon in the upper-right of the Photoshop workspace and type what you are looking for; in Figure 9-6, you see that the Frame tool is typed into the textbox. If you do not see the Frame tool in your tools panel, you can try this now too.

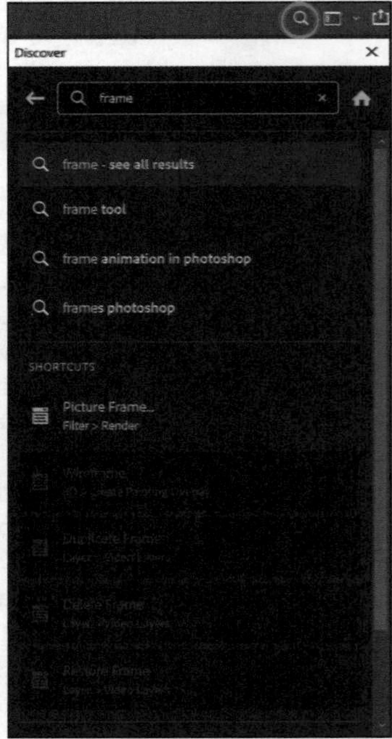

FIGURE 9-6: Find hidden tools and feature using the Search tool right in Adobe Photoshop.

Using the Frame Tool

The Frame tool works as a mask in that you can click and drag an area over a layer and your image will only show through within that area. Follow these steps to try it on your image:

1. **Open any image in Photoshop.**
2. **Make sure you have a layer in your layers panel. If your image indicates that it is a Background layer, click on it twice and click OK to convert it into a layer. You can also simply Alt-double-click (Windows) or Option-double-click (Mac OS) on it to turn it into a layer more quickly.**
3. **Select the Frame tool and then choose either a rectangle or elliptical frame. Click and drag an area for your frame in your image. (See Figure 9-7.) You can adjust the area after by using the handles. If you don't see the Frame tool, use the Search tool (see the previous section) to find it.**
4. **Adjust the placement of the image by clicking on the image in the middle of the frame area and clicking and dragging.**

   Note that a new layer mask was created when you used the Frame tool. If you no longer want the layer mask, simply right-click on the Frame thumbnail in the Layers panel and choose to Delete Frame. Choose Frame Only from the Adobe Photoshop dialog box that appears next.

FIGURE 9-7: Click and drag using the Frame tool in order to mask out the rest of your image.

Working with Artboards in Photoshop

Those of you familiar with Adobe Illustrator might be working with artboards already. Just like in Adobe Illustrator, Photoshop artboards provide the capability to build separate pages or screens within one document. This can be especially helpful if you are building multiple screens for a mobile application or small brochure.

You can think of an artboard as a special type of layer group created using the Layers panel. Its functionality might not be as intuitive in Photoshop as it is in Adobe Illustrator, but with a little practice you can get the hang of it. Follow these steps to see how you can make artboards on your own. If you want to follow along, open the image named `WaterSky` from the Book03_Photoshop folder that you downloaded in Book 1, Chapter 1:

1. **Open any image, or use the image named `WaterSky` that is located in the Book03_Photoshop folder.**
2. **Choose Window ⇨ Layers to show your Layers panel if it is not already visible.**
3. **From the Layers panel menu in the upper-right, choose New Artboard; a dialog box appears that allows you to choose from a variety of preset artboard sizes, as shown in Figure 9-8. In this example, the Set Artboard To preset is set to iPhone 8/7/6. Click OK to create an artboard at that size.**

   Note: While creating artboards, you can choose from a wide variety of preset sizes or define your own custom artboard size. Next, you add a text layer.

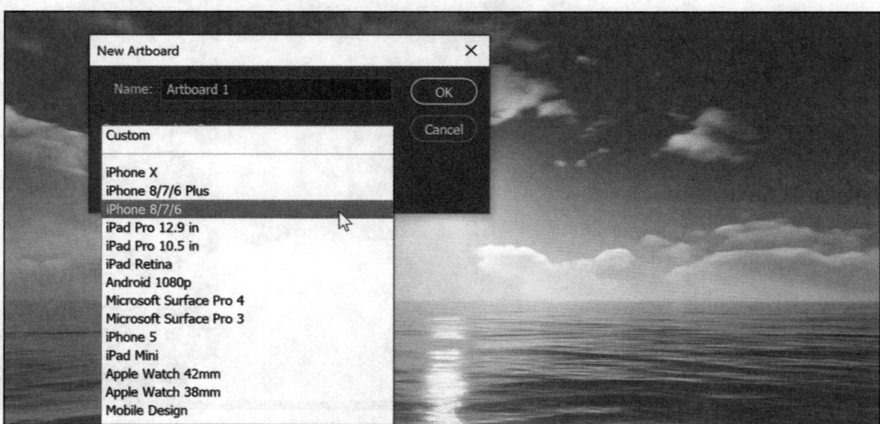

FIGURE 9-8: Add an artboard to your existing image.

4. **Select the Type tool and click and drag on the artboard to create a text area and add a new text layer.**
5. **Type the text** SKY & WATER. **Using the controls in the Options bar, set the type to the following properties:**
   - *Font:* Myriad Pro Bold, or any other available font.
   - *Size:* 50 pt
   - *Color:* Black
6. **Select the Commit button in the Options bar after you have set your type.**

TIP

You can press Ctrl+Enter (Windows) or ⌘+Return (Mac) in order to commit a type layer as well.

The result should be similar to what you see in Figure 9-9.

FIGURE 9-9: Add a text layer to your artboard.

7. **Because you will be adding other artboards, name your artboard by double-clicking on the default Artboard name in the Layers panel as shown in Figure 9-10.**

   In this example, the artboard was renamed Home.

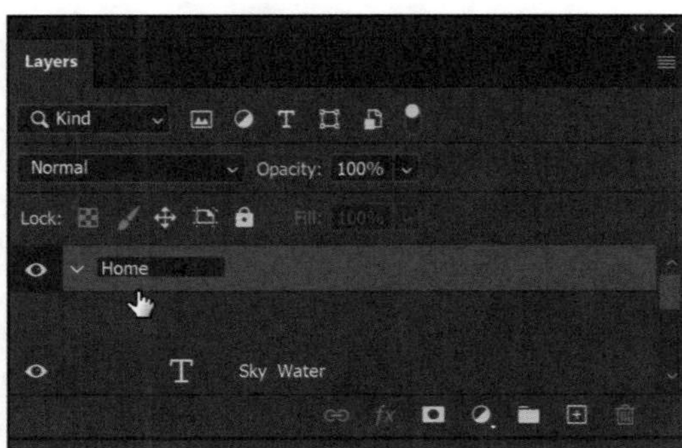

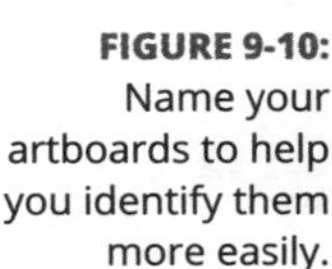

FIGURE 9-10: Name your artboards to help you identify them more easily.

As you see, you work with artboards just like you would work with any other Photoshop document. The difference comes in the next step, where you will find how to add additional artboards to your file.

Adding additional artboards

In this section, you create additional artboards. Perhaps you want a multi-screen mobile app created in Photoshop, or you just want to create multiple versions of one screen. For this section, you continue with the project started in the previous section.

1. **Click on the Move tool and select the hidden Artboard tool shown in Figure 9-11.**

   Plus sign icons appear on all sides of the artboard. You can click these plus sign icons to add additional artboards.

2. **Click the plus sign icon to the right of your existing artboard to see that a new blank artboard is added. (See Figure 9-11.)**

   Look in the Layers panel. Ensure that, just like layers, the newest artboard appears above the selected layer.

FIGURE 9-11: Add a blank artboard by clicking on the plus sign icon.

3. **Name the new artboard Beach by double-clicking on the default artboard name in the Layers panel.**

   Note: You can also use the Artboard tool to create your own custom sized artboard. Do this by clicking and dragging with the artboard tool on the Photoshop canvas. From the tool options bar, select a preset size from the Size pop-up menu, or just leave it in your custom size. Make sure not to click inside the artboard edge or you will activate the Move tool.

Adding content to the second artboard

Now, add an image and some text to the second artboard:

1. **Select the Move tool. You can also press V to select the Move tool.**

2. **Make sure that the Beach artboard is selected in the Layers panel and choose File➪Place Embedded. Navigate to the Book03_Photoshop folder and select the image named** `SunsetBeach` **and press Place. The image is added to the second artboard.**
3. **Click and drag on the corner points to proportionally scale the image larger and fill the screen.**

   When you are finished scaling the image press the Confirm checkmark in the upper-right of the Options bar.
4. **With the Move tool active, click and drag the image. Note that even though it is in the same document, it does not overlap or interfere with artwork on other artboards, as you see in Figure 9-12.**

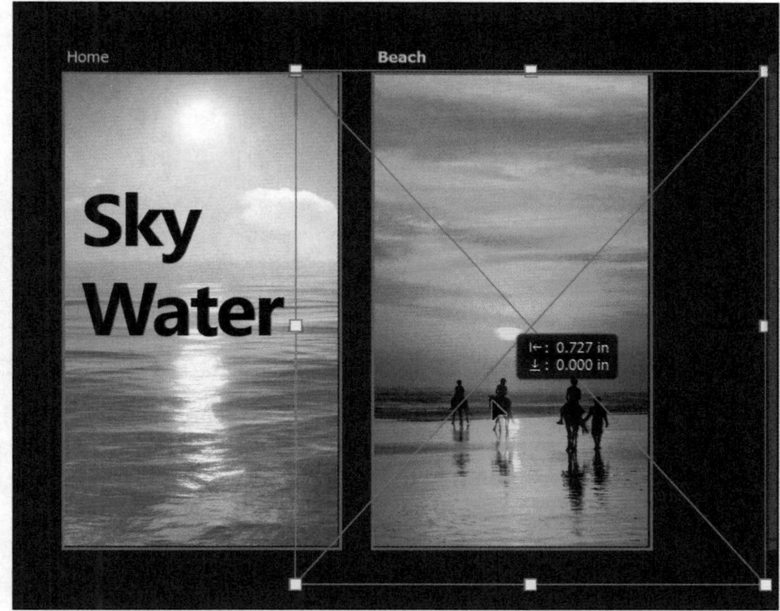

FIGURE 9-12: Reposition the content on one artboard without interfering with content on the other artboard.

Cloning from one artboard to another

Now, clone the text from the Home artboard to the Beach artboard:

1. **With the Move tool still selected, Ctrl-click (Windows) or ⌘-click (Mac) on the WATER & SKY text in the Home artboard. This activates that layer.**
2. **Hold down your Alt (Windows) or Option (Mac) key and click and drag the text from one artboard to the next.**

   If you hold down the Shift key while dragging, the image will remain aligned.

3. **Switch to the artboard and double-click on the newly cloned text to activate the text area and type** Fun Beaches**. You should have completed something similar to what you see in Figure 9-13.**

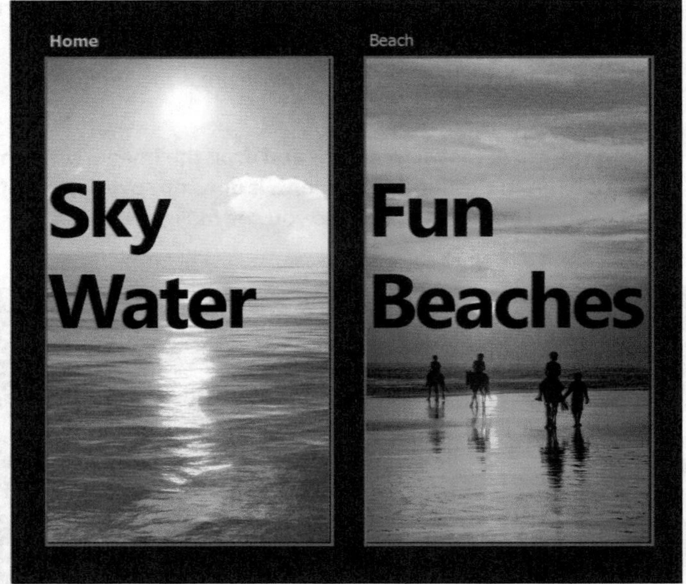

FIGURE 9-13: Two artboards with the text layer cloned to the second layer and then edited.

Moving elements from one artboard to another

To move an element from artboard to artboard, simply drag the elements from one artboard to another. When you move an element between artboards, Photoshop tries to position it in the same location relative to the ruler origins located in the upper-left corners of the artboards.

Exporting your artboards

You can save your file as a Photoshop file to maintain layers, but in this example you will export the file as a two-page PDF. You can use the File⇨Export feature to export your artboards as PNGs, JPEGs, PDFs, and more.

1. **Choose File⇨Export⇨Artboards to PDFs.**

   The Artboards to PDF dialog box appears.

2. **Choose a location and rename the file if you like.**

3. **Make sure that Multi-Page Document is selected under the Options. Also, check Reverse Page Order if you want your Home page to appear first in the exported PDF document. Then click Run.**

   When the export is complete you will receive a successful message alert.

Experimenting with 3D Files

As we mention earlier in this chapter, working with 3D files is beyond the scope of this book, but even a 3D novice can experiment with the 3D features in Photoshop CC. Follow these steps to try out some 3D features on your own:

1. **Open an existing Photoshop document.**

   To eliminate any confusion, it is best if this document simply has a Background layer and no additional layers.

2. **If you have an image open with multiple layers, select a layer to which you want to apply 3D perspective.**

3. **Select the Move tool.**
4. **Choose 3D⇨New Mesh from Layer⇨Mesh Preset⇨Cube Wrap.**

   If a Warning dialog box appears asking if you want to switch to the 3D workspace, click Yes.

   The image is applied to all sides of a cube. You've entered the 3D workspace, which includes a perspective plane. Also, note that the layer now has a 3D icon (looks like a 3D box) on the layer thumbnail in the Layers panel, as shown in Figure 9-14, indicating that it is now a 3D layer.

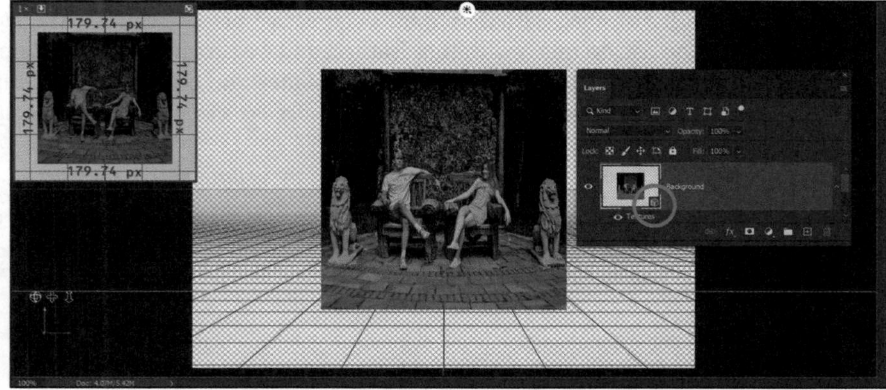

FIGURE 9-14: An icon in the layer indicates that this layer is now a 3D layer.

5. **Experiment with angle and positioning by clicking and dragging on the 3D layer.**

6. **Use the Object Orbit, Roll, Pan, Slide, and Zoom tools located in the Options bar, as shown in Figure 9-15, to change the positioning of the 3D layer.**

   Notice that these tools have additional options available on the Options bar across the top of your Photoshop document.

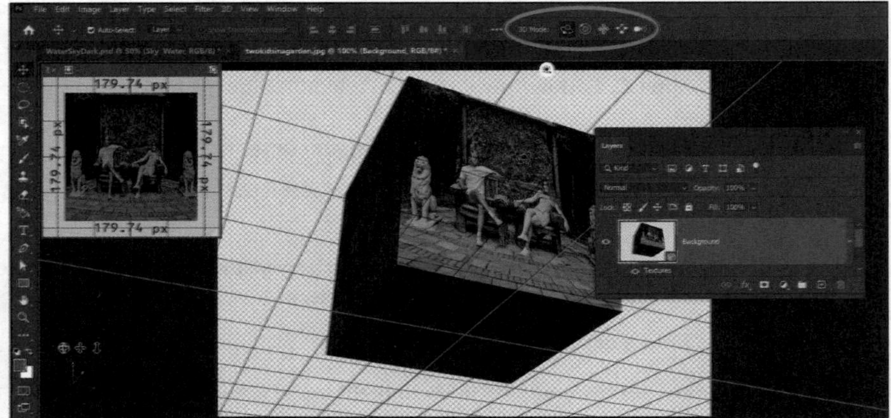

FIGURE 9-15: Click and drag with the Rotate the 3D Object tool to reposition the 3D layer.

Preserving Corrective Perspective with the Vanishing Point Feature

The incredible Vanishing Point feature lets you preserve correct perspective in edits of images that contain perspective planes, such as the sides of a building. You can do much with this feature, and we provide you with a simple introduction. Try experimenting with multiple planes and copying and pasting items into the Vanishing Point window for even more effects. Follow these steps:

1. **Open a file that you want to apply a perspective filter to.**

   If you don't have an appropriate image handy, you can find one in the Book03_Photoshop folder. For this example, the `VanishingPoint.jpg` image will work well.

2. **Create a new, blank layer by clicking the Create a New Layer button at the bottom of the Layers panel.**

   If you create a new layer every time you use Vanishing Point, the results appear on a separate layer, preserving your original image, because you can delete the result of the vanishing point filter and still retain the original layer.

3. **Choose Filter ⇨ Vanishing Point.**

   A separate Vanishing Point window appears. If you see an error message about an existing plane, click OK.

4. **Select the Create Plane tool and define the four corner nodes of the plane surface. In Figure 9-16, the plane is built on the missing stone wall on the left side of the doorway. If necessary, press Ctrl+– (Windows) or ⌘ +– (Mac) to zoom back to see the entire image.**

TIP

   Try to use objects in the image — for example, parts of the existing wall — to help create the plane.

   After the four corner nodes of the plane are created, the tool automatically is switched to the Edit Plane tool, which looks like an arrow.

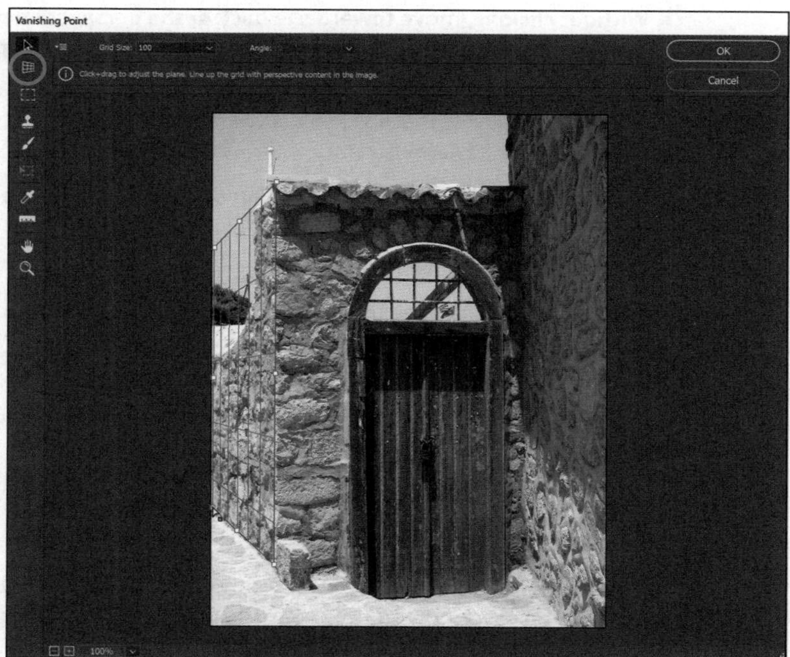

FIGURE 9-16: Create a perspective plane. Adjust it until it is blue, indicating that it is in proper perspective.

5. **Select and drag the corner nodes to make an accurate plane.**

   The plane grid should appear blue, not yellow or red, if it's accurate.

   After creating the plane, you can move, scale, or reshape the plane. Keep in mind that your results depend on how accurately the plane lines up with the perspective of the image.

TIP

   You can use your first Vanishing Point session to simply create perspective planes and then click OK. The planes appear in subsequent Vanishing Point sessions when you choose Filter ➪ Vanishing Point. Saving perspective planes is especially useful if you plan to copy and paste an image into Vanishing Point and need to have a readymade plane to target.

6. **Choose the Stamp tool in the Vanishing Point window and then select On from the Heal drop-down list on the Options bar.**
7. **With the Stamp tool still selected, cross over part of the area or part of the image you want to clone and Alt-click (Windows) or Option-click (Mac) to define it as the source to be cloned.**

   In the image `VanishingPoint`, the middle part of the stone wall was Alt/Option-clicked on to set it as the source for the cloning.

8. **Without clicking, move toward the back of the perspective plane (you can even clone outside the plane), and then click and drag to reproduce the cloned part of the image.**

   Notice in Figure 9-17 that it's cloned as a smaller version, in the correct perspective for its new location.

9. **Start from Step 7 and clone any region of an image closer to the front of the perspective pane.**

   The cloned region is now cloned as a larger version of itself.

   You can use the Marquee tool options (Feather, Opacity, Heal, and Move Mode) at any time, either before or after making the selection. When you move the Marquee tool, the Stamp tool, or the Brush tool into a plane, the bounding box is highlighted, indicating that the plane is active.

10. **Click OK.**

    To preserve the perspective plane information in an image, save your document in JPEG, PSD, or TIFF format.

In this chapter, you discovered many advanced layer capabilities. Read on to Chapter 10 to discover how to save your files for different uses and applications.

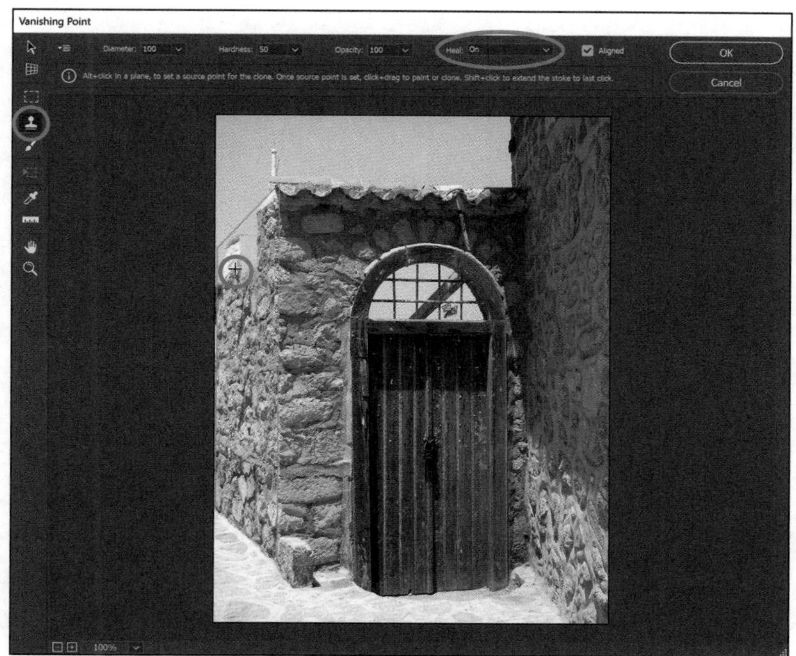

FIGURE 9-17: The stone is cloned in perspective using the Stamp tool.

» Preparing images for the web

» Discovering the color table

» Saving your settings

Chapter 10
Saving Photoshop Images for Print and the Web

It may seem like a minor task, saving your image, but you have to make some important decisions at this stage. Where will the image be viewed, onscreen or print? Will the image be large, or will it be the size of your thumbnail? Without selecting the correct settings and format, you may not be able to import your file or place it into other applications. If you save the file in the wrong format, you could delete valuable components, such as layers or saved selections. This chapter provides you with the necessary information to save the file correctly for both print and onscreen displays. Here you discover what the file format choices are and then progress into the Export feature (for saving in the GIF, JPEG, PNG, and WBMP file formats).

TIP

Saving files in the correct file format is important not only for file size, but also in support of different Photoshop features. If you're unsure about saving in the right format, save a copy of the file, keeping the original in the PSD format (the native Photoshop format). Photoshop alerts you automatically when you choose a format in the Save As or the Export As dialog box that doesn't support Photoshop features. When you choose a format that doesn't support some of the features you've used, such as channels or layers, a yield sign appears when a copy is being made.

With the capability of all Adobe applications (and even non-Adobe applications) to read native Photoshop (.psd) files, be wise and keep files in this native file format unless you have a compelling reason not to.

Backing Up with a Save

Before covering additional details about saving your files, keep in mind that you can set up Automatic Saving in your Photoshop preferences. This can be helpful if your software crashes or your computer shuts down unexpectedly. You can change these saving options by selecting Edit⇨Preferences⇨File Handling (Windows) Photoshop⇨Preferences⇨File Handling (Mac), as shown in Figure 10-1. Look for the drop-down menu that states Automatically Save Recovery Information Every: and select the length of time you would like to lapse between saving.

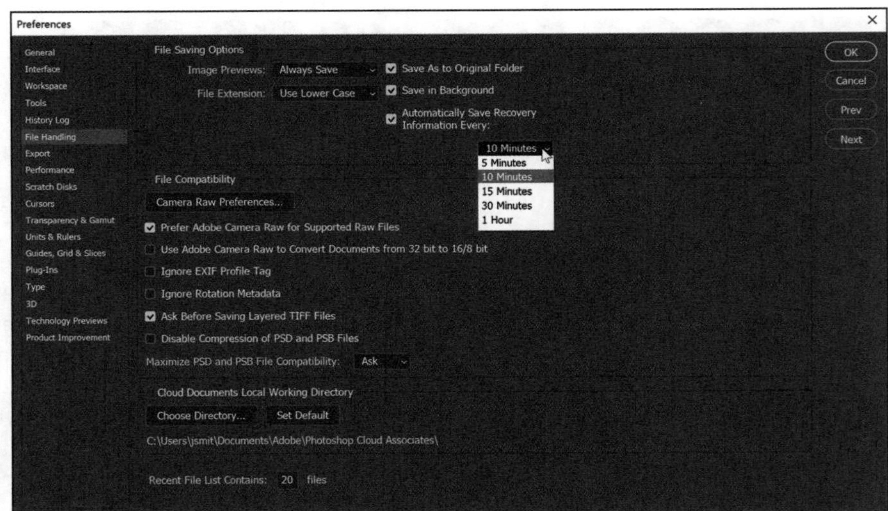

FIGURE 10-1: You can set up Photoshop to automatically save your files.

Choosing a File Format for Saving

When you choose File⇨Save for the first time (or you choose File⇨Save As to save a different version of a file), you see at least 18 different file formats to choose from in the Save As Type (Windows) or Format (Mac) drop-down list. We don't cover every format in this chapter (some are specific to proprietary workflows), but we show you which formats are best for the typical workflow you may face.

Wonderful and easy Photoshop PSD

If you're in an Adobe workflow (you're using any Adobe product), you can keep the image in the native Photoshop PSD format. By selecting this format, transparency, layers, channels, and paths are all maintained and left intact when placed in the other applications.

If compatibility with older versions of Photoshop is an issue, choose Edit⇨Preferences⇨File Handling (Windows) or Photoshop⇨Preferences⇨File Handling (Mac). Select Always from the Maximize PSD and PSB File Compatibility drop-down list. This choice saves a *composite* (flattened) image along with the layers of your document. (The PSB format is used for saving large Photoshop documents — in this case, *large* means they measure more than 30,000 x 30,000 pixels.)

REMEMBER

Leaving the Maximize PSD and PSB File Compatibility drop-down list set to Always creates a larger file. If file size is an issue, leave the drop-down list set to Ask and use the feature only when you need to open the Photoshop file in older versions of Photoshop.

Photoshop EPS

Many applications accept the Encapsulated PostScript (EPS) file format. It's used to transfer PostScript-language artwork between various applications. It supports vector data, duotones, and clipping paths. This means that text layers and other vector-based artwork will remain scalable.

When you choose to save in the EPS format, an EPS Options dialog box appears. Leave the defaults alone and click OK.

WARNING

Alter the settings in the EPS Options dialog box *only* if you're familiar with custom printer calibration or if you need to save your image to a specific screen ruling. Screen rulings (*lpi*, or lines per inch) are usually set in a page layout application, such as Adobe InDesign or QuarkXPress.

TIP

Even though PowerPoint accepts files saved from Photoshop in the EPS format, it is better to use the PNG format. The PNG format is designed for a better onscreen presentation and produces clearer results. Find out more in the later "PNG" section.

Photoshop PDF

If compatibility is an issue, save your file in the Photoshop PDF (Portable Document Format) format. PDF files are supported by more than a dozen platforms when viewers use Acrobat or Adobe Reader. (Adobe Reader is available for free at www.adobe.com.) What a perfect way to send pictures to friends and family! Saving a file in the Photoshop PDF format supports the capability to edit the image when you open the file by choosing File⇨Open in Photoshop.

If you're planning to send a layered file by email, choose Layer⇨Flatten Layers before choosing to save the file as a PDF. This command cuts the file size considerably.

TIFF

The Tagged Image File Format (TIFF) is a flexible bitmap image format that's supported by most image-editing and page-layout applications widely supported by all printers. TIFF supports layers and channels but has a maximum size of 4GB. We hope your files aren't that large!

DCS

The Photoshop Desktop Color Separation (DCS) 1.0 and 2.0 formats are versions of EPS that enable you to save color separations of CMYK (Cyan, Magenta, Yellow, Black) or multichannel files. Some workflows require this format, but if you've implemented spot color channels in an image, using the DCS file format is required to maintain them.

Choose the DCS 2.0 format unless you receive specific instructions to use the DCS 1.0 format — for example, for reasons of incompatibility in certain workflows.

Saving Images for the Web

There are many ways that you can save your art for the web or application design but the method that you choose depends upon your content. In this section, you discover the major formats used for website and application design that you can use within Photoshop, PNG, JPG, and GIF.

Ensuring that the image size is correct before you save the file for the web is a good practice. If you need to read up on resizing images, see Chapter 5 of this minibook. Generally speaking, you should resize the image to the right pixel dimensions. Choose Filter⇨Sharpen⇨Unsharp Mask to regain some of the detail that is lost in resizing the image, and then save the image for the web.

PNG

When you choose to save in the PNG (Portable Network Graphics) format, you have two options from which to choose — PNG-8 and PNG-24.

The PNG-8 format refers to palette variant, which supports only 256 colors, this allows the file format to be smaller in size than some other formats. PNG-8 can be a GIF substitute; see the "Saving a GIF" section, later in this chapter, for more details about that format.

PNG-24 saves 24-bit images that support anti-aliasing (the smooth transition from one color to another). These images work beautifully for continuous-tone images but are much larger than JPEG files. The truly awesome feature of a PNG file is that it supports 256 levels of transparency. In other words, you can apply varying amounts of transparency in an image, as shown in Figure 10-2, where the image shows through to the background.

FIGURE 10-2: A PNG-24 file with varying amounts of transparency.

There are several ways in which you can save a PNG image:

- **Choose File ➪ Quick Export as PNG:** This does not provide any additional save options other than the ability to change the name. If you want to set up preferences for your Quick Export, you can find settings by going to Edit ➪ Preferences ➪ Export (Windows) or Photoshop ➪ Preferences ➪ Export (Mac), as shown in Figure 10-3. A great format if you have an image set to the right size and don't need any other modifications.
- **Choose File ➪ Export Save For Web (Legacy).**
- **In the Preset drop-down menu, in the upper right of the Save for Web dialog box, you can choose PNG-24 or PNG-8 Dithered. Then press Save.**

Note: Photoshop can use dithering to mix the pixels of available colors to simulate missing colors.

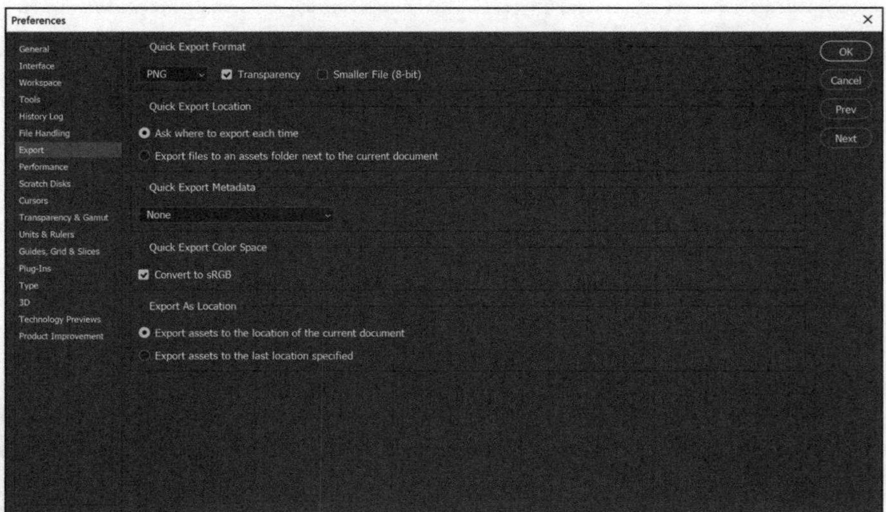

FIGURE 10-3: You can set up preferences for your Quick Export.

JPEG

JPEG (Joint Photographic Experts Group) is a good format for continuous-tone images — those with smooth transitions from one color to another, as in photographs — like the image shown in Figure 10-4.

FIGURE 10-4: Images with smooth transitions from one color to another are good candidates for the JPEG file format.

REMEMBER

The JPEG format is lossy, so you shouldn't save a JPEG and then open it, edit it, and save it again as a JPEG. Because the JPEG compression causes data to be lost, your image will eventually look like it was printed on the same texture as a paper towel. Save a copy of the file as a JPEG, keeping the original image in PSD format if you need to edit the image later, open the original PSD, make your changes, save the PSD, and then save a copy of the edited file as a JPEG.

The JPEG format does *not* support transparency, but you can cheat the system a bit by using matting.

A good image to save in the JPEG format is a typical photograph or illustration with lots of smooth transitions from one color to the next; this includes graphics that use gradients. To save an image as a JPEG, follow these steps:

1. **Choose File⇨Export⇨Save for Web (Legacy), and then click the 2-Up tab to view the original image (left) at the same time as the optimized image (right).**
2. **Select one of the JPEG preset settings from the Settings drop-down list.**

   You can choose Low, Medium, or High or customize a level between the presets by using the Quality slider.
3. **Leave the Optimized checkbox selected to build the best JPEG at the smallest size.**

   The only issue with leaving this checkbox selected is that some very old browsers don't read the JPEG correctly (not likely an issue for most viewers).
4. **Leave the Embed Color Profile checkbox deselected unless you're in a color-managed workflow and color accuracy is essential.**

   Selecting this checkbox increases the file size, and most people aren't looking for *exact* color matches from an image on the monitor.
5. **If you have to have the file size even smaller, use the Blur slider to bring down some detail.**

   It's funny, but one JPEG that has exactly the same pixel dimensions as another may vary in file size because the more detailed an image is, the more information is needed. So, an image of lots of apples will be larger than an image of the same size that has one apple in it. The Blur feature blurs the image (surprise!), so you may want to use it for only a low source image in Dreamweaver.
6. **(Optional) Select a matte color from the Matte drop-down list.**

   Because JPEG doesn't support transparency, you can flood the transparent area with a color you select from the Matte drop-down list. Select the color you're using for the background of your web page by selecting Other and entering the hexadecimal color in the lower portion of the Color Picker.
7. **Click Save.**

GIF

Some say that the way you pronounce *GIF* (Graphics Interchange Format) is based on the brand of peanut butter you eat. Is it pronounced like the peanut butter brand (Jif) or with a hard *g*, as in *gift*? Most people seem to pronounce it as "gift" (minus the T).

Use GIF if you have large areas of solid color, such as a logo like the one shown in Figure 10-5.

FIGURE 10-5: An image with lots of solid color makes a good GIF.

GIF isn't *lossy* (it doesn't lose data when the file is compressed in this format), but it reduces file size by using a limited number of colors in a color table. The lower the number of colors, the smaller the file size. If you've ever worked in Index Color mode, you're familiar with this process.

Transparency is supported by the GIF file format. Generally, GIF files don't do a good job on anything that needs smooth transitions from one color to another because of the format's poor support of *anti-aliasing*, which is the method Photoshop uses to smooth jagged edges. When an image transitions from one color to another, Photoshop produces multiple colors of pixels to create an even blend between the two colors.

Because anti-aliasing needs to create multiple colors for this effect, GIF files generally aren't recommended. In fact, when you reduce the size of a GIF file, you're more apt to see *banding* (an artifact that appears as streaks in the image) because the anti-aliasing can't take place with the limited number of colors available in GIF.

You can, of course, dramatically increase the number of colors to create a smoother transition, but then you risk creating monster files that take forever to download.

Saving a GIF

When you choose File⇨Save for Web (Legacy), you first see the available GIF options. They may be clearer to you if you have an image open (with lots of solid color in it).

To save a file for the web as a GIF, follow these steps:

1. **Choose File⇨Export⇨Save for Web (Legacy).**

   The Save for Web dialog box appears. This legacy saving method is used so that you can see results when changing options.

2. **At the top, click the 2-Up tab.**

   You see the original image on the left and the optimized image on the right (or top and bottom, depending on the image proportions).

   In the lower portion of the display, you see the original file size compared with the optimized file size, as well as the approximate download time. The download time is important. Nobody wants to wait around for a web page to load; most people don't wait more than ten seconds for an entire web page to appear, so try to keep an individual image's download time to a few seconds at the most. Remember that waiting for all images on a page to load can add up to a monstrous wait time for the viewer!

TIP

   Change the estimated download speed by choosing the Select Download Speed drop-down menu to the right of the file size information in the lower-left of the preview window.

3. **Select GIF 32 No Dither from the Preset drop-down list.**

   You may see a change already. As you can see in Figure 10-6, Photoshop supplies you with presets to choose from, or you can customize and save your own. In the example shown, the colors have been reduced down to 8. You will find out more about the color table in the next section.

FIGURE 10-6: Choose from presets, or create your own custom settings.

4. **Choose whether you want dithering applied to the image by selecting an option from the Specify the Dither Algorithm drop-down list.**

   This choice is purely personal. Because you may be limiting colors, Photoshop can use dithering to mix the pixels of available colors to simulate missing colors. Many designers choose the No Dither option because dithering can create unnatural color speckles in an image.

5. **If your image is on a transparent layer and you want to maintain that transparency on a web page, select the Transparency checkbox.**

Using the color table in the Save for Web dialog box

When you save an image in GIF using the Save for Web feature, you see the color table. This color table displays the limited color palette used in the image and is located on the right side of the Save for Web window. The color table is important because it enables you to see the colors used in the image and to customize the color table by using the options at the bottom.

You may want to customize your color table by selecting some colors to be web-safe and locking colors so that they're not bumped off as you reduce the number that's used.

To customize a color table, follow these steps:

1. **If your image has only a few colors that you want to convert to web-safe colors, choose the Eyedropper tool from the left of the Save for Web window and click the color (right on the image) in the Optimized view.**

   The sampled color is highlighted in the color table.

2. **Click the Web Safe button at the bottom of the color table, as shown in Figure 10-7.**

   When you cross over this button, the tooltip Shifts/Unshifts Selected Colors to Web Palette appears.

   A diamond appears, indicating that the color is now web-safe.

3. **Lock colors that you don't want to delete as you reduce the number of colors in the color table.**

   Select a color with the Eyedropper tool, or select it in the color table, and then click the Lock Color button. A white square appears in the lower-right corner, indicating that the color is locked.

   Obviously, if you lock 32 colors and then reduce the color table to 24, some of your locked colors are deleted. If you choose to add colors, those locked colors are the first to return. In Figure 10-7, you see that some of the colors were locked down when the color table was set to 16 colors. When the colors were reduced to 16 they were not deleted from the color table.

TECHNICAL STUFF

How is the color table created? Based on the color-reduction algorithm method you choose, the Save for Web feature samples the number of colors you indicate. If keeping colors web-safe is important, select the Restrictive (Web) option for the method; if you want your image to look better on most monitors but not necessarily to be web-safe, choose the Adaptive option.

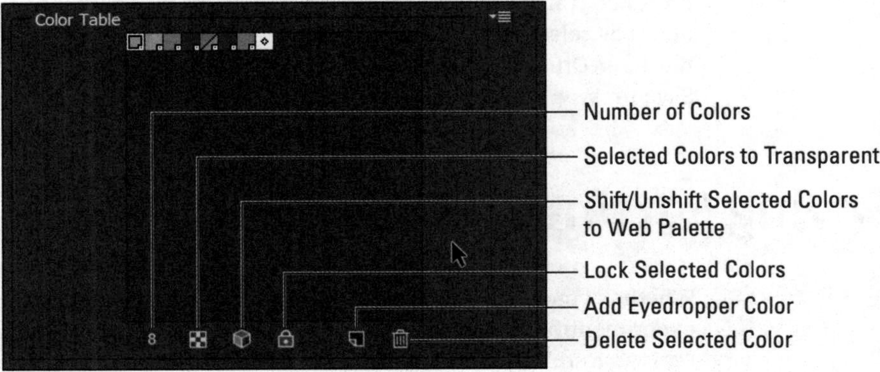

FIGURE 10-7: Customize colors by using the color table in the Save for Web window.

4. **Use the Colors drop-down list or enter a number to add or delete colors from the color table.**
5. **If your image uses transparency, select the Transparency checkbox near the top of the Save for Web dialog box.**

   REMEMBER

   Remember that transparency is counted as one of your colors in the color table.
6. **Select the Interlaced checkbox only if your GIF image is large (25K or larger).**

   Selecting this option causes the image to build in several scans on the web page — a low-resolution image that pops up quickly and is then refreshed with the higher-resolution image when the download is complete. Interlacing gives the illusion of the download going faster but makes the file size larger, so use it only if necessary.
7. **Click Save.**

   Now the image is ready to be attached to an email message or used in a web page.

Matte

Matting appears as a choice in the GIF, JPEG, and PNG format options. Matting is useful if you don't want ragged edges appearing around your images. Matting looks for pixels that are 50 percent or more transparent and makes them fully transparent; any pixels that are 49 percent or less transparent become fully opaque.

Even though your image might be on a transparent layer, it will have some "iffy" pixels — the ones that aren't sure what they want to be . . . to be transparent or not to be transparent. Choose a matte color to blend in with the transparent iffy pixels by selecting Eyedropper Color, Foreground Color, Background Color, White, Black, or Other (to open the Color Picker) from the Matte drop-down list in the Save for Web dialog box.

Saving Settings

Whether you're saving a GIF, JPG, or PNG file, you probably spent some time experimenting with settings to find what works best for your needs. Save selected options to reload later by saving the settings. Do so by clicking the arrow to the right of the Preset drop-down list. Select Save Settings from the list that appears and name your settings. Your named, customized settings then appear in the Preset drop-down list.

If you are busy creating designs for websites and applications, you will appreciate the Generate Image Assets feature. This feature allows you to name layers by adding extensions to layer names that will automatically determine the file format upon saving.

Follow these steps to try out this feature. You can use the image provided named Generate Assets in the Book03_Photoshop folder.

1. **Open the file named Generate Assets, or any other file that has multiple layers. Note that the GenerateAssets.psd file has three layers.**
2. **In the Layers panel, add the following extensions to the end of the layer names, as shown in Figure 10-8.**
   - Pinkflower**.gif**
   - Redflower**.jpg**
   - Purpleflower**.png**

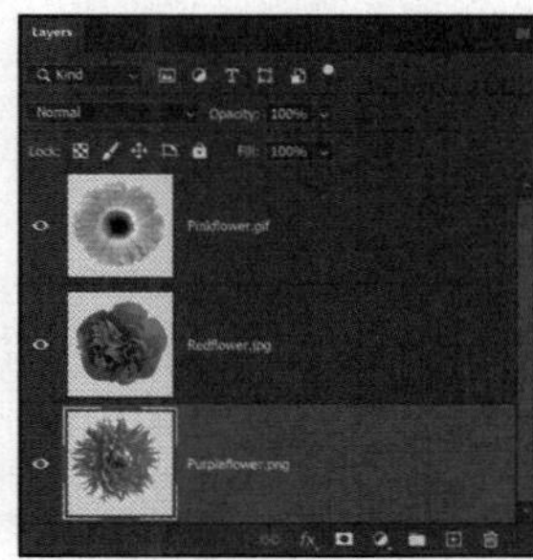

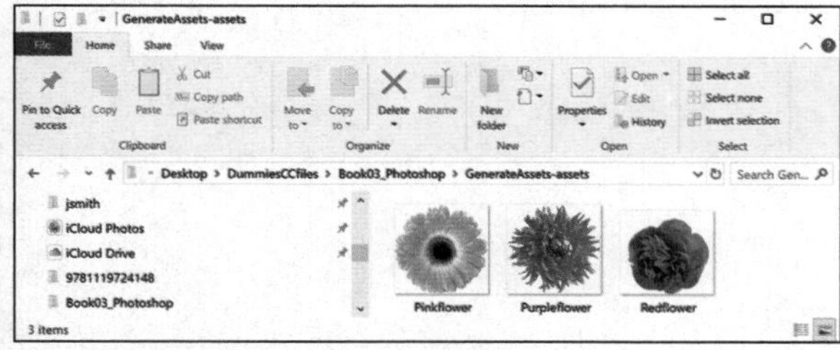

FIGURE 10-8: Add extensions to the end of the layers' names. After generating assets, a folder is created with the files in the format you specified.

3. **Now choose File⇨Generate⇨Image Assets.**

   You don't see it happening, but a new folder has been created in the same folder location that is named Generate Assets-assets. Inside this folder are the named layers, in the format that you indicated by using the extension you added. A great time-saving feature for those creating complex art for applications or the web.

4

InDesign CC

Contents at a Glance

» Creating new documents

» Looking at and setting up the workspace

» Creating your first publication

Chapter 1
Introducing InDesign CC

InDesign is a sophisticated page-layout program. You can use it to create professional-looking documents, including newsletters, books, and magazines. You can also use it to create documents for distribution on a tablet such as an iPad, and even documents that include interactivity or videos. InDesign has evolved into a tool that lets you publish content to just about any device or in print. For example, using InDesign you can create a document and distribute content in print and then add hyperlinks and video and export it to PDF or EPUB.

As powerful an application as InDesign is, you'd think it would be difficult to use, but you'll find that creating most basic documents is a snap. This minibook shows you how to use InDesign to make creative page layouts. In this chapter, you discover the InDesign interface and start your first publication.

Getting Started with InDesign CC

InDesign is used for creating page layouts that include type, graphics (such as fills and strokes), and images. The InDesign document you see in Figure 1-1 includes elements from Adobe Illustrator (logos) and Photoshop (images). If this file were to be exported as a PDF or an EPUB file, it could include video or interactive buttons.

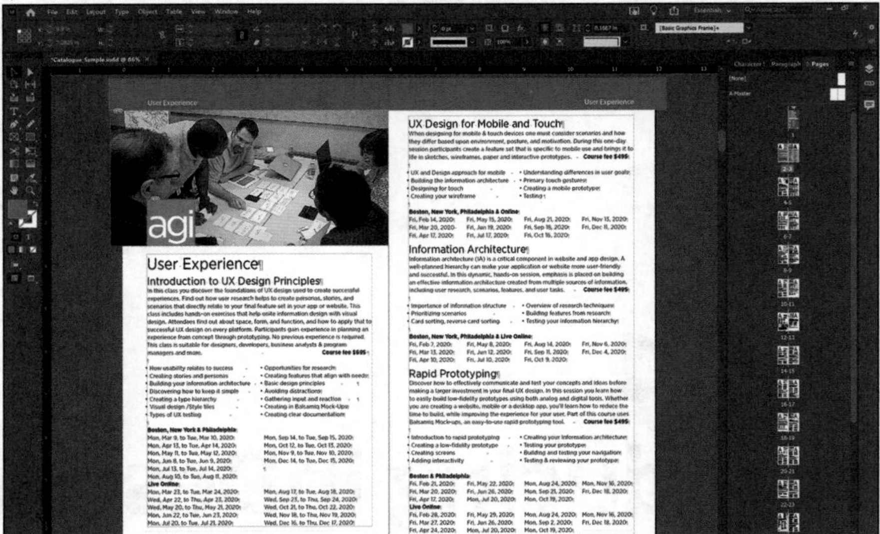

FIGURE 1-1: A sample page layout created using InDesign CC.

In the following sections, you get familiar with creating and opening documents in InDesign. In Chapters 2 through 8 in this minibook, you see how to add various elements to your pages.

Creating a new publication

After you launch InDesign, you can create a new InDesign document. Just follow these steps to create a new publication. Keep in mind that there are several methods that you can use to create a new document, and each may appear differently based upon you preferences. In this book the default preferences are referenced:

1. **In order to make sure that you have a consistent experience click the Home button in the upper-left of the InDesign workspace.**
2. **Click on the Create New button on the left side.**

   The New Document dialog box opens, as shown in Figure 1-2.
3. **Select whether you're designing a print, web, or mobile document from the header Intent menu. In this example Print is selected.**

   Depending upon your selection you are provided with common presets from which to choose from. You can also create your own custom sizes as well.

   Note: You can also access the New Document dialog box by choosing File ➪ New ➪ Document, or pressing Ctrl+N (Windows) or Command+ N (Mac).

TIP

If you want to preview your adjustments in this panel, select the Preview option at the bottom of the New Document window. Your updates in the steps that follow will adjust the appearance of the document in InDesign while you make your settings.

4. **Change the Units to a preferred measurement unit. For this example, Inches is selected. This can also be changed while working on your document.**

5. **For this example, confirm the Facing Pages checkbox has been selected to have the pages arranged as spreads with left and right pages.**

   With this option selected, pages in your document are arranged in pairs, so you have *spreads,* which are facing or adjacent pages in a layout. For example, you select this option if you're creating a publication that will be arranged like a book or magazine. If you deselect this option, pages are arranged individually, which is a good choice for a single-page flyer or a document with only a front and back side.

6. **Select a Blank Document Preset.**

   The Blank Document Preset should be set to the size of paper you intend to print on or the size at which the content will be displayed. You can also create your own custom sized document by typing values into the Width and Height text boxes.

   The Orientation setting changes from Portrait (tall) to Landscape (wide) based on the settings you enter in the Width and Height fields of the Page Size section.

TIP

The Intent selection you make determines the choices in the Blank Document Presets list. If you select Web for the Intent, you can choose from various screen resolutions, whereas if you select Mobile, you can select from various popular tablet sizes, such as the iPad, NOOK/Kindle Fire, or Android.

TIP

You can enter page sizes by typing the most common forms of measurement or just use the appropriate abbreviation in the size textboxes. For example, you enter **8 in** for 8 inches or **15 cm** for 15 centimeters. You can use most forms of measurement in all InDesign dialog boxes and panels; just make sure to specify the form of measurement you want to use. When creating web or mobile documents, these values switch to pixels.

7. **Choose a number for the columns on the page.**

   This step creates guides for columns that do not print or display in the completed project. These guides help you organize your pages as you create them. You can also enter a value in the Gutter field, which specifies the space between each of the columns. For more information about using columns in page layout, see Chapter 3 of this minibook. Again, this can be changed as needed when working on your document.

8. **Choose values for the page margins.**

   Notice the Make All Settings the Same button, which is a chain icon, to the right of the four text fields where you enter margin values. Toggle this button to set all margins to the same value, or set different values (broken chain icon).

   If you see Top, Bottom, Inside, and Outside, you're specifying margins for a page layout that has facing pages, which you specified earlier. If you see Top, Bottom, Left, and Right, you're creating a page layout without facing pages. The *inside* margins refer to the margins at the middle of the spread, and the *outside* margins refer to the outer left and right margins of a book or magazine. You can set the Inside setting to accommodate the binding of a book, which may need wider margins than the outside.

TIP

   If you use the same settings repeatedly, saving those settings as a preset is a good idea. Get your settings the way you want them, and then click the Save Document Preset button, located to the right of the document title and under the Preset Details header before you click Create. Enter a name for the preset, and then click Save Preset. After you save your preset, select Saved from the header menu (refer to the top of Figure 1-2), and you can select your saved preset whenever you create a new document.

9. **When you're finished, click Create.**

   After you click Create in the New Document dialog box, the new document is created with the settings you just specified.

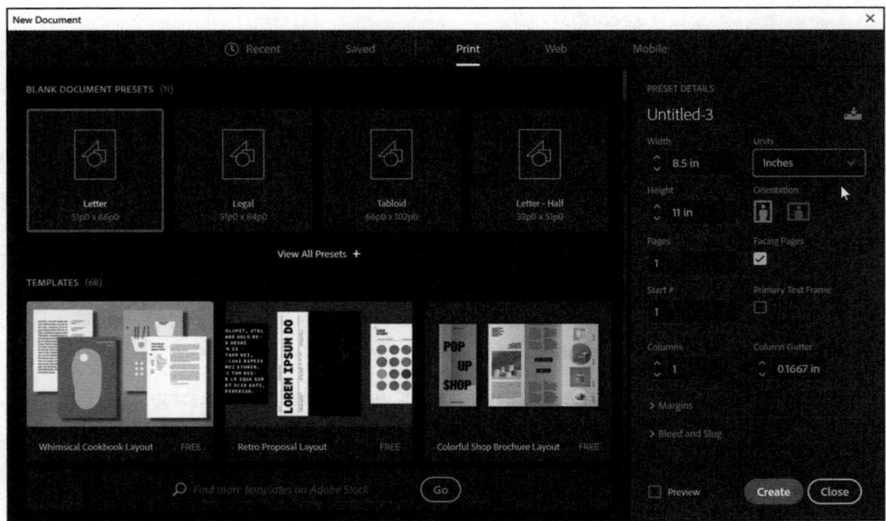

FIGURE 1-2: Setting up a new document with InDesign.

Margins, columns, orientation, and page size are discussed in more detail in Chapter 3 of this minibook.

Opening an existing publication

You may have InDesign files on your hard drive that you created or have saved from another source. To open existing InDesign documents (files that end with `.indd`), follow these steps:

1. **Choose File⇨Open.**

   The Open a File dialog box appears.

2. **Browse your hard drive and select a file to open.**

   If you want to use our example, choose the `SampleCatalog` document from the Book04_InDesign folder that you downloaded in Book 1, Chapter 1.

   Select a file by clicking the document's title. To select more than one document, press Ctrl (⌘ on the Mac) while you click the filenames.

3. **Click the Open button to open the file.**

   The file opens in the workspace.

Touring the Workspace

Just like the other applications in the Creative Cloud, InDesign has a standardized layout. Using panels that can be docked and a single-row Tools panel, you can keep much more space open in your work area.

The InDesign workspace, or *user interface*, includes a large number of tools and panels — but most users use only a few. You'll likely use several panels over and over again, so you should keep them easily accessible. In the default user workspace, many of these panels are already docked to the right. Figure 1-3 shows how the InDesign workspace layout looks when you open a new document. The Windows workspace is nearly identical to the Macintosh version of InDesign. If your workspace is not similar you can reset the defaults by selecting Window⇨Workspace⇨Essentials, or, if that is already active, choose Reset Essentials from further down in that same menu.

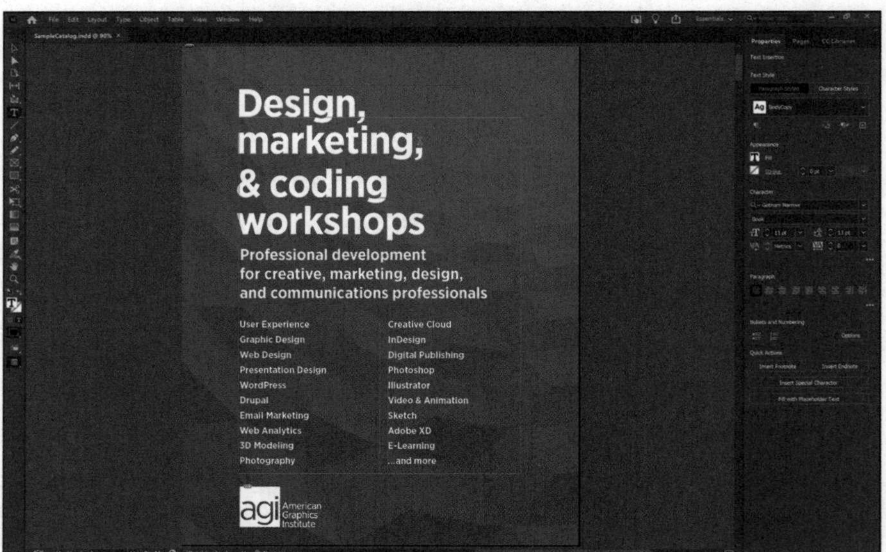

FIGURE 1-3: The InDesign default workspace.

Here are the elements that create the InDesign workspace:

- **Page:** The main area of the InDesign workspace is a page. It's the area that's printed or exported when you finish making a layout.
- **Master pages:** You can define how certain text elements and graphics appear in an entire document (or just portions of it) by using a master page. It's much like a template for your document because you can reuse elements throughout the pages. For example, if you have an element you want on each page (such as page numbering), you can create it on the master page. If you need to change an element on the master page, you can change it at any time, and your changes are reflected on every page that the master page is applied to. You access master pages in the Pages panel. (You find out more about master pages in Chapter 3 of this minibook.)
- **Spread:** A *spread* refers to a set of two or more pages that will be printed side by side. You usually see spreads in magazines and books when you open them — just like the book you're holding now. If your document has only a single page — front and back, or with only one side — you will not see a spread in the InDesign document window. InDesign will display only the one page, or, for a two-sided page, both pages (if you reduce the magnification).

» **Pasteboard:** The pasteboard is the area around the edge of a page. You can use the pasteboard to store objects until you're ready to put them into your layout. Pasteboards aren't shared between pages or spreads. For example, if you have certain elements placed on a pasteboard for pages 4 and 5, you can't access these elements when you're working on pages 8 and 9 — so each page or spread has its own pasteboard.

Tools

The Tools panel (See Figure 1-4) is where you find tools to edit, manipulate, or select elements in your document, as well as tools to make adjustments to pages. Simply use the cursor and click a tool to select it. See Figure 1-4 for the default Tools panel layout.

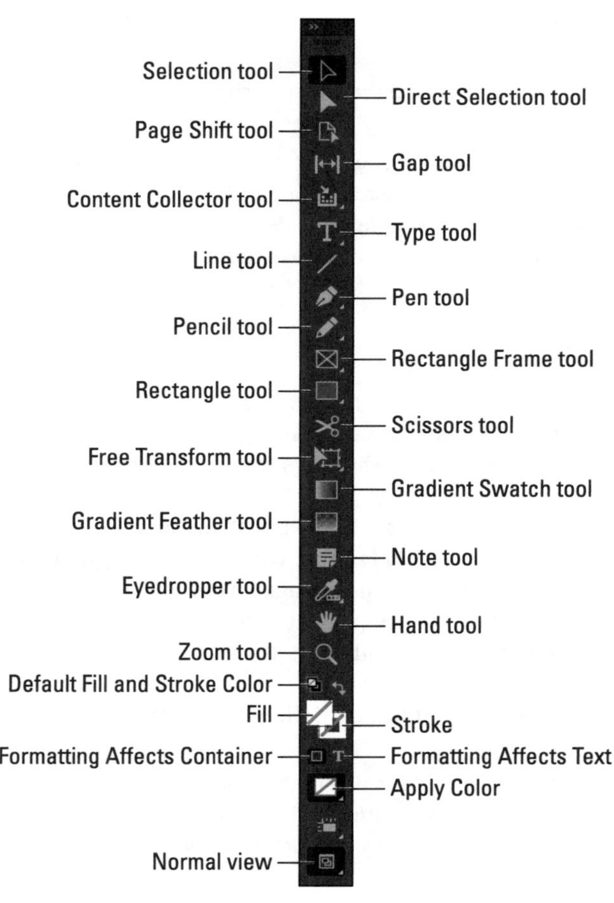

FIGURE 1-4: The Tools panel contains tools for creating, selecting, and editing elements and adjusting pages.

TIP

If you decide that a single row of tools just isn't for you, you can go back to an older version's Tools panel by clicking the two arrows in the gray bar at the top of the Tools panel. If you want to relocate the tools, click the bar at the top of the tools, under the double arrows, and drag the tools panel to a new location.

You find out more about these tools and how to use them in the related chapters of this minibook. For example, we discuss the drawing tools in Chapter 4 of this minibook.

With the tools in the Tools panel, you can do the following:

- » Create stunning new content on a page using drawing, frame, and text tools.
- » Select existing content on a page to move or edit.
- » View the page in different ways by moving (panning) and magnifying the page or spread.
- » Edit existing objects, such as shapes, lines, and text. Use the Selection tool to select existing objects so that you can change them.

When a tool has a small arrow next to the button's icon, more tools are hiding behind it. When you click the tool, and hold down the mouse button, a menu opens that shows you other available tools. While pressing the mouse button, move the cursor to the tool you want and release the mouse button after it's highlighted.

Menus

The menus on the main menu bar are used to access some of the main commands and control the user interface of InDesign. They also allow you to open and close panels used to edit and make settings for the publication.

InDesign menu commands such as New, Open, and Save are similar to most other applications you're probably familiar with. The InDesign menus also include commands that are especially used for page layout, such as Fill with Placeholder Text. For more information on using menus, see Book 1, Chapter 4. Remember to refer to the common commands and shortcuts that are also detailed in that chapter.

The InDesign main menu has the following options:

- » **File:** This menu includes some of the basic commands to create, open, and save documents. It also includes the Place command to import new content and many options to control document settings, exporting documents, and printing.

- **Edit:** You can access many commands for editing and controlling selections in this menu — such as copying and keyboard shortcuts. The dictionary and spell checker are on this menu, too.
- **Layout:** Use this menu to create guides. These options help you lay elements on the page accurately and properly align them. Use the menu to navigate the document's pages and spreads.
- **Type:** From this menu, you can select fonts and control characters in the layout. You can access the many settings related to text from this menu, which opens the associated panel where you make the changes.
- **Object:** You can modify the look and placement of objects on the page with this menu. Which options are available on this menu depends on which element you've selected in the workspace, such as a text frame or an image.
- **Table:** Use this menu to create, set up, modify, and control tables on the page.
- **View:** You can modify the view of the page from this menu, including zooming in and out, as well as work with guides, rulers, or grids to help you lay out elements.
- **Window:** Use this menu to open and close panels or switch between open documents.
- **Help:** This menu is where you can access the Help documents for InDesign.

Panels

In the default layout, you see a large area for the document. To the right of the document are several *panels* that snap to the edge of the workspace — panels that are attached to the edge of the workspace are considered *docked*. Panels are used to control the publication and edit elements on pages. Panels can be maximized, minimized, moved around, or closed altogether. As a default, you see the Properties panel. This panel provides you helpful information about whatever you have selected in the document.

To expand a panel, you can simply click the panel name, and it expands. The panels you expand are automatically collapsed again when a different panel is selected.

If you'd rather work with all panels expanded, simply click the left-facing double arrows on the gray bar above the panels. You can collapse all the panels again by clicking the right-facing double arrows on the gray bar above the expanded panels.

Even though some InDesign panels perform different functions, similar panels are grouped together depending on what they're used for. You can change the groupings by clicking and dragging a panel's tab into another grouping.

Some panels change when you're manipulating specific types of content on an InDesign page. Throughout Chapters 2 through 8 of this minibook, you discover these specific panels as you create layouts. For now, we briefly show you two general InDesign panels: Control and Pages.

Control panel

As a default, the Control panel, located across the top of the document window, is not visible as many of the Control panel options are included in the Properties panel off to the right. The Control panel can be made visible by selecting Window⇨Control. Just like the Properties panel, the Control panel is used to edit elements you have selected in InDesign, as shown for the Type tool in Figure 1-5. This panel is *context sensitive*, so it changes depending on which element you've selected on a page. For example, if you have selected text within a frame on the page, the Control panel displays options allowing you to edit the text. If you have a shape selected, the panel displays options allowing you to modify the shape.

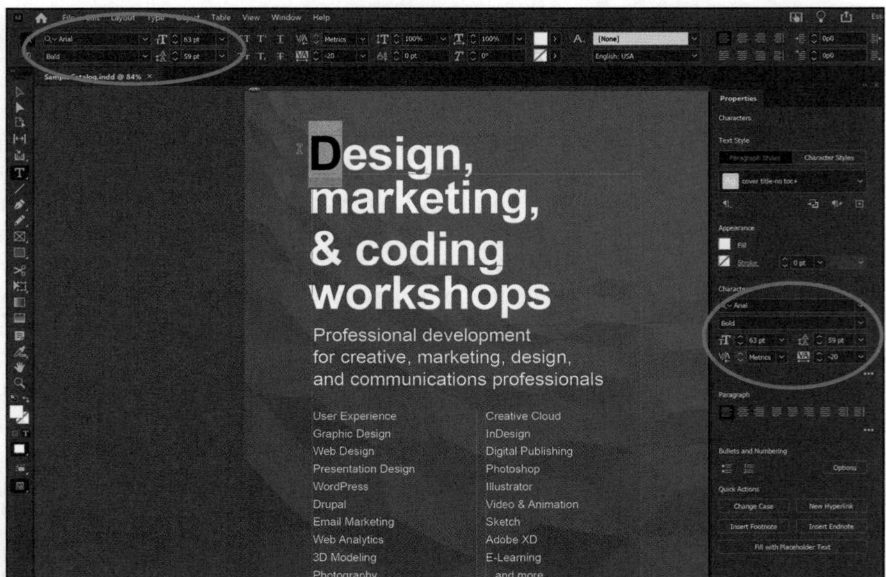

FIGURE 1-5: The Control and Properties panel, when the Type tool is active.

Figure 1-6 shows the Control and Properties panel when a frame is selected using the Selection tool. Note that you have controls that not only can change the frame, but also the text within when you use the Properties panel. Which method you use to make changes is up to you, but if you are a new user making a habit out of using the Properties panel would probably benefit you more.

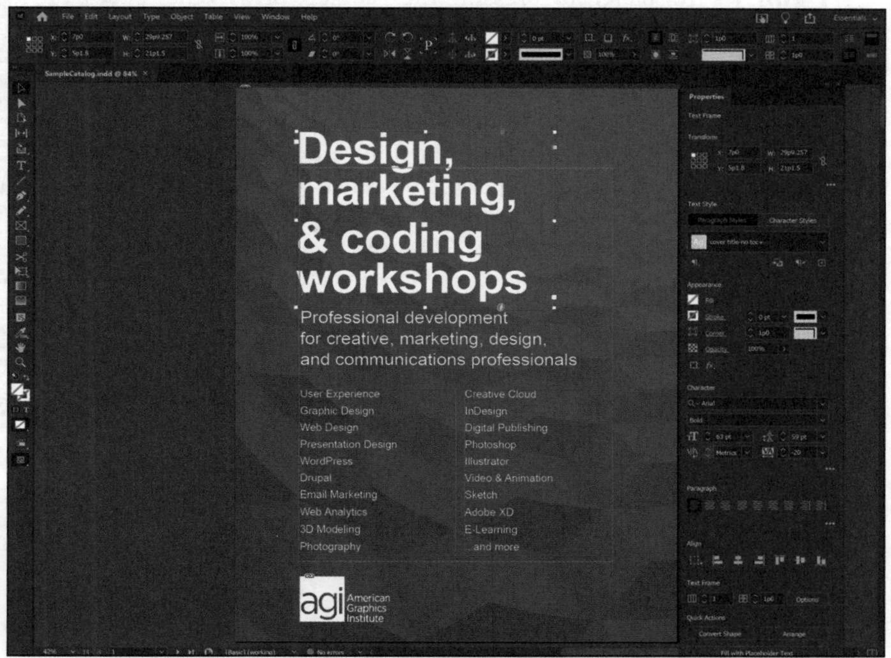

FIGURE 1-6: The Control panel, when a frame is selected using the Selection tool.

Pages panel

If you do not see the Pages panel, choose Window⇨Pages. You can control pages by using the Pages panel, as shown in Figure 1-7. This panel allows you to arrange, add, and delete pages in your document. If you are creating electronic documents, you can also use the Pages panel for creating alternative layouts for vertical and horizontal displays on a tablet. You can also navigate among pages with this panel, which we discuss further in Chapter 3 of this minibook.

TIP

You can also add and delete pages by choosing Layout➪Pages, and even use a keyboard shortcut to add pages, Ctrl+Shift+P (Windows) or ⌘+Shift+P (Mac).

REMEMBER

You can hide all open panels (including the Control panel) by pressing the Tab key; press Tab for them to return to view. In InDesign CC, you can leave tools and panels hidden and access them when you want by moving the cursor to the left or right side of the work area. When you hover your cursor over the thin vertical gray bar on either side, the tools or panels (depending on which side of the workspace you're in) reappear! By the way, they go away again after you leave the area.

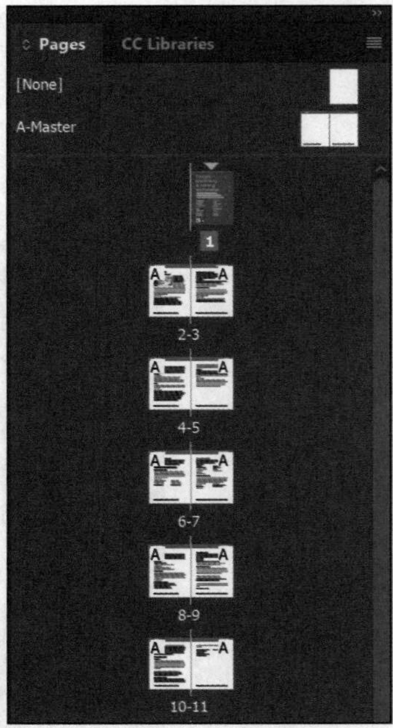

FIGURE 1-7: Use the Pages panel to add, delete, and move pages, as well as create alternative layouts for tablet displays.

You can navigate the document's pages by using the left- and right-arrow buttons on either side of the page number in the lower-left corner of the document window. You can also move to a specific page by entering a page number into the page field and pressing Enter/Return or by selecting the page from the drop-down list in the lower-left corner of the Document window.

Contextual menus

Contextual menus (or context menus) are menus that pop up when you right-click (Windows) or Control-click (Mac) the mouse. Contextual menus change depending on which element you click and which tool you're using. If you have no elements selected, the contextual menu opens for the overall InDesign document, allowing you to select options such as Zoom, Paste, Rulers, and Guides. If you have an element selected, your options include transforming, modifying, and editing the object.

REMEMBER

Contextual menus are context sensitive (hence the name!). Remember to select an element on the page before you right-click (Windows) or Control-click (Mac) to open the contextual menu. If you don't select the object first, the menu that displays is for the document instead of for an object.

You can find out more about editing and transforming elements in Chapters 2 and 3 of this minibook.

Setting Up the Workspace

Workspace settings are important because they help you quickly create the type of layout you need. Overall document settings control elements such as grids or guides that help you align elements on the page. Neither guides nor grids print when you print or publish your document.

Showing and hiding grids and guides

Grids and guides are onscreen lines that help you with your layout but, by default, don't appear when printing. A *document grid* is applied across the entire document page area. Use the document grid when you need to divide a document into sections to achieve your intended design. You can have objects on a page align to the document grid, which helps you accurately line up or space objects on your page.

Another type of grid is the *baseline grid*, which runs horizontally across the page. Use the baseline grid to make sure that text in different columns is aligned, thus creating a cleaner page layout.

REMEMBER

The *document grid* is used for aligning elements on the page, and the *baseline grid* is used for aligning the bottom of text across multiple columns.

- To show or hide the document grid, choose View⇨Grids & Guides⇨Show (or Hide) Document Grid.
- To show or hide the baseline grid, choose View⇨Grids & Guides⇨Show (or Hide) Baseline Grid. If you don't see the baseline grid, you may be zoomed out below the view threshold of 75%.

You can immediately see the difference between these two kinds of grids.

Guides can be placed anywhere on the page (or pasteboard) and are used to accurately position objects in a layout. Unlike grids, guides are typically created individually. Use them to align specific objects — such as the tops of several images — that appear across a page. Objects can snap to guides just like they can snap to a grid.

To create a guide and show or hide guides, follow these steps:

1. **Make sure that rulers are visible by choosing View⇨Show Rulers, or press Ctrl+ R (Windows) or Command+R (Mac).**

   Rulers appear in the workspace. If you already have rulers visible, the option View⇨Hide Rulers is on the View menu. Don't hide the rulers.

2. **Move the cursor to a horizontal or vertical ruler.**

   Make sure that the cursor is over a ruler.

3. **Click the ruler and drag the mouse toward the page.**

   A *ruler guide* shows on the page as a line.

4. **Release the mouse where you want the guide.**

   You just created a ruler guide!

5. **To hide the guide, choose View⇨Grids & Guides⇨Hide Guides.**

   This step hides the guide you created but doesn't delete it. You can make the guide reappear easily in the next step.

6. **To see the guide again, choose View⇨Grids & Guides⇨Show Guides.**

   The guide you created is shown on the page again.

You can edit the color of the ruler guide you created by positioning the mouse over it, clicking once to select it, and then right-clicking (Windows) or Control-clicking it (Mac) and selecting a new color from the Ruler Guides option.

You can find out more about the different kinds of guides and how to use them in page layout in Chapter 3 of this minibook.

You can also control the color of the guides and grid in your preferences. Access them by choosing Edit⇨Preferences⇨Grids (Windows) or InDesign⇨Preferences⇨Grids (Mac). You may need to scroll down through the Edit drop-down menu to find Preferences. When the Preferences dialog box opens, you can change the color and spacing of the lines. Click Guides & Pasteboards in the list on the left to change the color settings for guides. To change the color of your ruler guides, go to Layout⇨Ruler Guides, and then click the color drop-down menu to select a different color, or click the color box and select a custom color from the Color Picker. This changes the color of all future ruler guides.

Note: Change the color of your guides when no other guides are selected. Otherwise, all guides will change color.

Snapping to a grid or a guide

You can have elements on the page snap to a grid or a guide. Grid or guide snapping is useful so that you don't have to eyeball the alignment of several elements to one another, because they're precisely aligned to a grid or guide. To make sure that this setting is enabled, choose View⇨Grids & Guides⇨Snap to Document Grid or View⇨Grids & Guides⇨Snap to Guides. If these options are already selected, clicking them will turn them off.

Using Smart Guides

Give yourself an added hand when aligning objects on the InDesign page with Smart Guides. Illustrator and Photoshop users may be familiar with these interactive guides, but if you're not, read on to discover how you can take advantage of them.

You can experiment with this capability by creating two objects in an InDesign document. It doesn't matter which object or shape — any will do! You can also use page 2 and 3 in the SampleCatalog file we provided in the Book04_InDesign folder.

With the Selection tool, click and slowly drag one object in a circular motion around the other. You'll notice guides appear and disappear, indicating when the objects are aligned on the top, center, or bottom of the other object, as shown in Figure 1-8.

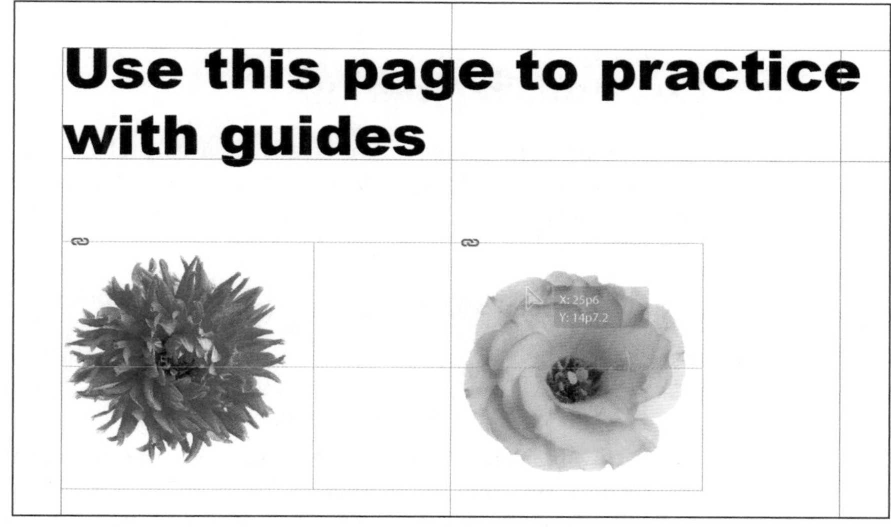

FIGURE 1-8: Click and drag one shape around another to see the interaction with Smart Guides.

As a default, pink guides appear when you align an object with the center of the page, as shown in Figure 1-9.

TIP

You can see a print preview of your document by clicking the Preview Mode button at the bottom of the Tools panel. Click and hold the Normal button to access the Preview Mode. When you click this button, all object bounding boxes, guides, and the grid disappear.

FIGURE 1-9: Know when your object is at the exact center of the page when a guide crosshair appears.

Customizing menus

InDesign has many menus and many choices under each menu. It's likely you'll need to use only some of these menu choices. If you discover that you're wading through many menu items to find the items you need, you can customize InDesign to hide menu items you don't use.

To customize your InDesign menus, follow these steps:

1. **Choose Edit⇨Menus.**

   The Menu Customization window opens.

2. **In the Application Menu Command column, click the triangle next to the menu(s) you want to customize.**

   For example, if you never expect to import XML, which is located under the File menu, click the triangle to the left of the File menu to display this menu item.

3. **Click the eye icon next to the menu item that you want to hide.**

   Or if you want to highlight a menu item, click the color column to the right of the eye icon to add a color to the menu choice.

4. **Click the Save As button at the top of the window to save the menu customization set. Enter a name for your customized menu set, click OK to save the name, and then click OK again to close the Menu Customization window.**

   The customized menus are saved. To return to the original default menus, or to further customize the menus, choose Edit⇨Menus at any time you're using InDesign.

Customizing the interface appearance

InDesign lets you adjust the overall appearance of the workspace to customize the color theme. You can set exactly how dark or light you would like the panels and pasteboard to be, or select between several predefined settings.

To customize the InDesign interface appearance, follow these steps:

1. **Choose Edit⇨Preferences⇨Interface (Windows) or InDesign⇨Preferences⇨Interface (Mac).**

   The Preferences window opens, with the Interface category active. The Appearance section at the top of this window is where you adjust settings that affect the user interface appearance.

2. **In the Appearance section, adjust the color theme for the user interface and pasteboard.**

   The default setting is Medium Dark, and you can switch between each setting and InDesign will preview your selection. Alternatively, you can enter a custom percentage setting to fine-tune the brightness setting.

3. **Select or deselect Match Pasteboard to Theme Color.**

   The option is selected by default, which means the pasteboard (the area around your document) is a darker shade. Deselect this option if you prefer a white pasteboard.

4. **Click OK to apply your changes.**

   The workspace appearance updates.

Saving a custom workspace

You've seen that InDesign has a number of panels. If you find that you're using some panels more than others, you can have InDesign remember the grouping of panels you use most frequently, including which ones are visible and which ones are hidden. InDesign calls this a *workspace*. The next time you want a certain group of panels open together, you can return to the workspace you previously saved.

The workspace isn't attached to a particular document, so you can have one workspace for editing text and another for working with a layout.

To save a custom workspace, follow these steps:

1. **Have the InDesign workspace configured in the way you want to save it — with any panels open that you might want to access together.**

   The open panels will be saved as a custom workspace.

2. **Choose Window⇨Workspace⇨New Workspace.**

   The New Workspace dialog box opens.

3. **Choose whether to save the panel locations or any menus that have been customized. Type a new name for the workspace in the Name text field.**

   Enter a name that reflects the type of work you do in that workspace, such as text editing or layout.

4. **Click OK.**

   The custom workspace is saved. The name you entered for your workspace is displayed on the Workspaces menu.

To access your workspace, choose Window⇨Workspace⇨*Your Workspace* (where *Your Workspace* is the name you gave the workspace in Step 3).

You can delete the workspace if you no longer want it saved. Simply choose Window⇨Workspace⇨Delete Workspace.

Working with Documents

After you're comfortable getting around the InDesign workspace, you're ready to begin working with a new document. After you've started working on a document, you should find out how to import content from other programs and to save that document on your hard drive. A lot of the content you use when creating layouts with InDesign is imported from other programs — both Creative Cloud applications like Photoshop and Illustrator, and non-Creative Cloud applications like Microsoft Word and Excel. You use InDesign to organize, modify, and integrate text and graphics into a layout. To begin, we show you the steps needed to import content and save new files.

We show you how to open new and existing documents earlier in this chapter, in the sections "Creating a new publication" and "Opening an existing publication."

TECHNICAL STUFF

You may also be working with a *template*, which is a layout you reuse by applying it to a document that requires a particular predesigned format. For example, a company may use a template for its official letterhead because every new letter requires the same page format and design. InDesign templates use the `.indt` file extension.

Importing new content

You can use many different kinds of content in an InDesign document because you can import many supported file types. You can import text, formatted tables, and graphics that help you create an effective layout. This capability makes integration with many different programs easy.

Follow these steps to import an image file into InDesign (in this example, we import a bitmap graphic file):

1. **Choose File➪New➪Document.**

   The New Document dialog box appears.

2. **Review the settings, make any changes depending upon the size and type of document you want to create, and then click Create.**

   Feel free to alter the settings to change the number of pages, page size, and intent before clicking Create. Afterwards, a new document opens.

3. **Choose File➪Place.**

   The Place dialog box opens, enabling you to browse the contents of your hard drive for supported files. If you were to select the Show Import Options checkbox, another dialog box opens before the file imports. Leave this option deselected for now.

4. **Click the file you want to import, and then click Open.**

   Certain files, such as bitmap photo, graphic, and PDF files, show a thumbnail preview at the bottom or to the right of the dialog box.

   When you click Open, the Place dialog box closes, and the cursor becomes an upside-down L with a thumbnail of the image you are placing.

5. **Click the location on the page where you want the upper-left corner of the imported file (for example, an image) to appear.**

   The imported file is placed on the page.

TIP

Click and drag to place the file into a specific frame size, or if you've created an empty frame on the page, clicking on inside the frame causes the object being imported — whether it's text or an image — to be placed inside the frame.

TIP

You can Ctrl-click (Windows) or ⌘-click (Mac) to place multiple files. After you select the images and click OK, each click places an image on the page, or you can hold down the Shift+Ctrl (Windows) or Shift+⌘ (Mac) while dragging a rectangle to have all selected images placed, spaced evenly, in a grid.

Note that when you're placing multiple images, you can see a thumbnail of each image before it's placed. You can also scroll through the loaded images by pressing the arrow keys on your keyboard.

For general information about importing and exporting using the Adobe Creative Cloud, check out Book 1, Chapter 5. For more information on importing different kinds of file formats, such as text, images, spreadsheets, and PDFs, see Chapters 2 and 3 in this minibook.

Viewing content

You can view elements in several different ways on your document's pages. For example, sometimes you need to see objects on a page close up so that you can make precise edits. InDesign offers several ways to navigate documents:

- **Scroll bars:** You can use the scroll bars to move pages around. The scroll bars are located below and to the right of the pasteboard. Click a scroll bar handle and drag it left and right or up and down.
- **Zoom:** Zoom in or out from the document to increase or decrease the display of your document. Select the Zoom tool (the magnifying glass icon) from the Tools panel and click anywhere on the page to zoom in. Press the Alt (Windows) or Option (Mac) key and click to zoom out.
- **Hand tool:** Use the Hand tool to move the page around. This tool is perhaps the best and quickest way to move pages around and navigate documents. Select the Hand tool by pressing the spacebar (when using any tool other than the Type tool), and then click and drag to move around the pasteboard.
- **Keyboard:** Press Ctrl++ (plus sign) (Windows) or ⌘ ++ (plus sign) (Mac) to zoom in using the keyboard; replace the plus sign with the minus sign to zoom out.

Saving your publication

Even the best computers and applications fail from time to time, so you don't want to lose your hard work unnecessarily. Saving a publication often is important so that you don't lose any work if your computer or software crashes or the power goes out.

To save a file, choose File⇨Save or press Ctrl+S (Windows) or ⌘ +S (Mac).

You may also want to save different versions of your files. You may want to do this if you're experimenting with different design options, for example, and you want to retain earlier versions of your files. To do this, use the Save As command, which makes creating different versions of documents easy.

REMEMBER

Choose File⇨Save before proceeding if you want the current document to save the revisions you've made since you last saved the file. All new additions to the document are made in the new version of the file.

To save a new version of the current document and then continue working on the new document, follow these steps:

1. **Choose File⇨Save As.**

   The Save As dialog box opens.

2. **Choose the directory you want to save the file in.**

3. **In the File Name text field, enter a new name for the document.**

   This step saves a new version of the file. Consider a naming scheme at this point. If your file is `myLayout.indd`, you might call it `myLayout02.indd` to signify the second version of the file. Future files can then increase the number for each new version.

4. **Click the Save button when you're finished.**

   This step saves the document in the chosen directory with a new name.

The File⇨Save As command is also used for other means. You may want to save your design as a template. After you create the template, choose File⇨Save As, and then choose InDesign CC Template from the Save As Type (Windows) or Format (Mac) drop-down list.

You can also choose File⇨Save a Copy. This command saves with a new name a copy of the current state of the document you're working on, but you then continue working on the original document. Both commands are useful for saving incremental versions of a project you're working on.

To find out more about working with files, go to Chapter 7 of this minibook.

a publication

- » Adding and importing text
- » Exploring text frame options
- » Changing paragraph settings
- » Editing with text editors and spell checking
- » Working with tables
- » Creating and editing text on a path

Chapter 2
Working with Text and Text Frames

Most of the documents you create contain text, so it's important to know how to format, style, and control text in your layouts. Text is made up of characters, and the characters are styled in specific fonts.

This chapter explains how to create, edit, and style text by using InDesign. You get started by editing and manipulating text placed inside *text frames* — containers on the page that hold text content. The most important concepts you can take away from this chapter are how to add text to documents and then change the text so that it looks the way you want on the page. In Chapter 3 of this minibook, find out how to create effective layouts that contain both text and graphics so that your audience is encouraged to read everything you create.

Understanding Text, Font, and Frames

Text is usually integral to a publication because it contains specific information you want or need to convey to an audience. Understanding the terminology that appears in the following pages is important. *Text* and *font* are quite different from each other:

- **Text:** The letters, words, sentences, and paragraphs making up content in the text frames in your publication.
- **Font:** The particular design forming a set of characters used to style text. You can find thousands of styles of fonts from many manufacturers, and many are included in programs you install on your computer. InDesign's Font menu lets you preview font faces and even set favorite fonts.

Frames resemble containers that are used to hold content. You can use two kinds of frames in a publication:

- **Text:** Contains text on the page in your InDesign document. You can link text frames so that text flows from one text frame to another, and you can have text wrap around graphic frames.
- **Graphic:** Holds an image that you place in your publication.

When you create frames using InDesign, they can contain either text or graphics — so the methods for creating both types of frames are identical. InDesign automatically changes frames to adapt to content, so you can use both the frame and shape tools for designing your layout and creating frames that will contain text or graphics.

Creating and Using Text Frames

Text frames contain any text you add to a publication. You can create a new text frame in many different ways. In InDesign, you can add text to creative shapes you draw, thereby changing them into text frames. Creating and using text frames in a publication is important because you typically use a lot of text. Throughout the following subsections, we show you how to create text frames in different, but important, ways using three different tools. If you need a guide to the tools, check out Chapter 1 of this minibook.

TIP

Text frames are sometimes automatically created when you import text into a publication. You find out how to do this in the "Importing text" section, later in this chapter.

Creating text frames with the Type tool

You can use the Type tool to create a text frame. If you use the Type tool and click the page, nothing happens unless you've first created a frame to hold the text. Here's how to create a text frame by using the Type tool:

1. **Select the Type tool in the Tools panel and place the tool over the page.**

   The Type tool cursor is an I-bar. Move the cursor to the spot where you want to place the upper-left corner of the text frame.

2. **Click and drag diagonally to create a text frame.**

   When you click, the mouse has a crosslike appearance. When you drag, an outline of the text frame appears, giving you a reference to its dimensions, as shown in Figure 2-1.

3. **Release the mouse button when the frame is the correct size.**

   The text frame is created, and an insertion point is placed in the upper-left corner of the frame. You can start typing on the keyboard to enter text or import text from another source. (See the later section, "Importing text.")

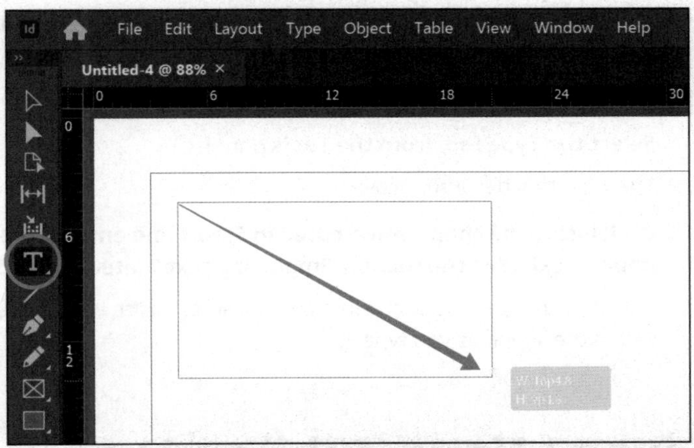

FIGURE 2-1: Using the Type tool, drag to create a text frame.

Working with Text and Text Frames

Creating text frames with the Frame tool

You can use the Frame tool to create frames that are rectangular, oval, or polygonal. Then, after you've placed the frame on the page, you can turn it into a text frame or use it as a graphic frame or simply a design object on the page. To create a new text frame with the Frame tool, follow these steps:

1. **Choose the Rectangle Frame tool from the Tools panel and click and drag diagonally on your InDesign page to create a new frame.**

   A new frame is created on the page.

2. **Select the Type tool and click inside the frame.**

   The X across the frame disappears, and the frame is now a text frame instead of a graphic frame.

3. **Choose the Selection tool and use it to move the text frame.**

   You can move the text frame if you click within the frame using the Selection tool and drag it to a new location.

Creating text frames from a shape

If you have an interesting shape that you've created with the drawing tools or copied and pasted from Illustrator, you can easily change the shape into a text frame so that it can be filled with text. Just follow these steps:

1. **Use the Pen tool, Pencil tool, or a Shape tool to create a shape with a stroke color and no fill. Or, copy and paste artwork from Illustrator.**

   A shape is created on the page that doesn't have a solid color for the fill.

2. **Select the Type tool from the Tools panel.**

   The Type tool becomes active.

3. **Click inside the shape you created in Step 1 and enter some text or import text. (See the section "Importing text," later in this chapter.)**

   This step changes the shape into a text frame. Notice how the text is confined within the shape as you type.

Adding Text to Your Publication

In the preceding sections' step lists, you find out how to add text by simply clicking in the text frame and typing new content, but you can also add text to publications in other ways. Doing so is particularly useful when you use other applications to create and edit documents containing text.

Importing text

You can import text you've created or edited using other software, such as Adobe InCopy or Microsoft Word or Excel. Importing edited text is a typical workflow activity when creating a publication, because dedicated text-editing software is often used to edit manuscripts before layout.

To import text into InDesign, follow these steps:

1. **Make sure no other objects are selected and then choose File ⇨ Place.**

   The Place dialog box opens. Choose an importable file (such as a Word document, an InCopy story, or a plain text file) by browsing your hard drive.

2. **Select a document to import and click the Open button.**

   The Place Text icon, the cursor arrow, and a thumbnail image of the text appear. Move the cursor around the page to the spot where you want the upper-left corner of the text frame to be created when the document is imported.

3. **Click to place the imported text.**

   This step creates a text frame and imports the text.

TIP

If you select a text frame *before* importing text, the text is automatically placed inside the text frame — so, in this case, you wouldn't have to use the cursor to place the text. If you choose the Selection tool you can move the text frame anywhere on the page after the text is added or resize the frame, if necessary.

Controlling text flow

Control the flow of the text by using these simple modifier keys while placing text:

- Choose File ⇨ Place, select the file you want to import, and click Open. Hold down the Shift key and, when the loaded cursor turns into a curvy arrow, click the document. The text is imported and automatically flows from one column to another or from page to page until it runs out. InDesign even creates pages, if needed.
- Choose File ⇨ Place, select the file you want to import, and click Open. Hold down the Alt (Windows) or Option (Mac) key. Then click and drag a text area. (Don't release the Alt key or Option key!) As you continue clicking and dragging additional text frames, your text flows from one text frame to another until you run out of copy.

TIP

If you select the Show Import Options checkbox in the Place dialog box, a second window appears in which you can choose to remove styles and formatting from text and tables. This action brings in clean, unformatted text to edit. On a Mac, select the Options button to access the Show Import Options checkbox.

Adding placeholder text

Suppose that you're creating a publication but the text you need to import into it isn't ready to import into InDesign — perhaps the text is still being created or edited. Rather than wait for the final text, you can use placeholder text and continue to create your publication's layout. *Placeholder text* is commonly used to temporarily fill a document with text. The text looks a lot like normal blocks of text, which is more natural than trying to paste the same few words repeatedly to fill up a text frame. However, placeholder text isn't in any particular language, because it's just being used as filler.

InDesign can add placeholder text into a text frame automatically. Here's how:

1. **Create a frame on the page by selecting the Type tool, clicking on the page and dragging diagonally to create a text frame.**
2. **Choose Type⇨Fill with Placeholder Text.**

   The text frame is automatically filled with characters and words, similar to the one shown in Figure 2-2.

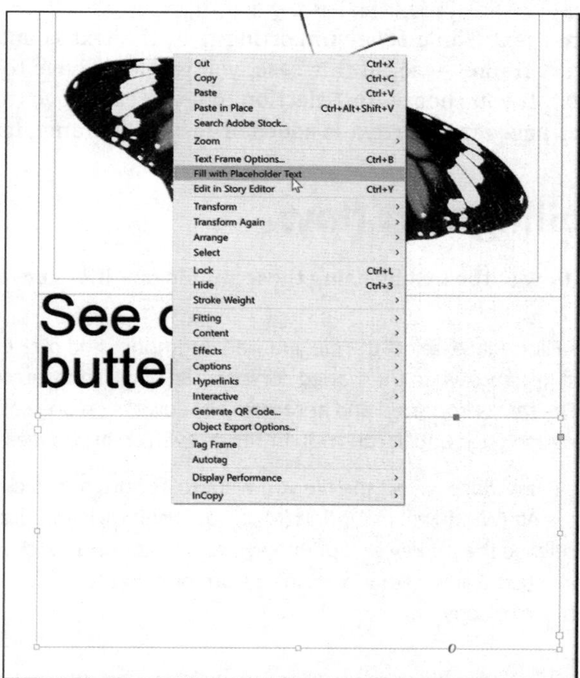

FIGURE 2-2: The text frame, filled with placeholder text.

Copying and pasting text

You can move text from another application into an InDesign publication by copying and pasting the text directly into InDesign. If you select and copy text in another program, you can paste it directly into InDesign from your computer's Clipboard. Here's how:

1. **Highlight the text you want to use in your publication and press Ctrl+C (Windows) or ⌘ +C (Mac) to copy the text.**

   When you copy text, it sits on the Clipboard (until it's replaced by something new), and you can transfer this information into InDesign.

2. **Open InDesign and press Ctrl+V (Windows) or ⌘ +V (Mac) to create a new text frame and paste the text into it.**

   A new text frame appears centered on the page with your selected text inside it.

TIP

You can also click in a text frame and press Ctrl+V (Windows) or ⌘ +V (Mac) to paste text from the Clipboard directly into an existing frame. You can do the same thing with an image.

All you need to do is double-click a text frame if you want to access or edit some text or type or paste it into the frame.

TIP

You can control what formatting you paste into your file by adjusting InDesign preferences. Select Edit⇨Preferences⇨Clipboard Handling (Windows) or InDesign⇨Preferences⇨Clipboard Handling (Mac), and then choose either All Information (swatches, styles, and so forth) or Text only.

Looking at Text Frame Options

In the previous sections of this chapter, we show you how to create text frames and enter text into them. In the following sections, we show you how to organize text frames in your publication and achieve the results you need. Controlling text frames so that they do what you need them to do is a matter of knowing how they work after you put text in them.

InDesign gives you a lot of control over the text in your publications. Changing text frame options allows you to change the way text is placed inside a frame. Changing these kinds of settings is sometimes important when you're working with particular kinds of fonts.

The text frame contextual menu contains many options for working with the text frame. You use this menu to perform basic commands, such as copy and paste, fill the text frame with placeholder text, make transformations, add or modify strokes, and change the frame type. Access the text frame contextual menu by right-clicking (Windows) or Control-clicking (Mac) a text frame with the selection tool. You can also find most of these options on the Type and Object menus.

Changing text frame options

To change text frame options that control the look of the text within the frame, follow these steps:

1. **Create a rectangular text frame on the page, select the frame with the Selection tool, and choose Object⇨Text Frame Options.**

   You can also press Ctrl+B (Windows) or ⌘ +B (Mac) or use the text frame's contextual menu to open the Text Frame Options dialog box.

   You can tell that a text frame is selected when it has handles around its bounding box.

   The Text Frame Options dialog box appears, showing you the current settings for the selected text frame.

2. **Select the Preview checkbox to automatically view updates.**

   Now any changes you make in the dialog box are instantly updated on the page, so you can make changes and see how they'll look before you apply them.

3. **In the Inset Spacing area of the dialog box, change the Top, Bottom, Left, and Right values.**

   These values are used to inset text from the edges of the text frame. The text is pushed inside the frame edge by the value you set. With the Make All Settings Same toggle, the chain symbol, enabled, you only need to enter a value into one of the four text boxes.

   You can also indent text, which we discuss in the section "Indenting your text," later in this chapter. You can choose in this dialog box how to align the text vertically (Top, Center, Bottom, or Justify). You can align the text to the top or bottom of the text frame, center it vertically in the frame, or evenly space the lines in the frame from top to bottom (Justify).

4. **When you finish making changes in this dialog box, click OK.**

   The changes you made are applied to the text frame.

Using and modifying columns

You can specify that the document contains a certain number of columns on the page when you create a new publication. Using columns allows you to snap new text frames to columns so that they're properly spaced on the page. You can even modify the size of the *gutter*, which is the spacing between columns.

You can also create columns within a single text frame by using the Text Frame Options dialog box. You can add as many as 40 columns in a single text frame. If you already have text in a frame, it's automatically divided among the columns you add. You can choose from three types of columns when creating a layout with text using InDesign:

- Use **Fixed Number** when you know exactly how many columns you want to appear in a text frame.
- Use **Fixed Width** when you know the exact width of columns that will appear in a text frame. If the text frame becomes larger or smaller, the number of columns may increase or decrease.
- Use **Flexible Width** if you want the width of columns to vary depending upon the size of the text frame. With Flexible Width, InDesign adds or reduces the number of columns as needed depending upon the width of the text frame.

The following steps show you how to add columns to a text frame on a page:

1. **Create a rectangular text frame on the page.**

   Use the Text or Frame tool to create the text frame. You can create columns in text frames that are rectangular, oval, or even freehand shapes drawn on the page.

2. **After you create the text frame, the cursor is automatically placed inside of the frame, and you can now enter some text.**

   You can type some text, paste text copied from another document, or add placeholder text by choosing Type⇨Fill with Placeholder Text.

3. **With the text frame still selected, choose Object⇨Text Frame Options.**

   In the Text Frame Options dialog box that opens, you may wish to select the Preview checkbox to immediately view the changes your settings make to the frame on the page.

4. **In the Columns section, change the value in the Number text field.**

   In this example, we entered **2** in the Number text field. The selected text frame divides the text in the frame into two columns. If you selected the Preview checkbox, when you click in a different text field in the dialog box, the text frame updates on the page to reflect the new value setting.

5. **Change the width of the columns by entering a new value in the Width text field.**

   The width of the columns is automatically set, depending on the width of the text frame you created. We entered **10** (picas) in the Width text field for this example. The text frame changes size depending on the width you set in this column.

6. **Change the value in the Gutter text field.**

   The gutter value controls the amount of space between columns. If the gutter is too wide, change the value in the Gutter text field to a lower number. We entered **0p5** in the Gutter text field for this example to change the gutter width to half a point.

7. **When you finish, click OK to apply the changes.**

   The changes are applied to the text frame you modified.

After you create columns in a text frame, you can resize the frame by using the handles on its bounding box, detailed in the later section, "Resizing and moving the text frame." The columns resize as necessary to divide the text frame into the number of columns you specified in the Text Frame Options dialog box. If you select the Fixed Column Width option from the drop-down menu, your text frames are always the width you specify, no matter how you resize the text frame. When you resize the text frame, the frame snaps to the designated fixed width.

You can also change the number of columns in the Control panel after selecting the text frame with the Selection tool, or by using the paragraph options in the Control panel while using the Type tool.

Modifying and Connecting Text Frames on a Page

Making modifications to text frames and then connecting them to other text frames in a publication so that the story can continue on a separate page is vital in most publications. You typically work with stories of many paragraphs that need to continue on different pages in the document.

When you have a text frame on the page, you need to be able to change the size, position, and linking of the frame. You need to thread (link) the frame to other frames on the page so that the text can flow between them — which is important if you're creating a layout that contains a lot of text.

If you paste more text content than is visible in the text frame, the text still exists beyond the boundaries of the text frame — so if you have a text frame that's 20 lines tall but you paste in 50 lines of text, the last 30 lines are cropped off. You need to resize the text frame or have the text flow to another frame in order to see the rest of the text you pasted. You can tell that the frame has more content when you see a small red plus sign (+) in a special handle in the lower-right of the text frame.

Resizing and moving the text frame

When creating layouts, you regularly resize text frames and move them around the document while you figure out how you want the page layout to look. You can resize and move a text frame by following these steps:

1. **Use the Selection tool to select a text frame on the page.**

   A bounding box with handles appears on the page. If the text frame has more text than it can show at the current size, a small red box with a red plus sign appears in the lower-right corner of the bounding box.

2. **Drag one of the handles to resize the text frame.**

   The frame updates automatically on the page while you drag the handles, as shown in Figure 2-3. Change the width or height by dragging the handles at the center of each side of the frame, or change the height and the width at the same time by dragging a corner handle. Note that in this example there is a red plus sign in the lower-right of the frame, indicating that the frame needs to be larger to fit the current text.

   Shift-drag a corner handle to scale the text frame proportionally.

FIGURE 2-3: Resize a text frame by dragging its handles.

3. **When you're finished resizing the text frame, click in the middle of a selected frame and move it around the page.**

   If you click within the frame once and drag it, you move the frame around the page. An outline of the frame follows the cursor and represents the spot where the frame is placed if you release the mouse button. Simply release the frame when you finish moving it.

REMEMBER

If you're using guides or grids on the page, the text frame snaps to them. Also, if you opened a document with columns, the text frame snaps to the columns when you drag the frame close to the column guidelines.

You can also use the Transform panel to change the location and dimensions of a text frame. If the Transform panel isn't already open, choose Window⇨Object and Layout⇨Transform to open it. Make sure the text frame is selected, then follow these steps:

1. **Change the values in the X and Y text fields.**

   Enter **1** in both the X and Y text fields to move the text frame to the upper-left corner of the page.

   The X and Y coordinates (location) of the text frame update to 1,1. The small square in the middle or along the edge of the text is the *reference point* of the text frame: The X and Y coordinates you set match the position of this point.

   Change the reference point by clicking any square in the reference point indicator in the upper-left corner of the control panel.

2. **Change the values in the W and H text fields.**

   For this example, we entered **35** (picas) in the W and H text fields. The text frame's width and height change to the dimensions you specify. Using the Transform panel to change the width and height is ideal if you need to set an exact measurement for the frame.

TIP

You not only can resize and move text frames, but also can change their shapes. Select a text frame and choose the Direct Selection tool from the Tools panel. You can then select the corners on the text frame and move them to reshape the text frame. Note that if the image is selected when you select the Direct Selection tool, you'll need to click off and then click on the frame to be able to reshape it.

Threading text frames

Understanding how to thread text frames together is important if you plan to build page layouts with a lot of text. *Threading* occurs when text frames are arranged so that the text in one frame continues in a second text frame. Threading is useful for most layouts because you can't always include all text in a single frame.

First, take a look at some of the related terminology because Adobe has given some special names to text frames that are linked. Figure 2-4 shows some of the concepts and icons we refer to in the following list:

- **Flowing:** Describes text starting in one frame and continuing in a second frame.
- **Threading:** Describes two text frames that have text flowing from the first frame to the second.
- **Story:** The name of a group of sentences and paragraphs in threaded text frames.
- **In port:** An icon on the upper-left side of a text frame's bounding box indicating that a frame is the first one in a story or has text flowing in from another frame. An In port icon has a story flowing into it if it contains a small arrow; otherwise, the In port icon is empty.
- **Out port:** An icon on the lower-right side of the text frame's bounding box indicating that a frame has text flowing out of it. The Out port icon contains a small arrow if the frame is threaded to another frame; an empty Out port icon signifies that the frame isn't connected to another text frame.

  If a text frame isn't connected to another frame and has *overset text* (more text than can be displayed in a text frame), the Out port shows a small red plus sign (+) icon.

Find a block of text that you want to thread (for best results, use one that has formed sentences as opposed to placeholder text), and then follow these steps:

1. **Copy some text on the Clipboard, such as from the InDesign Help files, a page loaded in a web browser window, or a Word, Notepad, or SimpleText document.**

   The type of content you're pasting doesn't matter. You only need to make sure the text is at least a few paragraphs long so that you have enough text to flow between frames.

   In Figure 2-4, you can see the text thread represented by a line connecting one text frame to another. InDesign shows you text threads if you choose View➪Extras➪Show Text Threads.

2. **Use the Type tool to create two text frames on a page.**

   The text frames can be above or beside one another, similar to the layout shown in Figure 2-5.

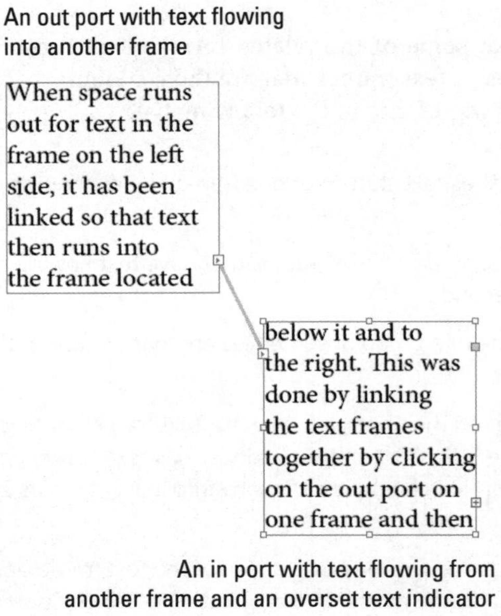

An out port with text flowing into another frame

When space runs out for text in the frame on the left side, it has been linked so that text then runs into the frame located

below it and to the right. This was done by linking the text frames together by clicking on the out port on one frame and then

An in port with text flowing from another frame and an overset text indicator

FIGURE 2-4: Threaded text between two linked text frames.

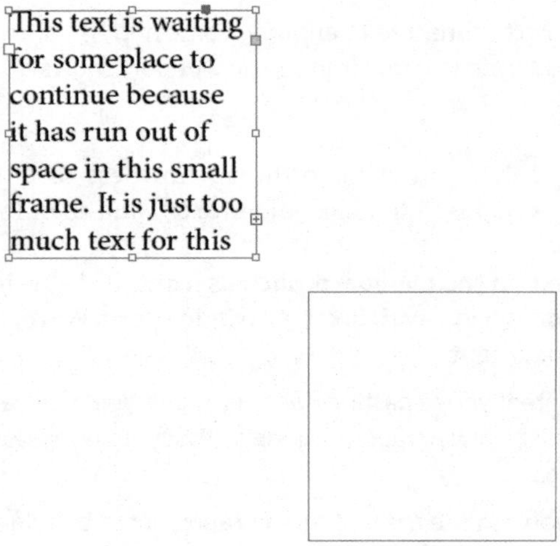

This text is waiting for someplace to continue because it has run out of space in this small frame. It is just too much text for this

FIGURE 2-5: Two frames on the page; the first contains text.

3. **Using the Text tool, click in the first text frame, which is above or to the left of the second text frame.**

   The blinking insertion point that appears in the first text frame lets you know that you can enter or paste text into the frame.

4. **Press Ctrl+V (Windows) or ⌘ +V (Mac) to paste the text into the text frame.**

   The text you've copied to the Clipboard enters into the frame. If you've pasted enough text, you see the overset text icon (a red plus sign) on the lower-right side of the text frame. (Refer to Figure 2-5.) If you don't see the overset text icon, use the Paste command a second time so that more text is entered into the frame.

5. **Click the overset text icon with the Selection tool.**

   The cursor changes to the loaded text icon so that you can select or create another text frame to thread the story.

6. **Move the cursor over the second text frame and click.**

   The cursor changes to the thread text icon when it hovers over the second text frame. When you click the second text frame, the two frames are threaded because the text continues in the second frame.

You can continue creating frames and threading them. You can thread them on the same page or on subsequent pages.

You can *unthread* text as well, which means that you're breaking the link between two text frames. You can rearrange the frames used to thread text, such as changing the page the story continues on when it's threaded to a second text frame. Break the connection by double-clicking the In port icon or the Out port icon of the text frame that you want to unthread. The frame is then unthreaded (but no text is deleted).

TIP

If your document doesn't have multiple pages in it, choose Layout ⇨ Pages ⇨ Insert Pages. Indicate the number of pages to add, and then click OK when you're finished. Now you can click through the pages using the Page Field control at the bottom of the workspace and create a new text frame into which you can link your text.

Adding a page jump number

If your document has multiple pages, you can add a *page jump number* (text that indicates where the story continues if it jumps to a text frame on another page) to an existing file. Before you start, make sure that a story threads between text frames on two different pages, and then follow these steps:

1. **Create a new text frame on the first page and type** continued on page.
2. **Use the Selection tool to select the text frame you just created.**

3. **Move the text frame so that it slightly overlaps the text frame containing the story (if needed, repeat Steps 1 and 4 from the preceding section, "Threading text frames," and create a new text frame with overset text).**

4. **Let InDesign know what text frame it's tracking the story from or to. Overlap the two text frames (and keep them overlapped), as shown in Figure 2-6, so that InDesign knows to associate these text frames (the continued-notice text frame and the story text frame) with each other.**

   TIP

   You can then *group* these two text frames (so that they move together). To do so, select both frames (Shift-click with the Selection tool to select both text frames), then choose Object⇨Group.

5. **Double-click the new text frame, which contains the text *continued on page.* Place the cursor at the spot where you want the page number to be inserted. (This is called the insertion point.)**

   The page number is inserted at the insertion point. Make sure that a space appears after the preceding character, so they do not run together. In this example, you would need a space after the word *page*.

6. **Choose Type⇨Insert Special Character⇨Markers⇨Next Page Number.**

   A number is added into the text frame. The number is sensitive to the location of the next threaded text frame, so if you move the second text frame, the page number updates automatically. This example will look like "continued on page 2."

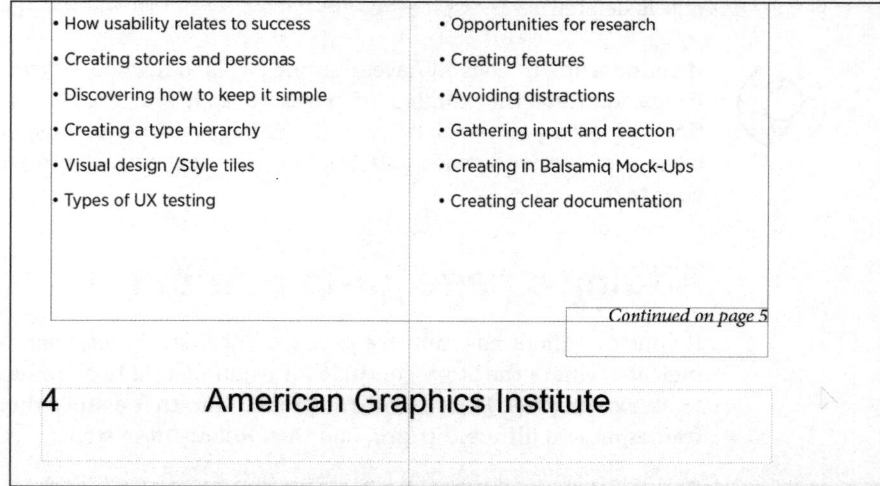

FIGURE 2-6: Slightly overlap the two text frames when creating a page jump, so that the page jump is associated with a specific story.

TIP

You can repeat these steps at the spot where the story is continued *from* — just choose Type➪Insert Special Character➪Markers➪Previous Page Number in Step 6 instead.

Understanding Paragraph Settings

You can change the settings for an entire text frame or a single paragraph in a text frame in several ways. You can use the Paragraph panel to make adjustments to a single paragraph or to an entire text frame's indentation, justification, and alignment. Open the Paragraph panel by choosing Window➪Type & Tables➪Paragraph.

TIP

If you want changes in the Paragraph panel to span across all text frames you create, don't select any paragraph or text frame before making the changes; instead, first select the entire text frame or frames on the page. Then the selections you make in the Paragraph panel affect all paragraphs in the selected text frames, not just one paragraph. If you want the selections you make in the Paragraph panel to affect just one paragraph within a text frame, select that paragraph first by using the Type tool, and then make your changes.

Indenting your text

You can indent a paragraph in a story by using the Paragraph panel. Indentation moves the paragraph away from the edges of the text frame's bounding box. Here's how to modify indentation:

1. **Create a text frame on the page and fill it with text.**

   You can fill the text frame by typing text, copying and pasting text, or inserting placeholder text by choosing Text➪Fill with Placeholder Text.

2. **Make sure that the insertion point is blinking in the text frame in the paragraph you want to change or use the Selection tool to select the text frame.**

3. **Choose Window➪Type & Tables➪Paragraph to open the Paragraph panel.**

   The Paragraph panel opens, showing the text frame's current settings. See Figure 2-7 to find out the name of each setting control.

TIP

4. **Change the value in the Left Indent text field and press Enter.**

   The larger the number, the greater the indent. You can specify the unit of measurement as you enter the text by entering *in* for inches or *pt* for points, using any forms of measurement InDesign supports.

5. **Change the value in the First Line Left Indent text field and press Enter.**

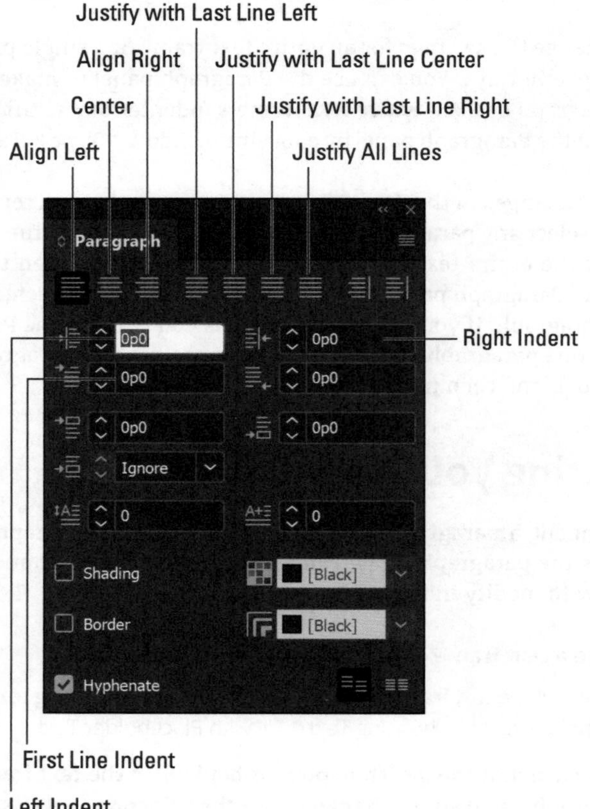

FIGURE 2-7: The Paragraph panel.

TIP

To change all paragraphs in a story, click the insertion point in a paragraph and choose Edit⇨Select All before changing settings.

Text alignment and justification

You can use the alignment and justification buttons in the Paragraph panel to format text frames:

» **Align** helps you left-, center-, or right-align text with the edges of the text frames.

» **Justification** lets you space text in relation to the edges of the text frame, and it lets you justify the final line of text in the paragraph.

To align or justify a block of text, select the text or the text frame and click one of the Align or Justify buttons. (Refer to Figure 2-7 to see the Align and Justify buttons in the Paragraph panel.)

Saving a paragraph style

Do you ever go to all the trouble of finding just the right indent, font, or spacing to use in your copy, just to find that you have to apply those attributes a hundred times to complete your project? Or, have you ever decided that the indent is too much? Wouldn't it be nice to change one indent text box and have it update all other occurrences? You can do this using paragraph styles in InDesign.

To create a paragraph style, follow these steps:

1. **Create a text frame, add text, and apply a first-line indent of any size you want.**

   Select some text — you don't have to select it all.

2. **Choose Window⇨Styles⇨Paragraph Styles.**

   The Paragraph Styles panel opens.

3. **From the Paragraph Styles panel menu, choose the new paragraph style icon on the bottom of the panel. It looks like a square with a plus sign inside.**

   Single-click the new style, Paragraph Style 1, and note that the Paragraph panel will show that every attribute is already recorded in this unnamed style.

   Double-click the new style, Paragraph Style 1, and the Paragraph Style Options dialog box opens. You don't have to do anything at this point other than name the style.

4. **Change the name from Paragraph Style 1 to something more appropriate, such as** BodyCopy, **and click OK.**

   Your style is created! After you click OK, the dialog box closes, and the new style is added to the Paragraph Styles panel list. You can modify the settings by double-clicking the style name in the Paragraph Styles panel. You can apply the style to other text frames by selecting the frame and clicking the style in the Paragraph Styles panel.

TECHNICAL STUFF

If you want to change an existing style, the Paragraph Style Options dialog box has several different areas in a large list on the left side. Select an item in the list to view and change its associated paragraph properties on the right side of the dialog box to update all instances of that paragraph style.

TIP

You can import paragraph styles from other documents or from a file on your hard drive, which is particularly useful when you need to use a particular set of styles for a template. To import paragraph styles, choose Load Paragraph Styles from the Paragraph Styles menu on the right side of the header. A dialog box prompts you to browse your hard drive for a file. Select the file to load, click Open, and then click OK to load the selected styles.

Editing Stories

Your documents likely contain all sorts of text, and some of that text may need to be edited. InDesign has a built-in story editor for editing text. This feature can be useful when it's inconvenient or impossible to open another text editor to make changes.

TIP

InDesign also integrates with another Adobe product: InCopy. It's a text editor that's similar to Microsoft Word but has integration capabilities with InDesign for streamlined page layout. If you work in IT or in editorial management and have some users who only write and others who only handle layout, you might want to have a look at InCopy as a possible text editor.

Using the story editor

The InDesign story editor lets you view a story outside tiny columns and format the text as necessary. To open the story editor to edit a piece of text, follow these steps:

1. **Find a piece of text that you want to edit and select the text frame with the Selection tool.**

   A bounding box with handles appears around the text frame.

2. **Choose Edit⇨Edit in Story Editor or use the keyboard shortcut Ctrl+Y (Windows) or ⌘+Y (Mac).**

   The story editor opens in a new window directly in the InDesign workspace.

3. **Edit the story in the window as necessary and click the Close button when you finish.**

   Your story appears in one block of text. Any paragraph styles you apply to the text in the story editor are noted in an Information pane on the left side of the workspace.

TIP

Notice in Figure 2-8 that you can see tables in the story editor. Click the small table icon to collapse and expand the table in the story editor.

help unite information design with visual design.
Attendees find out about space, form, and function, and
how to apply that to successful UX design on every
platform. Participants gain experience in planning an
experience from concept through prototyping. No previous
experience is required. This class is suitable for
designers, developers, business analysts & program
managers and more.
BodyCopy — Included in this course:
Bullet_2_column — • How usability relates to success • Opportunities for research
Bullet_2_column — • Creating stories and personas • Creating features
Bullet_2_column — • Discovering how to keep it simple • Avoiding distractions
Bullet_2_column — • Creating a type hierarchy • Gathering input and reaction
Bullet_2_column — • Visual design /Style tiles • Creating in Balsamiq Mock-Ups
Bullet_2_column — • Types of UX testing • Creating clear documentation
Row 1: This is a sample table
Row 2: January / February / March
Row 3: 10-11 / 13-14 / 12-13
Row 4
ClassName — UX Design for Mobile and Touch

Table icon

FIGURE 2-8: You can see text and tables in the story editor.

TIP

If you want to make text edits but would like to keep the original text unchanged, you can choose Type ➪ Track Changes ➪ Track Changes in Current Story. The original text remains in the story editor along with your marked changes, but in layout mode only the edited text appears. Later you (or an editor) can use the Track Changes command to accept or reject text edits using the Story Editor.

Checking for correct spelling

Typos and spelling errors are easy to make. Therefore, you must check for incorrect spelling in a document before you print or export it to a PDF. Here's how to check for spelling in InDesign:

1. **Choose Edit⇨Spelling⇨Check Spelling.**
2. **In the Check Spelling dialog box that appears, choose a selection to search from the Search drop-down list, and then click the Start button.**

   The spell check automatically starts searching the selection, story, or document.
3. **Choose from three options:**
   - Click the Skip button to ignore a misspelled word.
   - Select a suggested spelling correction from the list in the Suggested Corrections pane and click the Change button.
   - Click Ignore All to ignore any more instances of that word.

   The spelling is corrected in the text frame and moves to the next spelling error.
4. **Click the Done button to stop the spell check; otherwise, click OK when InDesign alerts you that the spell check is done.**

TIP

It is always a good practice to proofread your work manually, even when you use the spell check feature to adjust your writing. The spell checker won't reveal problems such as grammatical errors or word misuse.

Using custom spelling dictionaries

You can easily add words, such as proper nouns, to your dictionary by clicking the Add button.

You can create a user dictionary or add user dictionaries from previous InDesign versions, from files that other people have sent you, or from a server. The dictionary you add is used for all your InDesign documents.

Follow these steps to create your own, custom dictionary:

1. **Choose Edit⇨Preferences⇨Dictionary (Windows) or InDesign⇨Preferences⇨Dictionary (Mac).**

   The Preferences dialog box appears with the Dictionary section visible.

2. **From the Language drop-down list, select the language of your dictionary.**
3. **Click the New User Dictionary button below the Language drop-down list.**
4. **Specify the name and location of the user dictionary, and then click Save (Windows) or OK (Mac).**

TIP

If you want to see when a spelling error occurs without opening the Check Spelling dialog box, choose Edit⇨Spelling⇨Dynamic Spelling. Unknown words are then highlighted. To correct the spelling, right-click (Windows) or ⌘-click (Mac) and select the correct spelling from the contextual menu or add the word to your dictionary.

Using Tables

A *table* is made of columns and rows, which divide a table into cells. You see tables every day on television, in books and magazines, and all over the web. In fact, a calendar is a table: All the days in a month are shown down a column, every week is a row, and each day is a cell. You can use tables for many different tasks, such as listing products, employees, or events.

The following list describes the components of a table and how to modify them in InDesign:

- **Rows:** Extend horizontally across a table. You can modify the height of a row.
- **Columns:** Are vertical in a table. You can modify the width of a column.
- **Cells:** A text frame. You can enter information into this frame and format it like any other text frame in InDesign.

Creating tables

The easiest way to create a table is to have data ready to go. (Mind you, this isn't the only way.) But flowing in existing data is the most dynamic way of seeing what InDesign can do with tables.

Follow these steps to experiment with the table feature:

1. **Create a text area and insert tabbed copy into it.**

   The example uses dates for an event:

   Summer Events

June	July	August
1	2	3
4	5	6

   Notice that the text was simply keyed in by pressing the Tab key between every new entry and pressing the Return or Enter key for each new line. The text doesn't even need to be lined up.

2. **Select the text and choose Table⇨Convert Text to Table.**

   The Convert Text to Table Options dialog box appears. You can select columns in the window that appears or let the tabs in your text determine columns. You can find out more about table styles in the later section "Creating table styles."

   TIP

   You can assign a table style at the same time you convert text to a table.

3. **Click OK to accept the default settings and select the table.**

4. **Hold down the Shift key and use your mouse to click and grab the outside right border to stretch the table in or out.**

   The cells proportionally accommodate the new table size.

5. **Double-click and drag across the top three cells to highlight them, and then choose Table⇨Merge Cells to merge them.**

To create a new table without existing text, follow these steps:

1. **Create a new text frame with the Type tool.**

   The insertion point should be blinking in the new text frame you create. If it isn't blinking, or if you created a new frame another way, double-click the text frame so that the insertion point (I-bar) is active.

   Note that you don't have to have a text frame. You can choose Table⇨Create Table, and a text frame will be automatically created.

2. **Choose Table⇨Insert Table.**

3. **In the Insert Table dialog box that opens, enter the number of rows and columns you want to add to the table in the Rows and Columns text fields, and then click OK.**

   For example, we created a table with four rows and three columns.

Editing table settings

You can control many settings for tables. InDesign lets you change the text, fill, and stroke properties for each cell or for the table itself. Because of this flexibility, you can create fully customized tables to display information in an intuitive and creative way. In this section, we show you some basic options for editing tables.

To start editing table settings, follow these steps:

1. **Select the table you want to make changes to by clicking in a cell.**
2. **Choose Table⇨Table Options⇨Table Setup.**

   The Table Options dialog box opens with the Table Setup tab selected. The dialog box contains several tabs that contain settings you can change for different parts of the table.

   From the Table Setup tab, you edit the columns and rows, border, and spacing and specify how column or row strokes are rendered in relation to each other. For this example, we changed the number of rows and columns. Now we will change the table border weight to a 3-point stroke.
3. **Select the Preview checkbox at the bottom of the dialog box.**

   The Preview opens so that you can view the changes you make on the page while you're using the dialog box.
4. **Click the Row Strokes tab and change the options.**

   For this example, we selected Every Second Row from the Alternating Pattern drop-down list, changed the Weight option to 2, and changed the Color property for the first row to C=15, M=100, Y=100, K=0 (the CMYK equivalent of red).

   This step causes every second row to have a red, 2-point stroke. You can also click the Column Strokes tab to change the properties for column strokes. The two tabs work the same way.
5. **Click the Fills tab and change the options.**

   For this example, we selected Every Other Row from the Alternating Pattern drop-down list, changed the Color property to the same CMYK equivalent of red, and left the Tint at the default of 20 percent. This step changes the first row and every other row to a red tint.
6. **Click OK.**

   The changes you made in the Table Options dialog box are applied to the table.
7. **Click a table cell so that the insertion point is blinking.**

   The table cell is selected.
8. **Find an image you can copy to the Clipboard, and then press Ctrl+C (Windows) or ⌘ +C (Mac) to copy the image.**

9. **Return to InDesign and paste the image into the table cell by pressing Ctrl+V (Windows) or ⌘+V (Mac).**

   The image appears in the table cell, and the height or width (or both) of the cell changes based on the dimensions of the image. Make sure that the insertion point is active in the cell if you have problems pasting the image.

   Keep in mind that if you place a large image into a table cell that it will become overset and you won't be able to see the image. You can fix this issue by adjusting the size of the image in Photoshop before placing it or by clicking and dragging to create a frame for your image when you see the loaded cursor.

You not only can change the table itself, but also customize the cells within it. Choose Table⇨Cell Options⇨Text to open the Cell Options dialog box. You can also make changes to each cell by using the Paragraph panel. Similarly, you can change the number of rows and columns and their widths and heights from the Tables panel. Open the Tables panel by choosing Window⇨Type & Tables⇨Table.

InDesign lets you import tables from other programs, such as Excel. If you want to import a spreadsheet, choose File⇨Place. The spreadsheet is imported into InDesign as a table that you can further edit as necessary.

Creating table styles

If you've spent time customizing strokes, fills, and spacing for your table, you certainly want to save it as a style. Creating a table style lets you reuse your table setup for future tables. To create a table style, follow these steps:

1. **Make a table look the way you want.**

   The easiest way to create a table style is to complete the table setup and make a table look the way you want it at completion.

2. **Select the table.**

   Click and drag to select it with the text tool.

3. **Choose Window⇨Styles⇨Table Styles.**

   The Table Styles panel appears.

4. **Hold down the Alt (Windows) or Option (Mac) key and click the Create New Style button at the bottom of the Table Styles panel.**

   The New Table Style dialog box appears.

5. **Name the style and click OK.**

   Your table attributes are saved as a style.

If you want to edit table style attributes, you can simply double-click the named style in the Table Styles panel. (Make sure nothing is selected.)

Looking at Text on a Path

You can create some interesting effects with text on a path. Using the Type On a Path tool, you can have text curve along a line or shape. This feature is particularly useful when you want to create interesting titling effects on a page.

To create text on a path, follow these steps:

1. **Use the Pen tool to create a path on the page.**

   Create at least one curve on the path after you create it. (See Chapter 4 of this minibook to find out how to wield the Pen tool with confidence.)

2. **Click and hold the Type tool to select the Type On a Path tool.**
3. **Move the cursor near the path you created.**

   When you move the cursor near a path, a plus sign (+) symbol appears next to the cursor, and you can click and start typing on the path.

4. **Click when you see the + icon and type some text on the path.**

   An insertion point appears at the beginning of the path after you click, and you can then add text along the path. You select type on a path as you would normally select other text — by dragging over the text to highlight it.

To change properties for type on a path, select the text, and use the Type On a Path Options dialog box, which you open by choosing Type➪Type On a Path➪Options. In the Type On a Path Options dialog box, you can use effects to modify the way each character is placed on the path. You can also flip the text, change character spacing, and change character alignment to the path in the Align drop-down list or to the stroke of the path in the To Path drop-down list. Play with the settings to see how they affect your type. Click OK to apply changes; to undo anything you don't like, press Ctrl+Z (Windows) or ⌘ +Z (Mac).

TIP

To hide the path while keeping the text visible, set the stroke weight for the path to 0 pt.

IN THIS CHAPTER

» Working with and importing image files

» Selecting images on the page

» Knowing page layout settings

» Using text and graphics in your layouts

» Working with pages

» Using master pages and spreads

Chapter 3 Understanding Page Layout

This chapter shows you how to put graphics and text together so that you can start creating page layouts. Interesting and creative page layouts help draw attention to the pictures and words contained within the publication. An interesting layout motivates more members of the audience to read the text you place on a page.

Importing Images

You can add many kinds of image files to an InDesign document. Some of the most common are AI, PSD, PDF, JPG, PNG, and TIF. Images are imported into graphic frames. You can create the frames before importing, or if you don't have a frame, InDesign creates one for you instantly when you add the image to the page.

When you import an image into your InDesign document, the original image is still needed when you print or export the final document. Using the Link features you can keep track of the image and are notified if the original has been changed and needs to be updated. You also find additional settings at the time you import an image, which you access using the Image Import Options dialog box.

Follow these steps to import an image into your InDesign layout:

1. **Make sure that nothing on the page is selected.**

   If an object on the page is selected, click an empty area so that everything is deselected before you proceed.

2. **Choose File➪Place to place an image that you can access locally.**

   The Place dialog box opens, where you can browse your hard drive for image files to import. You can use this dialog box to import various kinds of files into InDesign, not just images.

3. **Select the image you want to import and click Open.**

   The Place dialog box closes, and the cursor displays a thumbnail of the image you selected.

   Note: You can also place files directly from CC Libraries. Choose File➪Place from CC Libraries. When the CC Libraries panel appears, you can select a saved library and then choose a file to place.

   TIP

   You can import multiple images at a time into an InDesign layout. Simply hold down the Ctrl (Windows) or ⌘ (Mac) key and select multiple files in the Place dialog box.

4. **Move the cursor to wherever you want the upper-left corner of the first image to be placed on the page, and then click the mouse.**

   If you've selected multiple images, you can use the left and right arrow keys to navigate the thumbnail images in your loaded cursor before clicking on the page. After you click on the page, the next image is placed, until there are no more images to place.

   Images are imported and placed into the publication inside a graphic frame. You can resize, move, and modify the image by using the Selection or Direct Selection tool or modify the frame and image together by using the Selection tool.

TIP

When placing multiple images, you can place all the selected images in one step — space them evenly in a grid by pressing the Shift+Ctrl (Windows) or Shift+⌘ (Mac) keys while dragging a rectangle using your mouse.

Don't worry if the image is imported and is too large for the layout or needs to be cropped. For more information about selecting graphic frames and modifying them, check out Chapter 4 of this minibook. To find out about importing and working with text and stories, see Chapter 2 of this minibook.

TIP

It's sometimes easier to create an empty graphic frame and then add an image to it than to import the image and create the frame at the same time. You can create an empty frame and even set fitting properties before you import an image — so that the image fits correctly at the time you import it. To set the fitting properties in a blank frame, choose Object➪Fitting➪Frame Fitting Options.

Importing PDFs

You can import PDF files to place them as images in InDesign layouts. When importing, you can preview and crop the pages by using the Place PDF dialog box. (Choose File➪Place, select the Show Import Options checkbox, and then click Open. On a Mac, you must click the Options button to be able to see the Show Import Options checkbox.) You import one page at a time, so you need to use the Forward and Back buttons displayed under the preview to select a page to place. Also, you can't import any video, sound, or buttons, and you can't edit the PDF after it's imported into InDesign — so it is more like importing a static image such as a JPEG file.

The Place PDF dialog box offers the following options:

- **Crop To:** You can crop the page you're importing using this drop-down list. Some options may be unavailable because they depend on what's in the PDF you're importing. The hatched outline in the preview shows you the crop marks.
- **Transparent Background:** Selecting this checkbox makes the PDF background transparent so that elements on the InDesign page show through. The PDF background is imported as solid white if this option isn't selected.

TIP

If your monitor is large enough for you to view your document window and folders on your computer, you can click and drag image files directly into your layout, bypassing the Place command altogether.

Importing other InDesign documents

You can place one InDesign document inside another. This feature might sound a bit odd, but it has many uses. For example, if you have a page from a book that you want to promote in a catalog, you can import an image of the book page to place into a catalog — all without converting the book page into some type of image format. This strategy not only removes a step, but also creates a higher-quality version of the image being placed into InDesign.

Here's how to take advantage of this feature:

1. **With a document open, choose File⇨Place or use the keyboard shortcut Ctrl+D (Windows) or ⌘+D (Mac).**

   The Place dialog box appears.

2. **Select the Show Import Options checkbox at the bottom of the Place dialog box.** On a Mac, you must click the Options button to be able to see the Show Import Options checkbox.)

3. **Navigate to an InDesign file and double-click to open it.**

   The Place InDesign Document dialog box appears, as shown in Figure 3-1, offering you the opportunity to choose which page or pages you want to place.

4. **Click OK.**

5. **Click the page to place the document.**

   If you're placing a document with multiple pages, click again to place each additional page.

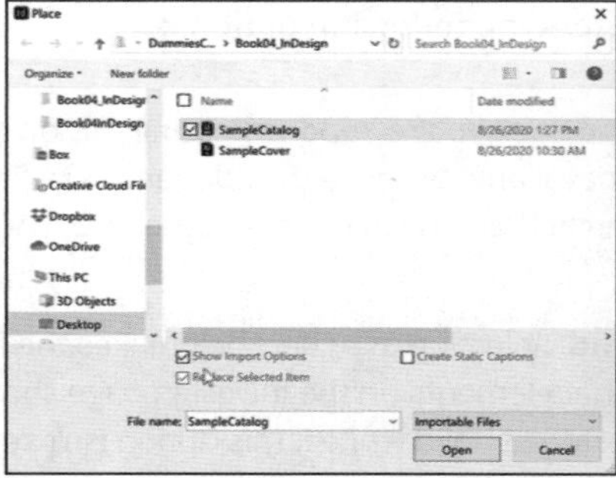

FIGURE 3-1: Choose pages to import in the Place InDesign Document dialog box.

Linking and Embedding Images

You can have images that you import either linked to your document or embedded in your document. Here's the difference between linking and embedding:

- **Linking:** The image that appears in the InDesign document is a preview of the image stored somewhere else on your computer or network. If the file you linked to your InDesign document is changed, it must be updated.

» **Embedding:** The image is copied into and saved within the InDesign document itself. It doesn't matter where the original file is located or whether you alter the file, because an embedded image is copied and saved directly within the InDesign document.

When you print or export a document, InDesign uses the linked images to generate the information necessary to create a high-quality printed document, a PDF file, or an image for posting on the web. InDesign keeps track of linked files and alerts you if any of them are moved or changed. You can update links for an image by selecting the image in the Links panel and choosing Update Link or Relink from the Links panel menu. You're prompted to find that file on your hard drive so that the file can be linked to the new location. And, if you send your InDesign file to someone else, make sure to also send its linked files along with the document. You can gain a better understanding of this process in Chapter 7 of this minibook.

REMEMBER

If you choose to embed images rather than link to images, your publication's file size increases because of the extra data that's being stored within it.

To find out which files are embedded or linked, look at the Links panel. (See Figure 3-2.) Choose Window⇨Links to open the panel and see whether any linked or embedded images are listed in the panel.

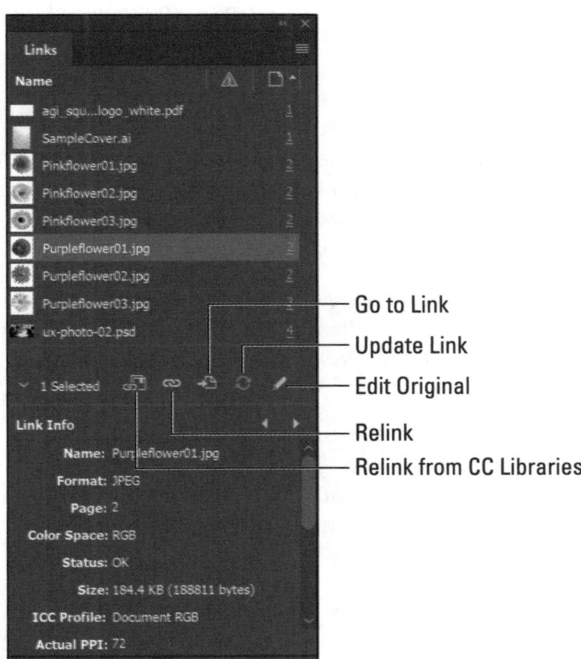

FIGURE 3-2: The Links panel keeps track of all images used in your documents.

You can embed a file by using the Links panel menu. Select the linked file from the list, then click the drop-down menu in the upper-right corner to access the menu. Select Embed Link if you want a linked file to be embedded within the document. Alternatively, choose Unembed Link from the Link panel menu to link a file rather than have it embedded in the document. We recommend linking to all images so that your files don't become too large and because it provides you with flexibility to manipulate the image files separately.

Setting Image Quality and Display

You can select quality settings that determine how images are displayed when they're part of an InDesign layout. These settings may help speed up your work if your computer is older or slower, or if you have many images. Displaying images at a higher resolution can give you a better idea of the finished print project and may avoid the need to print the project several times for proofing. These settings are applicable only to how you see the images while using InDesign to create a document — they don't impact the final printed or exported product.

To change the image display quality, choose Edit⇨Preferences⇨Display Performance (Windows) or InDesign⇨Preferences⇨Display Performance (Mac). You can then select one of the following settings from the Default View drop-down list:

- **Fast:** To optimize performance, the entire image or graphic is grayed out.
- **Typical (default):** This setting tends to make bitmaps look a little blocky, particularly if you zoom in. The speed of zooming in and out is increased if you select this option. InDesign uses a preview that it created (or that was already imported with the file) to display the image on the screen.
- **High Quality:** The original image is used to display onscreen. You can preview an accurate depiction of the final layout, but you may find that InDesign runs slowly when you use this option.

Notice the difference among these settings in Figure 3-3.

TIP

To change the display for an individual image, select the graphic frame and choose Object⇨Display Performance. Then choose one of the three options from the submenu.

FIGURE 3-3: From left to right: Fast Display, Typical Display, and High Quality Display.

Understanding Page Layout

Selecting Images

After you import an image into a document, you can select images in several different ways with the Selection or Direct Selection tools. It's useful to use the different methods depending upon whether you want to select and edit just the graphic frame or just the image inside it.

To select and then edit an image on the page, follow these steps:

1. **Place an image on a page by importing it into InDesign.**

   The image is placed within a graphic frame.

2. **With the Selection tool, drag one of the corner handles on the graphic frame in toward the center of the frame.**

   The graphic frame is resized, but the image isn't. The image appears to be cropped because you resized the graphic frame — though the image remains the same size within the frame, as shown in the center of Figure 3-4.

3. **Choose Edit⇨Undo or press Ctrl+Z (Windows) or ⌘+Z (Mac) to undo changes to the image.**

   The image returns to its original appearance on the page.

4. **Continue to use the Selection tool to click the center of the picture where a circle appears and move the picture within the frame.**

   The frame remains the same size, but its content is relocated.

FIGURE 3-4: The original image in a graphic frame (top); inside a resized frame (center); and repositioning an image within the frame (bottom).

5. **Switch to the Direct Selection tool, and then click within the image and drag to move the image within the graphic frame-bounding box.**

   A hand appears when you move the cursor over the graphic; when you move the image just past the edge of the graphic frame boundaries, that part of the image is no longer visible — it doesn't print and doesn't show when exported.

TIP

You can set the frame or image to resize by choosing Object⇨Fitting and selecting an option. You can even set this fitting before you place an image, which is especially helpful when creating templates. Resize before an image is placed by choosing Object⇨Fitting⇨Frame Fitting Options.

The Links panel can help you find images within documents, open images so that you can edit them, and view important information about selected images.

Manipulating Text and Graphics in a Layout

InDesign offers many tools that help you work with text and graphics together in a layout. From the tools in the Tools panel, to commands, to panel options, InDesign offers you an immense amount of control over the manipulation of graphics and text in a spread.

Page orientation and size

When you create a new document, you can set its page orientation and size. If you ever need to change your settings after you've created a document, choose File⇨Document Setup and change the following options, which affect all pages in your document:

- **Page Orientation:** Select either Landscape or Portrait. One of the first things you decide on when you create a new document is how to orient the pages. A *landscape* page is wider than it is tall; a *portrait* orientation is taller than it is wide.
- **Page Size:** Choose from many standardized preset sizes, such as Letter, Legal, and Tabloid. Alternatively, you can set a custom page size for the document. Make sure to properly set the page size so that it fits the kind of paper you need to print on, or screen size if you're publishing your work digitally.

TIP

You can also use the pages panel to adjust the size of individual pages and make some pages a different size. This topic is covered later in this chapter, in the section "Working with Pages and the Pages Panel."

Margins, columns, and gutters

Margins, columns, and gutters help divide a page for layout and confine its dimensions:

- **Margin:** The area between the edge of the page and the main printed area. Together, the four margins (top, bottom, left, and right) look like a rectangle around the page's perimeter. Margins don't print when you print or export the publication.
- **Column:** Divide a page into sections used for laying out text and graphics on a page. A page has at least one column when you start, which is between the margins. You can add column guides, represented by a pair of lines separated by a gutter area. Column guides aren't printed when you print or export the publication.
- **Gutter:** The space between two columns on the page. A gutter prevents columns from running together. You can define the gutter's width by choosing Layout ⇨ Margins and Columns; see Chapter 2 of this minibook for more information.

You can set margins and columns when you create a new document, which we discuss in Chapter 1 of this minibook. However, you can also modify margins and columns after the document has been created and specify different values for each page. You can modify the *gutter*, which is the width of the space between each column.

You can change margins and columns by setting new values in the Margins and Columns dialog box. Choose Layout ⇨ Margins and Columns, and then modify each individual page.

REMEMBER

Margins and columns are useful for placing and aligning elements on a page. These guides can have objects snap to them, enabling you to accurately align multiple objects on a page.

Using guides and snapping

Using guides when you're creating page layouts is a good idea because guides help you more precisely align elements on a page and position objects in the layout. Aligning objects by eyeballing them is difficult because you often can't tell whether an object is out of alignment by a small amount unless you zoom in to a large percentage.

REMEMBER

Make sure that snapping is enabled by choosing View➪Grids and Guides➪Snap to Guides. *Snapping* makes guides and grids useful. When you drag the object close to the grid, the object attaches to the guideline like it's a magnet. Aligning an object to a guide is easy after you create a guide, and you'll notice that InDesign displays temporary guides when you move an object near another object or near a guide.

Because guides are useful in creating a layout, check out the following kinds, available in InDesign (see Figure 3-5):

» **Column guides:** Evenly distribute a page into columns and can be used to align text frames in a document. These guides are set when you open a new document in InDesign that contains more than one column. Column guides can also be adjusted after a document is created by choosing Layout➪Margins and Columns.

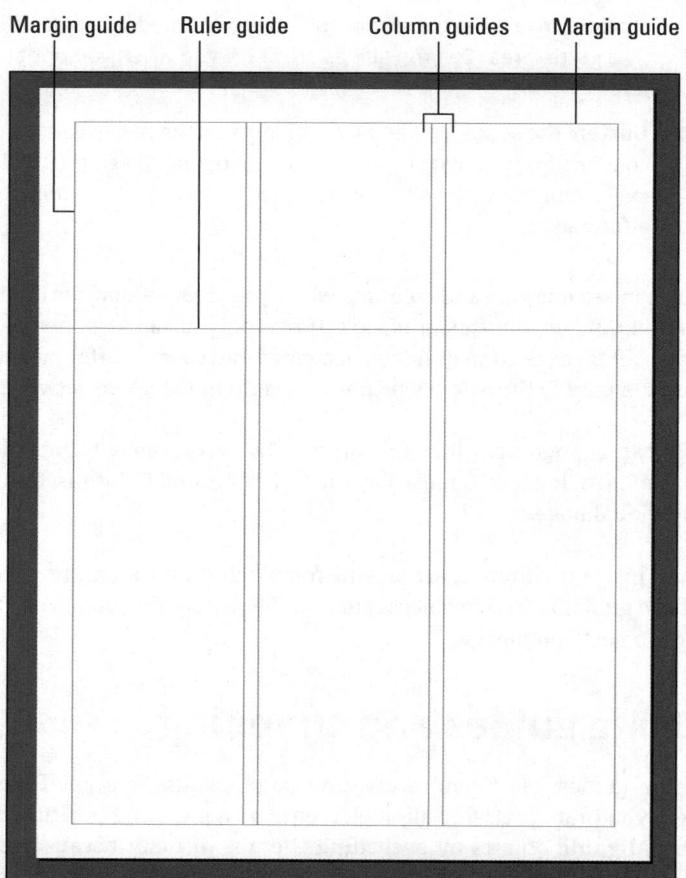

FIGURE 3-5: Column guides, margin guides, and ruler guides help create a layout.

- **Margin guides:** Define the area between the edge of the page and the main printable area. (We discuss these guides in the preceding section.)
- **Ruler guides:** Manually defined and can be used to align graphics, measure objects, or specify the location of a particular asset you want to lay out. See Chapter 1 of this minibook for details about adding ruler guides to the workspace.
- **Smart guides:** As previously mentioned, smart guides offer the capability to align objects on an InDesign page to other objects or even to the page. Smart Object alignment allows for easy snapping to page item centers or edges or page centers. In addition to snapping, smart guides give feedback to the user indicating the object to which you are snapping.
- **Liquid guides:** Liquid guides are useful when creating layouts that will be displayed on a variety of tablets. They are discussed more in Chapter 8 of this minibook.

To find out how to show and hide grids and guides, see Chapter 1 of this minibook.

TIP

With InDesign CC you can delete all guides at once by right-clicking the ruler (Windows) or Control-clicking the ruler (Mac) and selecting Delete All Guides on Spread from the contextual menu.

Locking objects and guides

You can lock in place elements such as objects and guides. This feature is particularly useful after you've carefully aligned elements on a page. Locking objects or guides prevents you from accidentally moving them from that position.

To lock an element, follow these steps:

1. **Use a drawing tool to create an object on a page, and then select the object with the Selection tool.**

   A bounding box with handles appears when the object is selected.

2. **Choose Object⇨Lock.**

   The object is locked in position. Now when you try to use the Selection or Direct Selection tools to move the object, it doesn't move from its current position.

To lock guides in place, follow these steps:

1. **Drag a couple of ruler guides to the page by clicking within a ruler and dragging toward the page.**

   A line appears on the page. (If rulers aren't visible around the pasteboard, choose View⇨Rulers.)

2. **Drag a ruler guide to a new location, if needed; when you're happy with the ruler guides' placements, choose View⇨Grids & Guides⇨Lock Guides.**

   All guides in the workspace are locked. If you try selecting a guide and moving it, the guide remains in its present position and cannot be moved. If you have any column guides on the page, they're locked as well.

TIP

Use layers in your publications for organizing different types of content, including guides. Layers are a lot like transparencies that lie on top of each other, so they can be used for stacking elements on a page. For example, you may want to stack graphics or arrange similar items (such as images or text) on the same layer. Each layer has its own bounding box color, which helps you determine which item is on which corresponding layer. For more information on layers in general, see Book 3, Chapter 8.

Merging Text and Graphics

When you have text and graphics together on a page, they should flow and work with each other to create an aesthetic layout. Luckily, you can work with text wrap to achieve a visual flow between text and graphics. In this section, you find out how to wrap text around images and graphics in your publications.

Wrapping objects with text

Images can have text wrapped around them, as shown in Figure 3-6. Wrapping is a typical feature of page layout in print and on the web. You can choose different text wrap options by using the Text Wrap panel, which you open by choosing Window⇨Text Wrap. Use the five buttons at the top of the panel to specify which kind of text wrapping to use for the selected object. Below the buttons are text fields where you can enter offset values for the text wrap. The fields are grayed out if the option isn't available.

You use the drop-down list at the bottom of the Text Wrap panel to choose from various contour options. The following list describes what happens when you click one of these buttons to wrap text around an object's shape:

- **No Text Wrap:** Use the default setting to remove any text wrapping from the selected object.
- **Wrap around Bounding Box:** Wrap text around all sides of the bounding box of the object.

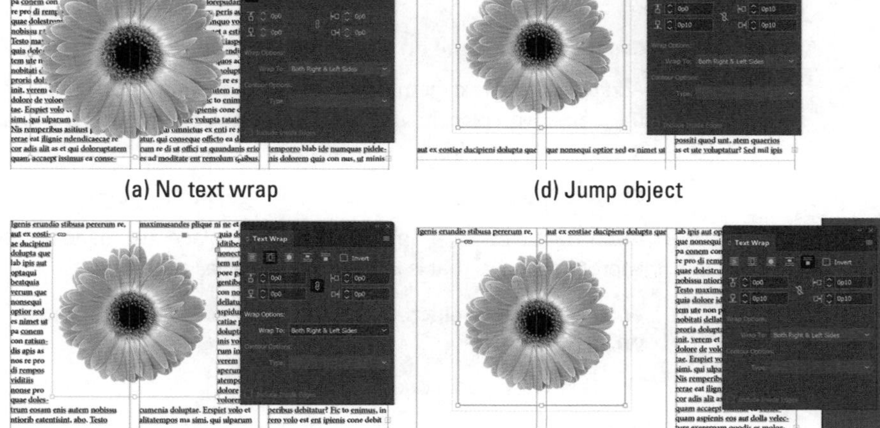

(a) No text wrap

(d) Jump object

(b) Wrap around bounding box

(e) Jump to next column

(c) Wrap around object shape

FIGURE 3-6: You can wrap text around images in InDesign in several ways.

- **Wrap around Object Shape:** Wrap text around the edges of an object.
- **Jump Object:** Make the text wrapping around the image jump from above the image to below it, with no text wrapping to the left or right of the object in the column.
- **Jump to Next Column:** Make text end above the image and then jump to the next column. No text is wrapped to the left or right of the image.
- **Offset:** Enter offset values for text wrapping on all sides of the object.
- **Wrap Options:** Select an option to determine on which sides of the object text will wrap.
- **Contour Options:** Select a contour from this drop-down list, which tells InDesign how the edges of the image are determined. You can choose from various vector paths or the edges to be detected around an object or image with transparency.

To add text wrapping to an object (a drawing or an image), follow these steps:

1. **Create a text frame on a page with a graphic.**

   Add text to the text frame by typing text, pasting text from elsewhere, or filling the frame with placeholder text. This text wraps around the image, so make sure that the text frame is slightly larger than the graphic frame you'll use.

2. **Choose File➪Place and place an image on this page.**

   If you want a good sample image you can place the image named LaptoGraphics.ai that is in the Book04_InDesign folder.

3. **With the image selected, choose Window➪Text Wrap to open the Text Wrap panel.**

   The Text Wrap panel opens.

4. **With the image still selected, click the Wrap around Object Shape button.**

   The text wraps around the image instead of hiding behind it.

5. **If you're working with an image that has a transparent background, select the Wrap Around Object Shape button to have InDesign find the edges for you.**

   The text wraps around the edges of the image, as in Figure 3-7. You can also push the text away from the image by increasing the Offset.

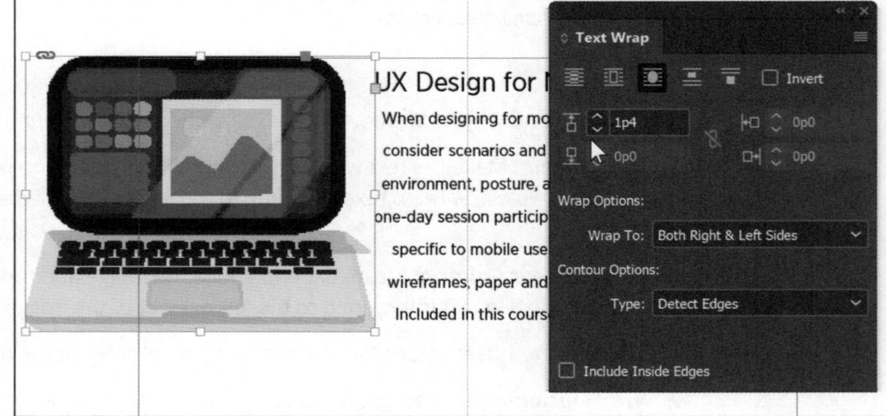

FIGURE 3-7: You can wrap text around an image with a transparent background.

Modifying a text wrap

If you've applied a text wrap around an object (as we show you how to do in the preceding section), you can then modify that text wrap. If you have an image with a transparent background around which you've wrapped text, InDesign created a path around the edge of the image; if you have a shape you created with the drawing tools, InDesign automatically uses those paths to wrap text around.

WARNING

Before proceeding with the following steps, be sure that the object uses the Wrap around Object Shape text wrap. (If not, open the Text Wrap panel and click the Wrap around Object Shape button to apply text wrapping.) Remember to choose Detect Edges if you're using an image with a transparent background.

To modify the path around an image with text wrapping, using the Direct Selection tool, follow these steps:

1. **Select the object by using the Direct Selection tool.**

   The image is selected, and you can see the path around the object.

2. **Click to select one of the anchor points on the text wrap path by using the Direct Selection tool; then click and drag the point to move it.**

   The path is modified according to how you move the point. (For more about manipulating paths, take a look at Chapter 4 of this minibook.) The text wrapping immediately changes, based on the modifications you make to the path around the object.

3. **Select the Delete Anchor Point tool from the Tools panel and delete an anchor point. This tool is hidden in the Pen tool.**

   The path changes again, and the text wrapping modifies around the object accordingly.

REMEMBER

You can also use the Offset values in the Text Wrap panel to determine the distance between the wrapping text and the edge of the object. Simply increase the values to move the text farther from the object's edge.

Working with Pages and the Pages Panel

The *page* is the central part of any publication — it's where the visible part of your publication is created. Navigating and controlling pages are important parts of working in InDesign. The Pages panel allows you to select, adjust, move, and navigate pages in a publication. When you use default settings, pages are created as *facing pages,* which means that they're laid out as two-page pairs, or spreads. Otherwise, pages are laid out individually. Whether a page is part of a spread or a single page is reflected in the Pages panel, the option is specified when you create a new document and can be changed in the Document Setup window.

The Pages panel, which you open by choosing Window➪Pages, also lets you add new pages to the document, duplicate pages, delete a page, or change the size of a page. The pages panel, shown in Figure 3-8, contains two main areas: the master pages (upper) section and the (lower) section containing the document's pages. Note that this image of the Pages panel was created by changing the panel view to Jumbo. If you would like to see your panels in a larger view simply choose Panel Options from the Panel menu in the upper-right of the Pages panel and then select Jumbo from the Pages Size drop-down menu.

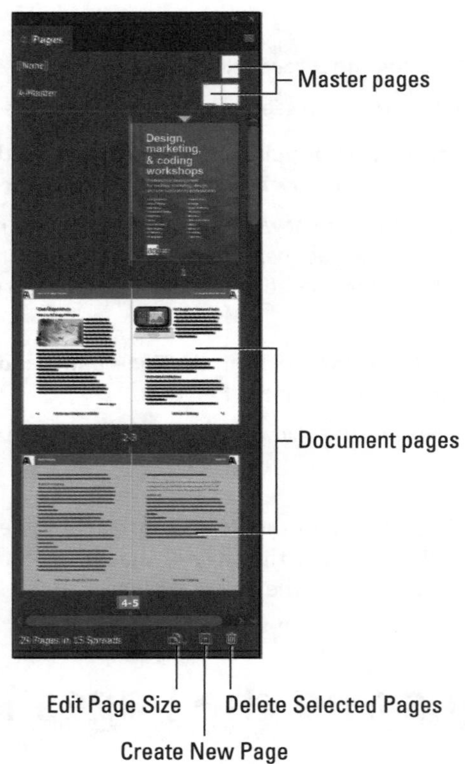

FIGURE 3-8: The Pages panel with page previews.

To discover more about master pages and how they differ from regular pages in your document, see the "Using Master Spreads in Page Layout" section, later in this chapter.

Selecting and moving pages

Use the Pages panel to select a page or spread in your publication. Select a page by clicking the page. If you Ctrl-click (Windows) or ⌘-click (Mac) pages, you can select more than one page at a time. The Pages panel also lets you move pages to a new position in the document: Select a page in the document pages area of the panel, and then drag it wherever you want to move the page. A small line and changed cursor indicate where the page will be moved. You can move a page so that it's between two pages in a spread; a hollow line indicates where you're moving the page. If you move a page before or after a spread, a solid line appears. Release the mouse button to move the page to the new location.

Adding and deleting pages

You can also add new pages to the publication by using the Pages panel. To add a new page, follow these steps:

1. **Choose Window⇨Pages to open the Pages panel.**

   The Pages panel opens.

2. **Click the Create New Page button.**

   A new page is added to the document.

   TIP

   Alt-click (Windows) or Option-click (Mac) the Create New Page button, and you can then specify the exact number of pages to add and the location of these new pages.

3. **Select a page in the Pages panel.**

   The selected page is highlighted in the Pages panel.

4. **Click the Create New Page button again.**

   A new page is added following the selected page.

TIP

To delete a page, select it in the Pages panel and click the Delete Selected Pages button. The selected page is removed from the document.

You can also add, delete, and move pages and more without the Pages panel by choosing from the Layout⇨Pages submenu.

Numbering your pages

When you're working with longer documents, adding page numbers before you print or export the publication is a good idea. You don't have to add them manually: A special InDesign tool lets you number pages automatically. This tool is particularly useful when you move pages around the document. You don't have to keep track of updating the numbering when you make these kinds of edits.

To number pages, follow these steps:

1. **Using the Type tool, create a text frame on the page where you want the page number to be added.**
2. **Choose Type➪Insert Special Character➪Markers➪Current Page Number.**

   The current page number appears in the text frame you selected. If you added the page number to a master page, the master page's letter appears in the field instead.

If you want page numbers to appear on all pages in the document, add the text frame to a master page. Remember that page numbers are added only to the pages in your document that are associated with that master page. If you want to add page numbers to the left and right sides of a book or magazine, you need to repeat this process on the left and right sides of the master pages. Remember that if you add a page number only on a document page — and not on a master — the page number is added to only that single page.

TIP

To modify automatic-numbering settings, choose Layout➪Numbering and Section Options. You can choose to have numbering start from a specific number or use a different style, such as Roman numerals.

Using Master Spreads in Page Layout

Master pages are a lot like templates you use to format page layouts. The settings, such as margins and columns, are applied to each layout that the master page is applied to. If you put a page number on a master page, the number also appears on every page that uses the layout. You can have more than one master page in a single publication, and you can choose which pages use a particular master page.

A master page, or spread, typically contains layout items that are used on many pages, such as page numbering, text frames, background images, or headings that are used on every page. By default, you cannot edit items that have been placed on a master page, on the pages that are using a particular master. You can, however, override or even completely detach these master page elements to modify them directly on a document page by pressing Ctrl+Shift (Windows) or ⌘ +Shift (Mac) while clicking on the master page element. This functionality is useful for special cases like changing the background of just one, or removing a page number.

REMEMBER

Master pages are lettered. The first master page is the A-Master by default. If you create a second master page, it's the B-Master by default. When you create a new publication, the A-Master is applied to all pages you initially open in the document. You can add pages at the end that don't have a master page applied to them.

Creating master pages and applying them to your publication enables you to create a reusable format for it, which can dramatically speed up your workflow when you put together documents with InDesign.

Creating a master spread

You may need more than one master page or master spread for a document. You may have another series of pages that need a unique format. In this situation, you need to create a second master page. You can create a master page or a master spread from any other page in the publication, or you can create a new one with the Pages panel.

To create a master page by using a page in the publication, do one of the following:

- Choose New Master from the Page panel's menu, and then click OK. A blank master page is created.
- Drag a page from the pages section of the panel into the master page section of the Pages panel. The document page turns into a master page.

WARNING

If the page you're trying to drag into the master pages section is part of a spread, select *both pages in the spread* before you drag it into the master pages section. You can drag individual pages into the master page section only if they're *not* part of a spread.

Applying, removing, and deleting master pages

After you create a master page, you can apply it to a page. You can also remove a page from a master page layout and delete a master page altogether:

- **To add master page formatting to a page or spread in a publication:** In the Pages panel, drag the master page you want to use from the master page section on top of the page you want to format in the document pages section. When you drag the master page on top of the page, it has a thick outline around it. Release the mouse button when you see this outline, and the formatting is applied to the page.

» **To remove any master page applied to a document page:** In the Pages panel, drag the None page from the master area in the Pages panel to that document page. You may need to use the scroll bar in the master pages area of the Pages panel to find the None page.

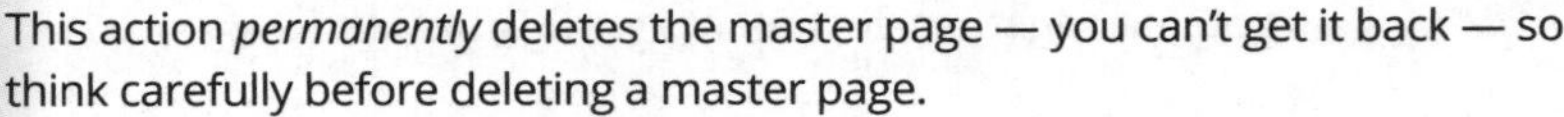

» **To delete a master page:** In the Pages panel, select the unwanted master page, and then choose Delete Master Page from the panel menu.

WARNING

This action *permanently* deletes the master page — you can't get it back — so think carefully before deleting a master page.

Changing individual page sizes

Using the Pages panel, you can change the size of individual pages in a document, which is useful if you have one page that folds out and is larger than others. Or, maybe you want to create a single document that includes a business card, an envelope, and a letterhead.

To change the size of individual pages using the Pages panel, follow these steps:

1. **In the Pages panel, click to select the page you want to modify.**
2. **Click the Edit Page Size button at the bottom of the Pages panel and select the new size. Be careful not to judge the size of a page by what it looks like in the Pages panel.**
3. **Repeat the process to adjust the size for any pages you want to modify.**

   When you're done editing the size of the pages, continue to work on their design and layout like you would work on any other pages. The only difference is that some pages in your document may be a different size.

TIP

You can also use the Pages panel to create alternative layouts for display on various tablet devices. These alternative layouts are discussed in Chapter 8 of this minibook.

» Drawing and editing shapes and paths

» Introducing corner effects

» Working with fills and layers

» Creating QR Codes

Chapter 4
Drawing in InDesign

Many of the tools you find in the InDesign Tools panel are used for drawing lines and shapes on a page. You can create anything from basic shapes to intricate drawings inside InDesign, instead of having to use a drawing program such as Illustrator. Even though InDesign doesn't replace Illustrator (see Book 5), which has many more versatile drawing tools and options for creating intricate drawings, InDesign is adequate for many drawing tasks. In this chapter, you discover how to use the most popular InDesign drawing tools and how to add colorful fills to illustrations.

Even if you don't think you'll be drawing using InDesign, you should still take a look at the sections "Modifying Frame Corners" and "Using Fills," later in this chapter, before jumping ahead.

Getting Started with Drawing

When you're creating a document, you may want drawn shapes and paths to be parts of the layout. For example, you may want to have a star shape for a yearbook page about a talent show or to run text along a path. Whatever it is you need to do, you can draw shapes and paths to get the job done.

Paths and shapes

Paths can take a few different formats. They can either be open or closed and with or without a stroke:

- **Path:** The outline of a shape or an object. Paths can be closed and have no gaps, or they can be open like a line on the page. You can draw freeform paths, such as squiggles on a page, freely by hand.
- **Stroke:** A line style and thickness that you apply to a path. A stroke can look like a line or like an outline of a shape.

Figure 4-1 shows the different kinds of paths and strokes you can create.

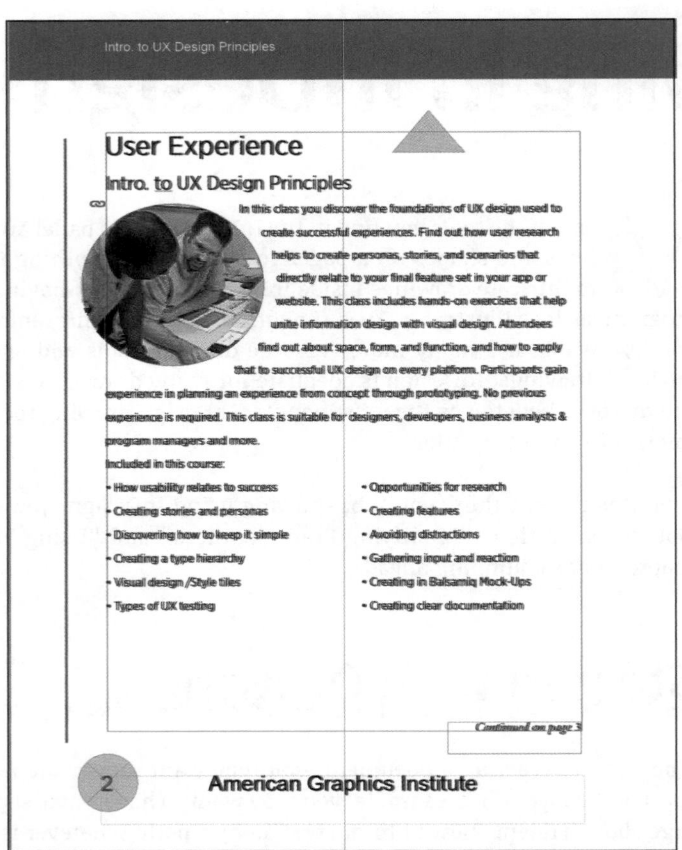

Intro. to UX Design Principles

User Experience

Intro. to UX Design Principles

In this class you discover the foundations of UX design used to create successful experiences. Find out how user research helps to create personas, stories, and scenarios that directly relate to your final feature set in your app or website. This class includes hands-on exercises that help unite information design with visual design. Attendees find out about space, form, and function, and how to apply that to successful UX design on every platform. Participants gain experience in planning an experience from concept through prototyping. No previous experience is required. This class is suitable for designers, developers, business analysts & program managers and more.

Included in this course:

- How usability relates to success
- Creating stories and personas
- Discovering how to keep it simple
- Creating a type hierarchy
- Visual design /Style tiles
- Types of UX testing
- Opportunities for research
- Creating features
- Avoiding distractions
- Gathering input and reaction
- Creating in Balsamiq Mock-Ups
- Creating clear documentation

Continued on page 3

2

American Graphics Institute

FIGURE 4-1: Different kinds of paths and strokes created in InDesign.

Paths contain points where the direction of the path can change. (You can find out more about points in the later section "Points and segments.") You can make paths by using freeform drawing tools, such as the Pen or Pencil tools, or by using the basic shape tools, such as Ellipse, Rectangle, Polygon, or Line.

The shape tools create paths in a predefined way so that you can make basic geometric shapes, such as a star or an ellipse. Just select the shape tool and drag the cursor onto the page, and the shape is automatically drawn. Creating shapes this way is a lot easier than trying to create them manually by using the Pen or Pencil tool. Figure 4-2 shows shapes drawn with the shape tools found in the Tools panel.

FIGURE 4-2: You can create many shapes with the basic shape tools.

You can change shapes into freeform paths, like those drawn with the Pencil or Pen tools. Similarly, you can make freeform paths into basic shapes. Therefore, you don't need to worry about which tool you initially choose.

TIP

We created the star and starburst shown in Figure 4-2 by double-clicking the Polygon tool and changing the options. Read more about the Polygon tool in the "Drawing Shapes" section, later in this chapter.

Keep in mind that you can also rough out a shape, like a triangle and then choose Object⇨Convert Shape and choose Triangle in order to correct your shape.

Points and segments

Paths are made up of points and segments:

- **Point:** Where the path changes somehow, such as a change in direction. Many points along a path can be joined with segments. Points are sometimes called *anchor points.* You can create two kinds of points:
  - *Corner points:* Have a straight line between them. Shapes such as squares and stars have corner points.
  - *Curve points:* Occur along a curved path. Circles or snaking paths have lots of curve points.
- **Segment:** A line or curve connecting two points — similar to connect-the-dots.

Figure 4-3 shows corner points and curve points joined by segments.

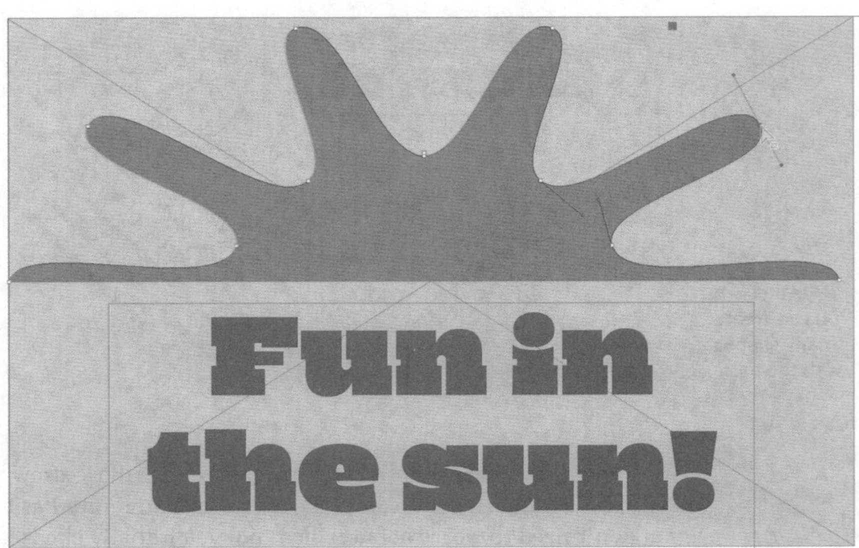

FIGURE 4-3: Points are joined by line segments.

Getting to Know the Tools of the Trade

The following sections introduce you to tools that you'll probably use most often when creating drawings in your publications. When you draw with these tools, you're using strokes and fills to make designs. The following sections show you what these common tools can do to help you create basic or complex illustrations in InDesign.

The Pencil tool

The Pencil tool is used to draw simple or complex shapes on a page. Because the Pencil tool is a freeform tool, you can freely drag it all over the page and create lines or shapes, instead of having them automatically made for you, such as when you use basic shape tools. The Pencil tool is an intuitive and easy tool to use. You find out how to use it in the later section "Drawing Freeform Paths."

The Pen tool

You use the Pen tool to create complex shapes on the page. The Pen tool works with other tools, such as the Add, Remove, and Convert Point tools. The Pen tool works by adding and editing points along a path, thereby manipulating the segments that join them.

Drawing with the Pen tool isn't easy at first. In fact, it takes many people a considerable amount of time to use this tool well. Don't become frustrated if you don't get used to it right away — the Pen tool can take some practice to get it to do what you want. You find out how to use the Pen tool in the "Drawing Freeform Paths" section, later in this chapter.

Basic shapes and frame shapes

Basic shapes are preformed shapes that you can add to a document by using tools in the Tools panel. The basic shape tools include the Line, Rectangle, Ellipse, and Polygon tools.

You can also draw these shapes and turn them into *frames* (containers that hold content in a document). You can use a frame as a text frame or as a graphic frame used to hold pictures and text. Draw a basic shape and then convert it to a graphic or text frame by choosing Object⇨Content⇨Text or Object⇨Content⇨Graphic. We discuss graphic frames and text frames in more detail in Chapter 2 of this minibook.

TECHNICAL STUFF

The frame and shape tools look the same and can even act the same. Both can hold text and images, but look out! By default, shapes created with the shape tools have a 1-point black stroke around them. Many folks don't see these strokes on the screen but later discover them surrounding their text boxes when they print. Stick with the frame tools, and you'll be fine.

Drawing Shapes

InDesign allows you to create basic shapes in a document. You can easily create a basic shape by following these steps:

1. **Create a new document by choosing File ⇨ New ⇨ Document.**
2. **When the New Document dialog box appears, select Print, choose a blank document preset size, and then click Create.**

   A new document opens.
3. **Select the Rectangle tool in the Tools panel.**
4. **Click anywhere on the page and drag the mouse diagonally.**

   When the rectangle is the dimension you want, release the mouse button. You've created a rectangle.

That's all you need to do to create a basic shape. You can also use these steps with the other basic shape tools (the Line, Ellipse, and Polygon tools) to create other basic shapes. To access the other basic shapes from the Tools panel, follow these steps:

1. **Click the Rectangle tool and hold down the mouse button.**

   A menu that contains all the basic shapes opens.
2. **Release the mouse button.**

   The menu remains open, and you can mouse over the menu items. Each menu item becomes highlighted when the mouse pointer is placed over it.
3. **Select a basic shape tool by clicking a highlighted menu item.**

   The new basic shape tool is now active. Follow the preceding set of steps to create basic shapes by using any of these tools.

TIP

To draw a square shape, use the Rectangle tool and press the Shift key while you drag the mouse on the page. The sides of the shape are all drawn at the same length, so you create a perfect square. You can also use the Shift key with the Ellipse tool if you want a perfect circle — just hold down Shift while you're using the Ellipse tool. Release the mouse before the Shift key to ensure that this constrain shape trick works!

Creating a shape with exact dimensions

Dragging on the page to create a shape is easy, but making a shape with precise dimensions using this method requires a few more steps. If you want to make a shape that's a specific size, follow these steps:

1. **Select the Rectangle tool or the Ellipse tool.**

   The tool is highlighted in the Tools panel.

2. **Click anywhere on the page, but don't drag the cursor.**

   This point becomes the upper-left corner of the Rectangle or Ellipse *bounding box* (the rectangle that defines the object's vertical and horizontal dimensions). After you click to place a corner, the Rectangle or Ellipse dialog box appears.

3. **In the Width and Height text fields, enter the dimensions for creating the shape.**

4. **Click OK.**

   The shape is created on the page, with the upper-left corner at the place where you initially clicked the page.

Using the Polygon tool

A *polygon* is a shape that has many sides. For example, a square is a polygon with four sides, but the Polygon tool enables you to choose the number of sides you want for the polygon you create. When you're using the Polygon tool, you may not want to create a shape with the default number of sides. You can change these settings before you start drawing the shape.

To customize the shape of a polygon, follow these steps:

1. **Make sure that you do not have any other shapes selected. Select the Polygon tool in the Tools panel by selecting the Rectangle tool and holding down the mouse button until the menu pops up.**

2. **Double-click the Polygon tool in the Tools panel.**

   The Polygon dialog box opens, as shown at the bottom of Figure 4-4.

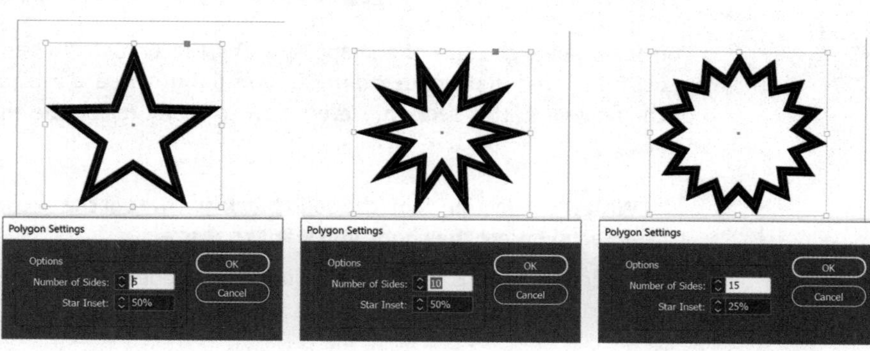

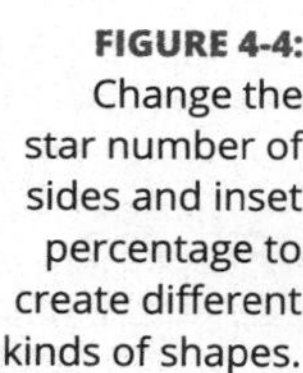

FIGURE 4-4: Change the star number of sides and inset percentage to create different kinds of shapes.

3. **In the Number of Sides text field, enter the number of sides you want the new polygon to have.**

   To create a star instead of a polygon, enter a number in the Star Inset text field for the percentage of the star inset you want the new shape to have.

   A higher percentage means that the sides are inset farther toward the center of the polygon, creating a star. If you want a regular polygon and not a star, enter **0** in the Star Inset text field. If you want a star, enter **50%**; for a starburst, enter **25%**.

4. **Click OK.**

5. **Move the cursor to the page and click and drag to create a new polygon or star.**

   Your new polygon or star appears on the page.

Figure 4-4 shows what a few different polygons and stars with different settings look like.

TIP

You can select a polygon after you have created and then double-click on the Polygon shape tool to change the sides and inset as well.

Editing Basic Shapes

You can edit basic shapes InDesign in order to create original shapes and craft exactly the kind of design you want. You aren't stuck with predetermined shapes, such as a square or an oval. You can make these forms take on much more custom shapes.

Changing the size with the Transform panel

You can change the size of a shape by using the Transform panel. The Transform panel is covered in more detail in Chapter 6, but you are introduced to the basics of the feature in this section. Here's how to easily resize the shapes you create in InDesign:

1. **With the Selection tool (the solid black arrow tool that's used to select objects), select the shape you want to resize.**

   When the shape is selected, a bounding box appears around it. You can see a selected shape in Figure 4-5.

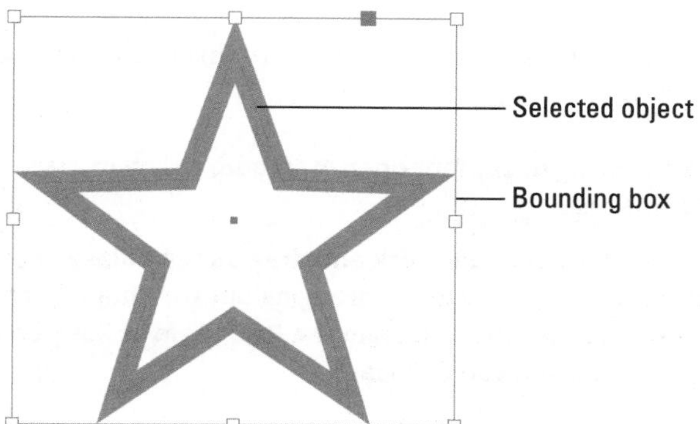

FIGURE 4-5: A selected shape with the bounding box visible.

2. **Open the Transform panel by choosing Window ➪ Object and Layout ➪ Transform.**
3. **In the Transform panel that appears, enter different number values in the W and H fields to change the size of the shape.**

   The shape automatically changes size on the page to the new size dimensions you specify in the Transform panel.

Changing the size with the Free Transform tool

Easily resize objects in InDesign by using the Free Transform tool.

To resize a selected object with the Free Transform tool, follow these steps:

1. **Make sure that *only* the object you want to resize is selected.**

   TIP

   Group multiple objects if you want to resize several objects simultaneously. To group objects, select one object and Shift-click to add it to the selection, and then press Ctrl+G (Windows) or ⌘ +G (Mac).
2. **Select the Free Transform tool.**
3. **Click any corner point and drag to resize the object.**

   TIP

   Hold down the Shift key while dragging to keep the objects constrained proportionally as you resize.

Creating your own custom shapes

No matter how good you are when drawing freehand, replicating that skill when creating custom shapes can be difficult. InDesign has the Pathfinder panel that can help you create those custom shapes. Try this exercise to experiment with the Pathfinder feature.

1. **Have an existing document open, or create a new InDesign document, any size is fine.**
2. **Select the Ellipse tool and click and drag out an Ellipse. If you like, you can hold down the Shift key when dragging out the shape to create a circle.** The Ellipse tool may be hidden in the Rectangle or Polygon tool, depending upon which one you used last.
3. **Select the Rectangle tool and click and drag out a rectangle. You can also hold down the Shift key while dragging if you would like to create a square.**
4. **Use the Select tool to overlap the shapes, as you see in Figure 4-6.**
5. **Make sure both shapes are selected. You can click on one and then Shift-click on the other.**
6. **Choose Window⇨Properties and click on Fill and change both shapes to any color.**
7. **In the Properties panel, look for the Pathfinder section and select Subtract; the frontmost object, the rectangle, is subtracted from the circle, as shown in Figure 4-6.**

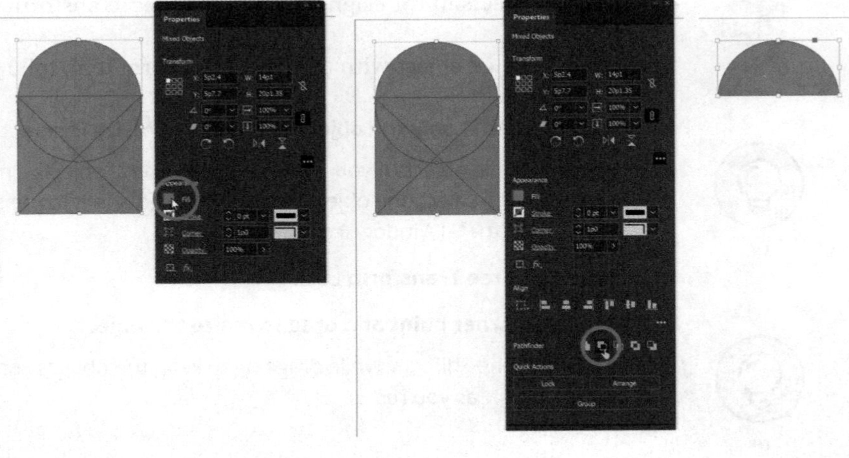

FIGURE 4-6: Overlap the shapes, change their fill and then choose a Pathfinder option such as Combine Shapes.

Changing the stroke of a shape

You can change the stroke of shapes you've created. The *stroke* is the outline that appears around the edge of the shape. The stroke can range from no stroke to a very thick stroke, and it's measured in point sizes. Even if a shape has a stroke set to 0 points, it still has a stroke — you just can't see it.

Follow these steps to edit the stroke of your shapes:

1. **Select a shape on the page.**

   A bounding box appears around the selected shape.

2. **Select a new width for the Stroke by using the Stroke Weight drop-down list in the Control panel.**

   As soon as a value is selected, the stroke automatically changes on the page. This number is measured in points. You use some of the other options in the following step list.

You can click in the Stroke text field and manually enter a numerical value for the stroke width. The higher the number you enter, the thicker the stroke. You can also change the style of the stroke from the Control panel by following these steps:

1. **With a basic shape selected, select the stroke type from the drop-down list in the Control panel and select a new line.**

   As soon as a new line is selected, the stroke automatically changes.

2. **Choose a new line weight from the Stroke Weight drop-down list.**

   For example, we chose 10 points. The shape updates automatically on the page.

TIP

If you want to create custom dashes, you can see more options by choosing Window⇨Stroke. Choose the panel menu in the upper-right corner of the panel and select Stroke Styles, and then select New. You can define the dash and gap size. Define custom dashes by clicking in the ruler and dragging the triangles to define the dash length. You can then click on the ruler to add additional dashes. Enter one value for an even dash or several numbers for custom dashes for maps, diagrams, fold marks, and more! Your custom stroke appears in the Stroke panel after you create it.

Add special ends to the lines with the Start and End drop-down lists. For example, you can add an arrowhead or a large circle to the beginning or end of the stroke. The Cap and Join buttons allow you to choose the shape of the line ends and how they join with other paths when you're working with complex paths or shapes.

Changing the shear value

You can change the shear of a shape by using the Transform panel. *Skew* and *shear* mean the thing: The shape is slanted, so you create the appearance of some form of perspective for the skewed or sheared element. This transformation is useful if you want to create the illusion of depth on a page.

Follow these simple steps to skew a shape:

1. **With a basic shape selected, choose Window ⇨ Object & Layout ⇨ Transform.**
2. **Select a value from the Shear X Angle drop-down list in the lower-right corner of the Transform panel.**

   After you select a new value, the shape skews (or shears), depending on what value you select. Manually entering a numerical value into this field also skews the shape.

Rotating a shape

You can change the rotation of a shape by using the Transform panel. The process of rotating a shape is similar to skewing a shape (see the preceding section):

1. **With a basic shape selected, choose Window ⇨ Object & Layout ⇨ Transform.**

   The Transform panel opens.
2. **Select a value from the Rotation Angle drop-down list.**

   After you select a new value, the shape rotates automatically, based on the rotation angle you specified. You can also manually enter a value into the text field.

Drawing Freeform Paths

You can use different tools to draw paths. For example, you can use the Pencil tool to draw freeform paths. These kinds of paths typically look like lines, and you can use the Pencil and Pen tools to create simple or complex paths.

Using the Pencil tool

The Pencil tool is perhaps the easiest tool to use when drawing freeform paths. (See Figure 4-7.)

Follow these steps to get started:

1. **To create a new document, choose File ➪ New ➪ Document ➪ Print, and then click Create in the New Document dialog box that appears.**

2. **Select the Pencil tool in the Tools panel. Make sure you have defined a stroke weight and color for your stroke.**
3. **Drag the cursor around the page.**

   You've created a new path by using the Pencil tool.

FIGURE 4-7: This freeform drawing was created with the Pencil tool.

Using the Pen tool

Using the Pen tool is different from using the Pencil tool. When you start out, the Pen tool may seem a bit complicated — but after you get the hang of it, using it isn't hard after all. The Pen tool uses points to create a particular path. You can edit these points to change the segments between them. Gaining control of these points can take a bit of practice.

To create points and segments on a page, follow these steps:

1. **Close any existing documents and create a new document by choosing File ➪ New ➪ Document.**
2. **Select Print, then click Create in the New Document dialog box.**

   A new document opens with the default settings.

3. **Select the Pen tool in the Tools panel. Make sure a stroke weight and color are defined before using this tool.**
4. **Click anywhere on the page, and then click a second location.**

   You've created a new path with two points and one segment joining them.
5. **Ctrl-click (Windows) or ⌘ -click (Mac) an empty part of the page to deselect the current path.**

   After you deselect the path, you can create a new path or add new points to the path you just created.

Drawing in InDesign

6. **Using the Pen tool, add a new point to a selected segment by selecting the segment, then hovering the mouse over the line and clicking.**

   A small plus (+) icon appears next to the Pen tool cursor. You can also do the same thing by selecting the Add Anchor Point tool (located on the menu that flies out when you click and hold the Pen icon in the Tools panel).

7. **Repeat Step 6, but this time click a new location on a line segment and drag away from the line.**

   This step creates a curved path. The segments change and curve depending on where the points are located along the path. The point you created is a *curve* point.

For more information about working with the Pen tool, check out Book 5, Chapter 4.

Editing Freeform Paths

Even the best artists sometimes need to make changes or delete parts of their work. If you've made mistakes or change your mind about a drawing, follow the steps in this section to make changes.

To change a path segment, make sure the segment is deselected, then select a point with the Direct Selection tool. When a point is selected, it appears solid; unselected points appear hollow.

TIP

You can automatically select the Direct Selection tool by pressing the A key.

All you need to do to select a point is click the point itself. Then you can use the handles that appear when the point is selected to modify the segments. Follow these steps:

1. **Select the Direct Selection tool (or press the A key) from the Tools panel, and then click a point.**

   The selected point appears solid. If you select a curve point, handles extend from it.

   A curved point and a corner point edit differently when you select and drag them. Curve points have handles that extend from the point, but corner points don't.

2. **Drag the point where you want it; to edit a curve point, click a handle end and drag the handle left or right.**

   The path changes, depending on how you drag the handles.

Suppose that you want to make a corner point a curve point. You can do just that with the Convert Direction Point tool. To understand how the Convert Direction Point tool works best, you should have a path that contains both straight and curved segments. Follow these steps to change a corner point into a curved point and vice versa:

1. **Select the Convert Direction Point tool.**

   This tool resides on a menu under the Pen tool in the Tools panel. Hold down the mouse button over the Pen tool icon until a menu appears; select the Convert Direction Point tool from the menu.

2. **Click a curved point with the Convert Direction Point tool.**

   The point you click changes into a corner point, which changes the path's appearance.

3. **Click and drag a corner point with the Convert Direction Point tool.**

   The point is modified as a curved point. This step changes the appearance of the path again.

This tool is handy when you need to alter the way a path changes direction. If you need to manipulate a point in a different way, you may need to change its type by using the Convert Direction Point tool.

Modifying Frame Corners

You can use corner effects on basic shapes to customize the shape's look. Corner effects are useful for adding an interesting look to borders. You can be quite creative with some of the shapes you apply effects to or by applying more than one effect to a single shape. Here's how to create a corner effect on a rectangle:

1. **Select the Rectangle tool and create a new rectangle anywhere on the page.**

   Hold the Shift key when using the Rectangle tool if you want to create a square.

2. **With the Selection tool, select the shape, and then choose Object ⇨ Corner Options.**

   The Corner Options dialog box opens. If you prefer to use the Properties panel, choose Window ⇨ Properties and click on the Corners button to see the same Corner Options window.

3. **Select the type of corner to apply, then click OK.**

   The corner option is applied to the shape. To apply the corner option to all corners of the shape, make sure the Make All Settings the Same chain icon is selected.

TIP

To adjust corner effects visually, use the Selection tool to click and select a frame. With the frame selected, click the yellow box that appears toward the top of the right edge of the frame. After clicking the yellow box, each corner handle becomes yellow and can be dragged horizontally left or right.

Using Fills

A fill is located inside a path. You can fill paths and shapes with several different kinds of colors, transparent colors, or even gradients. Fills can help you achieve artistic effects and illusions of depth or add interest to a page design.

You may have already created a fill. The Tools panel contains two swatches: one for the stroke (a hollow square) and one for the fill (a solid box). (Refer to Chapter 2 of this minibook to locate the Fill and Stroke boxes.) If the Fill box contains a color, your shape has a fill when it's created. If the Fill box has a red line through it, the shape is created without a fill.

Creating basic fills

You can create a basic fill in several different ways. To create a shape with a fill, follow these steps:

1. **Create a new shape on the page.**

   Select a shape tool and drag on the page to create a shape. The shape is filled with the fill color you chose.

2. **If you do not already see the Properties panel choose Window⇨Properties.**
   Click on Fill, choose your color from the swatches, Color Picker as shown in Figure 4-8, or Gradient.

3. **Select a color in the Color panel.**

   You can enter values into the RGB (Red, Green, Blue) or CMYK (Cyan, Magenta, Yellow, Black) fields manually or by using the sliders. Alternatively, you can use the Eyedropper tool to select a color from the Color Picker at the bottom of the Color panel.

TIP

   The Fill box in the Tools panel is updated with the new color you've selected in the Properties panel.

TIP

As in the other Creative Cloud applications, you can create tints of a color built with CMYK by holding down the Shift key while dragging any color's slider. All color sliders then move proportionally.

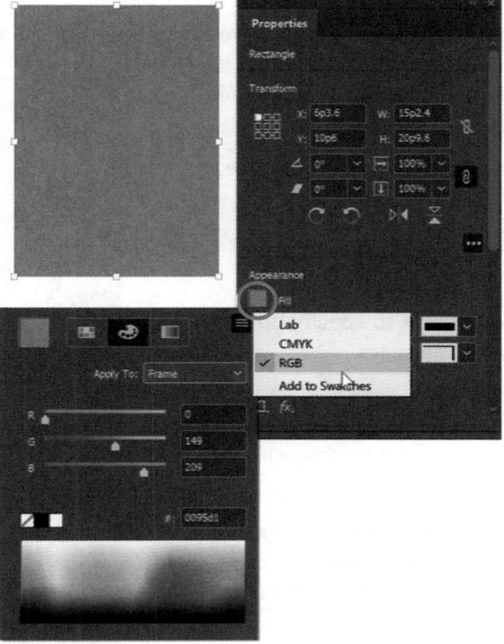

FIGURE 4-8: Choose to fill using swatches, the Color Picker, or with a gradient. Click on the Panel menu to change the color mode; this example is set for RGB color.

You can also choose to use color swatches to select a fill color by using the Swatches panel. (Choose Window⇨Color⇨Swatches to open the Swatches panel.) Create a new color swatch (of the present color) by clicking the New Swatch button at the bottom of the panel. Double-click the new swatch to add new color properties by using sliders to set CMYK color values or by entering numbers into each text field.

Perhaps you already have a shape without a fill and you want to add a fill to it, or maybe you want to change the color of an existing fill. Select the shape and click on the Fill color in the Windows⇨Properties panel and select a new color from the list of Swatches.

TIP

You can drag and drop a swatch color to fill a shape on a page, even if that shape isn't selected. Open the Swatches panel by choosing Window⇨Color⇨Swatches, and then drag the color swatch over to the shape. Release the mouse button, and the fill color is applied automatically to the shape.

Making transparent fills

Fills that are partially transparent can create some interesting effects for the layout of your document. You can set transparency to more than one element on the page and layer those elements to create the illusion of depth and stacking.

Follow these steps to apply transparency to an element on the page:

1. **Using the Selection tool, select a shape that has any fill color on the page.**

   A bounding box appears around the selected shape.

2. **If your Properties panel is not visible choose Window ⇨ Properties.**

3. **Use the Opacity slider to specify how transparent the shape appears.**

   Click the arrow to open the slider or click in the text field to manually enter a value using the keyboard. The effect is immediately applied to the selected shape.

4. **Click open the Apply Effects to Object menu to choose which element you want to apply the transparency to, as shown in Figure 4-9.**

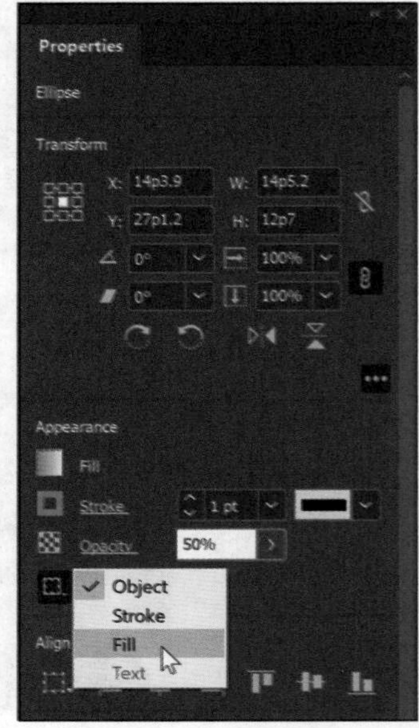

FIGURE 4-9: You can apply transparency to the entire object, the fill, stroke, or text if you have a text object selected.

Filling with gradients

A *gradient* is the color transition from one color (or no color) to a different color. Gradients can have two or more colors in the transition.

Gradients can add interesting effects to shapes, including 3D effects. Sometimes you can use a gradient to achieve glowing effects or the effect of light hitting a surface. The two kinds of gradients available in InDesign are radial and linear; the linear gradient is shown in Figure 4-10 and described in this list:

- **Radial:** A transition of colors in a circular fashion from a center point radiating outward
- **Linear:** A transition of colors along a straight path

You can apply a gradient to a stroke or a fill or even to text. To apply a gradient to a stroke, simply select the stroke instead of the fill.

WARNING

Even though you can apply a gradient to the stroke of live text, you'll create a printing nightmare — use these features sparingly!

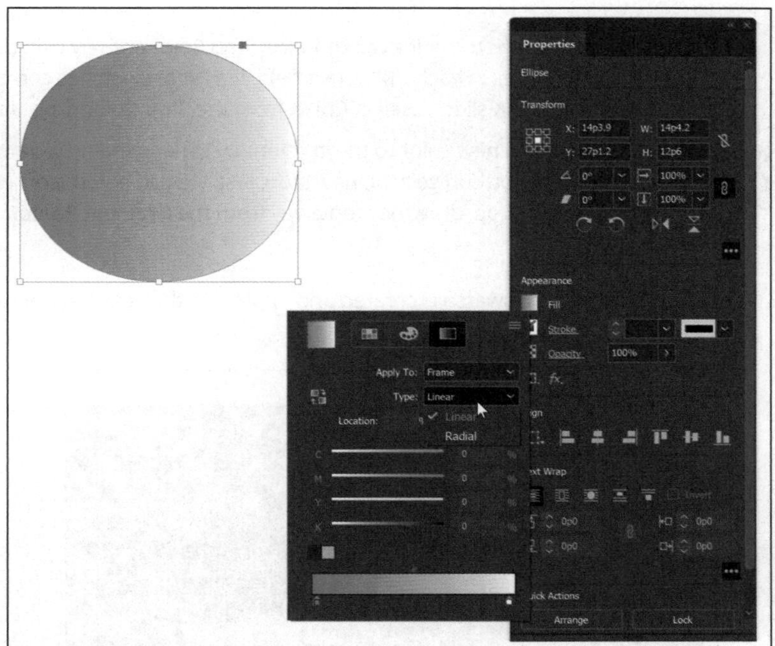

FIGURE 4-10: An example of a linear gradient.

Here's how to add a gradient fill to a shape:

1. **With the Selection tool, select the object you want to apply a gradient to and then choose Window ⇨ Color ⇨ Swatches.**

   The Swatches panel opens.

2. **Choose New Gradient Swatch from the Swatches panel drop-down menu.**

   The New Gradient Swatch dialog box opens. (See Figure 4-11.)

3. **Type a new name for the swatch in the Swatch Name field.**

   Sometimes, giving the swatch a descriptive name, such as one indicating what the swatch is being used for, is helpful.

4. **Select Linear or Radial from the Type drop-down list.**

   This option determines the type of gradient the swatch creates every time you use it. We selected Radial from the drop-down list. (Refer to Figure 4-11.)

5. **Manipulate the gradient stops, below the Gradient Ramp, to position each color in the gradient.**

   Gradient stops are the color chips located below the Gradient Ramp. You can move the diamond shape above the Gradient Ramp to determine the center point of the gradient. You can select each gradient stop to change the color and move them around to edit the gradient. When the gradient stops are selected, you can change the color values in the Stop Color area by using sliders or entering values in each CMYK text field. If a list of swatches comes up and you want the CMYK sliders, select CMYK from the Stop Color drop-down menu.

TIP

You can add a new color to the gradient by clicking the area between gradient stops. Then you can edit the new stop, just like you edit the others. To remove the gradient stop, drag the stop away from the Gradient Ramp.

6. **Click OK.**

The gradient swatch is created and applied to the selected object.

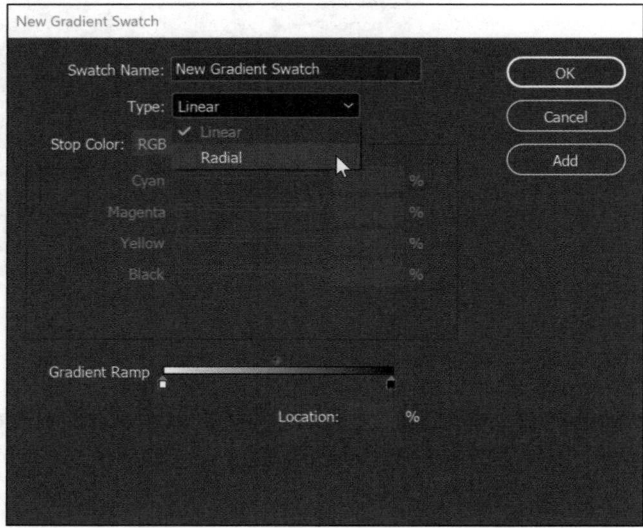

FIGURE 4-11: The New Gradient Swatch dialog box.

REMEMBER

To edit a gradient, double-click the gradient's swatch in the Swatches panel. This step opens the Gradient Options dialog box, where you can modify the settings made in the New Gradient Swatch dialog box.

Removing fills

Removing fills is even easier than creating them. Follow these steps:

1. **Select the shape with the Selection tool.**

   A bounding box appears around the shape.

2. **Click the Fill box in the Tools panel.**
3. **Click the Apply None button, located below the Fill box.**

The button is white with a red line through it. The fill is removed from the selected shape, and the Fill box changes to No Fill. You also see the None fill on the Swatches and Color panels.

If you're using a single-row Tools panel, you don't see the Apply None button unless you click and hold down the Apply Gradient (or Apply Color) button.

Adding Layers

Layers are like transparent sheets stacked on top of one another. If you add layers to a drawing, you can create the appearance that graphics are stacked on top of one another. The Layers panel allows you to create new layers, delete layers you don't need, or even rearrange layers to change the stacking order. You can even duplicate an InDesign file, and then add or remove layers to create an alternative version of the document.

Here's how to work with layers in InDesign:

1. **Open the Layers panel by choosing Window ⇨ Layers.**

   This panel allows you to create, delete, and arrange layers.

2. **Draw a shape on the page using a shape tool.**

   Create the shape anywhere on the page and make it large enough so that you can easily stack another shape on top of part of it.

3. **Click the Create New Layer button, on the bottom of the Layers panel, to create a new layer.**

   A new layer is stacked on top of the selected layer and becomes the active layer.

   TIP

   Double-click a layer to give it an appropriate name or, even better, hold down the Alt (Windows) or Option (Mac) key and click the Create New Layer button to open the Layer Options dialog box before the layer is created.

   WARNING

   Make sure that the layer you want to create content on is selected before you start modifying the layer. You can tell which layer is selected because it's always highlighted in the Layers panel. You can easily add content to the incorrect layer if you don't check this panel frequently. (If you inadvertently add an item to the wrong layer, you can always cut and paste items to the correct layer.)

4. **Make sure that a shape tool is still selected, and then create a shape on the new layer by dragging the cursor so that part of the new shape covers the shape you created in Step 2.**

   The new shape is stacked on top of the shape you created in Step 2.

For more information about working with layers, refer to Book 3, Chapter 8.

Creating QR Codes

You can use InDesign to create and modify QR code graphics. *QR codes* are a form of barcode that can store information such as words, numbers, URLs, or other forms of data. The user scans the QR code using her camera and software on a device, such as a smartphone, and the software makes use of the encoded data; for example, it opens a browser with the encoded URL, identifies a product, tracks a package, and so on. InDesign lets you encode hyperlinks, text, text messages, email messages, or business cards.

InDesign treats QR codes as graphics, so you can scale and modify the colors like other artwork in your documents. You can even copy and paste them into Illustrator!

You add a QR code to an empty frame in InDesign. Follow these steps to add one to your document:

1. **Click and drag the Rectangle Frame tool on the page to create an empty frame.**
2. **With your new frame selected, choose Object ⇨ Generate QR Code.**

   The Generate QR Code dialog box appears, with the Content tab active.
3. **Select what type of data to encode using the Type menu.**

   Select Web Hyperlink, Plain Text, Text Message, Email, or Business Card. The content area below this menu updates depending on what Type you choose.
4. **Enter the data to encode in your QR code.**

   The area below the Type menu updates with text boxes to enter your data in. For example, if you chose Email, you enter an address, subject, and message to send. Or if you chose Business Card, you need to fill in Name, Organization, addresses, and so on.
5. **Click the Color tab and choose a color swatch.**

   This sets the color for your QR code, affecting how it displays on the page. You can modify the color after placing the QR code on the page by changing the Fill and Stroke attributes of the selected frame.
6. **Click OK.**

   The QR code is added to the selected frame. If you need to edit it further, select the text frame and choose Object ⇨ Edit QR Code.

» Looking at color controls and models

» Discovering how to use swatches and swatch libraries

Chapter 5
Understanding Color

Color is an important part of design. Advertisements often rely on color to relay brands or effective messages — think of the package delivery company based on brown or a soft drink company known for its red cans and bottles. Color can enhance your message and, when used consistently, helps create a brand identity. When you use color, you want it to be accurate, whether you're printing or viewing the color online. In this chapter, you find out some of the fundamental aspects of working with color and the basic instructions on how to prepare documents for printing.

Selecting Color with Color Controls

You have several different color modes and options to choose when working in InDesign. In this minibook, you find out how to add color to drawings with the Color panel. In this section, you use the Color panel to choose colors and apply them to elements on your page. You also discover how to save colors as swatches so you can easily reuse them.

You should use swatches whenever possible because they use named colors that can be matched by others more easily. A swatch can have exactly the same appearance as any color you choose that's unnamed, but a swatch establishes a link between the color on the page and the name of a color, such as a Pantone color number. Discover more about these kinds of color in the later section "Using Color Swatches and Libraries."

You can use these color controls for choosing colors for selections in a document:

- **Stroke color:** Choose colors for strokes and paths in InDesign. A hollow box represents the Stroke color control.
- **Fill color:** Choose colors for filling shapes. A solid square box represents the Fill color control.

  In the Tools panel, you can toggle between the Fill and Stroke color controls by clicking them. Alternatively, you can press X on the keyboard to toggle between selected controls.
- **Text color:** When you're working with text, a different color control becomes active. The Text color control is visible and displays the selected text color. Text can have both the stroke and fill colored.

To apply colors to selections, you can click the Apply color button below the color controls in the Tools panel. Alternatively, you can select and click a color swatch.

REMEMBER

The default colors in InDesign are a black stroke and no fill color. Restore the default colors at any time by pressing D. This shortcut works while using any tool except the Type tool.

Understanding Color Models

You can use any of three kinds of color models in InDesign: CMYK (Cyan, Magenta, Yellow, Black), RGB (Red, Green, Blue), and LAB colors (Lightness and *A* and *B* for the *color*-opponent dimensions of the color space). A *color model* is a system used for representing each color as a set of numbers or letters (or both). The best color model to use depends on how you plan to print or display your document:

- If you're creating a PDF that will be distributed electronically and probably not printed, use the RGB color model. RGB is how colors are displayed on a computer monitor.
- You must use the CMYK color model if you're working with *process color.* Instead of having inks that match specified colors, you have four ink colors layered to simulate a particular color. Note that the colors on the monitor may differ from the ones that are printed. Sample swatch books and numbers can help you determine which colors you need to use in a document to match colors printed in the end.

» If you know that the document needs to be printed by professionals who determine what each color is before it's printed, it doesn't matter whether you use CMYK, RGB, or Pantone colors. In many cases, CMYK colors give you a more accurate representation of what can be printed, because RGB colors are often more vivid and bright on your monitor.

Using Color Swatches and Libraries

The Swatches panel and swatch libraries help you choose, save, and apply colors in your documents. The colors you use in a document can vary greatly. For example, one publication you make with InDesign may be for a newsletter that has only two colors; another may be a catalog printing with CMYK along with a separate spot color. For each document, you should customize the available swatches so you can work more efficiently.

The Swatches panel

You can create, apply, and edit colors from the Swatches panel. In addition, you can also create and save solid colors. Choose Window⇨Color⇨Swatches to open or expand the Swatches panel.

To create a new color swatch to use in a document, follow these steps:

1. **Click the panel menu drop-down located in the upper-right corner of the Swatches Panel; choose New Color Swatch.**

   The New Color Swatch dialog box opens.

2. **Leave the Name with Color Value checkbox selected or deselect to name your new color swatch.**

   The colors in the Swatches panel are named by their color values as a default.

   The name you choose is displayed next to the color swatch when it's entered into the panel.

3. **Choose the color type from the Color Type drop-down list.**

   Are you using a spot color, Pantone, for example, or Process, which is used when printing a color with Cyan, Magenta, Yellow, and Black (CMYK)?

4. **Choose the color mode.**

   From the Color Mode drop-down list, select a color mode. For this example, we use CMYK. Many of the other choices you see are prebuilt color libraries for various systems.

5. **Create the color by using the color sliders.**

   Note that if you start with Black, you have to adjust that slider to the left to see the other colors.

WARNING

   If you choose a spot color, such as Pantone, you are presented with a list of available color swatches instead of the color sliders.

6. **Click OK or Add.**

   Click Add if you want to continue adding colors to the Swatches panel or click OK if this color is the only one you're adding. The color or colors are added to the Swatches panel. If you clicked Add, click OK when you're done adding color swatches.

TIP

You can make changes to the swatch by selecting it in the Swatches panel and then choosing Swatch Options from the panel menu, or by simply double-clicking it in the Swatches panel.

Swatch libraries

Swatch libraries, also known as color libraries, are standardized sets of named colors that help you because they're the most commonly and frequently used sets of color swatches. You can avoid trying to mix your own colors, which can be a difficult or tedious process to get right. For example, InDesign includes a swatch library for Pantone spot colors and a different library for Pantone process colors. These libraries are quite useful if you're working with either color set. If you skipped over it, you may want to review the earlier section "Understanding Color Models," where we explain the difference between spot and process colors.

To choose a swatch from a swatch library, follow these steps:

1. **Choose New Color Swatch from the Swatches panel menu.**

   The New Color Swatch dialog box opens.

2. **Choose the color type you want to work with from the Color Type drop-down list.**

   Choose from Process or Spot Color types.

   Process will create your color from a build of CMYK colors (Cyan, Magenta, Yellow, and Black).

Spot adds a new solid color to your document. A professional printer using a CMYK press can add your specific ink color using a spot color.

Note: K stands for black in CMYK, as it relates to black being the Key color.

3. **Select a color library from the Color Mode drop-down list.**

   The drop-down list contains a list of color swatch libraries to choose from, such as Pantone Process Coated or TRUMATCH. After choosing a swatch set, the library opens and appears in the dialog box. For this example, we chose PANTONE+ Solid Coated, which is standard. If you're looking for the standard numbered Pantone colors, this set is the easiest to choose from. The Pantone Solid-Coated library of swatches loads.

4. **Pick a swatch from the library.**

   If you were provided with a Pantone number, type it into the Pantone textbox. Most organizations have set Pantone colors that they use in order to keep their color consistent no matter where it is printed or displayed. You can also scroll and click a swatch in the library's list of colors, shown in Figure 5-1.

WARNING

Your monitor doesn't provide the most accurate representation of a color or show you how a color will print. If color accuracy is critical, you should consider purchasing a color swatch book from Pantone, such as the Pantone Color Bridge Set. For more details about this guide, go to `www.pantone.com` and search for *Color Bridge*.

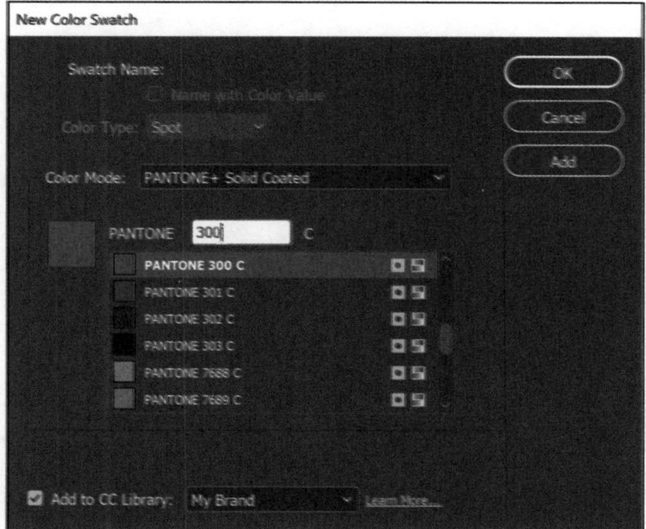

FIGURE 5-1: Choose a color from the swatch library to add it to the Swatches panel.

5. **Click the Add button.**

   This step adds the swatch to your list of color swatches in the Swatches panel. You can add as many color swatches as you like.

6. **When you finish adding swatches, click the OK button.**

   After you add a new color, the swatch is added to the list of swatches in the swatches panel and is ready to use in your project. Look in the swatches panel to see the newly added colors.

» **Transforming objects with the Free Transform tool**

» **Rotating and scaling objects**

» **Shearing and reflecting objects**

» **Adding a clipping path**

» **Aligning and distributing objects in a layout**

Chapter 6
Clipping Paths, Alignment, and Object Transformation

In this chapter, you discover several different ways to manipulate and arrange objects on a page. You find out how to use the Transform panel and other tools in the Tools panel to transform objects on page layouts. You can make the same transformation to an object several different ways in InDesign, so we show you a couple different ways to do the same job.

Aligning and distributing objects and images helps you organize elements logically on a page. In this chapter, you find out how to align objects by using the Align panel. In Chapter 4 of this minibook, we touch on paths. This chapter provides more information about clipping paths. We show you how to create a new path to use as a clipping path for an image in your document.

Working with Transformations

Chapter 4 of this minibook shows you how to transform graphic objects by using the Free Transform tool. You can manipulate objects in InDesign in many ways. You can transform an object by selecting an individual object and choosing the Transform panel, which is available by choosing Window➪Object & Layout➪Transform, or by using the Free Transform tool to visually adjust objects. Transforming a size of a selected object can also be completed using the Properties panel.

Looking at the Transform panel

The Transform panel and Properties panel, shown in Figure 6-1, are both useful when changing the scale, rotation, or skew of selected objects. You can choose from a range of values for some of these modifiers or manually set your own by typing them. Since the Properties panel is more accessible, it will be covered in more detail first.

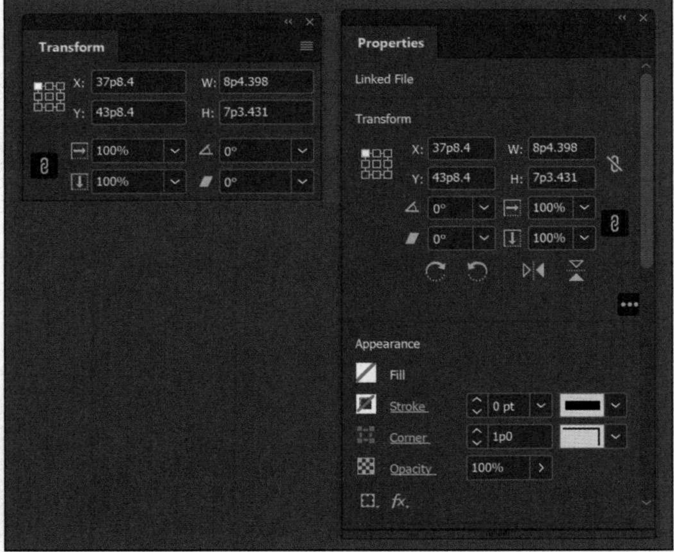

FIGURE 6-1: Both the Transform and Properties panel make it easy to resize, rotate, and reposition selected objects.

Both the Transform and Properties panels offer the following information and functionality:

» **Reference point:** Indicates which handle is the reference for any transformations you make. For example, if you reset the X and Y coordinates, the reference point is set to this position. In Figure 6-2, the reference point is in the top-left corner, as indicated by the solid square.

- **Position:** Change these values to reset the X and Y coordinate positions of the selected object.
- **Size:** The W and H text fields are used to change the current dimensions of the object.
- **Scale:** Enter or choose a percentage from the Scale X Percentage and Scale Y Percentage drop-down lists to scale (resize) the object on either of these axes.
- **Constraining proportions:** Click the Constrain Proportions button to maintain the current proportions of the object being scaled.
- **Shearing:** Enter or choose a negative or positive number to modify the shearing angle (skew) of the selected object.
- **Rotation angle:** Set a negative value to rotate the object clockwise; a positive value rotates the object counterclockwise.

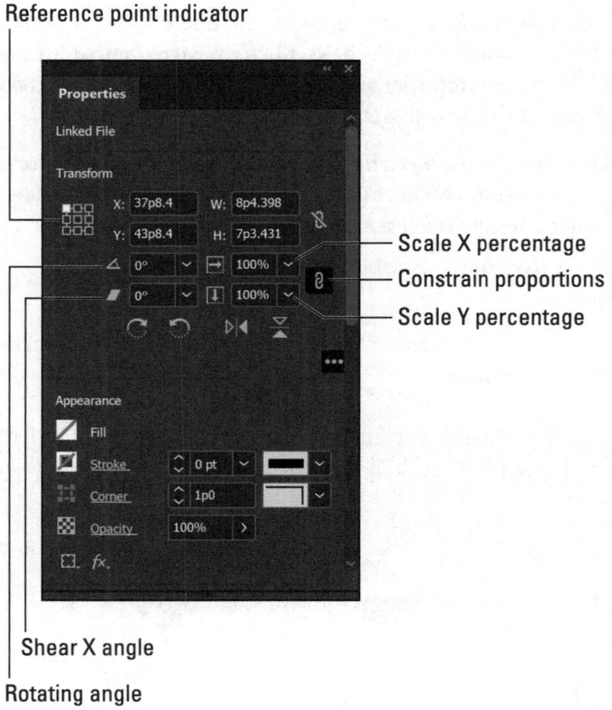

FIGURE 6-2: The Properties panel has the tools you need for resizing your images.

When you're scaling, shearing, or rotating an object in your layout, it transforms based on the reference point in the Transform panel. For example, when you rotate an object, InDesign considers the reference point to be the center point of the rotation.

Click a new reference point square in the Transform panel to change the reference point of the graphic to the equivalent bounding box handle of the selected object.

Using the Free Transform tool

The multipurpose Free Transform tool lets you transform objects quickly. Using the Free Transform tool, you can move, rotate, shear, reflect, and scale objects. If you would like to experiment with SVG Scalable Vector Images, you can follow these steps to create a file to use.

1. **Create a new file using File⇨New⇨Document. When the New Document window appears, choose any size. In this example, Print was selected and then the document size Letter was double-clicked.**
2. **Choose File⇨Place and navigate to the Book04_InDesign folder and open the folder named Farm-Images. Click on the topmost image and then hold down the Shift key and click on the last image. All the images have been selected. Click the Open button.**
3. **All the files are ready to be placed using the loaded cursor. Simply click once to see that the first image is placed. Continue to click, each time placing an image until the cursor is no longer loaded.**

   You now have many scalable illustrations to work with as you navigate this chapter.
4. **Select any illustration and then click on the Free Transform tool located in your Tools panel.**

The functions of the Free Transform Tool are represented in InDesign by different cursors, as shown in Figure 6-3.

FIGURE 6-3: Different cursors indicate options for using the Transform tool.

TIP

If you are worried about the resolution of your images, check to see if they are displaying using the High-Quality Display. You can check this by right-clicking on your image and selecting Display Performance from the contextual menu. See Figure 6-4.

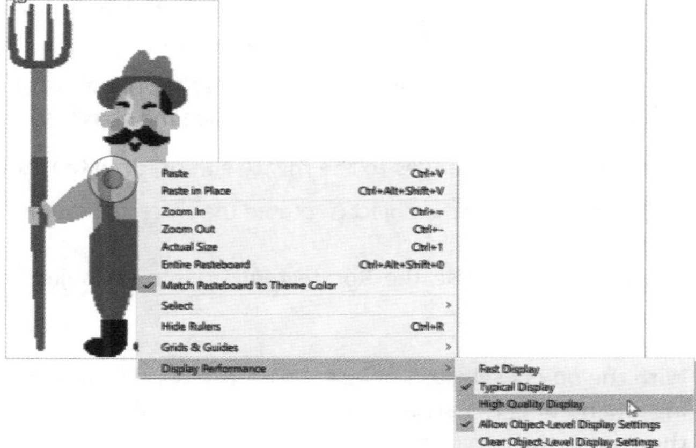

FIGURE 6-4: Change your display performance if you are worried about the resolution of your selected image.

To move an object by using the Free Transform tool, follow these steps:

1. **Use the Selection tool to select an object on the page.**

   You can use an object that's already on the page or create a new shape by using the drawing tools. When the object is selected, you see handles around its edges.

2. **Select the Free Transform tool from the Tools panel.**

   The cursor changes to the Free Transform tool.

3. **Move the cursor over the middle of the selected object.**

   The cursor changes its appearance to indicate that you can drag to move the object. (Refer to Figure 6-2.) If you move the cursor outside the edges of the object, the cursor changes when other tools, such as rotate, scale, and shear, become active.

4. **Drag the object to a different location.**

   The object is moved to a new location on the page.

Rotating objects

You can rotate an object by using the Free Transform tool, the Rotate tool, or the Transform panel. Use the panel to enter a specific degree that you want the object to rotate. The Free Transform tool lets you visually manipulate the object on the page.

To rotate an image by using the Free Transform tool, follow these steps:

1. **Select an object on the page with the Selection tool.**

   Handles appear around the edges of the object. You can rotate any object on the page.

2. **Select the Free Transform tool in the Tools panel and move it near the handle of an object outside the bounding box.**

   The cursor changes when you move it close to the handle of an object. For rotation, you must keep the cursor just outside the object.

3. **When the cursor changes to the rotate cursor, drag to rotate the object.**

   Drag the cursor until the object is rotated the correct amount.

Alternatively, you can use the Rotate tool to spin an object by following these steps:

1. **With the object selected, select the Rotate tool in the Tools panel by clicking and holding the arrow in the lower corner of the Free Transform tool, then move the cursor near the object.**

   The cursor looks similar to a crosshair.

2. **Click the cursor anywhere on the page near the object.**

   The point that the object rotates around is set on the page.

3. **Drag the cursor outside the object.**

   The object rotates around the reference point you set on the page. Hold the Shift key if you want to rotate in 45-degree increments.

You can also rotate objects by using the Transform panel. Here's how:

1. **Select an object on the page with the Selection tool.**

   The bounding box with handles appears around the selected object.

2. **If the Transform panel isn't open, choose Window ⇨ Object & Layout ⇨ Transform.**

   The Transform panel appears.

3. **Select a value from the Rotation Angle drop-down list or click the text field and enter a percentage.**

   The object rotates to the degree you set in the Transform panel. Negative angles (in degrees) rotate the image clockwise, and positive angles (in degrees) rotate the image counterclockwise.

Scaling objects

You can scale objects by using the Transform panel (refer to Figure 6-1), the Scale tool, or the Free Transform tool. Use the Transform panel to set exact width and height dimensions that you want to scale the object to, just as you can set exact percentages for rotating.

To scale an object by using the Free Transform tool or the Scale tool, follow these steps:

1. **Select an object on the page.**

   A bounding box appears around the object.

2. **Select the Free Transform tool or the Scale tool from the Tools panel.**
3. **Move the cursor directly over a corner handle.**

   The cursor changes into a double-ended arrow. (Refer to Figure 6-3.)

4. **Drag outward to increase the size of the object; drag inward to decrease the size of the object.**

   TIP

   If you want to scale the image proportionally, hold down the Shift key while you drag.

5. **Release the mouse button when the object is scaled to the correct size.**

To resize an object using the Transform panel, select the object and enter new values into the W and H text fields in the panel. The object then resizes to those exact dimensions.

Shearing objects

Shearing an object means that you're skewing it horizontally, slanting it to the left or right. A sheared object may appear to have perspective or depth because of this modification. You use the Shear tool to create a shearing effect, as shown in Figure 6-5.

Follow these steps to shear an object:

1. **Select an object on the page.**

   The bounding box appears around the object that's selected.

2. **Choose the Shear tool in the Tools panel by clicking and holding the Free Transform tool.**

   The cursor changes so that it looks similar to a crosshair. Click the corner of the object that you want to shear from, and a crosshair appears (refer to Figure 6-3).

FIGURE 6-5: The original image is on the left, and the sheared image is on the right.

3. **Click anywhere above or below the object and drag.**

   The selected object shears depending on which direction you drag. Press the Shift key while you drag to shear an object in 45-degree increments.

TIP

To shear objects with the Free Transform tool, begin dragging a handle and then hold down Ctrl+Alt (Windows) or ⌘ +Option (Mac) while dragging.

You can also enter an exact value into the Transform panel to shear an object. Select the object and then enter a positive or negative value in the panel representing the amount of slant you want to apply to the object.

TIP

You can apply shear by choosing Object ➪ Transform ➪ Shear to display the Shear dialog box.

Reflecting objects

You can reflect objects to create mirror images by using the Transform panel menu. The menu provides several additional options for manipulating objects.

Follow these steps to reflect an object:

1. **Select an object on the page and then choose Window ➪ Object & Layout ➪ Transform to open the Transform panel or open it from the Window menu.**

   The object's bounding box and handles appear. The Transform panel shows the current values of the selected object.

2. **Click the panel menu in the Transform panel.**

   The menu opens, revealing many options available for manipulating the object.

3. **Select Flip Horizontal from the Transform panel menu.**

   The object on the page flips on its horizontal axis. You can repeat this step with other reflection options in the menu, such as Flip Vertical.

TIP

You can also reflect objects with the Free Transform tool by dragging a corner handle past the opposite end of the object. The object reflects on its axis.

Understanding Clipping Paths

Clipping paths allow you to create a path that crops a part of an image based on the path, such as removing the background area of an image. This shape can be one you create using InDesign, or you can import an image that already has a clipping path. InDesign can also use an existing alpha or mask layer, such as one created using Photoshop, and treat it like a clipping path. Clipping paths are useful when you want to block out areas of an image and have text wrap around the leftover image.

You can create a clipping path directly in InDesign by using a drawing tool, such as the Pen tool. You use the tool to create a shape and then paste an image into this shape on the page. If you expect you'll want to reuse the clipping path and the image in the future, create the path in Photoshop and save it as part of the image instead. See Book 3, dedicated to Photoshop, for more details on how to do this.

Here's a quick way to remove the background of an image using InDesign:

1. **Choose File ➪ Place and browse to locate and open an image.**

2. **With the Pen tool, create a path right on top of the image.**

   The path should be created so that it can contain the image.

3. **With the Selection tool, click to select the image and then choose Edit ➪ Cut.**
4. **Select the shape you created in Step 1 and choose Edit ➪ Paste Into.**

   The image is pasted into the selected shape you drew with the Pen tool.

If an image placed into your InDesign layout already includes a clipping path, you can choose to use it by selecting Show Import Options at the time you place the image into your layout. On the Mac, click on the Options button to be able to check the Show Import Options checkbox.

If that image contains multiple paths, you can choose which path to use by selecting Object ➪ Clipping Path ➪ Options, and then selecting either Alpha Channel or Clipping Path from the Type menu, depending upon what is used in the image.

If an image you import into your layout was photographed against a solid light colored or white background — such as a picture of a product for a catalog that was taken against a white background — you can have InDesign create a clipping path for that image. Here's how:

1. **Find an image that has a solid background and place it into your document. Using the Selection tool, click to select the image.**
2. **Choose Object ➪ Clipping Path ➪ Options.**

3. In the Clipping Path dialog box, click the Type drop-down menu and choose Detect Edges. Adjust the threshold slider until the background surrounding the image disappears.

4. Click OK.

The path is created so that the image is visible and the background has been removed. The image can be placed over any other object, and it will reveal any items behind it.

TIP

For a clipping path to reveal objects behind it, the object with the clipping path needs to be at the top-most layer or the top-most object within its layer. You can check this by selecting the object and choosing Object ⇨ Arrange ⇨ Bring to Front and also using the layers panel, available by choosing Window ⇨ Layers.

Arranging Objects on the Page

This section covers the additional ways you can arrange objects, which gives you more control over the placement of elements in your document.

Aligning objects

In InDesign CC, you can align visually without the need for any extra tools or panels. If you keep Smart Guides activated (they're on by default), when you use the Selection tool to select and move objects around a page, guides appear automatically. These guides appear when the selected object is aligned with other objects on the page or with the page itself. If viewing these pesky guides starts to bother you, choose Edit ⇨ Preferences ⇨ Guides & Pasteboard (Windows) or InDesign ⇨ Preferences ⇨ Guides & Pasteboard (Mac) and turn off the four options underneath the Smart Guide Options heading.

You can also align objects on a page by using the Align panel: Choose Window ⇨ Object & Layout ⇨ Align. This panel gives you control over the way elements align to one another or to the overall page. The Align panel has many buttons to control selected objects. Mouse over a button to see its tooltip describe how that button aligns elements.

TIP

If you're not sure what each button does after reading the associated tooltip, look at the icon on the button. The icon is sometimes helpful in depicting what the Align button does to selected objects.

Here's how to align elements on the page:

1. **Select several objects on the page with the Selection tool.**

   Hold the Shift key while clicking each object to select several objects.

   Each object is selected when you click it on the page. If you don't have a few objects on a page, quickly create a couple new objects by using the drawing tools.

2. **Choose Window ⇨ Object & Layout ⇨ Align.**

   The Align panel opens.

3. **Select the kind of alignment you want to apply to the selected objects.**

   Try clicking the Align Vertical Centers button. Each selected object aligns to the vertical center point on the page.

Distributing objects

In the previous section, you find out how to align a few objects on a page, but what if the objects you're aligning aren't distributed evenly? Maybe their centers are lined up, but there's a large gap between two of the images and a narrow gap between the other ones. In that case, you need to *distribute* objects and align them. Distribute objects on the page to space them relative to the page or to each other in different ways. Here's how:

1. **Select objects on a page that are neither aligned nor evenly distributed by using the Selection tool while holding the Shift key.**

   The objects are selected when you click each one. All objects you select will be aligned to each other on the page.

2. **If the Align panel isn't open, choose Window ⇨ Object & Layout ⇨ Align.**

   The Align panel opens.

3. **Click the Distribute Horizontal Centers button and then click the Align Vertical Centers button directly above it on the Align panel.**

The selected objects are distributed evenly and aligned horizontally on the page.

Don't forget about the cool Multiple Place feature: It lets you distribute and align on the fly! Try this handy option to place several images at one time:

1. **Choose File ⇨ Place.**
2. **Press Ctrl (Windows) or ⌘ (Mac), select multiple images, and then click the Open button.**

3. **Before clicking to place the images, hold down Ctrl+Shift (Windows) or ⌘ +Shift (Mac). Note that you might have to wait a moment while the grid loads.**

   The cursor appears as a grid, as shown in Figure 6-6.

4. **Click and drag to create the rectangle that you want your images aligned to and distributed within.**

   The images are aligned and distributed automatically, as shown in Figure 6-7.

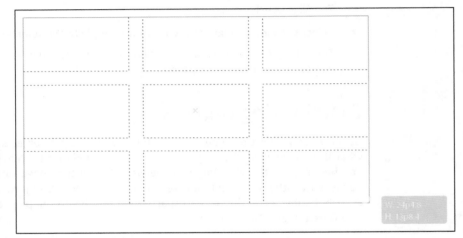

FIGURE 6-6: Place multiple images by holding down Ctrl+Shift (Windows) or ⌘ +Shift (Mac) and dragging.

FIGURE 6-7: Images are aligned and distributed automatically when placed.

» Packaging documents for delivery

» Understanding file formats

» Exporting to different file formats

» Printing documents

Chapter 7
Exporting to PDF and Printing

In addition to printing, you can also export InDesign documents into several different kinds of file formats. In this chapter, you take a closer look at the preparations for delivering a file to a print vendor, as well as many of the different kinds of files you can create electronically from an InDesign document. Before you print a document, it's good to make sure you have everything you need, so we start by looking at the preflight option.

Preflight: Preparing Your Documents for Printing

The *Preflight* feature in InDesign helps confirm that everything you need to print your document is ready and available. It also provides information about the document you're printing, such as listing its fonts, print settings, and inks used. Using Preflight can help you determine whether your InDesign document has unlinked images or missing fonts before printing it. Here's how to use the Preflight option:

1. **If you don't have a file to work with, you can access the file named `Preflight` in the Book04_InDesign folder in the DummiesCCfiles folder located at `https://www.agitraining.com/dummies`.**

   This page intentionally has missing fonts and images.

2. **With the file still open, choose Window ⇨ Output ⇨ Preflight.**

   The Preflight panel opens.

3. **Make sure the On checkbox is selected and the Profile is set to [Basic] (working).**

   These settings provide a basic preflight of the document, confirming that all images and fonts are available for printing.

4. **If any errors are noted in the document, click the page number listed to the right of the error in the preflight window to view the error.**

   Common errors, such as missing fonts and overset text (text that doesn't fit completely into a text frame) are all listed in the Preflight panel, and can generally be corrected easily with a little time. It's much better to fix errors before you print them!

TIP

When the Preflight option is turned on, a small green or red circle appears along the bottom-left corner of the document window. Green indicates that the document is preflighted and no errors are reported, whereas red indicates a possible error. You can simply click the errors drop-down along the bottom-left corner of the Document window to open the Preflight panel and investigate any possible errors.

If you are using the example `Preflight` file that we supplied, you will see that an image is missing, and that there is overset text. Follow these steps to remove the errors:

1. **Choose Window⇨Links top open the Links panel. Note that the image Runonbeach is missing. This file is in the Images folder in the DummiesCCFiles folder.**

2. **Click on the Relink icon, as shown in Figure 7-1, and navigate to DummiesCCFiles folder and then the Images folder.**

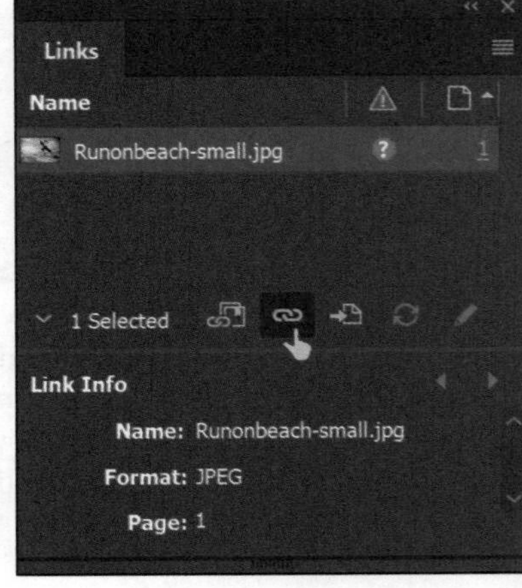

FIGURE 7-1:
The Relink icon.

3. **Select the Runonbeach image and click OK.**

   The error is now repaired.

4. **To remove the text overflow, simply expand the textbox by dragging the middle handle at the bottom. Do this until the text fits.**

   Note: If you wanted the text to continue on to another page you could also click on the red plus icon in the lower-right corner of the text box. When the cursor becomes loaded with text you can click and drag to create a new text area, or simply click on an existing text area. The text flow then continues from the overflow into the new textbox.

Packaging Your Documents

If you need to deliver your documents to a print service provider or to another designer, you'll want to provide everything that provider needs to continue working on the InDesign file. This is where the Package command is helpful. Package gathers copies of all images and fonts used in the document along with a copy of the InDesign document itself into a single folder that you can easily deliver to a print service provider or to a colleague. You can even use this option to archive completed jobs, to make sure that all necessary elements are stored together. Here's how:

1. **Choose File ➪ Package.**

   The Package dialog box opens. The Summary screen shows you all current images and fonts in the document, based on an analysis of that document.

2. **Click Fonts in the list on the left side of the dialog box.**

   Any fonts in your document are listed on this screen. Select fonts from this list and click the Find Font button to discover where they're located. Click the More Info button to see the path to the font. These fonts are saved directly into the package folder when you finish.

3. **Click Links and Images in the list on the left side of the dialog box.**

   The Links and Images screen lists the images within your document. If any images aren't properly linked, your document is incomplete and prints with pictures missing. Find the image, update it, and repair links before packing the file.

4. **When you're finished, click the Package button at the bottom of the dialog box.**

   Your document and all its associated files are saved into a folder. You're given the opportunity to name the folder, specify a location on your hard drive, and add instructions for printing.

TIP

If you're taking a file to a professional print service provider, you can give the vendor your original InDesign document or you can create a high-quality PDF file suitable for printing. It's a good idea to ask your print service provider about the file type it prefers to receive. By providing an InDesign file, you give the printing service a chance to fine-tune the document before printing, and if you provide a PDF, the document has limited capabilities for editing. Different print vendors have their preferences so communication is important.

Understanding File Formats

The kind of file you decide to create by exporting depends on your needs. The first thing to determine is where you'll use the exported file. For example, you might need to

- Put an image of your InDesign document, or page, on the web.
- Send an entire document to someone who doesn't have InDesign, but wants to receive it by email.
- Import the content into a different program, such as Adobe Animate or Illustrator.
- Take a particular kind of file somewhere else to print it.

Exporting InDesign documents lets you make them "portable" so that they can be used in different ways — such as on the web or in another program. You can choose from the many file formats InDesign supports, and you can control many settings related to the files you create.

Some of the file formats you can create from InDesign are listed in Table 7-1.

REMEMBER

JPEG, PDF, and EPS files can be exported from InDesign and then imported into other software programs. You can export these images for use in print after they're imported into a different graphics program, or you can use the images on the web. It all depends on how you set up the document for export and the settings you use.

After determining which file format to export your file in, learn how to export the file and the different kinds of settings you can control when doing so. The rest of this chapter shows you how to export different file types from InDesign.

TABLE 7-1 File Formats

File Format	Description
JPEG (Joint Photographic Experts Group), PNG (Portable Network Graphics)	Either of these commonly used formats for compressed images is a good choice for creating a picture of an InDesign page to post on a website. This is only a picture and is good for something like a thumbnail image representing a document. Don't use this if you want someone to actually read the InDesign file on the web — for that you'll want to export to PDF. This is simply a picture preview of what a document page looks like.
EPS (Encapsulated PostScript)	A self-contained image file that contains high-resolution printing information about all the text and graphics used on a page. This format is commonly used for high-quality printing when you need to have an image of an InDesign page used within another document — such as a picture of a book cover created with InDesign that needs to appear in a promotional catalog — so that you can use an EPS of the book cover in your layout.
XML (Extensible Markup Language)	Lets you separate the content from the layout so that all the content on a page can be repurposed and used in different ways — online or in print. Corporations commonly use XML for storing their product data when they have a large amount of information — such as thousands of items in a catalog.
PDF (Portable Document Format)	Used to exchange documents with users on different computer systems and operating systems. This format is used extensively for distributing files such as e-books and brochures. You may need to distribute the file to a wide audience or to a service provider for printing. Anyone who has installed Adobe Reader (also known as Acrobat Reader) on a computer can view your document. You can export PDFs designed for printing by selecting PDF (Print). Select PDF (Interactive) if your PDF contains hyperlinks or movies and will be distributed online instead of being printed.
Rich or Plain text (text files)	Can include formatting (Rich) or plain text only (Plain). A text file is a simple way to export content. If you need the text from your document to incorporate or send elsewhere, you can export it as Plain (Text Only), Tagged, or Rich Txt. If you need to send a document to someone who doesn't have InDesign, exporting it as text may be a good option.
EPUB	Use the e-pub file format to create electronic books that can be read using an electronic book reader, including devices such as the nook or iPad, or using any e-book reader software. With additional conversion, this file can also be read on the Amazon Kindle.
HTML	Use the HTML file format to export a document as HTML (HyperText Markup Language), which is the language used by web browsers. Documents exported using this option typically require some HTML editing for formatting and design, which you can do using Adobe Dreamweaver.
IDML	Use the InDesign Markup Language option to export InDesign documents in a format that can be read by earlier versions of InDesign. For example, a colleague is using InDesign CS6, but you are using InDesign CC. Export the document using the IDML option so that your colleague can successfully open it in InDesign CS6. If you simply save the file, the older version of InDesign is not able to open the document.

Exporting Publications

You can export publications from the Export dialog box. After you open it by choosing File⇨Export, you can choose the file format, name, and location. After specifying a name and location and a format to export to, click Save. A new dialog box opens, where you can make settings specific to the file format you picked. We discuss in the following sections some of the most common file formats you're likely to use for export.

Exporting PDF documents for printing

Create a PDF file of your document if you want to make sure that viewers see exactly what you have created — even if they don't have InDesign. A PDF file also limits the editing capabilities, making it unlikely your document will be changed. If you choose to export a PDF document, you have many options to customize the document you're exporting. You can control the amount of compression in the document, the marks and bleeds it has in InDesign, and its security settings. Here's how to export to PDF:

1. **Choose File⇨Export.**

   The Export dialog box opens.

2. **Choose a location in which to save the file and then enter a new filename.**

   Browse to a location on your hard drive using the Save In drop-down list (if you're using Windows) and name the file in the File Name text field. If you're using a Mac, name the file in the Save As text field and select a location from the Where drop-down list.

3. **Select Adobe PDF (Print) from the Save As Type (Windows) or Format (Mac) drop-down list at the bottom of the Export window.**

4. **Click Save.**

   The Export Adobe PDF dialog box appears with the General options screen open, as shown in Figure 7-2.

5. **Choose a preset from the Adobe PDF Preset drop-down list.**

   These presets are easy to use. If you're familiar with Adobe Acrobat and the Adobe Distiller functions, they're the same. (For more detailed information about what each setting does, see Book 6.)

TIP

   The presets on the Adobe PDF Preset drop-down list automatically change the individual export settings of a document. For example, you can select Smallest File Size from the list if you're displaying your work online or select High Quality Print if you plan for the PDF to be printed on home printers. Select Press Quality if you intend to have the PDF professionally printed.

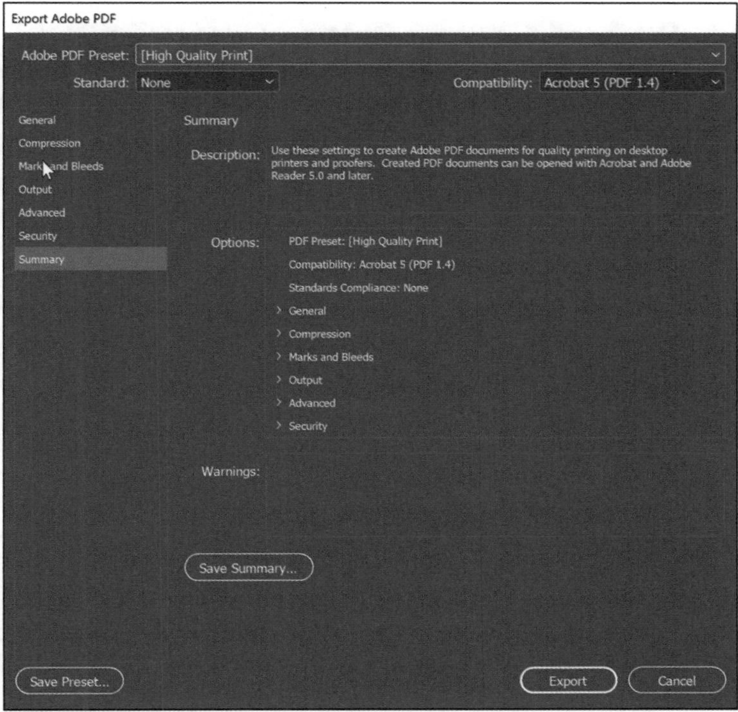

FIGURE 7-2: Choose the export options to export a file in the PDF format.

6. **Leave the Standard drop-down list at None.**

   Leave it at None unless, of course, you know about the PDF/X standards used for sharing advertisements. (For more on PDF/X, see Book 6.)

7. **Select a range of pages to export by typing the start page (and then a hyphen) and the end page in the Range text box.**

   You can also export nonconsecutive pages by separating the page numbers with a comma.

   By default, all pages are exported.

8. **Choose a compatibility setting for the PDF from the Compatibility drop-down list.**

   Compatibility settings determine which kind of reader is required in order to view the document. Setting compatibility to Acrobat 5 (PDF 1.4) ensures that a wide audience can view your PDF files. Some older PDF readers may not be able to interpret certain features in your document if you choose compatibility for a higher version. Setting this value to Acrobat 5 or later ensures that anyone who has installed Adobe Reader or Acrobat in the past ten years will be able to view the PDF file.

9. **Choose whether to embed thumbnails and whether to optimize the document, and then choose which kinds of elements to include in the file by selecting the checkbox to the left of the options in the Include section.**

   Other settings specify the inclusion of bookmarks, links, and other elements in the file. Unless you've added any of these elements, you don't need to worry about selecting these options. You may want to embed thumbnail previews, but Acrobat creates thumbnails automatically when the file is opened, so this can add to the file size unnecessarily.

TECHNICAL STUFF

   Click Security in the list on the left of the Export Adobe PDF dialog box to open the Security screen, where you can specify passwords to open the document. You can also choose a password that's required to print or modify the PDF file.

10. **Click the Export button to export the file.**

    The file is saved to the location you specified in Step 2.

Exporting EPS files

From InDesign, you can export EPS files, which are useful for importing into other programs. EPS files are single-page graphics files, which means that each exported InDesign page is saved as a separate EPS file.

TIP

You do not need to export an EPS file to place one InDesign file into another! If you're creating classified pages or any page that contains other InDesign pages, you can save yourself a few steps by simply choosing File⇨Place and selecting the InDesign file. InDesign can import InDesign files into a layout, or even PDFs — so the only reason to export to EPS is to create a picture for an older software or database that doesn't work with newer file formats.

Here's how to export EPS files:

1. **Choose File⇨Export.**

   The Export dialog box opens.

2. **Select a location on your hard drive to save the EPS files, enter a new filename, and select EPS from the Save As Type (Windows) or Format (Mac) drop-down list; click Save.**

   The Export EPS dialog box opens.

3. **Choose a page or range of pages to export.**

   Select the All Pages option to export all pages or select the Ranges option and enter a range of pages. If you want spreads to export as one file, select the Spreads radio button.

If you're creating more than one EPS file (for example, exporting more than one page of your InDesign document), the file is saved with the filename, an underscore, and then the page number. For example, page 7 of a `cats.indd` file would be saved as `cats_7.eps` in the designated location.

4. **From the Color drop-down list, select a color mode; from the Embed Fonts drop-down list, select how to embed fonts.**

   From the Color drop-down list, select Leave Unchanged to retain the color mode you're using for the InDesign document. You can also change the color mode to CMYK (Cyan, Magenta, Yellow, Black), Gray (grayscale), or RGB (Red, Green, Blue). For more information on color modes, flip back to Chapter 5 of this minibook.

   From the Embed Fonts drop-down list, select Subset to embed only the characters used in the file. If you select Complete, all fonts in the file are loaded when you print the file. Selecting None means that a reference to where the font is located is written into the file.

5. **Choose whether you want a preview to be generated for the file by choosing from the Preview drop-down list.**

   A *preview* (a small thumbnail image) is useful if an EPS file can't be displayed. For example, you're browsing a library of images and you see a small thumbnail image of the EPS file; so whether or not you use the image or open it on your computer, you can see what the file looks like. From the Preview drop-down list, you can select TIFF to generate a preview; select None if you don't want a preview to be created.

6. **Click the Export button to export the files.**

   The files are saved to the location you designated in Step 2.

Exporting JPEG and PNG files

You can export JPEG and PNG files from an InDesign document. These file formats are commonly used in web publishing and are useful if you need to place a picture of an InDesign document or page online. You can export a single image on a selected page, or even entire pages and spreads as a JPEG or PNG. JPEG and PNG files allow you to effectively compress full color or black-and-white images, which is useful if you need a picture of an InDesign page to appear on the web.

To export a JPEG or PNG image, follow these steps:

1. **Select an object on a page or make sure that no object is selected if you want to export a page or spread.**
2. **Choose File ➪ Export.**

   The Export dialog box opens.

3. **Type a filename, locate the spot where you want to save the file on your hard drive, and select either JPEG or PNG from the Save As Type (Windows) or Format (Mac) drop-down list; click Save.**

   The Export JPEG or PNG dialog box opens.

4. **If you want to export a page, select the Pages radio button and enter the page number in the Range text field; if you want to export the selected object, make sure that the Selection option is selected.**

   The Selection option is available only if a selection was made in Step 1.

5. **Choose an image quality and format to export by selecting from the Image Quality and Format Method drop-down lists.**

   The Image Quality drop-down list controls the amount of compression that's used when you export a JPEG or PNG file. Choose from these options:

   - *Maximum:* Creates an image with the largest file size and highest quality
   - *High:* Creates an image with a larger file size and high quality
   - *Medium:* Creates an image with medium file size and medium quality
   - *Low:* Creates a smaller file of lesser quality because it includes less image information

   If you select the Baseline format from the Format Method drop-down list, the entire image has to be downloaded before it's displayed in a web browser. Select Progressive to show the image in a progressively complete display as it downloads in a web browser.

   The Format Method options do not appear if you are creating a PNG image.

6. **Click the Export button.**

   The file is exported and saved to the location you specified in Step 3.

Exporting text files

You can extract text from an InDesign document so that it can be edited or used elsewhere. The text formats vary slightly depending on the text in your document.

To export text, follow these steps:

1. **Select the Text tool from the toolbox and select some text within a text frame in your document, or place the cursor within a text frame where you want to export all the text.**

   The cursor must be in a text frame in order to export text.

2. **Choose File ➪ Export.**

   The Export dialog box opens.

3. **Enter a filename, select a location to save the file in, and select Text Only from the Save As Type (Windows) or Format (Mac) drop-down list; click Save.**

   The Text Export Options dialog box opens.

4. **Choose a platform and encoding for the export.**

   Select either PC or Macintosh from the Platform drop-down list to set the PC or Mac operating system compatibility. Select an encoding method for the platform you choose from the Encoding drop-down list; you can choose either Default Platform or Unicode.

   The universal character-encoding standard *Unicode* is compatible with major operating systems. *Encoding,* which refers to how characters are represented in a digital format, is essentially a set of rules that determines how the character set is represented by associating each character with a particular code sequence.

5. **Click the Export button.**

   The file is exported and saved to the location you specified in Step 3.

Printing Your Work

You can print your work from an InDesign document in many different ways, with many kinds of printers and processes. You can either use your home or office printer, or you can take your work to a professional printing service. Both printers and printing services vary in the quality of production they can offer you.

The following sections look at the different ways you can set up a document for printing and the kinds of issues you may encounter during this process.

Understanding bleed

If you want an image or span of color to go to the edge of a page, without any white margins, you *bleed* it off the edge of the document. *Bleeding* extends the print area slightly beyond the edge of the page into the area that will be cut and removed during the printing process. (Professional printers sometimes print your documents on a larger page size than necessary, and then trim off the extra paper.) You can bleed your own work, too. While you're designing, you can turn on crop and bleed marks to show where the page needs to be trimmed to make sure that the image bleeds far enough off the edge of the page area. We cover this topic in the next section.

Doing it yourself: Printing and proofing at home or the office

Because many InDesign documents are designed to be professionally printed, the available print settings are quite robust. Just because they are robust doesn't make them complicated, though. We demystify the settings for you here:

Choose File⇨Print to open the Print dialog box. Choose the printer you intend to use from the Printer drop-down menu. If you don't have a printer, choose Adobe PDF from the Printer drop-down menu to follow along with the steps below.

Many printing options are available in the list on the left side of the Print dialog box. Clicking an item causes the dialog box to change, displaying the settings you can change for the selected item. Click each of the items below to better understand these common printing options:

- **General:** Set the number of copies of the document you want to print and the range of pages to print. You can select the Reverse Order checkbox to print from the last to first page. Select an option from the Sequence drop-down list to print only even or odd pages instead of all pages. If you're working with spreads that need to be printed on a single page, select the Spreads radio button.
- **Setup:** Define the paper size, orientation (portrait or landscape), and scale. You can scale a page so that it's as much as 1,000 percent of its original size or as little as 1 percent. You can (optionally) constrain the scale of the width and height so that the page remains at the same ratio. The Page Position drop-down list is useful when you're using paper that's larger than the document you've created. This option helps you center the document on larger paper.
- **Marks and Bleed:** Turn on or off many of the printing marks in the document, such as crop, bleed, and registration marks. For example, you may want to show these marks if a bleed extends past the boundaries of the page and you need to show where to crop each page. You see a preview of what the page looks like when printed, and you can select options to print page information (such as filename and date) on each page.

REMEMBER

- **Output:** Choose how to print pages — for example, as a separation or a composite, using which inks (if you're using separations), or with or without trapping. InDesign can separate and print documents as plates (which are used in commercial printing) from settings you specify. Depending on the printer you have selected, you may or may not have separations as an option.

- **Graphics:** Control how graphics and fonts in the document are printed. The Send Data drop-down list controls bitmap images and specifies how much of the data from these images is sent to the printer. Here are some other options available when printing:
  - *All:* Sends all bitmap data. Use this for the highest quality printing, but it may take a long time for your jobs to print if you're using high-resolution images.
  - *Optimized Subsampling:* Sends as much image data as the printer can handle. Use this if you are having difficulty with printing high-quality images when using the All option.
  - *Proxy:* Prints lower-quality images mostly to preview them.
  - *None:* Prints placeholder boxes with an X through them. This option lets you quickly print a proof of a layout to just get a quick understanding of a page design.
- **Color Management:** Choose how you want color handled when it's output. If you have profiles loaded in your system for your output devices, you can select the profiles here. Color management is more commonly used by commercial print vendors and vendors with critical color needs. It's a complex topic — entire books are available on the subject of color management — and we don't have space to cover it in depth here.
- **Advanced:** Determine how you want images to be sent to the printer. If you are unfamiliar with Open Prepress Interface (OPI), you can leave this setting at the default. Also known as image-swapping technology, the OPI process allows low-resolution images inserted into InDesign to be swapped with the high-resolution version for output. It's a highly specialized setting used by commercial print vendors — such as catalog print vendors — that work with a large number of images.

  Flattening needs to be addressed if you use a drop shadow, feather an object in InDesign, or apply transparency to any objects, even if they were created in Photoshop or Illustrator.

  Use the preset Medium Resolution for desktop printers and High Resolution for professional press output.
- **Summary:** You can't make modifications but you can see a good overview of all your print settings.

TIP

After you finish your settings, click the Save Preset button if you want to save the changes you've made. If you think you may print other documents with these settings repeatedly, using the Save Preset feature can be a great timesaver.

After you click the Save Preset button, the Save Preset dialog box opens, where you can enter a new name to save the settings. The next time you print a document, you can select the saved preset from the Print Preset drop-down list in the Print dialog box.

Click the Print button at the bottom of the Print dialog box when you're ready to print the document.

» Alternate layouts for varying devices

» Liquid Layout rules

» Exporting digital books to EPUB

» Sharing for review online

» Publishing online

Chapter 8

Creating Digital Documents Using EPUB and Publish Online

In addition to creating printed documents, InDesign can also be used for creating digital documents and electronic books that are shared in digital format. InDesign includes tools for creating and sharing digital versions of flyers, brochures, and books without needing to use any type of coding.

These digital formats let others view your documents, even if they do not have the InDesign software. Digital documents can be shared across a variety of devices, from phones and tablets to traditional computers. InDesign can be used to create different kinds of digital files depending upon the type of content you'll be sharing and how it will be used. In this chapter, we discuss three different ways to share digital documents from InDesign:

- **Publish Online:** To share documents that can be viewed while readers are online and have Internet access, you can use this option.

- **EPUB:** The standard for electronic books that can be viewed online and offline is EPUB. If your content is text-heavy, the EPUB reflowable file format is what you should use. If your document is highly stylized and designed, the EPUB fixed layout is a better alternative and allows for animation and interactivity.
- **PDF:** If a document needs to retain its exact format and be shared with users across many different types of devices and operating systems, the PDF standard is a useful option.

In this chapter, you discover how to create documents for distribution using all three of these digital publishing formats.

Choosing the Right Digital Format

When distributing files digitally, you'll want to choose the best file format depending upon any requirements you have for the document or the type of content it contains. See Table 8-1 to help determine the best formats for files you wish to share digitally.

TABLE 8-1 File Formats for Digital Distribution

Documents Type	PDF	Publish Online	EPub Reflowable	EPub Fixed Layout
Requires formatting that should not change	X			
Contains Interactivity or multimedia		X		X
Long content, intended to read like a book			X	X
Must be downloadable and accessible offline	X		X	X

Planning Layouts for Digital Distribution

Before sharing a document digitally it's useful to consider the types of devices that viewers will be using, what they will need to do with the file, and the orientation that will be easiest to use when reading. For example, your printed documents may be printed on a letter size paper, but digitally they may need to be viewed on tablet or phone held horizontally. This planning allows you to create and distribute documents that will better meet the needs of your audience.

As part of this planning process, if a document is to be shared only in digital formats, you can design the document's layout for specific devices or orientations, such as an iPad horizontal layout which can be selected at the time you create a new document.

Adapting Print Documents to Share Digitally

If you have a file that was originally created for print format and you also want to share it digitally, you may want to adapt it for electronic distribution so that it is better suited for onscreen viewing. To do this, you will save a copy of the original InDesign document and work on the duplicated document as you adapt it to better meet the needs of a digital audience.

To convert an existing print document to a digital document layout, follow these steps:

1. **Open the existing document and choose File ⇨ Save As and name the copy of the document to reflect that it is for digital distribution. Then click Save.**

   Choose a name that relates to your intended distribution. Consider including terms such as Online, or eBook to the filename so you can distinguish the electronic version from the print version of the InDesign files.

   TIP

   If you're creating an EPUB document for a book that primarily uses text and that will be viewed on many readers, after Step 1 you can go directly to Step 6, because you do not need to specify layout details or size details for specific devices. For all other documents, including EPUB documents that require highly formatted layouts, continue with Steps 2 through 5.

2. **If you would like to have InDesign start the process of adjusting the size and positioning of objects for the new layout, choose Layout ⇨ Margins and Columns, click the Enable Layout Adjustment checkbox, then click OK.**

   If you prefer to adjust the layout manually for the new document size, skip this step. If you are creating different sized versions of a document for publishing to different destinations, see Creating liquid layout rules for alternative layouts in this chapter as an alternative.

3. **Choose File ⇨ Document Setup.**

4. **In the Document Setup window, choose Mobile from the Intent drop-down menu for documents to be shared with smaller screens, or choose Web for larger screens, then choose the desired Page Size and orientation (vertical or horizontal) and click OK.**
5. **Scroll through the document and make manual adjustments to the layout and confirm you are comfortable with the automatic adjustments. Use the Selection tool to move objects and use the Pages panel to add pages as needed, then choose File ⇨ Save to save these changes.**
6. **Choose Window ⇨ Articles to open the Articles panel. Use the Selection tool to drag items from your layout into the Articles panel, defining the order in which they will be displayed in an EPUB or HTML file.**

   If you have something that you don't want to include in the digital version of the book, don't place it into the Articles panel or delete it from the layout. If you have images that are specific to an area of text, make sure that those images are anchored to the text. This ensures that the images appear near the text, as opposed to at the end of the story.
7. **Save your document again after placing all elements to be exported into the Articles panel.**

Creating liquid layout rules for alternative layouts

The Liquid layout capability lets you use InDesign more efficiently when creating various sized layouts for a single project. It often allows for less manual work adapting the content between versions of a document, such as a printed and online version. You use liquid layouts in combination with another capability known as alternative layouts. Use these capabilities in the design process before converting your documents to different sizes. The process starts by creating liquid layout rules for how objects on page should be adjusted to a new layout. The liquid layout rules specify how content will be adjusted when page sizes are changed. After liquid layout rules are established for each page, then the page size can be changed or alternate layouts can be created.

Liquid layouts and alternate layouts are optional when designing for different page sizes. They help adapt content for different destinations, but are not required to prepare documents for digital publishing. If you prefer to not use automated tools to adapt the content on your pages to different size versions of your documents, you can skip this section.

To create Liquid Layouts, follow these steps:

1. **Choose the Page tool and click on a page to identify the page to which liquid layout rules will be applied.**
2. **Choose Layout ➪ Liquid Layout to open the Liquid Layout panel.**
3. **From the Liquid Layout panel, choose a Liquid Page Rule for each page as shown in Figure 8-1. Choose from these options:**

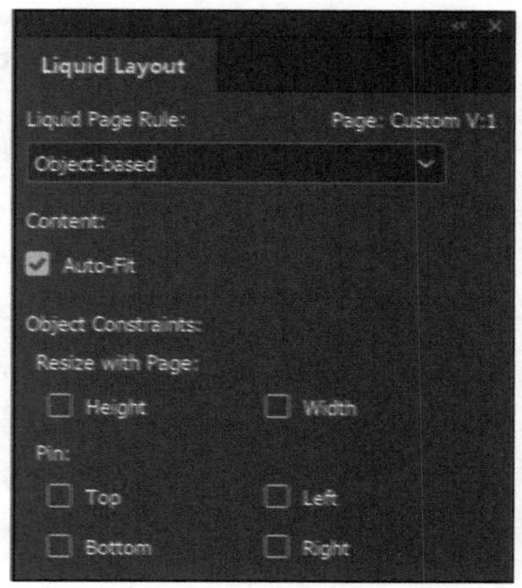

FIGURE 8-1: Create a Liquid Layout.

   - *Scale:* The entire page is resized with all elements scaling proportionally. This may leave white space around the sides of a portrait page converted to horizontal, or white space around the top and bottom for a landscape page converted to portrait. This does not fill the space of the new layout.
   - *Re-center:* Content is not resized although it is centered in the new page size.
   - *Guide-based:* Content adapts to special liquid guides that you can create by choosing the Page tool then clicking and dragging ruler guides from either the vertical or horizontal page ruler.
   - *Object-based:* Each object can have specific rules established by choosing this option then using the selection tool to select each page object and define the rules. The height or width of objects can be adjusted or remain constant, and the location of each object can be set to the top, bottom or side of the page.

4. **Repeat the process of selecting pages, as described in Step 1, and specifying rules for the pages. You can choose a different rule for each page of the document depending on the content.**
5. **Test liquid layout functionality by choosing the Page tool and dragging the page handles along the outside edges of the page. Notice how the objects change location based on the liquid layout settings as the page size changes.**

Creating alternate layouts

After you have indicated how you want your pages to change using the Liquid Layout rules, you can create separate horizontal and vertical layouts for your document, as well as layouts for different size versions of your document.

To create an alternative layout for a document, follow these steps:

1. **From the Layout menu, choose Layout ⇨ Create Alternate Layout to create a new alternative layout for your current layout.**

   For example, if you have a portrait (vertical) layout for print, you may want to also create a landscape (horizontal) version of the layout for onscreen use.

2. **Name the alternate layout and specify the page attributes outlined below, and then click OK, as shown in Figure 8-2.**

   - *Name:* Enter a name for the alternate layout.
   - *From Source Pages:* Specify if you wish to create an alternate layout from only a single page, a range of pages, or all pages.
   - *Page Size:* Set the dimensions for the alternate layout along with the orientation.
   - *Liquid Page Rule:* To use the rules set on each page select Preserve Existing; otherwise specify a rule to apply to any pages for which an alternate layout is being created.
   - *Link Stories:* Select this to create a link between any text in the original layout and the text duplicated in the new layout. If text in the original layout is revised, this option makes it easier to update the text in the new layout via the Links panel.
   - *Copy Text Styles to New Style Group:* Select this if you are using text styles and wish to create a duplicate of the styles so they can be varied in the alternate layout without affecting the original styles.
   - *Smart Text Reflow:* This option removes manual text formatting such as forced line breaks that were applied to the original source layout.

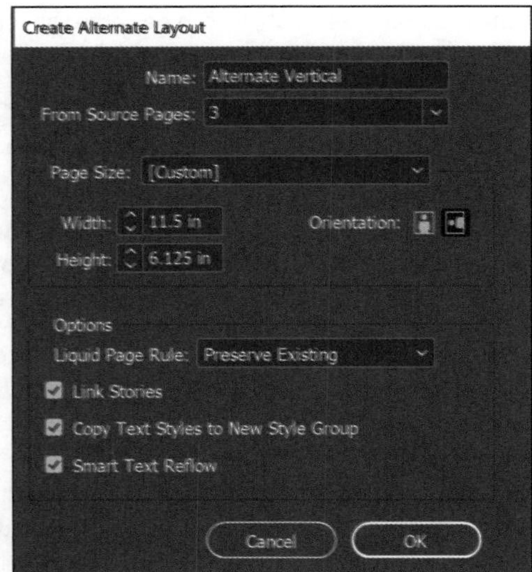

FIGURE 8-2: Create an alternate layout.

3. **Locate the alternate layout in the Pages Panel and make adjustments to page content as needed.**

   While the Liquid Layout starts the process of making content fit into a new page size or orientation, it does not replace your role as a designer. You will like need to finalize the design and review any changes to your new alternate layout after it is created.

REMEMBER

   Notice that frames that were duplicated onto the alternate layout may contain a link icon. This indicates that the text and images on the alternate layout are linked to the original. Changes made to the items in the original layout are reflected on the alternate layout. If an item in the alternate layout is different from the original layout, a yellow triangle appears instead of the link symbol. Double-click the yellow triangle, and the item in the alternate layout is then synchronized with the original layout. Any changes made to an alternate layout will not be reflected in the original layout.

Adding interactivity to digital documents

You can add interactive features to your digital files using several options. Depending upon the format of your final export, some of this functionality can be used. For example, buttons and forms are useful for creating interactive PDFs, while Animations are useful if you wish to use the Publish Online option.

Use these InDesign capabilities to add interactivity to your digital documents:

- **Animation panel:** Use the Animation panel to apply animation to objects on your page. For example, you can have an object fade in as the page appears, or fade out and disappear. Access this panel by choosing Window ➪ Interactive ➪ Animation, and apply animation to individual objects by selecting them and choosing an animation option from the Preset drop-down menu in the Animation panel.
- **Timing panel:** Use the Timing panel to modify animations you have created using the Animation panel. Access this panel by choosing Window ➪ Interactive ➪ Timing. Use the Event drop-down list to specify when an object's animation occurs, and use the Delay option to specify if the animation should wait before it occurs.
- **Buttons and Forms:** Use this panel to convert text or images to buttons that become interactive when the InDesign document is exported to certain formats such as modern EPUB or PDF. Access this panel by choosing Window ➪ Interactive ➪ Buttons and Forms. Buttons can have actions associated with them, so that something occurs when they are clicked. For example, when clicked, a button can play a movie or sound, navigate to a specific page, or play an animation.

Exporting Digital Books as EPUB

Digital books that are distributed through Apple, Google and many other digital book sellers are shared in the EPUB file format for use on readers such as the iBooks app, Nook, Kobo reader, or Sony eReader. Amazon's Kindle reader also has digital books start in the EPUB format, although the files used on the Kindle undergo an additional conversion process as the books are made into Amazon's proprietary format. The bottom line if you want to create electronic books: You need to know about creating EPUB files. InDesign is able to convert your files to the EPUB format, although you will need to make some decisions regarding the specific EPUB format you wish to use: either reflowable or fixed layout. Here you will discover how to create an EPUB file for your digital books using InDesign and which format is right for your books.

Preparing EPUB (Reflowable) Books

The EPUB – Reflowable format is best for books that are text-heavy, such as a novel or a biography. To prepare a book file for exporting to EPUB, do the following:

1. **Make sure any images used in the book are anchored to the portion of the text that relates to the picture. Do this by using the following steps:**

   A. *Choose the Selection tool.*

   B. *Click an image.*

   C. *Use the Edit ⇨ Cut command, and then switch to the Type tool.*

   D. *Click at the location where the image should remain near and use the Edit ⇨ Paste command.*

   The exact location of the image in the InDesign layout is less important, as the EPUB will take instructions and lay out the file based upon where the image is anchored to the text.

2. **Open the Articles panel by choosing Window ⇨ Articles.**
3. **Click on any text frames in the document and drag them to the Articles panel.**
4. **Move the stories up or down in the Articles panel to specify their sequence when the EPUB is created.**

   The order in which content flows in a reflowable EPUB file is dictated by the Articles panel.

Exporting EPUB (Reflowable) Books

The Exporting an EPUB converts the file from an InDesign format into a format the digital book reader can use. To convert a book into an EPUB reflowable format:

1. **With an InDesign document open, choose File ⇨ Export.**

   The Export dialog box appears.

2. **From the Save as Type (Windows) or Format (Mac) drop-down list, select EPUB (Reflowable). Find a location on your computer or network for the document to be saved. Enter a name for the EPUB file in the File Name portion of the Export dialog box.**

3. **Click the Save button.**

4. **In the EPUB Reflowable Layout Export Options dialog box set the options that are suitable for your book, as shown in Figure 8-3, which shows the General tab, one of eight tabs for specifying EPUB options.**

   Here is an overview of essential settings in the General tab when creating an EPUB:

   - *Version:* By default, the EPUB 2.0.1 option is selected. This allows readers on older devices to be able to view your reflowable content. If sharing files that will only be read on more modern devices and readers, this can be changed to EPUB 3.0.

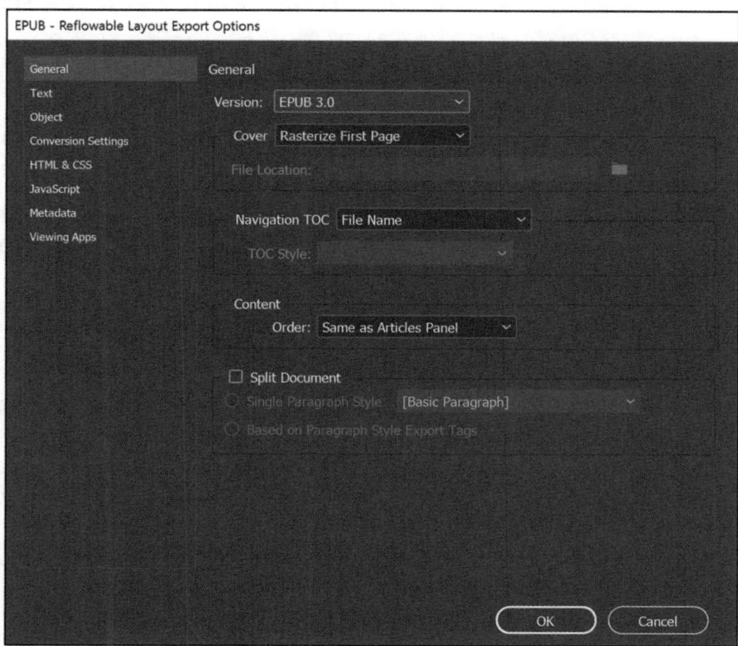

FIGURE 8-3: The Reflowable Layout Export options.

- *Cover:* If you want InDesign to generate the cover image displayed for the EPUB, choose whether it should use the first page (Rasterize First Page) of the document or choose a specific image file to use as a cover for your book if you prefer. The cover image is displayed in the digital library of book readers or within books stores.
- *Navigation:* Choose Multi Level (TOC Style) if you want InDesign to automatically generate a table of contents that will be used by readers to more easily navigate the EPUB file.
- *Content:* For Order, choose Same as Articles Panel. This flows your content based on how you added your images and stories into the Articles panel as discussed in the previous section.

5. **Additional Options available in the Text and Objects tab, on the left side of the dialog box.**

   You can specify additional options for how your text (bullets and numbering lists) and graphics (fixed or relative size) will be displayed under the Text and Objects tabs.

6. **Click the Conversion Settings tab, on the left side of the dialog box.**

   In the Conversion Settings, specify the resolution that images should be set at when created for the EPUB. If you choose Automatic for the format, InDesign will determine whether to save an image as a JPG, GIF, or PNG.

7. **Make these adjustments as needed, before creating the EPUB:**

   - *Metadata:* Select the Metadata tab. Document metadata is used by EPUB readers to describe the book title, publisher, ISBN, and other information. You can enter the Publisher here and the ISBN number for your book in the Identifier text box. Additional document metadata can be entered in the Document Information field. Access this by choosing File ➪ File Info.
   - *HTML & CSS:* Select the HTML & CSS tab. If you have created Cascading Style Sheets (CSS) to control the formatting for your EPUB files, you can add a style sheet, by selecting Add Style Sheet.
   - *Viewing Apps:* Select the app you wish to use on your computer to view EPUB files after they have been created.

TIP

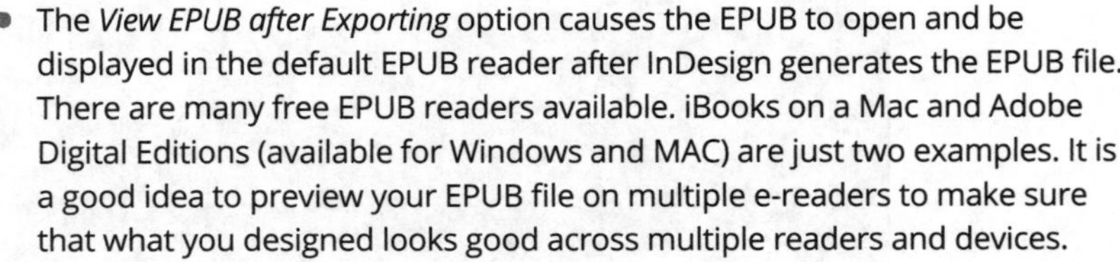

- The *View EPUB after Exporting* option causes the EPUB to open and be displayed in the default EPUB reader after InDesign generates the EPUB file. There are many free EPUB readers available. iBooks on a Mac and Adobe Digital Editions (available for Windows and MAC) are just two examples. It is a good idea to preview your EPUB file on multiple e-readers to make sure that what you designed looks good across multiple readers and devices.

TIP

8. **Click the OK button.**

The EPUB is generated. If you selected the View EPUB after Exporting checkbox, the EPUB file opens using your default EPUB reader.

TECHNICAL STUFF

If the EPUB Options dialog box looks similar to the HTML Export window, it is because EPUBs actually use HTML as their foundation. Electronic book readers are essentially specialized browsers designed to display books formatted using HTML and CSS. All the HTML and CSS content for an EPUB file is contained within the compressed folder named with the EPUB file extension.

Exporting EPUB (Fixed-Layout)

If your book is heavily designed and has very graphic layout, or contains any interactive elements added using InDesign, you will want to use the EPUB (Fixed-Layout) option. This is useful for cookbooks, picture books, or any other book with specific formatting that should be retained when converted to a digital book format. The steps to export your document and many of the options are similar to those used for the EPUB (Reflowable). Here is the process for creating a Fixed Layout EPUB from InDesign:

1. **With an InDesign document open, choose File ⇨ Export.**

   The Export dialog box appears.

2. **From the Save as Type (Windows) or Format (Mac) drop-down list,** select EPUB (Fixed-Layout) and then find a location on your hard drive or network for the document to be saved. Enter a name for the EPUB file in the File Name portion of the Export dialog box.

3. **Click the Save button.**

4. **In the EPUB Fixed Layout Export Options** dialog box, specify the General options for your book, as shown in Figure 8-4. Here is an overview of essential settings in the General tab when creating a Fixed Layout EPUB:

   - *Version:* By default, the EPUB 3 option is selected and no other version can be chosen if you want a fixed layout.
   - *Navigation:* Choose Multi Level (TOC Style) if you want InDesign to automatically generate a table of contents that will be used by readers to more easily navigate the EPUB file.

     However, if you are exporting a graphic heavy book such as a children's book, you can choose none and the table of contents will not be re-created in the sidebar of the reader.

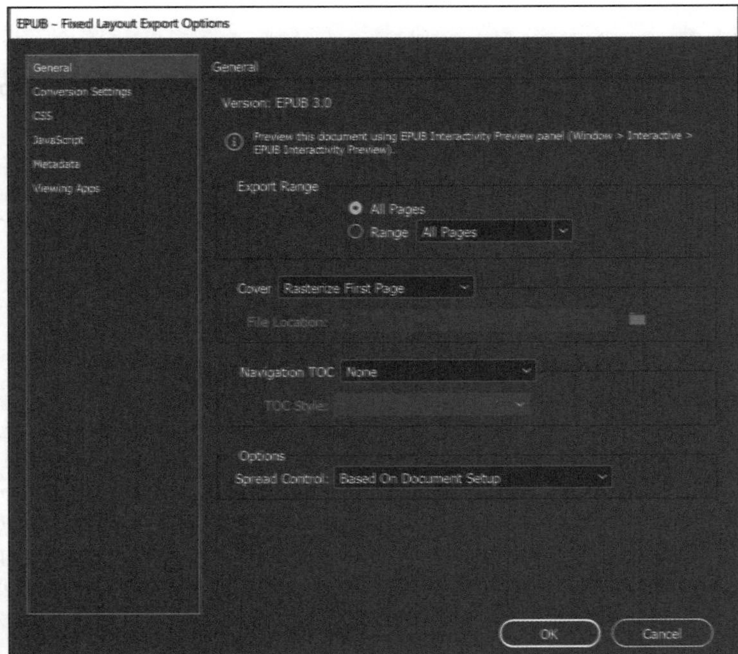

FIGURE 8-4: The Fixed Layout Export options.

- *Options:* For Spread Control, depending on how you created your document, using Facing Pages or not, you can control how these pages are displayed in the reader. If you choose Based on Document Setup, InDesign will output your pages as either single pages or spreads, depending on how you defined your pages in the Document Setup Dialog Box.

5. **Make any changes needed for Conversion Settings, CSS, JavaScript, Metadata, and Viewing Apps as discussed in the For Additional Export Layout Tab options; see the definitions for Export Reflowable.**
6. **Click OK to save.**

TECHNICAL STUFF

Both EPUB file formats are created using HTML and CSS. The HTML and CSS from an EPUB Reflowable is easily changed using a web editor such as Dreamweaver or Sigil. However, if you need to make a change to an EPUB (Fixed Layout) file, it is best if you go back to the original InDesign file, make your changes and reexport your document. The HTML and CSS for fixed layout EPUBs is rather complex and difficult to make changes to.

Publish Online

You can publish your InDesign documents online so that they can be viewed within a web browser. Publish Online works by creating a digital version of your document that can be viewed using any modern web browser connected to the Internet like Chrome, Firefox, or Safari. Documents published online require an Internet connection although you can allow readers to download a static PDF version of the document. This publishing option allows for more interactive capabilities such as forms, buttons, and animations. Documents shared using Publish Online are uploaded to Adobe's server; you receive a URL that you can share for others to view the file. Because the files reside on Adobe's servers, this option is not ideal if you want content to be placed on your own website.

To publish an InDesign document online:

1. **With the InDesign document open that you would like to publish online, choose File ➪ Publish Online or click the Export icon along the top-right side of the interface and select Publish Online.**
2. **In the Publish Your Document Online interface, enter a Title and Description for your document, as shown in Figure 8-5.**

   If your document contains left and right pages that should be kept together, such as an image displayed across two side-by-side pages, choose the Export as Spreads option; otherwise, use the Single Page option. Selecting the Allow Viewers to Download Your Document as PDF option enables a copy of the file to be downloaded for offline viewing by readers. If you do not want viewers to embed your document on websites that are not yours, select the Hide the Share and Embed options checkbox.

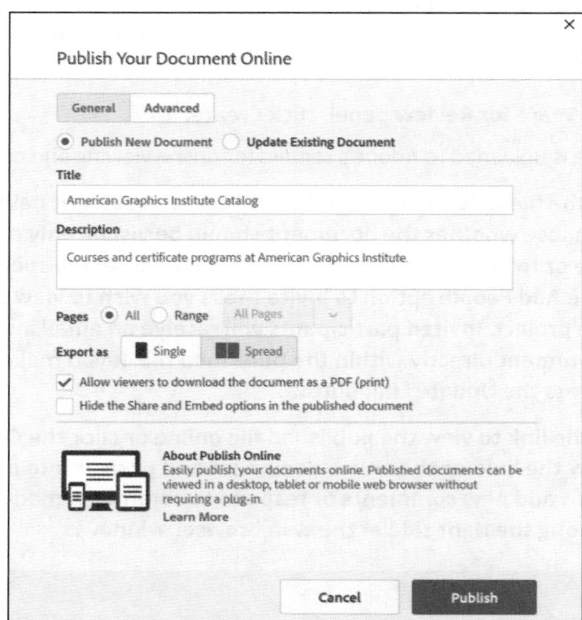

FIGURE 8-5: The Publish Your Document Online options.

3. **Click the Advanced button in the Publish Your Document Online interface and specify whether to publish all pages or only specific pages.**

   You can also specify which page or image should be used to represent the document as a small image thumbnail. If desired, you can fine-tune the quality of the images and, if you have made the option of a PDF download available, you can also choose the quality of the PDF, which will affect the file size of the PDF document.

4. **Click Publish to publish your document online.**

   InDesign provides a progress bar as the document is converted and uploaded to Adobe's servers. After the document is uploaded, you can click View Document to open your web browser and see it online, or copy the URL address to share the published document.

Share for Review

If you need to collaborate or receive feedback on your work, the Share for Review option may be useful. It creates an online version of your document, similar to Publish Online, with fewer options when publishing and adds capabilities for reviewers to provide comments.

To share an InDesign document for review:

1. **With the InDesign document open that you would like to share for review, click the Export icon along the top-right edge of the interface and select Share for Review.**
2. **In the Share for Review panel, click Create.**

   The file is uploaded to Adobe's servers for online viewing and commenting.

3. **After the file uploads, specify the title of the document being reviewed and choose whether the document should be visible only to certain people or to anyone with the link. If you limit access to specific people, use the Add People option to invite those you wish to view and comment on the project. Invited participants will receive an email invitation and can comment directly within the published file. If you make changes, you can press the Update Link button.**
4. **Click the link to view the published file online or click the Clipboard icon to copy the link for the online document and paste it into a web browser. You can add new comments or respond to comments made by collaborators along the right side of the web browser window.**

5. **Return to the InDesign document you have shared for review. Any comments made online are shared directly in the InDesign document within the Review panel. If the Review panel is not open, choose Window ⇨ Comments ⇨ Review. Within the review panel, click the eye icon to show or hide comments within the InDesign document. As you update the document to address comments, select the ellipse (. . .) adjacent to a comment and change the status of the comment.**

Sharing documents for review online is useful when all reviewers have access to the Internet while making comments. If a reviewer needs access to a document while offline, then creating a PDF may be a better option. Comments added to a PDF version of a document using the Adobe Acrobat annotation tools can be imported into InDesign using the PDF Comments panel which you can open by choosing Window ⇨ Comments ⇨ PDF Comments.

5 Illustrator CC

Contents at a Glance

- » Looking around the Document window
- » Opening and creating documents
- » Checking out the tools and panels
- » Changing your view
- » Zooming in and out

Chapter 1 Discovering Illustrator CC

Adobe Illustrator is generally used to create vector logos, illustrations, maps, packages, labels, signage, art for applications, websites, and more. (See the nearby "Vector graphics" sidebar for more information.)

This chapter gets you started with Illustrator and helps you understand when Illustrator is the tool best suited for creating your art.

Deciding When to Use Illustrator CC

How do you draw a line in the sand and decide to create graphics in Illustrator rather than in Photoshop? The following questions can help you make up your mind:

- » Does the graphic need to be scaled? Will it be used in sizes that you cannot predict at this time?

» Is the artwork technical in nature? Are precise measurements and placement of elements critical?

» Does the artwork require the use of spot colors, using a system such as the Pantone Matching System (PMS)?

» Will the artwork be used frequently in different formats, such as print and onscreen in apps or websites?

» Is the artwork illustrative with many solid colors and occasional gradients and vignettes? (If you have a photograph, or an illustration that contains many variations of tonal values, use Photoshop for your editing.)

If you answered yes to any of these questions, it is probably best to create and edit your artwork in Illustrator.

Keep in mind that by using Illustrator, you gain these benefits:

» **Illustrator can save and export graphics into most file formats.** By choosing to save or export, you can create a file that can be used in most other applications. For instance, Illustrator files can be saved as `.bmp`, `.jpg`, `.pdf`, `.svg`, `.tiff`, and `.png` files, to name a few.

» **Illustrator files are easily integrated into other Adobe applications.** You can save Illustrator files in their native format and open or place them in other Adobe applications such as InDesign, Photoshop, and Adobe XD. You can also save Illustrator artwork as Portable Document Format (`.pdf`) files, which lets anyone using the free Acrobat Reader software open and view the file but maintains editing capabilities when the file is opened later in Illustrator.

» **Illustrator artwork is reusable because the resolution of vector artwork isn't determined until output.** In other words, the higher the quality of your printer, the higher quality of the output. Illustrator graphics are quite different from the bitmap images you create or edit in Photoshop, where resolution is determined as soon as you scan, take a picture, or create a new bitmap (created from pixels) document.

» **Illustrator has limitless scalability.** You can create vector artwork in Illustrator and scale it to the size of your thumb or to the size of a barn, and either way, it still looks good. See the nearby sidebar "Vector graphics" for more information.

VECTOR GRAPHICS

Vector graphics are made up of lines and curves defined by mathematical objects called *vectors*. Because the *paths* (the lines and curves) are defined mathematically, you can move, resize, or change the color of vector objects without losing quality in the graphic.

Vector graphics are *resolution-independent:* They can be scaled to any size and printed at any resolution without losing detail. On the other hand, a predetermined number of pixels creates bitmap graphics, so you can't *scale* (resize) them easily — if you scale them smaller, you throw out pixels; if you scale them larger, you end up with a blocky, jagged picture.

The following figure shows the differences between an enlarged vector graphic (left; notice the smooth edges) and an enlarged bitmap graphic (right; note the jagged edges). Many designers create company logos as vectors to avoid problems with scaling: A vector graphic logo maintains its high-quality appearance at any size.

Creating a New Document

In Illustrator CC, you may see a start screen when you launch Adobe Illustrator, as shown in Figure 1-1.

To create a new document in Illustrator, click the New button on this screen.

Note: If you prefer not to see this start screen, you can choose to turn it off by unchecking Show the Home Screen When No Documents Are Open button in Edit ⇨ Preferences ⇨ General (Windows) or Illustrator ⇨ Preferences ⇨ General (Mac).

1. **If you do not see the start screen, choose File ⇨ New.**

   The New Document dialog box appears, as shown in Figure 1-2. You use it to determine the new document's profile, size, measurement unit, color mode, and page orientation as well as the number of artboards (pages) you want in the document.

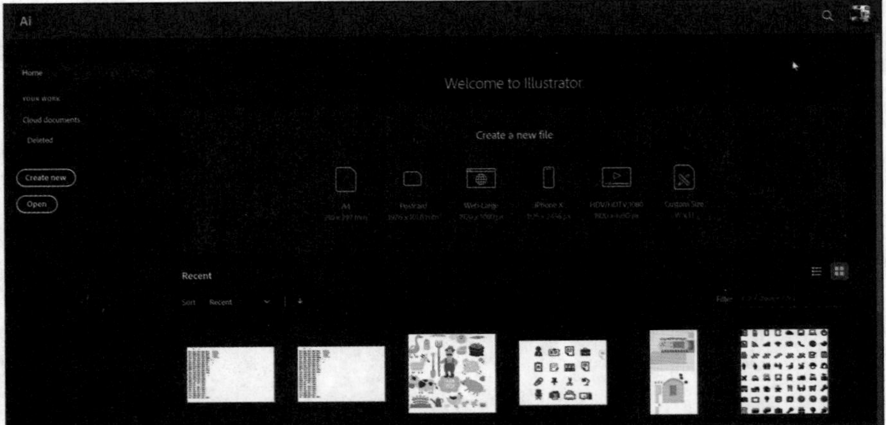

FIGURE 1-1: The initial start screen in Illustrator allows you to see recent files or start a new file.

If you are working with the default preferences, you see a large New Document window that allows you to choose a new document specific to the artwork you need to create, Mobile, Web, Print, Film & Video, Art & Illustration. Each of these choices comes with preset details that you can edit on the right side of the New Document dialog box. These options also come with templates that you are free to use in order to get a jump-start on your design.

TIP

Some users prefer to see the less cumbersome legacy New Document Window shown in Figure 1-3. You can set your preferences to always see the legacy New Document window by choosing Edit ➪ Preferences ➪ General (Windows) or Illustrator ➪ Preferences ➪ General (Mac) and checking the Use legacy "File New" interface option.

2. **In this instance, scroll down to the bottom of the Settings section on the right of the New Document window and select More Settings. (You have to scroll past the Advanced Options section to see the More Settings button.)**

   The legacy New Document appears. This is the workflow that is referenced in the following steps.

3. **Enter a name for your new file in the Name text field.**

   You can determine the name of the file now or later when you save the document. This can be done when you initially save your file as well. In this example, the new file is named `IllustratorExercise`.

4. **Choose a profile from the New Document Profile drop-down list.**

   Selecting the correct profile sets up preferences, such as resolution and colors. If you choose a Print profile, your basic color swatches are built from CMYK colors (cyan, magenta, yellow, and black), and your settings default to a higher resolution. For the web, your color swatches are built from RGB colors (red, green, blue), and your default resolution is set to a lower amount.

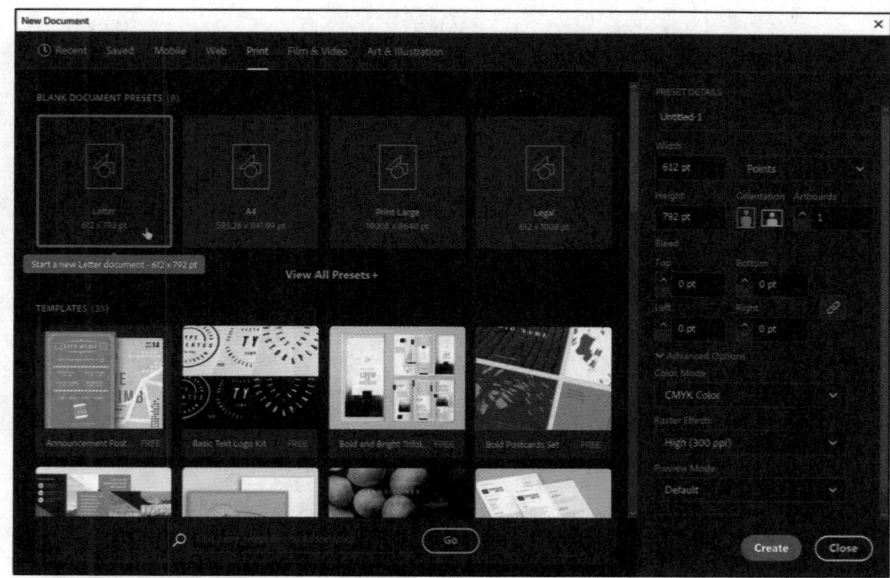

FIGURE 1-2: The New Document interface offers options for different types of files and templates.

You can choose from the following default profiles: Print, Web, Mobile, Film and Video, Art & Illustration, Devices, and Basic RGB.

Based upon the profile that you select, you will be provided with certain size options. In this example, Print is selected as the profile and Letter is selected as the size. In Figure 1-3, the Web profile is selected.

5. **In the Number of Artboards text box, you can enter the number of artboards you want in the document.**

   If you want a single-page document, leave this setting at 1. Creating a document with multiple artboards is discussed later in this chapter in the section "Creating multiple artboards."

6. **Enter in the Spacing text box the amount of space to leave between artboards.**

   Don't bother making changes in this section unless you selected multiple artboards. To make pages abut, enter **0** (zero) or enter additional values if you want a little space between each artboard. If you're adding artboards, you can enter in the Columns text box the number of columns of artboards you want arranged in the document. Keep in mind that this can be changed later.

7. **Select from the Size drop-down list or type measurements in the Width and Height text fields to set the size of the document page.**

   You can select from several standard sizes in the Size drop-down list or enter your own measurements in the Width and Height text fields. As previously mentioned, these size options are dependent upon your profile selection.

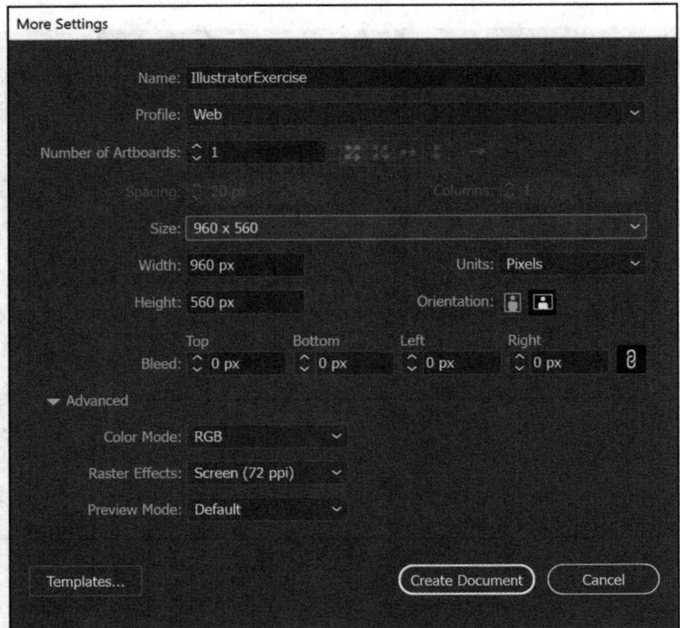

FIGURE 1-3: The legacy New Document window.

8. **Select from the Units drop-down list to select the type of measurement you're most comfortable with.**

   Your selection sets all measurement boxes and rulers to the increments you choose: points, picas, inches, millimeters, centimeters, or pixels.

9. **Pick the orientation for the artboard.**

   The *artboard* is your canvas for creating artwork in Illustrator. You can choose between *Portrait* (the short sides of the artboard on the top and bottom) and *Landscape* (the long sides of the artboard on the top and bottom).

10. **Add values in the Bleed text boxes, if necessary.**

    A *bleed value* is the amount of image area that extends beyond the artboard. To print from edge to edge, enter a value for the bleed. Keep in mind that most desktop printers need a *grip area,* which is the part of the page that a printer needs to grab on to the paper. It typically forces any image area near the edge of a page to not print. Bleeds are typically used in jobs to be printed on a press.

11. **When you're finished making selections, click Create Document.**

    One or more Illustrator artboards appear.

Try a Template

TIP

Want to start with a pre-created template? Simply choose File➪New from Template and double-click on the Blank Templates Folder. You are then provided with templates for various documents such as business cards and standard banner ad sizes. Double-click on the one you want to open and a new file with the appropriately sized artboards is opened.

Opening an Existing Document

It's a good idea to familiarize yourself with the workspace before starting to work in Adobe Illustrator, but to do this you should have a document open. In this section, you jump right in by opening an existing document. Here's how:

1. **Launch Adobe Illustrator CC.**
2. **Choose File➪Open and select a file in the Open dialog box.**

   For this exercise, you can open one of your own files or one of the sample files located in the DummiesCCfiles folder that you can download from `www.agitraining.com/dummies`. If you choose to use the samples provided, open the sample file called `SportsCover.ai`.

   Keep this file open so that you can follow along with the explanation of the workspace in the next section.

Taking a Look at the Document Window

To investigate the work area (shown in Figure 1-4) and become truly familiar with Illustrator, have a document open, like the `SportsCover.ai` file you were directed to in the preceding section, or create a new document. In the Illustrator work area, a total of 227 inches in width and height helps create your artwork (and all artboards). That's helpful, but it also leaves enough space to lose track of your illustration. The following list explains the tools you work with as you create artwork in Illustrator:

- **Imageable area:** If you're planning to print your Illustrator document, choose View➪Show Print Tiling. A dotted line appears that marks the printing area on the page. Many printers can't print all the way to the edges of the paper, so the imageable area is determined by the printer you set in the Print dialog box. If you don't want to see the print tiling, choose View➪Hide Print Tiling.

You can move the imageable area around on your page by using the Print Tiling tool. See the nearby sidebar, "The Print Tiling tool," for more on this tool.

- **Artboard:** This area, bounded by solid lines, represents the entire region that can contain printable artwork. By default, the artboard is the same size as the page, but it can be enlarged or reduced. The U.S. default artboard size is 8.5 x 11 inches, but it can be set as large as 227 x 227 inches. If you're using the default settings for Illustrator, the edge of the artboard is indicated by a transition from a white to a medium gray canvas.
- **Scratch area:** This area outside the artboards extends to the edge of the 227-inch square window. The scratch area represents a space on which you can create, edit, and store elements of artwork before moving them onto the artboard. Objects placed onto the scratch area are visible onscreen, but they don't print. However, objects in the scratch area appear if the document is saved and placed as an image in other applications.

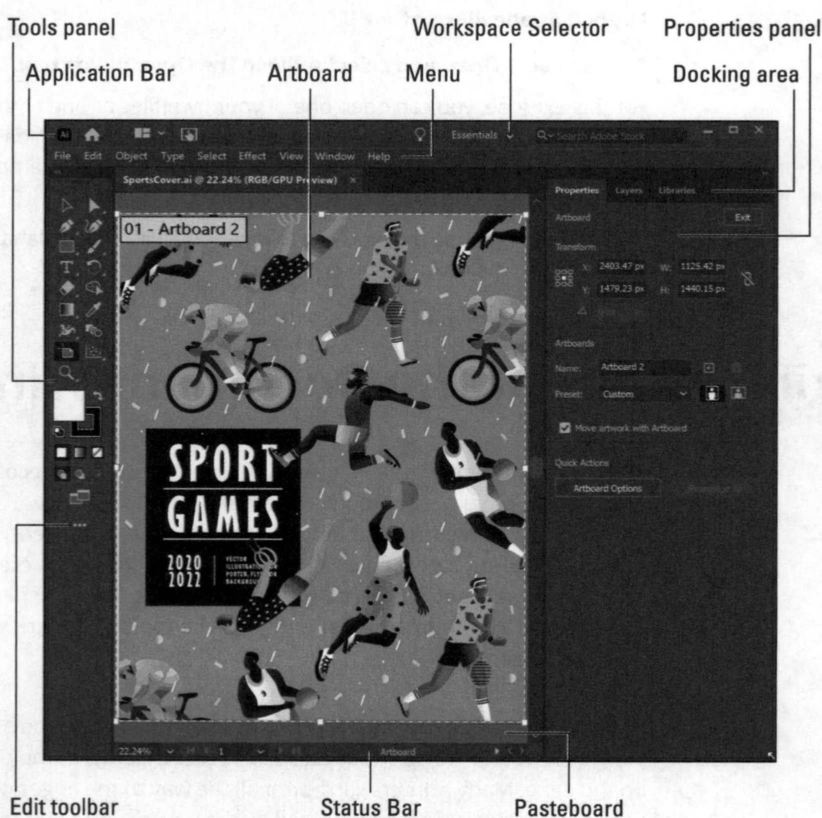

FIGURE 1-4: The Illustrator work area.

TIP

If you're having a difficult time seeing your items in the scratch area, choose Edit ➪ Preferences ➪ User Interface (Windows) or Illustrator ➪ Preferences ➪ User Interface (Mac) and select White for the Canvas color.

The Adobe Illustrator workspace includes many items such as toolbars, panels, menu items and more:

- *Application Bar:* Access menus and application controls.
- *Menu Bar:* File, Edit, and so forth. Illustrator functions are available from the drop-down menus.
- *Workspace Selector:* Change which panels, tools, and other options you see in Illustrator. The default workspace is Essentials.
- *Tools Panel:* Tools used to create and edit images, artwork, and page elements.
- *Artboard:* The visible work area in Illustrator. Keep in mind that you can have multiple artboards.

THE PRINT TILING TOOL

Use the Print Tiling tool to move the printable area of your page to a different location. For example, if your printer can print only on paper that's 8.5 x 11 inches or smaller but the page size is 11 x 17, you can use the Print Tiling tool (a hidden tool accessed by holding down the mouse button on the Hand tool) to indicate which part of the page you want to print. Follow these steps to use the Print Tiling tool:

1. **When adjusting page boundaries, choose View ➪ Fit All in Window so that you can see all your artwork.**
2. **Hold down the mouse button on the Hand tool to select the hidden Print Tiling tool.**

   The pointer becomes a dotted cross when you move it to the active window.
3. **Position the mouse pointer over the artboard and click and drag the page to a new location.**

   While you drag, the Print Tiling tool acts as though you're moving the page from its lower-left corner. Two gray rectangles are displayed. The outer rectangle represents the page size, and the inner rectangle represents the printable area of a page. You can move the page anywhere on the artboard; just remember that any part of a page that extends past the printable area boundary will not print.

* *Paste Board:* Objects and images can be created in the scratch area, but will not be printed or exported.
* *Properties panel:* The main panel for changing properties of selected objects.
* *Docking area:* A place that can hold your various panels, whether expanded or not.
* *Edit Toolbar:* Select Edit Toolbar to find any tools that may not be visible in your current view.

Navigating the Work Area with Zoom Controls

You can navigate the work area efficiently by using the Hand tool and the various zoom controls. You can change the magnification of the artboard in several ways, including using menu items, the Zoom tool, and keyboard commands. It will help to have a sample file open while practicing navigating your workspace. Use the `SportsCover.ai` example that was referenced earlier in this chapter, or open an existing file of your own. Choose the method you feel most comfortable with:

* **Hand tool:** Scroll around the Document window by using the scroll bars or the Hand tool. The Hand tool lets you scroll by dragging. You can imagine that you are pushing a piece of paper around on your desk when you use the Hand tool.

  Hold down the spacebar to temporarily access the Hand tool while any tool (except the Type tool) is selected. Holding down the spacebar while the Type tool is selected gives you only spaces!
* **View menu:** Using the View menu, you can easily select the magnification you want by using Zoom In, Zoom Out, Fit in Window (especially useful when you're lost in the scratch area), and Actual Size (provides a 100 percent view of your artwork).

* **Zoom tool:** Using the Zoom tool, you can click the Document window to zoom in; to zoom out, Alt-click (Windows) or Option-click (Mac). Double-click on the Zoom tool to quickly resize the Document window to 100 percent. Control which elements are visible when using the Zoom tool by clicking and dragging over the area you want zoomed into.

* **Keyboard shortcuts:** The zooming keyboard shortcuts make sense and are easy to remember. Table 1-1 lists the most popular keyboard shortcuts to change magnification.

TIP

The shortcuts in Table 1-2 require a little coordination to use, but they give you more control in your zoom. While holding down the keys, drag from the upper-left corner to the lower-right corner of the area you want to zoom to. A marquee appears while you're dragging; when you release the mouse button, the selected area zooms to the size of your window. The Zoom Out command does not give you much control; it simply zooms back out, much like the commands in Table 1-1.

If you are trying out these keyboard shortcuts, select an object in the `SportsCover.ai` file, to see how the zoom is related to the selected item.

TABLE 1-1 Magnification Keyboard Shortcuts

Command	Windows Shortcut	Mac Shortcut
Actual Size	Ctrl+1	⌘+1
Fit in Window	Ctrl+0 (zero)	⌘+0 (zero)
Zoom In	Ctrl++ (plus)	⌘++ (plus)
Zoom Out	Ctrl+– (minus)	⌘+– (minus)
Hand tool	Spacebar	Spacebar

TABLE 1-2 Zoom Keyboard Shortcuts

Command	Windows Shortcut	Mac Shortcut
Zoom In to Selected Area	Ctrl+spacebar+drag	⌘+spacebar+drag
Zoom Out	Ctrl+Alt+spacebar	⌘+Option+spacebar

Taking Advantage of Artboards

As you already discovered, the page itself is referred to as the *artboard*, in the `SportsCover.ai` file there is only one artboard. The options available to you when using artboards include the following:

* You can have multiple artboards in one document, creating essentially a multipage document.

- You can name and rearrange artboards by using the Artboard panel.
- You can copy entire artboards, allowing you to quickly create multiple versions for your artwork.
- You can use artboards to define a print area, eliminating the need to use the Print Tiling tool.

Creating multiple artboards

You can start off right away with multiple artboards, or add them as you need them. In order to create a new document with multiple artboards, you can follow these steps:

1. **Launch Adobe Illustrator CC and choose File ⇨ New.**

   The New Document dialog box appears where you can specify the number and size of your artboards by clicking on the More Settings button.

2. **Specify the number of artboards to start with by entering a number in the Number of Artboards text box.**

   In our Figure 1-5, four artboards are shown.

3. **You can then specify how to arrange the artboards, by clicking the grid or row arrangement icon to the right of the Number of Artboards text box.**

   Using these grid boxes, you can

   - Specify how many rows and columns to use.
   - Change the direction of the layout from left to right or right to left.

4. **Enter an amount in the Spacing text box to determine the distance between artboards.**

   Enter **0** (zero) if you want the artboards to butt against each other, or a higher value if you want some space between them.

5. **Click OK to create your new document. The document will not have multiple sized artboards or content.**

After you create multiple artboards, choose Window ⇨ Artboards to see a panel with artboards listed individually, as shown in Figure 1-6. In the next section you discover how you can enhance your artboard experience by working with this panel. In this example the artboards have default names; you can double-click on an artboard name in the Artboard panel and change the name if you like.

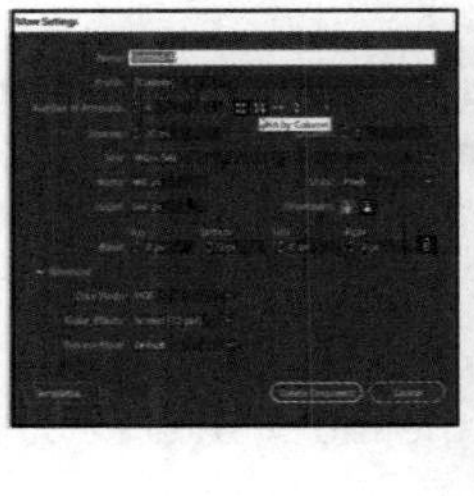

FIGURE 1-5: Add multiple artboards right from the start. A document created using four artboards.

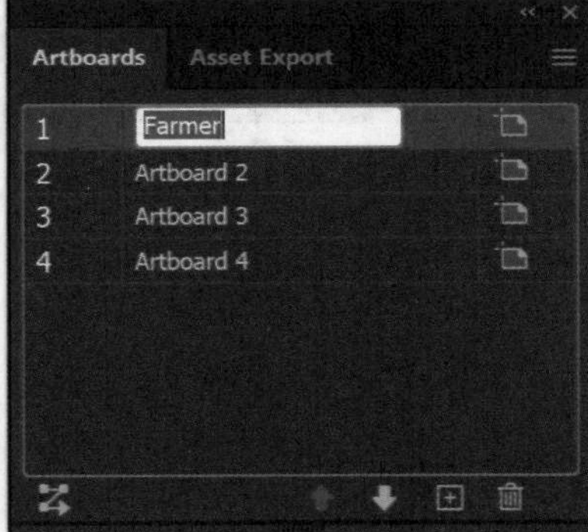

FIGURE 1-6: The Artboard panel helps to keep your artboards organized, and you can rename the artboards too.

Using the Artboard Panel

The Artboard panel adds additional capabilities for your artboards. Here are some things you can do to artboards from the Artboard panel:

- **Navigate to a specific artboard.** Just double-click the artboard's number.
- **Rearrange artboards.** Drag the artboards inside the Artboard panel to reorganize their stacking order.
- **Delete an artboard.** Drag it to the Trash icon.
- **Copy an existing artboard.** Drag it to the New Artboard icon.
- **Create a new artboard.** Click the New Artboard icon.

- **Edit additional artboard options, such as size and orientation, by selecting Artboard Options from the panel menu.** The Artboard options appear as you see in Figure 1-7. Keep in mind that each artboard can be a different size than others in the same document.
- **Rename an artboard.** Double-click the Artboard name right in the Artboards panel, type a new name, and then press the Tab or Enter key.

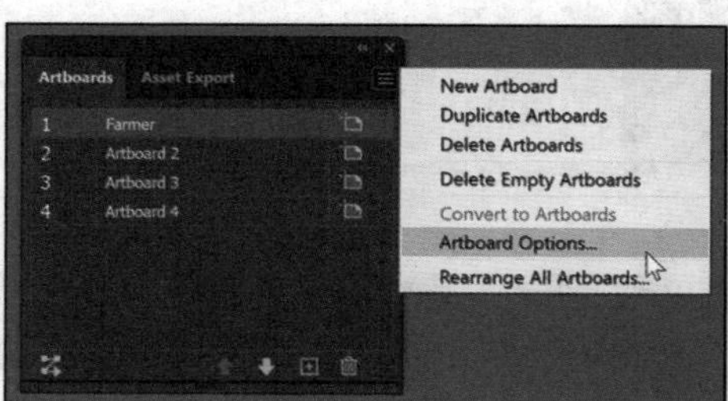

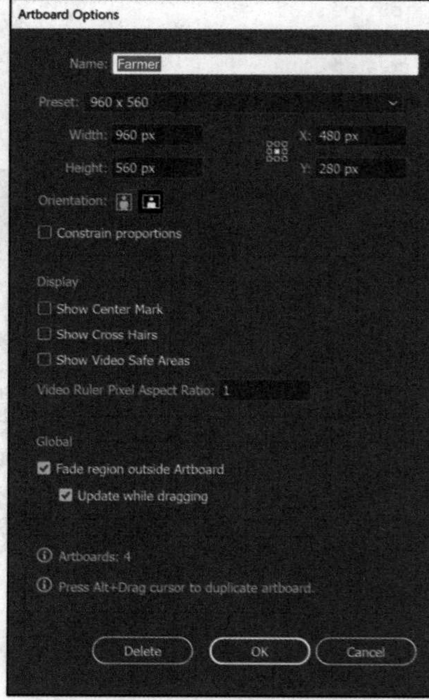

FIGURE 1-7: Create new or edit existing artboards by using the Artboard Options dialog box.

Printing a document with multiple artboards

Pay close attention before you print a document with multiple artboards, or else you may print needless pages. To control which artboards are printed, follow these steps:

1. **Choose File ⇨ Print.**

   The Print dialog box appears, as shown in Figure 1-8.

2. **Click the arrows underneath the Preview box in the lower-left corner to preview the artboards.**

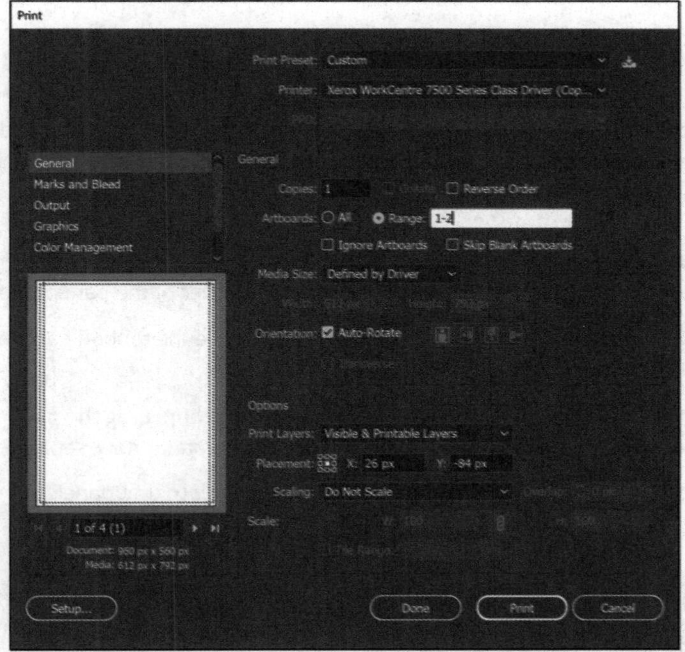

FIGURE 1-8: Printing a document with multiple artboards.

3. **After you decide which artboards to print, enter them into the Range text box.**

   To print all artboards, select the All radio button; otherwise, enter a consecutive range such as **1–3** in the Range text box. You can also print nonconsecutive pages by separating them with commas — enter **1, 3, 4** to print only artboards 1, 3, and 4.

4. **Click Print to print the selected artboards.**

Now that you have a general understanding of the artboards you will have the opportunity to review how workspaces can be used to make your job of creating illustrations easier.

Checking Out the Panels

The standardized interface in Adobe Creative Cloud is a useful boost for users because the Illustrator panel system is similar to all other products in the Creative Cloud. This consistency makes working and finding tools and features easier.

The panels you see as a default are docked together. To *dock* a panel means to attach the panel in the docking area for organizational purposes.

You can arrange panels to make them more helpful for production. You may choose to have only certain panels visible while working. Here's the lowdown on using Illustrator panels:

- To see additional options for each panel (because some options are hidden), click the panel menu in the upper-right corner of the panel. (See Figure 1-9.)
- To move a panel group, click and drag above the tabbed panel name, or on the rightmost part of the panel.
- To rearrange or separate a panel from its group, drag the panel's tab. Dragging a tab outside the docking area creates a new, separate panel.
- To move a tab to another panel, drag the tab to that panel.

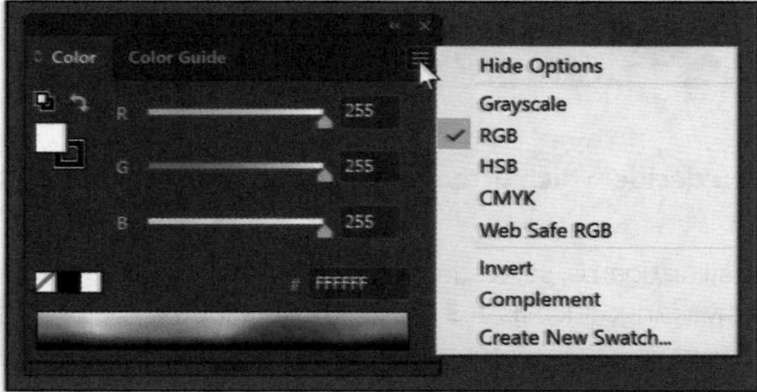

FIGURE 1-9: Many panels have additional menu options.

Look out for those panels — they can take over your screen! Some panels can be resized. To change the size of a panel, drag the lower portion of the panel Read the next section to find out how to keep panels in order.

Getting to Know Your Workspace

Depending upon the type of work you are creating you may choose to switch your Workspace. Workspaces changes the visibility of your panels and tools. As mentioned in the previous section, you can choose to hide or show, drag panels out of the docking area, or group them with other panels. This is a feature common to many of the apps in the Creative Cloud. Workspaces offer you the opportunity to use preconfigured workspaces or save your own custom combinations.

Workspaces can be handy when you just want to clean up a messy workspace, or when you want the panels or tools to be focused on a particular task, for instance, creating art for the web.

You can quickly change to a different workspace by using the Workspace Switcher located up in the upper-right of your screen, as shown in Figure 1-10.

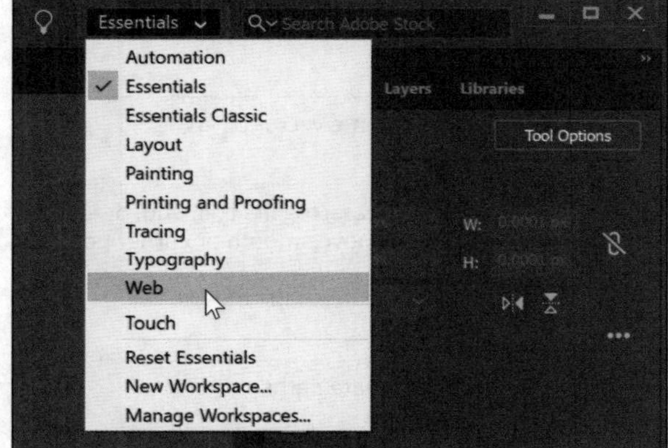

FIGURE 1-10: Change your workspace by using the Workspace Switcher.

Becoming Familiar with the Tools

As you begin using Illustrator, you'll find it helpful to be familiar with its tools. Tools are used to create, select, and manipulate objects in Illustrator. The tools should be visible as a default, but if not, you can access them by choosing Window⇨Tools. Depending upon your workspace, you may or may not see the tools that we list in this section. If you want to follow along with the default tools shown here, choose Essentials from the Workspace Switcher in the upper-right of your workspace. If you are already using the Essentials workspace and need to tidy things up, choose Reset Essentials.

TIP

Can't find a tool that you need? Click on the Edit tool (. . .) button at the bottom of the tools panel to find tools that may be hidden in your workspace.

Table 1-3 lists many the tools that we show you how to use throughout this minibook. Hover the cursor over the tool in the Tools panel to see the name of the tool in a tooltip. In parentheses on the tooltip (and noted in Column 2 of Table 1-3) is the keyboard command you can use to access that tool. When you see a small triangle at the lower-right corner of the tool icon, it contains additional, hidden tools. Select the tool and hold the mouse button to see it.

TABLE 1-3 **Illustrator CC Tools**

Icon	Tool or Keyboard Command	Task	See Minibook Chapter
	Selection (V)	Activate objects	2
	Direct Selection (A)	Activate individual points or paths	2
	Group Selection (A) Hidden in Direct Selection tool	Select grouped items	2
	Pen (P)	Create vector paths	4
	Curvature Tool (Shift+~)	Create, toggle, edit, add, or remove smooth or corner points	4
	Rectangle tool (M)	Create shape objects	3
	Paint Brush (B)	Create paths	n/a
	Type (T)	Create text	5
	Rotate (R)	Rotate objects	9
	Eraser (Shift+E)	Erase vector paths	4
	Shape Builder (Shift+M)	Combine, edit, and fill shapes	3
	Gradient (G)	Modify gradients	8
	Mesh (U) Hidden in Gradient tool	Create a gradient mesh	10
	Eyedropper (I)	Copy and apply attributes	8
	Width tool (Shift+W)	Adjust strokes to variable widths	9
	Blend (W)	Create transitional blends	10

Icon	Tool or Keyboard Command	Task	See Minibook Chapter
	Artboard tool (Shift+O)	Add or edit artboards	1
	Graph (J)	Create graphs	n/a
	Zoom (Z)	Increase and decrease the onscreen view	n/a

Table 1-4 lists additional tools that you may find useful. Many of these are also referenced throughout this minibook. Remember that if you can't find a tool, click on Edit Tools (. . .) at the bottom of the Tools panel and you will find them there.

TABLE 1-4 Illustrator CC Tools

Icon	Tool or Keyboard Command	Task	See Minibook Chapter
	Magic Wand (Y)	Select based on similarity	2
	Lasso (Q)	Select freehand	2
	Line Segment (\)	Draw line segments	5
	Shaper (Shift+N)	Create shapes	n/a
	Scale (S)	Enlarge or reduce objects	9
	Free Transform (E)	Transform objects	9
	Perspective Grid (Shift+P)	Provide perspective plane	11
	Symbol Sprayer (Shift+S)	Spray symbols	10
	Slice (Shift+K)	Create HTML slices	n/a
	Hand tool (H)	Push document window	1

As you become more efficient, you may find it helpful to reduce the clutter on your screen by hiding all panels except the ones necessary for your work. You can save your own panel configuration by choosing Window ⇨ Workspace ⇨ New Workspace. Choose Window ⇨ Workspace ⇨ Essentials to return to the default workspace. You may need to reset the Essentials workspace if it still doesn't look like the default. Choose Window ⇨ Workspace ⇨ Reset Essentials.

Changing Views

When you're working in Illustrator, precision is important, but you also want to see how the artwork looks. Whether for the web or print, Illustrator offers several ways in which to view your artwork:

» **Preview and Outline views:** By default, Illustrator shows Preview view, where you see colors, stroke widths, images, and patterns as they should appear when printed or completed for onscreen presentation. Sometimes this view can become a nuisance, especially if you're trying to create a corner point by connecting two thick lines. At times like this, or whenever you want the strokes and fills reduced to the underlying structure, choose View ⇨ Outline. You are switched to the Outline view where you now see the outline of the illustration, as shown in Figure 1-11. Colors, strokes, and fills are invisible in this view, taking away distractions when precision is important.

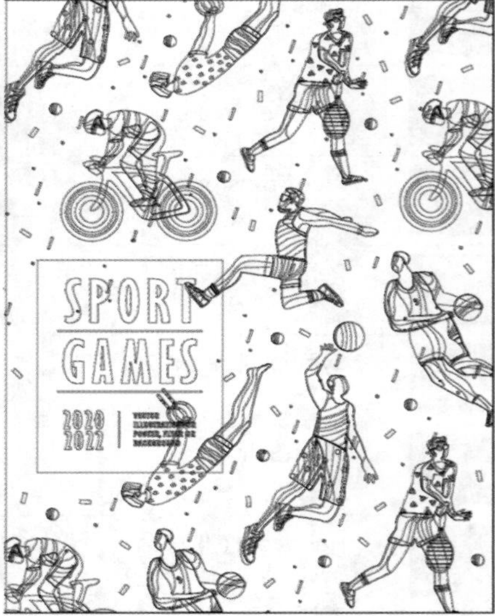

FIGURE 1-11: Preview view (left) and Outline view (right).

» **Pixel view:** If you don't want to be surprised when your artwork appears in your web browser, use Pixel view. This view, shown in Figure 1-12, maintains the vectors of your artwork but shows how the pixels will appear when the image is viewed onscreen, as though it's on the web.

FIGURE 1-12: See how your artwork translates into pixels in Pixel view.

TIP

Pixel view is helpful for previewing the look of text onscreen — some fonts just don't look good as pixels, especially if the text is small. In Pixel view, you can review fonts until you find one that's easily readable as pixels.

» **Overprint view:** For people in print production, the Overprint preview can be a real timesaver. Choose Window ⇨ Attributes to bring up the Attributes panel, which you can use to set the fill and stroke colors to overprint. This view creates additional colors when printing and aids printers when trapping abutting colors.

Trapping is the slight overprint of a lighter color into a darker color to correct for press *misregistration* — that is, when the printed colors don't align exactly. When several colors are printed on one piece, the likelihood of perfect alignment is slim! Setting a stroke to Overprint on the Window ⇨ Attributes panel is one solution. With Overprint selected, the stroke is overprinted on the nearby colors. This mixing of color produces an additional color but is less obvious to the viewer than a white space created by misregistration. Select Overprint (View ⇨ Overprint Preview) to see the result of overprinting in Overprint view, as shown in Figure 1-13. If you would like to try this feature you will need a file that is in the CMYK mode (cyan, magenta, yellow, black). You can also use the image named `SportsCoverCMYK.ai` that is located in the Book05_Illustrator folder.

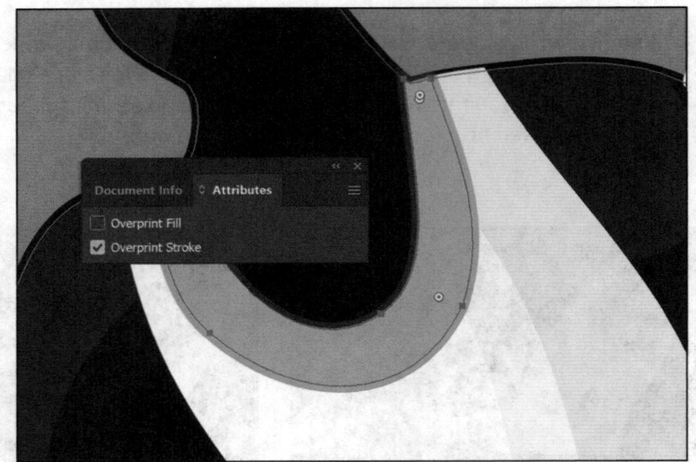

FIGURE 1-13: Choose to see the Overprint value using the Attributes panel.

tools

» Working with a selection

» Grouping and ungrouping selections

» Constraining movement and cloning objects

Chapter 2

Using the Selection Tools

If someone has been coaching you in using Adobe Illustrator, you may have heard the old line "You have to select it to affect it," meaning that if you want to apply a change to an object in Illustrator, you must have that object selected or no change will occur. Although making selections may sound simple, it can become tricky when you're working on complicated artwork. This chapter shows you the tools that help save you editing time when working in Adobe Illustrator.

Getting to Know the Selection Tools

In the following sections, you take a quick tour of anchor points (integral to the world of selections), the bounding box, and, of course, the selection tools.

TIP

See our website at `www.agitraining.com/dummies` to find files that you can practice with in this section or create your own simple document by following the instructions in the next section.

Giving selections a try

Before reading any further, give "select it to affect it" a try by creating a file and making simple changes to it. To follow along, create a new file of any size and add shapes to the artboard. If you need help getting through that task, read the steps below:

1. **Select the Rectangle tool and then Click and drag on the artboard from the upper-left corner of what will be your rectangle to the lower right.**
2. **Click and hold down on the Rectangle tool to select the hidden Ellipse tool, and also click and drag on the artboard from the upper-left to the lower-right. You now have two shapes on your artboard.**
3. **Click on the rectangle with the Selection tool and hold down the Alt (Windows) or Option (MACOS) key and drag to clone the rectangle. You now have three shapes on your artboard.**

Now that you have some shapes to work with, you can try out the main selection tools in Illustrator. You will have the opportunity to investigate these tools in more detail as you go through this chapter.

1. **Using the Selection tool, this is the solid arrow tool at the top of your tools panel. Click and drag any shape to see that you can move its position.**
2. **Click on any other shape to see that the past selection turns off and the other object is selected.**
3. **If you do not see the Properties panel choose Window⇨Properties and click on Fill. Select any color from the Swatches that appear, the shape is changed to the new color. See Figure 2-1.**
4. **Click on another shape and then click Stroke in the Properties panel and change the color of the border around the shape.**

   If you have a hard time seeing your stroke, change the value of 1, at the right of the color you selected, to a higher number. Note that any changes affect only the selected object.

5. **Click on the artboard where there are no objects. Now no objects are selected.**
6. **Select the Direct Selection tool and click and drag around the corner of one of the rectangle shapes, as shown in Figure 2-2. Using the Direct Selection tool only selects and activates this one corner.**
7. **Click and drag that corner point to see that you can now edit that shape.**

Now that you see how Illustrator works with selections and the properties of the selected object you can read on to find other important details.

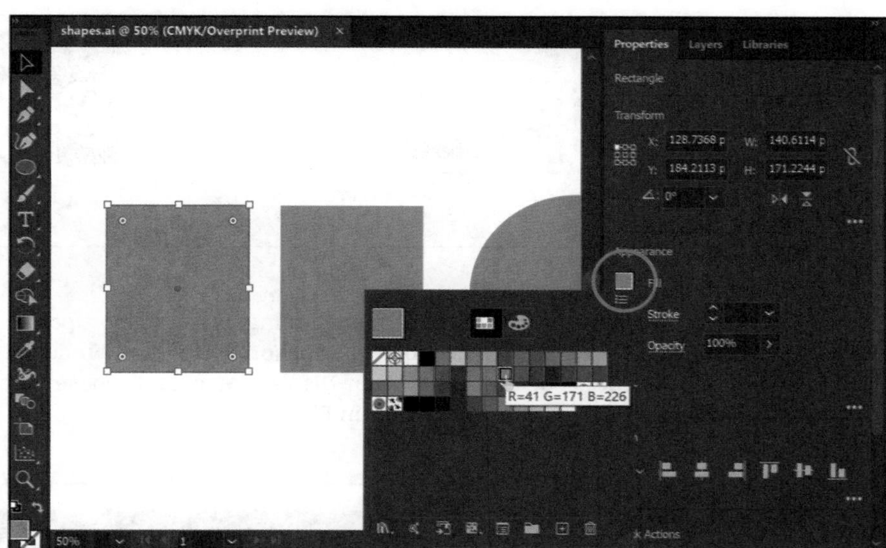

FIGURE 2-1: Select an object and then change the Fill color.

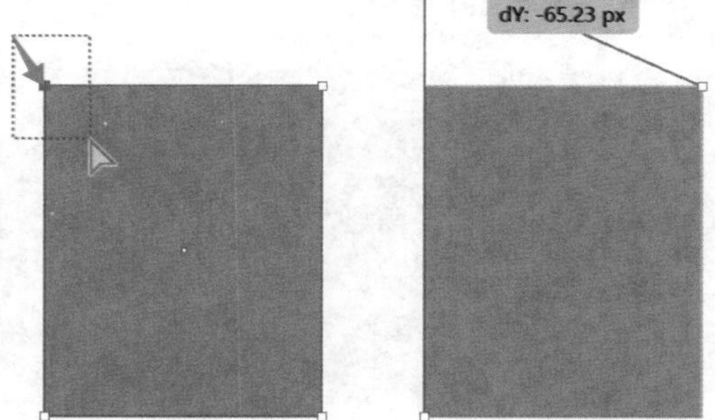

FIGURE 2-2: The Selection tool selects the entire object; the Direct Selection tool can select specific points.

Anchor points

It helps to understand how Illustrator works with *anchor points,* which act like handles and can be individually selected and moved to other locations. You essentially use the anchor points to drag objects or parts of objects around the workspace. After you place anchor points on an object, you can create strokes or paths from the anchor points.

You can select several anchor points at the same time, as shown in Figure 2-3, or only one. Selecting all anchor points in an object lets you move the entire object without changing the anchor points in relationship to one another. You just did this in the last section when you used the Selection tool. You can tell which anchor points are selected and active because they appear as solid boxes.

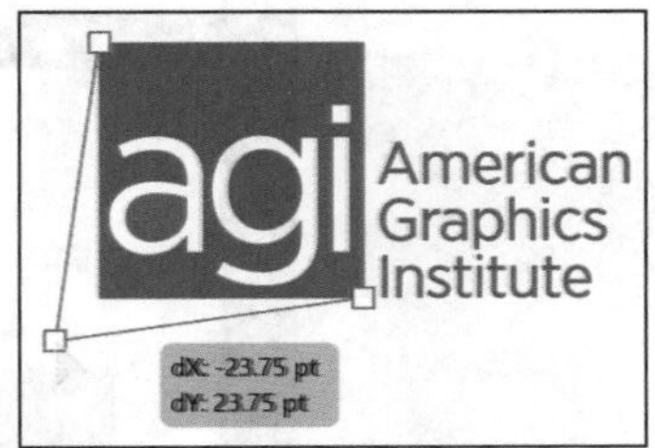

FIGURE 2-3: All anchor points selected, and one anchor point is selected.

If you are having a hard time seeing your anchor points you can choose Edit⇨Preferences⇨Selection & Anchor Display (Windows) or Illustrator⇨Preferences⇨Selection & Anchor Display (Mac) and choose to increase the size of your anchor points, as you see in Figure 2-4.

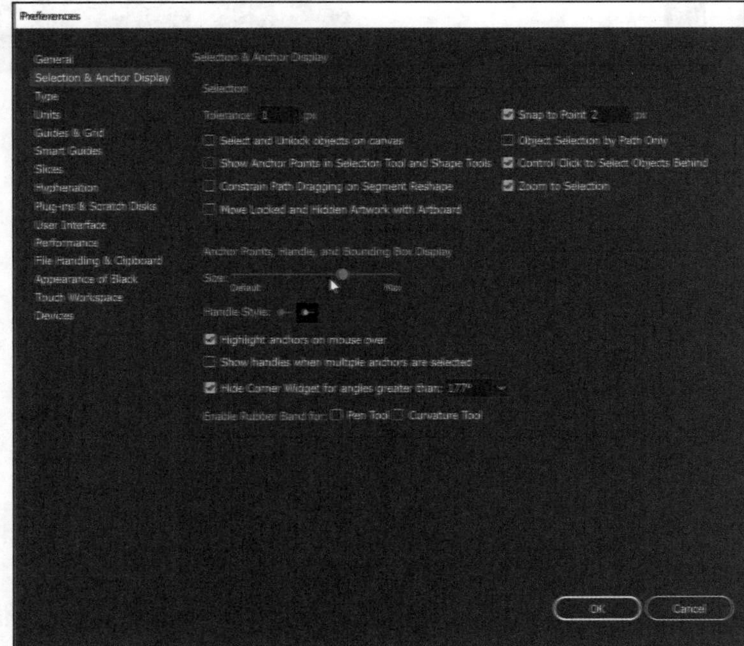

FIGURE 2-4: Increase the size of your anchor points if you are having a hard time seeing them.

Bounding boxes

As a default, Illustrator shows a bounding box when an object is selected with the Selection tool. This feature can be helpful if you understand its function but confusing if you don't know how to use it.

By dragging the handles, you can use the bounding box for quick transforms, such as scaling and rotating. To rotate, you pass the mouse cursor (without clicking) outside a handle until you see a rotate symbol and then drag.

If the bounding box gets in your way turn off the feature by choosing View⇨Hide Bounding Box.

Introducing all the selection tools

Illustrator CC offers five main selection tools. To see these selection tools without having to edit your toolbar, open Window⇨Workspace and choose Essentials Classic. This workspace is specifically created for those who have worked in previous versions of Illustrator and want to have the same tools available. Keep in mind that you can use the Edit Toolbar button at the bottom of the Tools panel and create your own custom workspace. As you can see in Figure 2-5, the arrow in the upper left of the Tools panel has been clicked in order to change the Tools panel to two columns.

FIGURE 2-5: The Essential Classic workspace. To edit the visible tools, click the Edit Toolbar button at the bottom.

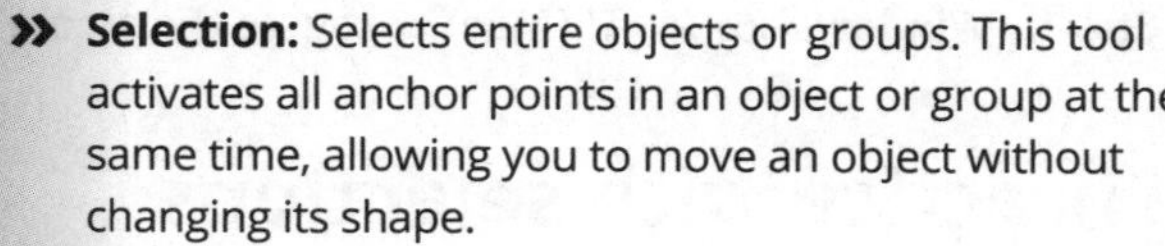

- **Selection:** Selects entire objects or groups. This tool activates all anchor points in an object or group at the same time, allowing you to move an object without changing its shape.

- **Direct Selection:** Selects individual points.

- **Group Selection:** Hidden in the Direct Selection tool in the Tools panel and used to select items within a group. This tool adds grouped items as you click objects in the order in which they were grouped. This selection tool becomes more useful to you as you find out about grouping objects in Illustrator.

- **Magic Wand:** Use the Magic Wand tool to select objects with like values, such as fill and stroke colors, based on a tolerance and stroke weight. Change the options of this tool by double-clicking it.

- **Lasso:** Use the Lasso tool to click and drag around anchor points you want to select.

You can select an object with the Selection tool by using one of three main methods:

- Click the object's path.
- Click an anchor point of the object.
- Drag a marquee around part or all of an object. (In a later section named "Using a marquee to select an object," we discuss using the marquee method.)

Working with Selections

After you gain an understanding about the basics of selections, you'll probably be eager to try some techniques. In the following sections, you see the basics: Make a selection, work with anchor points and the marquee, make multiple selections, and of course, save your work.

Smart guides, turned on by default in Illustrator CC, can help you make accurate selections. These guides are visible as you're drawing. They display names such as `anchor point` and `path`, and they highlight paths when you're lined up with endpoints or center points. You can come to love these helpful aids, but if you don't want to see them, simply choose View⇨Smart Guides or press the keyboard shortcut Ctrl+U (Windows) or ⌘ +U (Mac) to toggle the smart guides off and on.

More practice with selections

To work with selections, you need to have something on the page in Illustrator. Follow these steps to make a selection:

1. **Create a new page in Adobe Illustrator. (Any size or profile is okay.)**

   Alternatively, you can open an existing illustration; see Chapter 1 of this minibook for instructions on how to download the CCDummies files. A good sample to work with would be `SchoolIcons.ai`, located in the Book05Illustrator folder. Skip to Step 3 if you're working with an existing illustration.

2. **If you're starting from a new page, create an object to work with.**

   For example, click and hold down the Rectangle tool to select the Star tool. Then click and drag from the upper left to the lower right to create a star shape.

Exact size doesn't matter, but make it large enough that you can see it. To start over, choose Edit⇨Undo or press Ctrl+Z (Windows) or ⌘+Z (Mac).

As a default, all shapes start with a black stroke and a white fill. If yours isn't black and white, press D, which changes the selected object to the default colors.

TIP

You can see the width and height of your object while you click and drag. If you don't want those values to display, choose Edit⇨Preferences⇨Smart Guides (Windows) or Illustrator⇨Preferences⇨Smart Guides (Mac) and deselect the Measurement Labels checkbox.

3. **Using the Selection tool, click the object to make sure it's active.**

   All anchor points are solid, indicating that they're active. You see, as a default, many additional points you can use to transform your selected object. If you don't see all your individual points, turn off the Bounding Box feature by selecting View⇨Hide Bounding Box. The Bounding Box feature makes it easier to select and transform objects, so, if you like that feature, make sure that you go back and select Show Bounding Box again when you are done.

4. **Click and drag the shape to another location.**

   All anchor points travel together.

5. **When you're finished relocating your selection, use one of these three methods to deactivate your selection:**
   - Choose Select⇨Deselect.
   - Ctrl-click (Windows) or ⌘-click (Mac) anywhere.
   - Use the keyboard shortcut Ctrl+Shift+A (Windows) or ⌘+Shift+A (Mac).

TIP

To select an object behind another, use the select-behind keyboard shortcut. Simply place the cursor over the area where you know that the object (to be selected) is located and press Ctrl-click (Windows) or ⌘-click (Mac OS). (You may have to do this twice to ensure you select the object behind.)

Selecting an anchor point

You can also deselect all active anchor points and then make just one anchor point active. Follow these steps:

1. **Choose Select⇨Deselect to make sure the object isn't selected.**
2. **Select the Direct Selection tool (the white arrow) from the Tools panel.**
3. **Click one anchor point.**

   Only one anchor point (the one you clicked) is solid, and the others are hollow, as shown in Figure 2-6.

FIGURE 2-6: Select only one anchor point and click and drag it.

4. **Click and drag that solid anchor point with the Direct Selection tool.**

   Only that solid anchor point moves. If you hold the Shift key down while dragging it will constrain your movement to 45, or 90 degrees.

TIP

Note that an anchor point enlarges when you cross over it with the Direct Selection tool. This enlargement helps those of us who typically have to squint to see where the anchor points are positioned.

Using a marquee to select an object

Sometimes you can more easily surround the object you want to select by dragging the mouse to create a marquee. Follow these steps to select an object by creating a marquee:

1. **Choose the Selection tool.**
2. **Click outside the object and drag over a small part of it, as shown in Figure 2-7.**

   The entire object becomes selected.

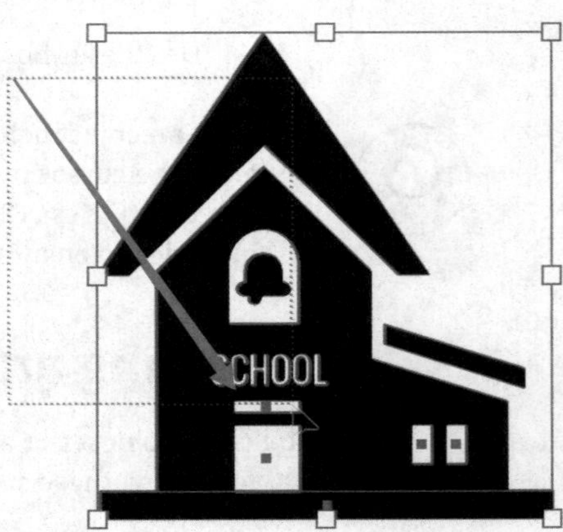

FIGURE 2-7: Select an entire object.

You can also select only one anchor point in an object by using the marquee method:

1. **Choose Select⇨Deselect to make sure the object isn't selected, and then choose the Direct Selection tool.**

2. **Click outside a corner of the object and drag over only the anchor point you want to select.**

   Notice that only that anchor point is active, which can be a sight-saver when you're trying to select individual points. (See Figure 2-8.)

FIGURE 2-8: Select individual anchor points.

TIP

You can use this method to cross over just the two top points or side anchor points to activate multiple anchor points as well.

Selecting multiple objects

If you have multiple items on a page, you can select them by using one of these methods:

- **Select one object or anchor point, and then hold down the Shift key and click another object or anchor point.** Depending on which selection tool you're using, you select either all anchor points on an object or additional objects (Selection tool) or additional anchor points only (Direct Selection tool).

  You can use the Shift key to deactivate an object as well. Shift-click a selected object to deselect it.

- **Choose Select⇨All or press Ctrl+A (Windows) or ⌘+A (Mac).**
- **Use the marquee selection technique and drag outside and over the objects.** When you use this technique with the Selection tool, all anchor points in the objects are selected; when using the Direct Selection tool, only the points you drag over are selected.

Saving a selection

Spending way too much time trying to make your selections? Illustrator comes to the rescue with the Save Selection feature. After you have a selection that you may need again, choose Select⇨Save Selection and name the selection. The selection now appears at the bottom of the Select menu. To change the name or delete the saved selection, choose Select⇨Edit Selection. This selection is saved with the document.

Grouping and Ungrouping

Keep objects together by grouping them. The Group function is handy when you're creating something from multiple objects, such as a logo. Using the Group function, you can ensure that all objects that make up the logo stay together when you move, rotate, scale, or copy it.

Creating a group

Follow these steps to create a group:

1. **If you aren't already working with an illustration that contains a bunch of objects, create several objects on a new page — anywhere, any size.**

   For example, select the Rectangle tool and click and drag on the artboard several times to create additional rectangles.

2. **Select the first object with the Selection tool and then hold down the Shift key and click a second object.**
3. **Choose Object➪Group or press Ctrl+G (Windows) or ⌘ +G (Mac).**
4. **Choose Select➪Deselect, and then click one of the objects with the Selection tool.**

   Both objects become selected.

5. **While the first two objects are still selected, Shift-click a third object.**
6. **With all three objects selected, choose Object➪Group again.**

   Illustrator remembers the grouping order. To prove it, choose Select➪Deselect to deselect the group and switch to the Group Selection tool. (Hold down the mouse button on the Direct Selection tool to access the Group Selection tool.)

7. **With the Group Selection tool, click the first object; all anchor points become active. Click the first object again; the second object becomes selected. Click the first object again, and the third object becomes selected.**

   The Group Selection tool activates the objects in the order you grouped them. After you group the objects, you can treat them as a single object.

To ungroup objects, choose Object➪Ungroup or use the keyboard shortcut Ctrl+Shift+G (Windows) or ⌘ +Shift+G (Mac). In a situation where you group objects twice (because you added an object to the group, for example), you have to choose Ungroup twice.

Using Isolation mode

When you use *Isolation mode* in Illustrator, you can easily select and edit objects in a group without disturbing other parts of your artwork. With the Selection tool, simply double-click a group, and it opens in a separate Isolation mode, where all objects outside the group are dimmed and inactive. Do the work you need to do on the group and exit from Isolation mode by clicking the arrow to the left of Group in the upper-left corner of the window, shown in Figure 2-9.

Press the Esc key to be released from isolation mode.

FIGURE 2-9: By double-clicking on an object you enter In Isolation mode. This helps you edit group contents without disturbing other artwork.

Manipulating Selected Objects

In the following list, you discover a few other cool things you can do with selected objects:

- **Move selected objects:** When an object is selected, you can drag it to any location on the page, but what if you only want to nudge it a bit? To nudge an item one pixel at a time, select it with the Selection tool and press the left-, right-, up-, or down-arrow key to reposition the object. Hold down the Shift key as you press an arrow key to move an object by ten pixels at a time.
- **Constrain movement:** Want to move an object over to the other side of the page without changing its alignment? Constrain something by selecting an object with the Selection tool and dragging the item and then holding down the Shift key before you release the mouse button. By pressing the Shift key mid-drag, you constrain the movement to 45-, 90-, or 180-degree angles!
- **Clone selected objects:** Use the Selection tool to easily *clone* (duplicate) an item and move it to a new location. To clone an item, simply select it with the Selection tool and then hold down the Alt key (Windows) or Option key (Mac). Look for the cursor to change to two arrows (see Figure 2-10) and then drag the item to another location on the page. Let go of the mouse before you let go of the Alt or Option key. Notice that the original object is left intact and that a copy of the object has been created and moved.

FIGURE 2-10: Drag the double arrow to clone an object.

- **Constrain the clone:** By Alt-dragging (Windows) or Option-dragging (Mac) an item and then pressing Shift, you can clone the item and keep it aligned with the original.

REMEMBER

Don't hold down the Shift key until you're in the process of dragging the item; otherwise, pressing Shift deselects the original object.

After you clone an object to a new location, try this neat trick to create multiple objects the same distance apart from each other by using the Transform Again command: Choose Object⇨Transform⇨Transform Again or press Ctrl+D (Windows) or ⌘+D (Mac) to have another object cloned the exact distance as the first cloned object. (See Figure 2-11.) Keep in mind that you must use the Transform again immediately after you make the clone. We discuss transformations in more detail in Chapter 9 of this minibook.

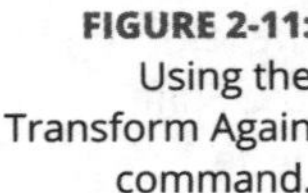

FIGURE 2-11: Using the Transform Again command.

- **Use the Select menu:** By using the Select menu, you can gain additional selection controls, such as choosing Select⇨Inverse, which allows you to select one object and then turn your selection inside out. Also, choosing the Select⇨Same option lets you select one object and then select additional objects on the page based on similarities in color, fill, stroke, and other special attributes.

Take advantage of the Select Similar button on the Control panel to easily access that feature. Notice that in Figure 2-12 you can hold down the arrow to the right of the Select Similar button to choose which similarities should be considered to make the selection.

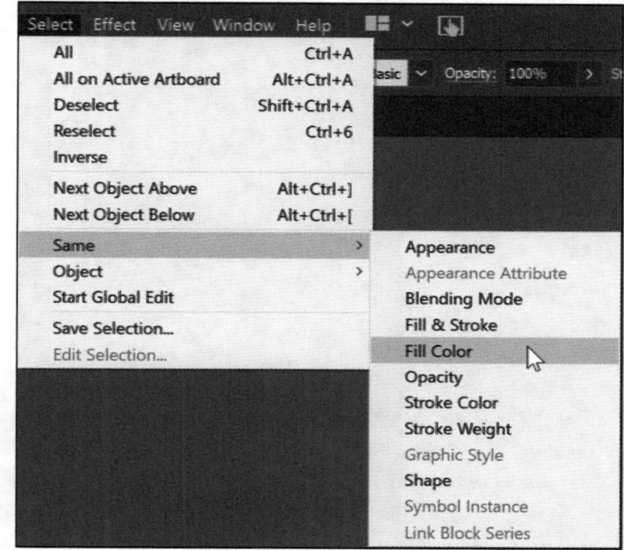

FIGURE 2-12: Define the similarities you want your selections based on.

» Resizing shapes

» Using the Pathfinder panel

» Using the Shape Builder tool

Chapter 3
Creating Basic Shapes and Beyond

Basic shapes, such as squares, circles, polygons, and stars, are used in all types of illustrations. With the right know-how and with the use of the right tools, you can easily create any shape you want for your artwork. In this chapter, you find out how to use the shape tools to create shapes based on precise measurements, and how to combine shapes to create custom shapes. You also learn how to edit and resize your shapes.

The Basic Shape Tools

As a default in the Essentials workspace, the only visible shape tool in the Tools panel is the Rectangle tool. If you click and hold down that tool, you can access hidden tools such as the Rounded Rectangle, Ellipse, Polygon, and Star tools, shown in Figure 3-1.

TIP

You can tear off this tool set so that you don't have to find the hidden shapes later. Click and hold the Rectangle tool and drag to the arrow on the far-right side, and then release the mouse button. These tools are now on a free-floating toolbar that you can drag to another location.

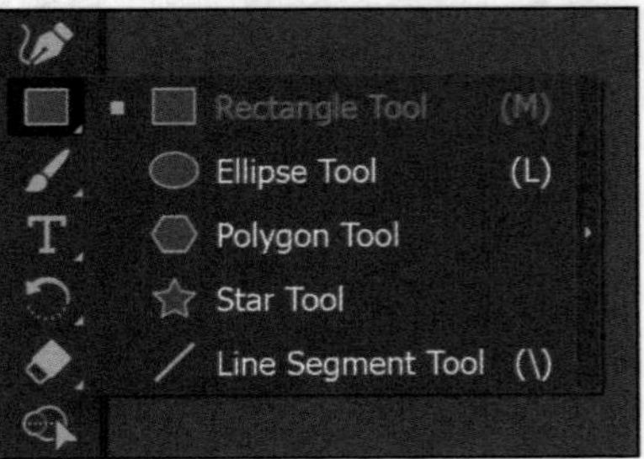

FIGURE 3-1:
Basic shape tools.

Creating rectangles and ellipses

Rectangles and ellipses are the most fundamental shapes you can create, and if you have been following the chapters in order, you have created one or two already. To create a rectangle shape freehand, select the Rectangle tool and simply click on the artboard where you want the shape to begin. Then drag diagonally toward the opposite side: Drag your mouse the distance you want the shape to be in size, and release the mouse button. (See Figure 3-2.) You can drag up or down. You do the same to create a circle or oval by using the Ellipse tool.

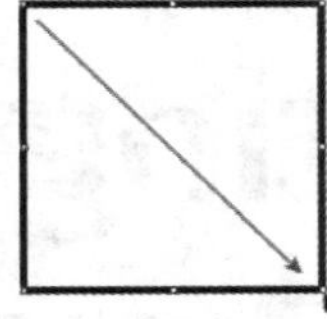

FIGURE 3-2:
Click and drag diagonally to create a shape.

After you create the shape, adjust its size and position by using the Selection tool. Reposition the shape by clicking the selected object and dragging. Resize the object by grabbing a handle and adjusting in or out. To adjust two sides together, grab a corner handle. To resize a shape proportionally, Shift-drag a corner handle.

Making an exact size? If you already know what size you need, don't bother clicking and dragging to create your shape. Simply select your shape, Rectangle, Ellipse, or other shape and then click and release on the artboard. Make sure that you do not click and drag! When you release a dialog box appears, as you see in Figure 3-3. Put your measurements in and press OK.

TIP

If your increments are not what you expected, you can change them right in the shape textbox. Type your measurement followed by the following in order to make sure you get what you expect. (See Table 3-1.)

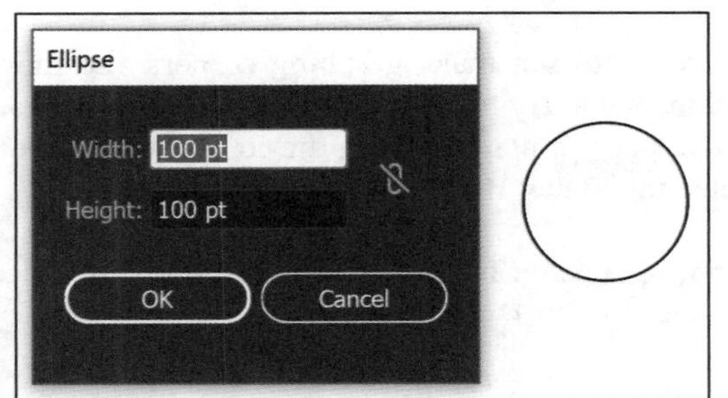

FIGURE 3-3: Click and release to enter exact dimensions for your shape.

TABLE 3-1 **Automatically Changing Measurement Increments**

Type this	For this measurement
in or "	Inches
Mm	Millimeters
Cm	Centimeters
Pt	Points
Px	Pixels

Working with the Live corners feature

Add rounded corners to all your default shapes by using the Corner Widgets. This can be completed visually by selecting and dragging any one of the corner widgets that you see in a shape. You can see in Figure 3-4 how a corner is being rounded by dragging the lower-right corner widget toward the center of the rectangle. You can also round corners of other shapes such as polygons. If you deselect your shape, and then select a corner with the Direct Selection tool and click and drag.

If you don't see the corner widgets, go to View⇨Show Bounding Box.

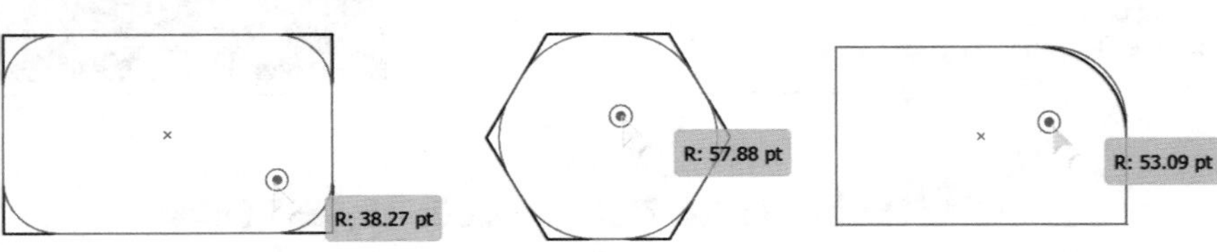

FIGURE 3-4: Use corner widgets to round corners of your shapes. Use the Direct Selection tool to round only selected corners.

Note: The corner widgets can make grabbing corners and moving objects more difficult. For instance, you try to move an object and you instead click and drag a corner widget. If you do not plan on using the corner widgets you can toggle them off and on by selecting View➪Hide or Show Corner Widget.

If you require more precision in your determination of rounded corner size you have a couple of options.

- With a rectangle selected choose Window➪Properties. Click on the More Options button in the Transform section of the panel. Note that you can add and change the corner radius directly in this panel.
- Double-click on any corner widget. If you have the Selection tool you see the Transform panel on the left in Figure 3-5. Here you can either link all four corners to the same size, or press the link button to convert it to unlink. Enter different values for each corner point.
- If you have the Direct Selection tool active and double-click on a specific corner, you see the Corner dialog box on the right of Figure 3-5. Here you can enter a specific value for that point, and even change the roundness of the corner.

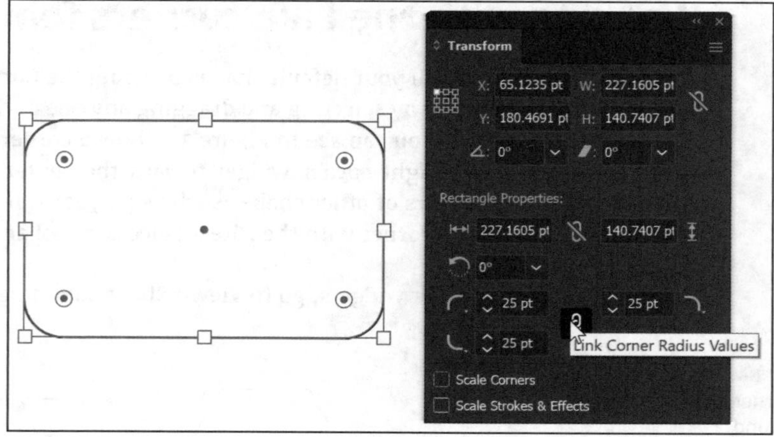

FIGURE 3-5: Making precision changes to your rounded corners.

The Rounded Rectangle tool

By default, you might not see the Rounded Rectangle tool, but if you are planning on using it frequently you can add it to your toolbar by clicking the Edit Toolbar button at the bottom of the Tools panel. Scroll until you find the Rounded Rectangle tool and click and drag it to your tools panel.

With this tool, you can click once on the artboard and enter your size dimensions, including how large you want your corner radius.

When using the Rounded Corner Rectangle you can change the rounded corner visually by pressing the up and down keys on your keyboard *while* you're dragging out the Rounded Rectangle shape on the artboard.

The smaller the value, the less rounded the corners; the higher the value, the more rounded the corners. Be careful: You can round a rectangle's corners so much that it becomes an ellipse!

Using the Polygon tool

You create stars and polygons in much the same way as you create rectangles and ellipses. Select the Polygon tool and click and drag from one corner to another to create the default six-sided polygon shape. You can also select the Polygon tool and click once on the artboard to change the Polygon tool options in the Polygon dialog box.

You can change the polygon shape by entering new values in the Radius and Sides text fields, as shown in Figure 3-6. The radius is determined from the center to the edge of the polygon. The value for the number of sides can range from 3 (making triangles a breeze to create) to 1,000. Whoa — a polygon with 1,000 sides would look like a circle unless it was the size of Texas!

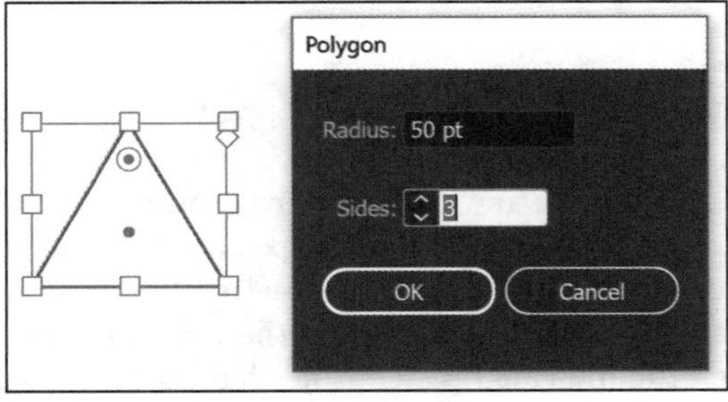

FIGURE 3-6: Click once on the artboard to enter exact dimensions for your polygon.

Using the Star tool

To create a star shape, select the Star tool from the Tools panel. (Remember that it may hide under other shape tools.) If you click the artboard once to open the Star dialog box, you see three text fields in which you can enter values to customize your star shape:

- » **Radius 1:** Distance from the outer points to the center of the star
- » **Radius 2:** Distance from the inner points to the center of the star
- » **Points:** Number of points that comprise the star

The closer together the Radius 1 and Radius 2 values are to each other, the shorter the points on your star. In other words, you can go from a starburst to a seal of approval by entering values that are close in the Radius 1 and Radius 2 text fields, as shown in Figure 3-7.

Scale to put as many images side by side as you can.

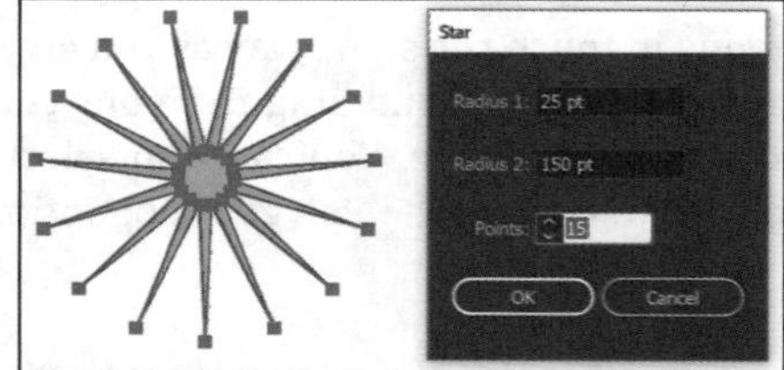

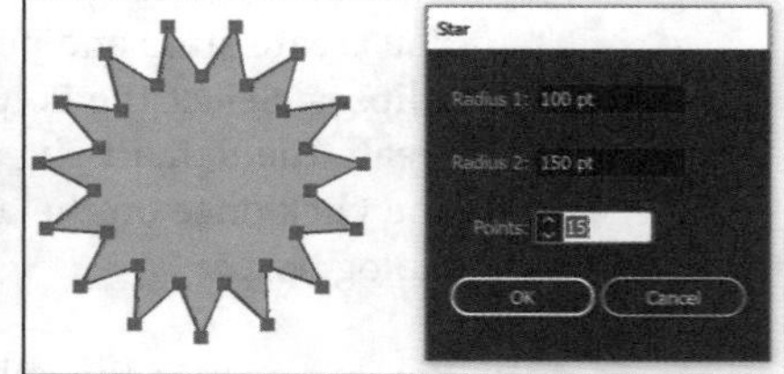

FIGURE 3-7: Radius 1 and Radius 2 are closer to each other in the star on the bottom.

TIP

When creating a Polygon or a Star you can change the amount of sides visually by pressing the up and down keys on your keyboard *while* you're dragging out the Polygon or Star shape on the artboard.

Resizing Shapes

You often need a shape to be an exact size (for example, 2 x 3 inches). After you create a shape, the best way to resize it to exact measurements is to use either the Transform section of the Properties panel or the Transform panel, shown in Figure 3-8. Have the object selected and then choose Window⇨Properties or Transform to open the indicated panel. Note that on these panels you can enter values to place an object in the X and Y fields as well as enter values in the width (W) and height (H) text fields to determine the exact size of an object.

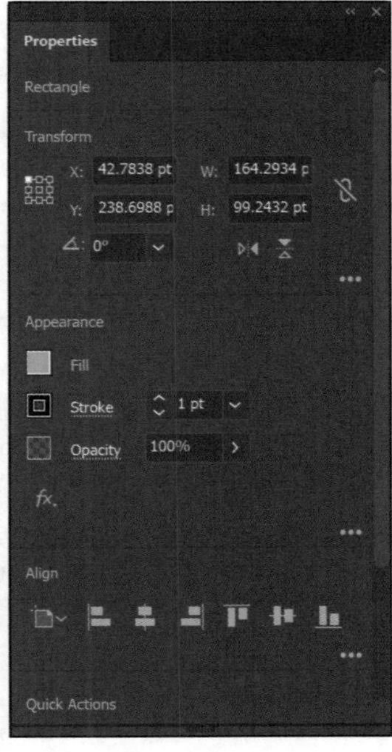

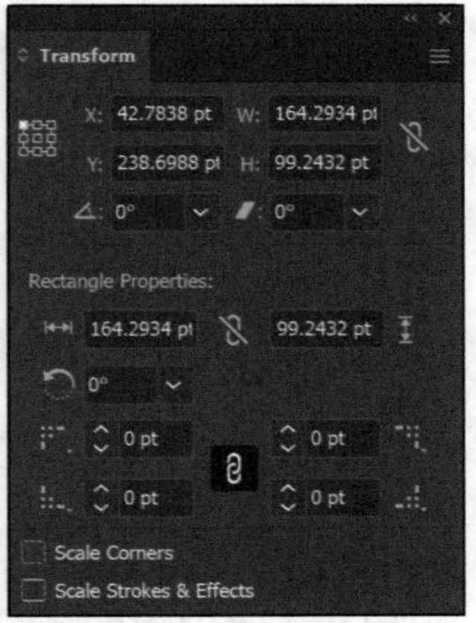

FIGURE 3-8: Precisely set the size of a shape using the Properties and Transform panels.

In many Adobe Illustrator panels, you may see measurement increments consisting of points, picas, millimeters, centimeters, or inches, which can be confusing and maybe even intimidating. But you can control which measurement increments to use.

TIP

Show rulers by choosing View⇨Rulers⇨Show Rulers or press Ctrl+R (Windows) or ⌘ +R (Mac). Then right-click (Windows) or Control-click (Mac) the ruler to change the measurement increment to an increment you're more familiar with. Using the contextual menu that appears, you can change the measurement increment directly on the document.

REMEMBER

If you don't want to bother creating a freehand shape and then changing its size, select the Shape tool and click the artboard. The Options dialog box specific to the shape you're creating appears, and you can type values into the Width and Height text fields.

If you accidentally click and drag, you end up with a tiny shape on your page. Don't fret. Simply get rid of the small shape by selecting it and pressing the Delete key, and then you can try again.

Tips for Creating Shapes

The following simple tips can improve your skills at creating basic shapes in Illustrator:

- Press and hold the Shift key while dragging with the Rectangle or Ellipse tool to create a perfect square or circle. (See Figure 3-9.) This trick is also helpful when you're using the Polygon and Star tools — holding down the Shift key constrains them so that they're straight.
- Create a shape from the center out by holding down the Alt (Windows) or Option (Mac) key while dragging. Hold down Alt+Shift (Windows) or Option+Shift (Mac) to pull a constrained shape out from the center.
- When creating a star or polygon shape by clicking and dragging, if you keep the mouse button down, you can then press the up- or down-arrow key to interactively add or subtract points or sides to your shape.

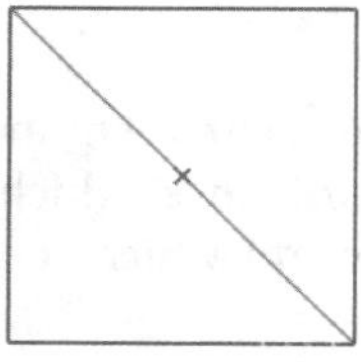
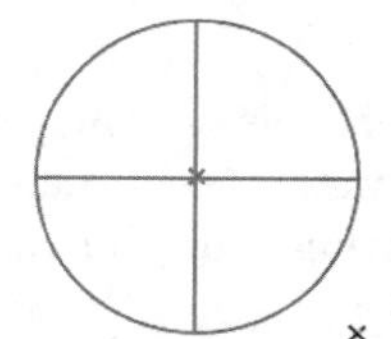
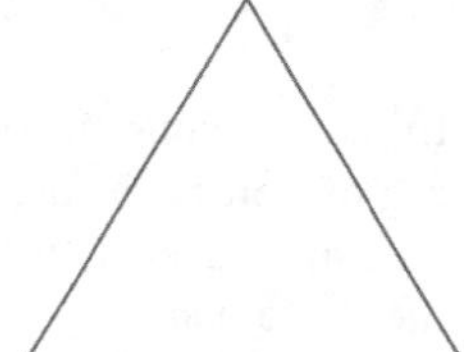

FIGURE 3-9: Use the Shift key to constrain a shape while you create it. In this example a rectangle, Ellipse and Polygon set to 3 sides is shown.

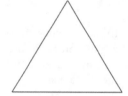

Making some pies

You can take an ellipse and convert it into a pie or a wedge. Do this by selecting an ellipse and dragging one of the pie widgets to create a pie shape, as shown in Figure 3-10.

You can also go to the Transform section of the Properties panel, click the More Options button to reveal additional options. Click the Invert Pie button (double arrows) to create a wedge shape. To reset a pie shape back to an ellipse, double-click one of the pie widgets.

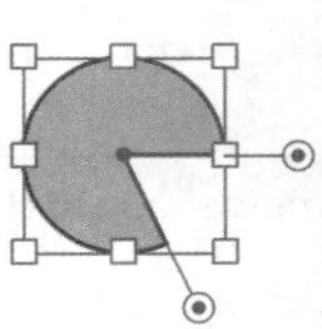

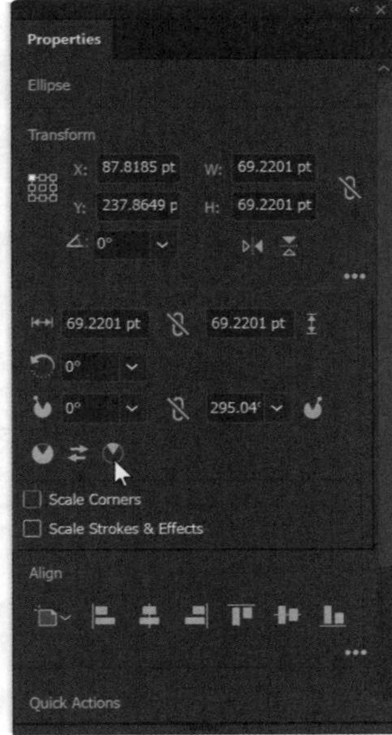

FIGURE 3-10: Use the Pie widget to create pies and wedges from your selected Ellipse.

Creating advanced shapes

At times, it may be wise to use advanced tools in Illustrator to create unique shapes. The Pathfinder panel is an incredible tool you can use to combine, knock out (eliminate one shape from another), and even create shapes from other intersected shapes.

You use the Pathfinder panel, shown in Figure 3-11, to combine objects into new shapes. To use the Pathfinder panel, choose Window➪Pathfinder.

Across the top row of the Pathfinder panel are the Shape modes, which let you control the interaction between selected shapes. You can choose from the Shape modes listed in Table 3-2.

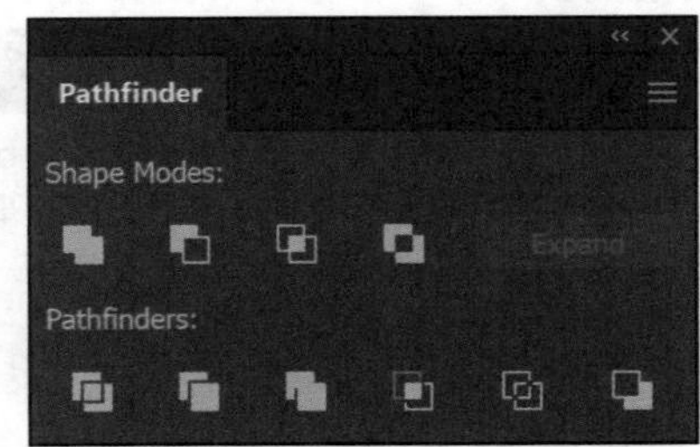

FIGURE 3-11: Combine objects into new shapes.

TIP

If you like the result from using Exclude Overlapping Shapes mode, you can also create a similar effect by selecting several shapes and choosing Object➪Compound Path➪Make. This command "punches" the topmost shapes from the bottom shape. If your results are less than impressive, pick a fill color for your object that is not white.

TABLE 3-2 Shape Modes

Button	Mode	What You Can Do with It
	Unite	Unite selected shapes into one shape.
	Minus Front	Cut out the topmost shape from the underlying shape.
	Intersect Shape Areas	Use the area of the topmost shape to clip the underlying shape as a mask would.
	Exclude Overlapping Shape Areas	Use the area of the shape to invert the underlying shape, turning filled regions into holes and vice versa.

TIP

If you are not ready to permanently combine shapes, you can keep the option of editing the individual shapes by holding down the Alt or Option key when you select a Shape Mode.

Using the Pathfinders

***Pathfinders* are the buttons at the bottom of the Pathfinder panel that let you create new shapes from overlapping objects. Table 3-3 summarizes what each Pathfinder does.**

TABLE 3-3 The Pathfinders

Button	Mode	What You Can Do with It
	Divide	Divide all shapes into their own individual shapes. This tool is quite useful when you're trying to create custom shapes.
	Trim	Remove the part of a filled object that's hidden.
	Merge	Remove the part of a filled object that's hidden. Also, remove any strokes and merge any adjoining or overlapping objects filled with the same color.
	Crop	Delete all parts of the artwork that fall outside the boundary of the topmost object. You can also remove any strokes. If you want strokes to remain when using this feature, select them and choose Object➪Path➪Outline Stroke.
	Outline	Divide an object into its shape's line segments or edges. This tool is useful for preparing artwork that needs a trap for overprinting objects.
	Minus Back	Delete an object that's behind the front-most object.

Using the Shape Builder tool

One of our favorite tools is the Shape Builder tool. Using the Shape Builder tool, you can intuitively combine, edit, and fill shapes on your artboard. Follow these steps to create your own unique shape using the Shape Builder tool:

1. **Create several overlapping shapes.**
2. **Select the shapes that you want to combine.**
3. **Select the Shape Builder tool and then click and drag across the selected shapes, as shown on the left in Figure 3-12.**

   The selected shapes are combined into one shape, as shown on the right in Figure 3-12.

   The Shape Builder tool also enables merging objects, breaking overlapping shapes, subtracting areas, and more.

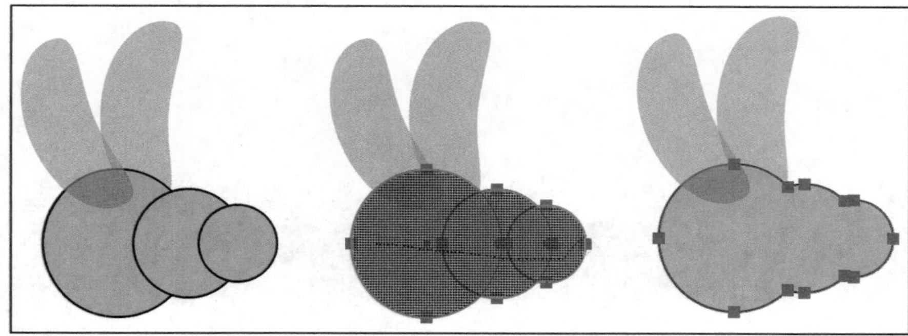

FIGURE 3-12: Create several shapes to use with the Shape Builder tool.

4. **Create another shape that overlaps your new combined shape.**
5. **Using the Selection tool, select both shapes.**
6. **Select the Shape Builder tool again.**
7. **Hold down the Alt (Windows) or Option (Mac) key and click and drag across the newly added shape, as shown in Figure 3-13.**

   It's subtracted from the underlying shape.

Coloring fills and strokes is easier now, too. When you're finished making your shape, you can use the Live Paint Bucket tool that is hidden in the Shape Builder tool to intuitively fill your shape with color.

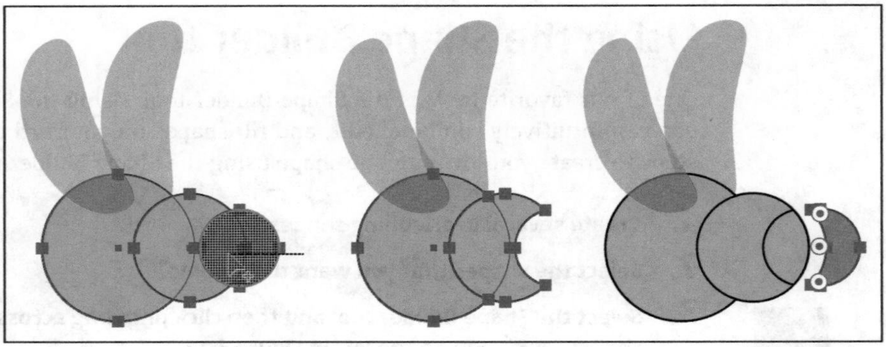

FIGURE 3-13: Alt/Option and drag to subtract one shape from another shape. Simply click a shape to divide or separate it from another shape.

» Creating paths, closed shapes, and curves

» Using the hidden Pen tools

» Tracing some artwork

» Using the Image Trace feature

» Placing images in Illustrator CC

Chapter 4
Using the Pen Tool and Integrating Images

You've seen illustrations that you know are made from paths, but how do you make your own custom paths? In this chapter, we show you how to use the Pen tool to create paths and closed shapes.

REMEMBER

Using the Pen tool requires more coordination than using other Illustrator tools. Fortunately, Adobe Illustrator includes features to help make using the Pen tool a little easier. After you master the Pen tool, the possibilities for creating illustrations are unlimited. Read this chapter to build your skills using one of most popular tools in the creative industry: the Pen tool's Bézier curve capabilities.

Pen Tool Fundamentals

You can use the Pen tool to create all sorts of elements, such as straight lines, curves, and closed shapes, which you can then incorporate into illustrations:

- **Bézier curve:** Originally developed by Pierre Bézier in the 1970s for CAD/CAM operations, the Bézier curve (shown in Figure 4-1) became the underpinnings of the entire Adobe PostScript drawing model. You can control the depth and size of a *Bézier curve* by using direction lines.

FIGURE 4-1: Bézier curves are controlled by direction lines.

- **Anchor point:** You can use anchor points to control the shape of a path or an object. Anchor points are created automatically when using shape tools. You can manually create anchor points by clicking from point to point with the Pen tool.
- **Direction line:** These lines are essentially the handles you use on curved points to adjust the depth and angle of curved paths.
- **Closed shape:** When a path is created, it becomes a closed shape when the start point joins the endpoint.
- **Simple path:** A *path* consists of one or more straight or curved segments. Anchor points mark the endpoints of the path segments.

In the next section, we show you how to control the anchor points.

Creating a straight line

A basic function of the Pen tool is to create a simple path. For this next set of steps create a new Illustrator document.

You can create a simple, straight line with the Pen tool by following these steps:

1. **Press D or click the small black-and-white color swatches at the bottom of the Tools panel.**

   Pressing D reverts to the default colors of a black stroke and a white fill. With black as a stroke, you can see your path more clearly.

   TIP: Pressing D to change the foreground and background colors black and white also works in Photoshop and InDesign.

2. **Click the Fill swatch, at the bottom of the Tools panel, to ensure that the Fill swatch is in front of the Stroke swatch, and then press the forward slash (/) key to change the fill to None.**
3. **Open a new blank page and select the Pen tool.**

   Notice that when you move the mouse over the artboard, the Pen cursor appears with an asterisk (*) beside it, indicating that you're creating the first anchor point of a path.

4. **Click the artboard to create the first anchor point of a line.**

   The asterisk disappears.

   WARNING: Avoid dragging the mouse or you'll end up creating a curve rather than a straight segment.

5. **Click anywhere else on the document to create the ending anchor point of the line.**

   Illustrator creates a path between the two anchor points. Essentially, the path looks like a line segment with an anchor point at each end. (See Figure 4-2.)

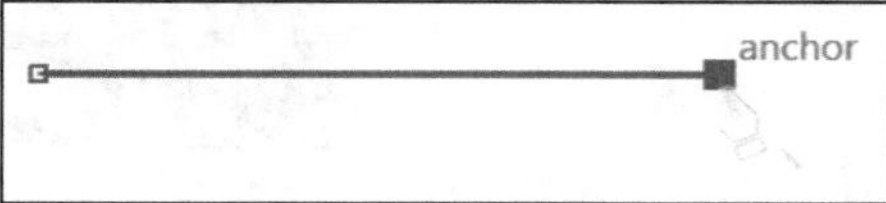

FIGURE 4-2: A path connected by two anchor points.

To make a correction to a line you created with the Pen tool (as described in the preceding step list), follow these steps:

1. **Choose Select⇨Deselect to make sure that no objects are selected.**
2. **Select the Direct Selection tool from the Tools panel.**

   Notice the helpful feature that enlarges the anchor point when you pass over it with the Direct Selection tool.

TURNING OFF THE RUBBER BAND FEATURE

As a default, the Pen tool uses a rubber band feature that hints where your path will go as you reposition your cursor. This feature is distracting to some users. If you prefer to not have that feature enabled, choose Edit ⇨ Preferences (Windows) or Illustrator ⇨ Preferences (Mac) and select the Selection and Anchor Display option. Uncheck the Pen option at the bottom of the window, as shown in the figure below.

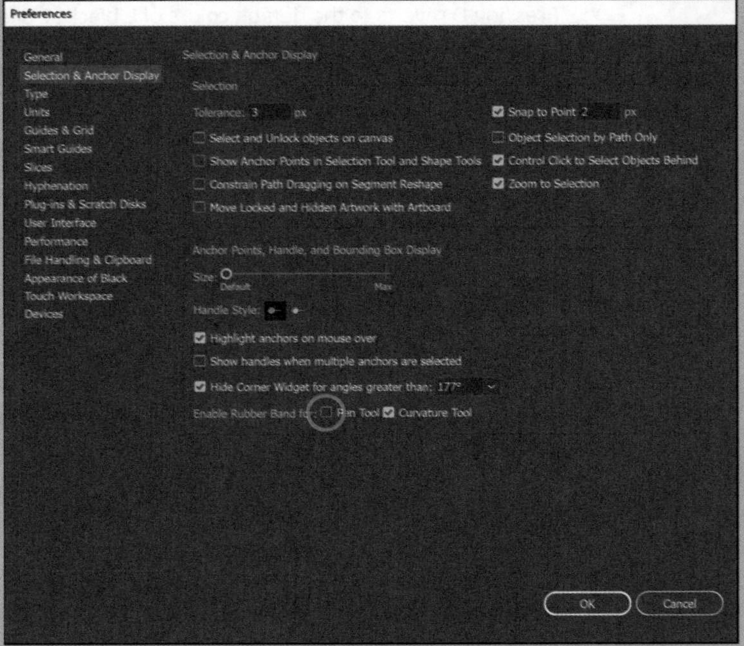

3. **Click an anchor to select one point on the line.**

   Notice that the selected anchor point is solid and the other is hollow. *Solid* indicates that the anchor point you clicked is active, whereas *hollow* is inactive.

4. **Click and drag the anchor point with the Direct Selection tool.**

   The selected anchor point moves, changing the direction of the path while not affecting the other anchor point.

REMEMBER

Use the Direct Selection tool (press A to use the keyboard shortcut to select the Direct Selection tool) to make corrections to paths.

Make sure that only the anchor point you want to change is active. If the entire path is selected, all anchor points are solid. If only one anchor point is selected, all but that one point will be hollow.

Creating a constrained straight line

In this section, we show you how to create a real straight line — one that's on multiples of a 45-degree angle. Illustrator makes it easy; just follow these steps:

1. **Select the Pen tool and click the artboard anywhere to place an anchor point.**
2. **Hold down the Shift key and click another location to place the ending anchor point.**

   Notice that when you're holding down the Shift key, the line snaps to a multiple of 45 degrees.

REMEMBER

Release the mouse button before you release the Shift key or else the line pops out of alignment.

Creating a curve

In this section, you see how to use the Bézier path to create a curved segment. We don't guarantee that you'll love this process — not at first, anyway. But after you know how to use a Bézier path, you'll likely find it useful. To create a Bézier path, follow these steps:

1. **Starting with a blank artboard, select the Pen tool and click the artboard anywhere to place the first anchor point.**
2. **Click someplace else to place the ending anchor point — don't let go of the mouse button — and then drag the cursor until a direction line appears.**

   If you look closely, you see that anchor points are square and that direction lines have circles at the end, as shown in Figure 4-3.

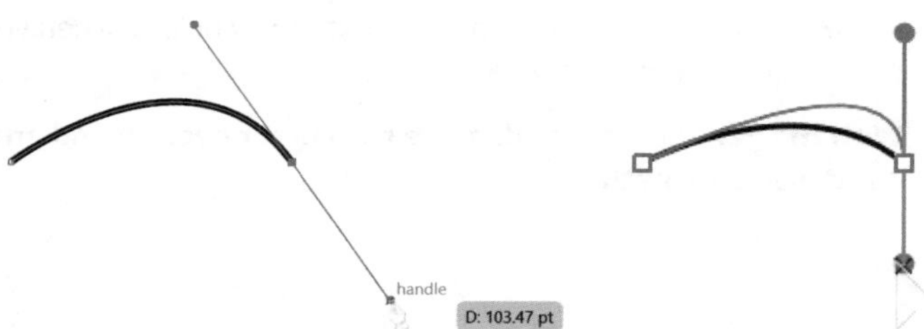

FIGURE 4-3: Click and drag with the Pen tool to create a curved path. You can then click on the directional line to adjust the curve.

3. **Drag the direction line closer to the anchor point to flatten the curve; drag farther away from the anchor point to increase the curve, as shown in Figure 4-3.**
4. **When you're happy with the curve, release the mouse button.**

You've created an *open path,* or a path that doesn't form a closed shape. We show you in the next section how to reconnect to the starting point of the path to make a closed shape.

To alter a curved segment after you create it, follow these steps:

1. **Choose Select⇨Deselect to ensure that no objects are selected.**
2. **Choose the Direct Selection tool and click the last anchor point created.**

   If the direction lines aren't already visible, they appear.

   TIP

   If you have difficulty selecting the anchor point, drag a marquee around it with the Direct Selection tool.
3. **Click precisely at the end of one of the direction lines; drag the direction line to change the curve.**

Reconnecting to an existing path

Creating one segment is fine if you want just a line or an arch. But if you want to create a shape, you need to add more anchor points to the original segment. If you want to fill your shape with a color or a gradient, you need to close it, which means that you need to eventually return to the starting anchor point.

To add segments to your path and create a closed shape, follow these steps:

1. **Create a segment (straight or curved).**

   We show you how in the preceding sections of this chapter.

   You can continue from this point, clicking and adding anchor points until you eventually close the shape. For this example, you deselect the path so that you can discover how to continue adding to paths that have already been created. Knowing how to edit existing paths is extremely helpful when you need to make adjustments to artwork.
2. **With the Pen tool selected, move the cursor over an end anchor point on the deselected path.**

3. **Click when you see the Pen icon with a forward slash to connect your next segment.**

   The forward slash indicates that you're connecting to this path.

4. **Click someplace else to create the next anchor point in the path; drag the mouse if you want to create a curved segment.**

5. **Click to place additional anchor points, dragging as needed to curve those segments.**

   Remember that you want to close this shape, so place the anchor points so that you can eventually come back to the first anchor point.

   The shape shown in Figure 4-4 is a result of adding several linked anchor points.

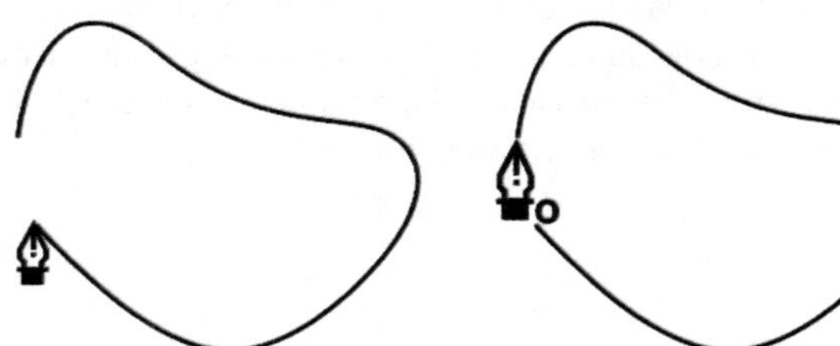

FIGURE 4-4: Adding anchor points to create a shape. Click when the close circle appears to create a closed shape.

6. **When you return to the first anchor point, move the cursor over it and click when the close icon (a small, hollow circle) appears, as shown in Figure 4-4.**

   The shape now has no endpoints.

Controlling curves

After you feel comfortable creating curves and paths, take control of those curves so that you can create them with a greater degree of precision. The following steps walk you through the manual method for changing the direction of anchor points and reveal a helpful keyboard shortcut to make controlling paths a little more fluid. At the end of this section, we introduce new tools that you may also want to take advantage of to help you get control of the Pen tool.

To control a curve, follow these steps:

1. **Create a new document and then choose View⇨Show Grid to show a series of horizontal and vertical rules that act as guides.**

   If it helps, use the Zoom tool to zoom in to the document.

2. **With the Pen tool, click an intersection of any of these lines in the middle area of the page to place the initial anchor point and drag upward.**

   Let go but don't click when the direction line has extended to the horizontal grid line above it, as shown in Figure 4-5(a).

3. **Click to place the second anchor point on the intersection of the grid directly to the right of your initial point; drag the direction line to the grid line directly below it, as shown in Figure 4-5(b).**

TIP

   If you have difficulty keeping the direction line straight, hold down the Shift key to constrain it.

4. **Choose Select⇨Deselect to deselect your curve.**

   Congratulations! You've created a controlled curve. In these steps, you created an arch that's going up, so you first clicked and dragged up. Likewise, to create a downward arch, you must click and drag down. Using the grid, try to create a downward arch like the one shown in Figure 4-5(c).

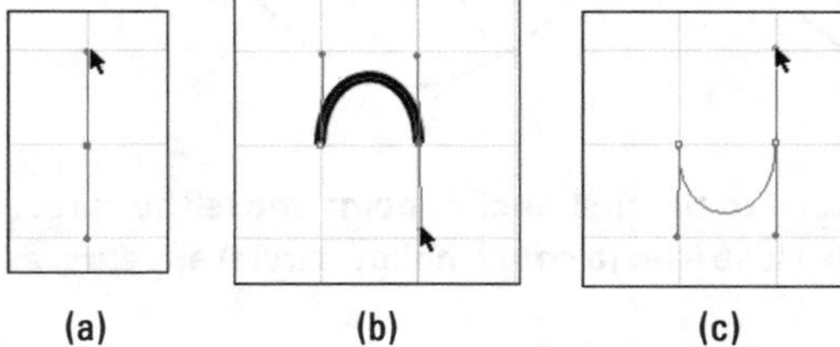

FIGURE 4-5: Creating a controlled Bézier curve.

Creating a corner point

To change the direction of a path from being a curve to a corner, you have to create a *corner point*, shown on the right in Figure 4-6. A corner point has no direction lines and allows for a sharp directional change in a path.

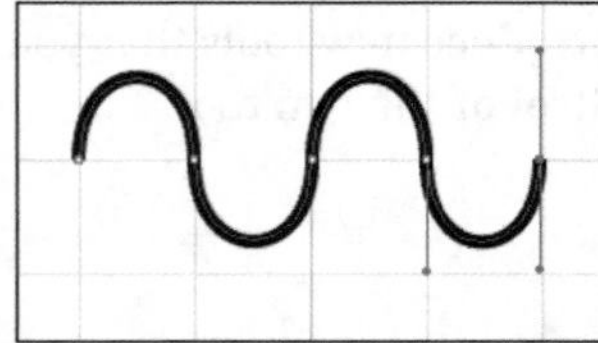

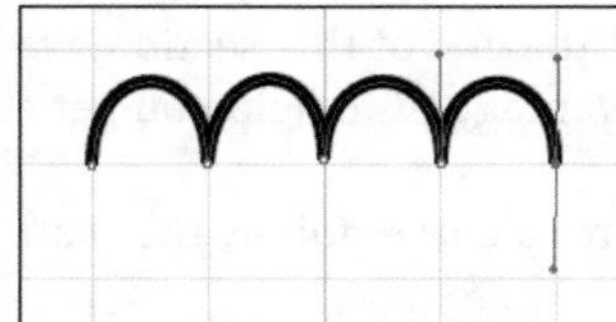

FIGURE 4-6: Smooth versus corner points.

You can switch from the Pen tool to the Convert Anchor Point tool to change a smooth anchor point into a corner point, but that process is time-consuming. An easier way is to press the Alt (Windows) or Option (Mac) key — the Pen tool temporarily changes into the Convert Anchor Point tool — while clicking the anchor point.

To change a smooth anchor point into a corner point by using the shortcut method, follow these steps:

1. **Create an upward arch.**

   We show you how in the preceding section, "Controlling curves." (Refer to Figure 4-7[b]).

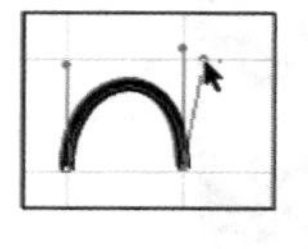

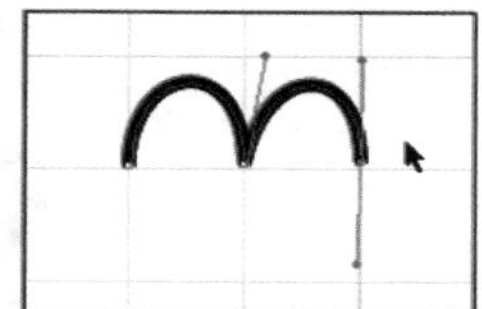

FIGURE 4-7: Converting from smooth to corner.

2. **Hold down the Alt (Windows) or Option (Mac) key and position the cursor over the last anchor point (the last point that you created with the Pen tool).**

3. **When the cursor changes to a caret (that's the Convert Anchor Point tool), click and drag until the direction line is up to the grid line above, as shown on the left in Figure 4-7.**
4. **Release the Alt (Windows) or Option (Mac) key and the mouse button, move the cursor to the grid line to the right, and click and drag down.**

Additional Pen Tools

Hold down the Pen tool icon in the Tools panel to access additional tools: the Add Anchor Point, Delete Anchor Point, and Convert Anchor Point tools, shown in Table 4-1. In the preceding section, we show you how to create a corner point by using the shortcut method, by pressing the Alt (Windows) or Option (Mac) key to access the Convert Anchor Point tool. You may feel more comfortable switching to that tool when you need to convert a point, but switching tools can be more time-consuming.

TIP

Even though you can use a hidden tool to delete and add anchor points, Illustrator automatically does this as a default when you're using the Pen tool. When you move the cursor over an anchor point by using the Pen tool, a minus icon appears. To delete that anchor point, simply click. Likewise, when you move the cursor over a part of the path that doesn't contain anchor points, a plus icon appears. Simply click to add an anchor point.

TABLE 4-1 **The Hidden Pen Tools**

Icon	Tool
	Pen
	Add Anchor Point
	Delete Anchor Point
	Convert Anchor Point

If you prefer to use the tools dedicated to adding and deleting anchor points, choose Edit⇨Preferences⇨General (Windows) or Illustrator⇨Preferences⇨General (Mac); in the Preferences dialog box that appears, select the Disable Auto Add/Delete checkbox. Then, when you want to add or delete an anchor point, select the appropriate tool and click the path.

Adding tools to help make paths

Some Pen tool modifiers are available in the Properties panel in Illustrator CC. You can take advantage of them for many Pen tool uses, but using keyboard shortcuts to switch the Pen tool to its various options is probably still faster. If you're resistant to contorting your fingers while trying to create a path, you should appreciate these tools.

To see the Properties panel tools, select the Pen tool and start creating a path. Notice that the Properties panel has a series of buttons available, shown in Figure 4-8.

Using the Eraser tool

If you haven't discovered the Eraser tool, you will wonder how you got along without it! Using the Eraser tool, you can quickly remove areas of artwork as easily as you erase pixels in Photoshop by stroking with your mouse over any shape or set of shapes.

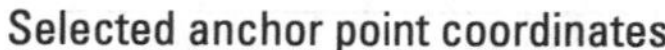

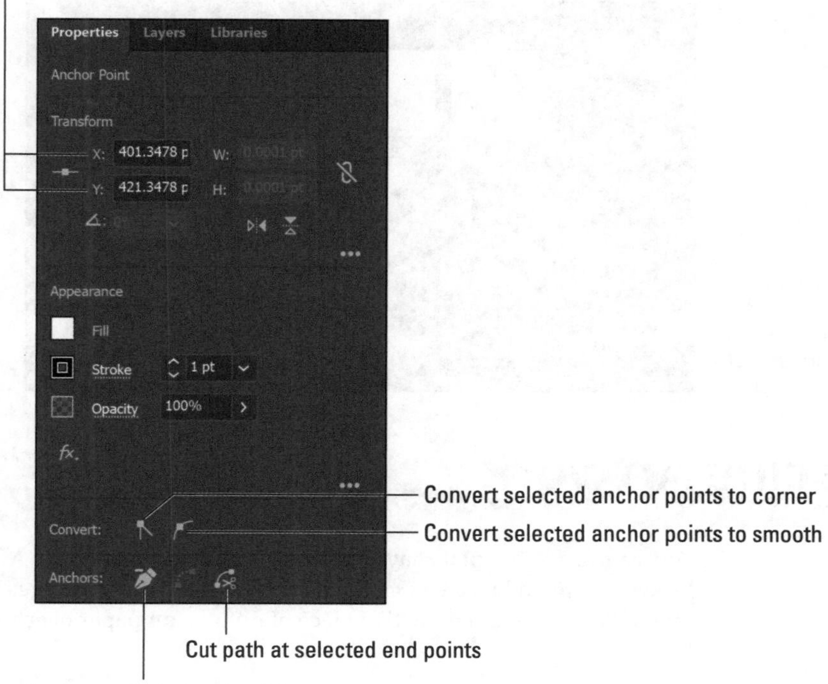

FIGURE 4-8: Properties panel tools for easy editing.

New paths are automatically created along the edges of the erasure, even preserving its smoothness, as shown in Figure 4-9.

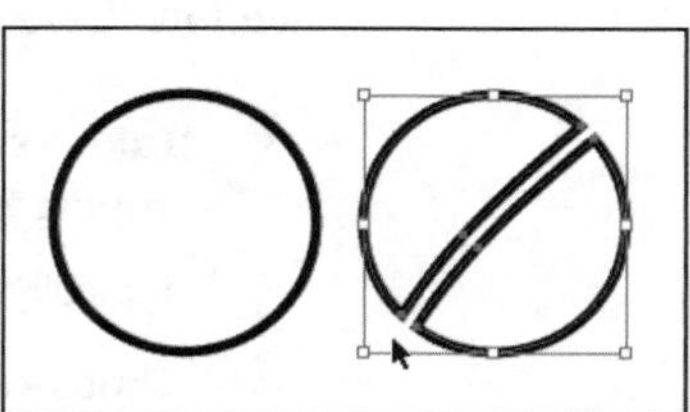

FIGURE 4-9: The Eraser tool deletes sections of a path.

By double-clicking the Eraser tool, you can define the diameter, angle, and roundness of your eraser. (See Figure 4-10.) If you're using a drawing tablet, you can even set Wacom tablet interaction parameters, such as Pressure and Tilt.

If you want to erase more than a single selected object, use Isolation mode to segregate grouped objects for editing. Remember that in order to enter this mode, you simply double-click a group of items. You can then use the eraser on all objects in that group at one time without disturbing the rest of your design.

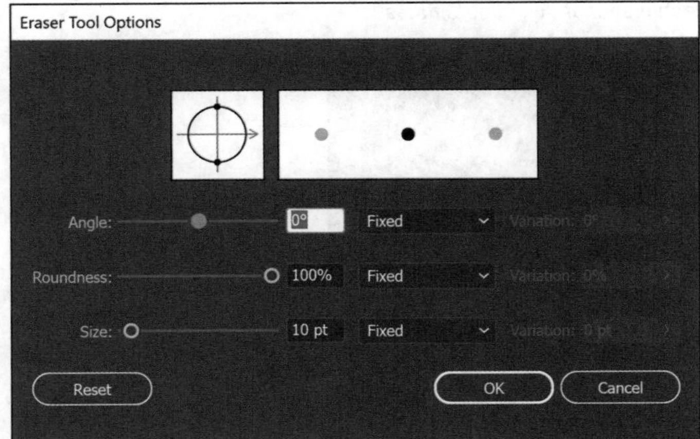

FIGURE 4-10: Double-click the Eraser tool to set various tool options.

Tracing Artwork

You can use a template layer to trace an image manually. A *template layer* is a locked, dimmed layer you can use to draw over placed images with the Pen tool, much like you would do with a piece of onion skin paper over the top of an image.

Creating a template layer

Follow these steps to create a template layer:

1. **Take a scanned image or logo and save it in a format that Illustrator can import from your image-editing program, such as Photoshop.**

   Typically, you save the image as an .eps, a .tif, or a native .psd (Photoshop) file.

2. **Choose File⇨Place to open the Place dialog box.**

3. **In the Place dialog box, locate the saved image; then select the Template checkbox and click Place.**

   On the Mac OS, you must select the Options button in order to be able to check the Template checkbox.

   Note that the Template checkbox is located at the bottom of the dialog box.

   Selecting the Template checkbox tells Illustrator to lock down the scanned image on a layer. Essentially, you can't reposition or edit your image.

   After you click Place, a template layer is automatically created for you, and another layer is waiting for you to create your path. The newly created top layer resembles a piece of tracing paper that has been placed on top of the scanned image.

4. **Re-create the image by tracing over it with the Pen tool.**
5. **When you're done, turn off the visibility of the placed image by clicking the Visibility icon to the left of the template layer.**

   You now have a path you can use in place of the image, which is useful if you're creating an illustration of an image or are digitally re-creating a logo.

For more about layers, check out Chapter 7 of this minibook.

Keep practicing to become more comfortable with clicking and dragging, flowing with the direction line pointing the way you want the path to be created; everything will fall into place.

Using Image Trace

With the Image Trace feature, you can take raster (bitmap) artwork and automatically trace it to convert it into vector artwork. This means that you can take scans of sketches, illustrations, and even photographs, and convert them into vector artwork that can be edited in Illustrator. The Image Trace feature is not totally new — in previous versions it was known as Live Trace — but with CC it has been greatly improved. Additional options have been added, and you can easily access those options and "test" your trace using the Image Trace panel. Follow these steps to try out this improved feature:

1. **Choose File⇨Place and place a scan or raster illustration that you want to convert to vector paths.**

   Immediately after placing, choose Window⇨Image Trace.
2. **Click the Preset drop-down menu to select a preset most similar to the artwork you want to convert.**

   Many of the presets are defined according to the type of artwork that you are tracing — such as a sketch, technical drawing, or photograph. (See Figure 4-11.) The 3 Colors, 6 Colors, and 16 Colors options are useful when converting a photograph to a vector image. If you have complex art, you may receive a message that it will take more time for this to process.

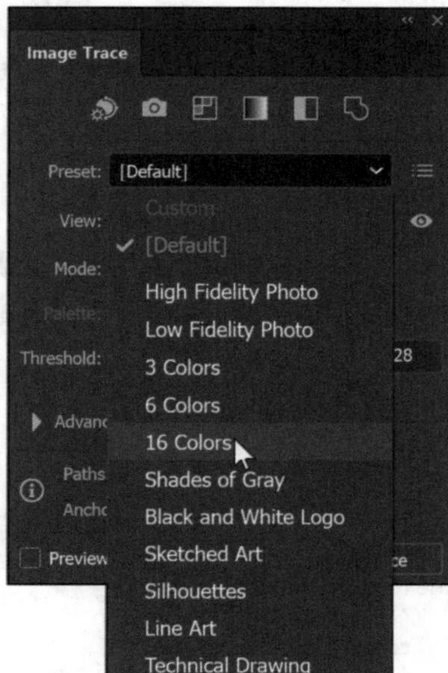

FIGURE 4-11: Select a preset that best fits your placed image.

3. **Expand the Advanced section and note the additional path options available there:**

- **Paths** determines how accurate a representation of the original art the path trace will be. The higher the percentage, the more accurate the tracing. (See Figure 4-12.) Be careful with this feature — if you set the percentage too high, you can end up with more anchor points than necessary. By default, this setting is at 50%, which should work for most of your artwork.
- **Corner** determines how the Image Trace traces corners. A lower percentage produces a more rounded corner; a higher amount a sharper corner. (See Figure 4-13.) Again, the default setting (75%) works for most cases.
- **Noise** determines the minimum details size Image Trace can reproduce. The default of 100 pixels tends to work well. This provides you an area that will be recognized by the Image Trace feature of about 10 x 10 pixels. You might have to play with this setting a bit to see what works best for your image.

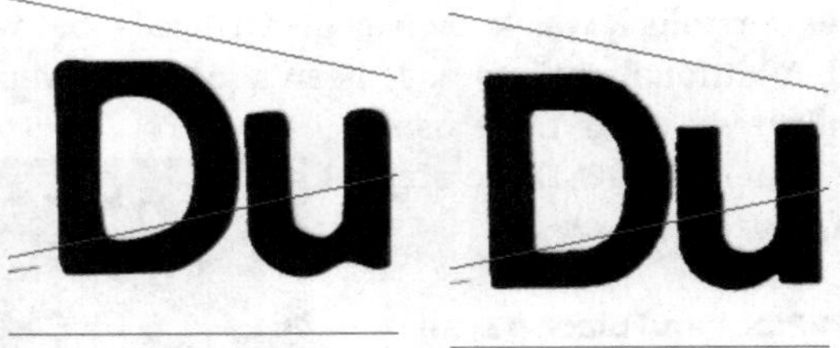

FIGURE 4-12: The image on the left has a low path tolerance of 0%. The image on the right has a high path tolerance of 100%.

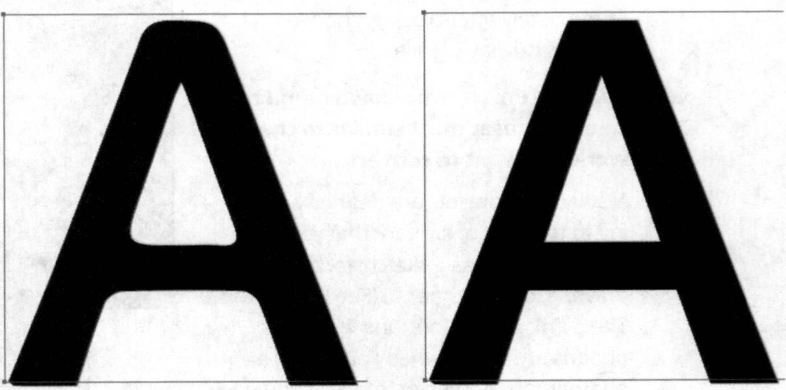

FIGURE 4-13: On the left, a traced image with the corner setting at 0%. On the right, a traced image with a corner setting of 100%.

Changing the view

Can't see the results of your trace? Keep in mind that you can change the view. Just select the View drop-down menu in the Image Trace panel. Using this you can turn off and on the view for the original image as well as the tracing results.

Other Details You Should Know about Placing Images

In the preceding section, you discover how to place an image as a template. But what if you want to place an image to be used in an illustration file? Simply choose File⇨Place.

Click an image once to see its Link checkbox. On the Mac OS, click the Options button in order to see this information. If you keep the checkbox selected, the image is linked to the original file, which is helpful if you plan to reference the file several times in the illustration (it saves file space) or edit the original and have it update the placed image in Illustrator. This option is usually selected by people in the prepress industry who want to have access to the original image file. Just remember to send the image with the Illustrator file if it's to be printed or used someplace other than on your computer.

TIP

You can unembed an image by selecting it in the Links panel (Window⇨Links) and then choosing Unembed from the panel menu, as shown in Figure 4-14.

If you are worried about keeping all your placed files associated with your Illustrator file, you can use File⇨Package in Illustrator CC. This feature gives you the option to create a folder that contains your linked files, fonts, and a report about your file, as you see in Figure 4-15.

If you deselect the Link checkbox, the image is embedded into the Illustrator file. This option builds the image data into the Illustrator file. With an embedded image you won't need to remember where the original file is stored, but it does make for a larger Illustrator file. Now that you have the ability to choose Unembed from the Links panel menu, you can retrieve the image data and save it for future editing if you need to. In certain instances, such as when you want an image to become a symbol (see Chapter 10 of this minibook), the image will have to be embedded, but most functions work with both linked and unlinked files.

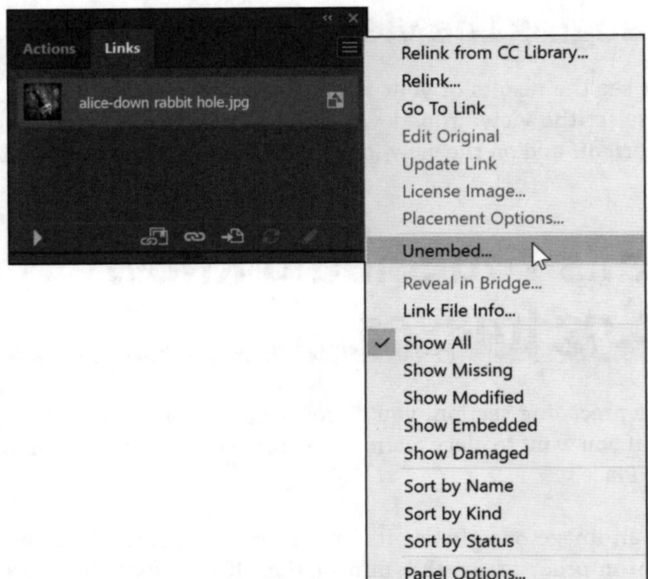

FIGURE 4-14: You can unembed an embedded image by using the Links panel options.

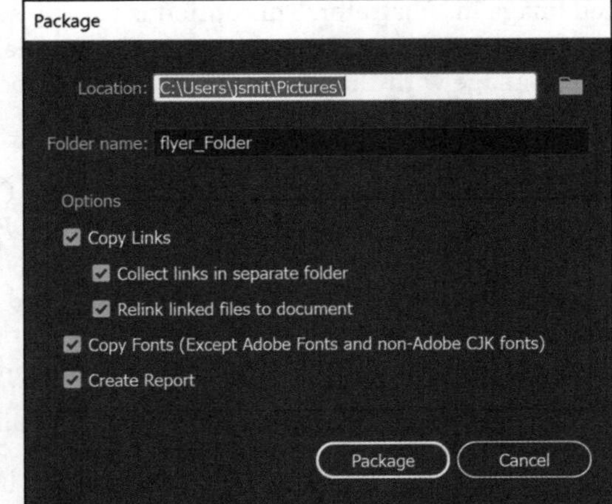

FIGURE 4-15: Collect images and fonts and create a report using the Package feature.

» Getting to know text areas

» Manipulating text along paths and within shapes

» Assigning font styles

» Discovering the Character, Control, and Paragraph panels

» Saving time with text utilities

Chapter 5
Using Type in Illustrator

One of Illustrator's strengths is manipulating text. Whether you're using Illustrator to create logos, business cards, or type to be used on the web or any number of other screen devices, you have everything you need to create professional-looking text.

In this chapter, you discover the Type tools and a few basic (and more advanced) text-editing tricks that you can take advantage of. You then find out about other text tools, such as the Character and Paragraph panels. At the end of this chapter, you get the quick-and-dirty lowdown on the Illustrator text utilities. These utilities can save you loads of time, so don't skip that section.

Working with Type

You can create all sorts of artwork with type, from creating a single line of text to placing text along paths, and wrapping text around objects. In this section, you start with simple text entry and then progress into more complicated text functions.

Click and hold the Type tool to see the hidden type tools, as shown in Figure 5-1. The different tools give you the capability to be creative and also accommodate formatting for foreign languages. Keep in mind that depending upon you selected workspace you may see additional tools. For this chapter Window ⇨ Workspace ⇨ Typography was selected for the workspace.

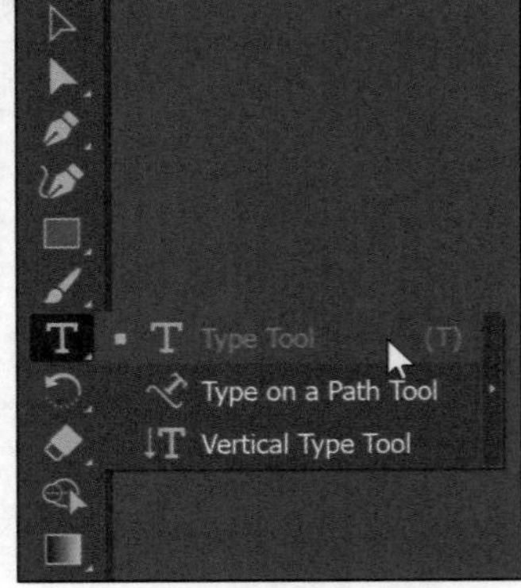

FIGURE 5-1: The Type tools.

Creating text areas

A *text area* is a region that you define. Text, when inserted in this region, is constrained within the shape. To create a text area, click and drag with the Type tool. *Lorem Ipsum* fills the area.

Note: Adobe Illustrator inserts placeholder text for you as a default. If you would prefer that your textbox stay empty, choose Edit ⇨ Preferences ⇨ Type (Windows) or Illustrator ⇨ Preferences ⇨ Type (Mac OS), and then uncheck the Fill New Type Objects With Placeholder Text checkbox.

TIP

As you create and finish typing in a text area, you may want to quickly click and drag to create a new text area elsewhere on your artboard. You do have two options to create multiple text boxes quickly on your artboard:

- Choose Select ⇨ Deselect and then create another area.
- Hold down the Ctrl (Windows) or ⌘ (Mac) key, and click anywhere on the artboard outside of the active text area. By clicking, you deactivate the current text box so that you can click and drag out a new text area.

Creating a line of text

To create a simple line of text, select the Type tool and click the artboard. As a default, a textbox filled with Lorem Ipsum text appears. It will disappear when you start typing. With this method, the line of type goes on forever (even beyond the end of the Scratch area) until you press Enter (Windows) or Return (Mac) to start a new line of text. This excess length is fine if you just need short lines of text for callouts or captions, for example, but it doesn't work well if you're creating a label or anything else that has large amounts of copy.

WARNING

Many new users click and drag an ever-so-small text area that doesn't allow room for even one letter. If you accidentally do this, switch to the Selection tool, delete the active type area, and then click to create a new text insertion point.

Flowing text into an area

Select the Type tool and then drag on the artboard to create a text area. The cursor appears in the text area; text you type flows automatically to the next line when it reaches the edge of the text area. You can also switch to the Selection tool and adjust the width and height of the text area with the handles.

TIP

Need an exact size for a text area? With the Type tool selected, drag to create a text area of any size. Then choose Window➪Properties panel. Type an exact width measurement in the W text field and an exact height measurement in the H text field.

Dealing with text overflow

Watch out for excess text! If you create a text area that's too small to hold all the text you want to put into it, a red plus sign appears in the lower-right corner, as shown in Figure 5-2.

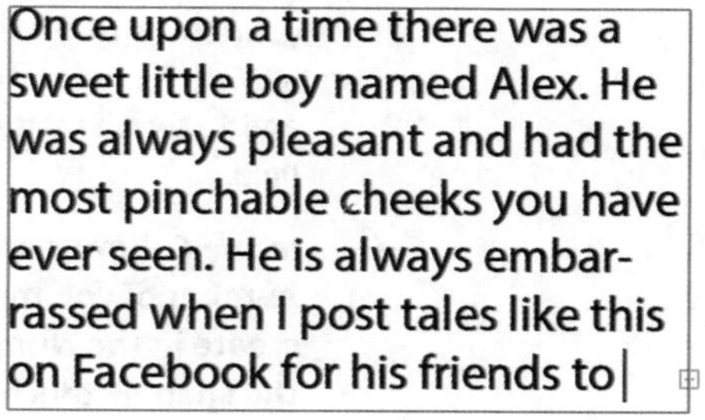

Once upon a time there was a sweet little boy named Alex. He was always pleasant and had the most pinchable cheeks you have ever seen. He is always embar-rassed when I post tales like this on Facebook for his friends to

FIGURE 5-2:
The plus icon indicates that text is overflowing.

When Illustrator indicates that you have too much text for the text area, you have three options:

- Make the text area larger by switching to the Selection tool and dragging the handles.
- Make the text smaller until you no longer see the overflow indicator.
- *Thread* this text area (link it to another), which is a topic covered later in this chapter, in the "Threading text into shapes" section.

Creating columns of text with the Area Type tool

The easiest and most practical way to create rows and columns of text is to use the area type options in Adobe Illustrator. This feature lets you create rows and columns from any text area. You can have only rows, only columns (much like columns of text in a newspaper), or even both.

1. **Select the Type tool and drag on the artboard to create a text area.**
2. **Choose Type⇨Area Type Options.**

   The Area Type Options dialog box appears, as shown in Figure 5-3. At the end of this section, a list explains all options in the Area Type.
3. **In the Area Type Options dialog box, enter a width and height in the Width and Height text fields.**

   The Width and Height text fields contain the height and width of your entire text area. In Figure 5-3, `302 pt` is in the Width text field and `184 pt` is in the Height text field.

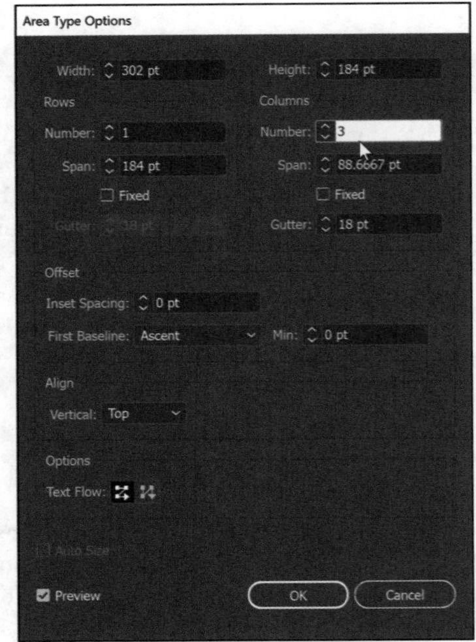

FIGURE 5-3: The Area Type Options dialog box lets you create columns of text.

4. **In the Columns area, enter the number of columns you want to create in the Number text field, the span distance in the Span text field, and the gutter space in the Gutter text field.**

   The *span* specifies the height of individual rows and the width of individual columns. The *gutter* is the space between columns and is automatically set for you, but you can change it to any value you like.
5. **Click OK.**

Note that you can also access the Area Type Options by selecting the Open Area Type Options button in the Area Type section of the Properties panel.

When you create two or more columns of text from the Area Type Options dialog box, text flows to the next column when you reach the end of the previous column, as shown in Figure 5-4.

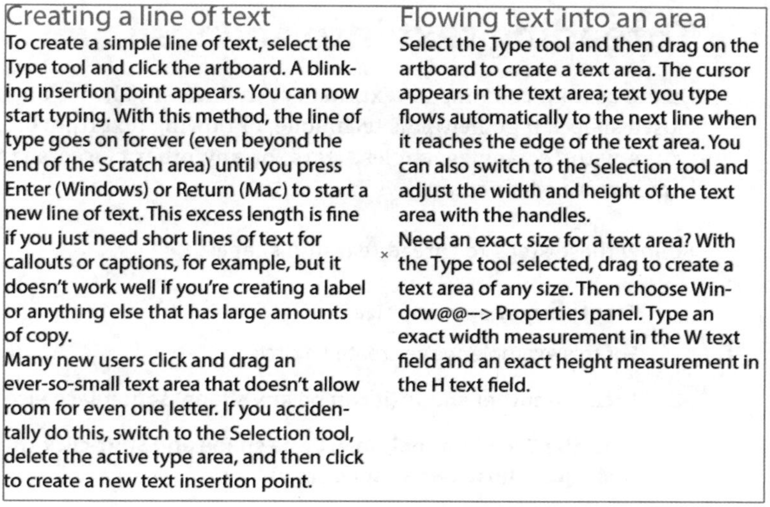

Creating a line of text

To create a simple line of text, select the Type tool and click the artboard. A blinking insertion point appears. You can now start typing. With this method, the line of type goes on forever (even beyond the end of the Scratch area) until you press Enter (Windows) or Return (Mac) to start a new line of text. This excess length is fine if you just need short lines of text for callouts or captions, for example, but it doesn't work well if you're creating a label or anything else that has large amounts of copy.

Many new users click and drag an ever-so-small text area that doesn't allow room for even one letter. If you accidentally do this, switch to the Selection tool, delete the active type area, and then click to create a new text insertion point.

Flowing text into an area

Select the Type tool and then drag on the artboard to create a text area. The cursor appears in the text area; text you type flows automatically to the next line when it reaches the edge of the text area. You can also switch to the Selection tool and adjust the width and height of the text area with the handles.

Need an exact size for a text area? With the Type tool selected, drag to create a text area of any size. Then choose Window@@–> Properties panel. Type an exact width measurement in the W text field and an exact height measurement in the H text field.

FIGURE 5-4: One column of text flows into the next.

TIP

The following list breaks down the options available in the Area Type Options dialog box (refer to Figure 5-3):

- **Width and Height:** The present width and height of the entire text area.
- **Number:** The number of rows and/or columns that you want the text area to contain.
- **Span:** The height of individual rows and the width of individual columns.
- **Fixed:** Determines what happens to the span of rows and columns if you resize the type area. When this checkbox is selected, resizing the area can change the number of rows and columns but not their width. Leave this option deselected if you want to resize the entire text area and have the columns automatically resize with it.
- **Gutter:** The empty space between rows or columns.
- **Inset Spacing:** The distance from the edges of the text area.
- **First Baseline:** Where you want the first line of text to appear. The default Ascent option starts your text normally at the top. If you want to put in a fixed size, such as 50 points from the top, select Fixed from the drop-down list and enter **50 pt** in the Min text field.
- **Text Flow:** The direction in which you read the text as it flows to another row or column. You can choose to have the text flow horizontally (across rows) or vertically (down columns).

Threading text into shapes

Create custom columns of text that are in different shapes and sizes by threading closed shapes together. This technique, of flowing text from one shape to another, works with rectangles, circles, stars, or any other closed shape and can lead to some creative text areas.

Follow these steps to thread text into shapes:

1. **Create any shape, any size.**

   For this example, we've created a circle.

2. **Create another shape (it can be any shape) someplace else on the page.**
3. **With the Selection tool, select one shape and Shift-click the other to make just those two shapes active.**
4. **Choose Type ➪ Threaded Text ➪ Create.**

   A threading line appears, as shown in Figure 5-5, indicating the direction of the threaded text.

5. **Select the Type tool, click the top of the first shape to start the threading, and start typing.**

   Continue typing until the text flows over into the other shape.

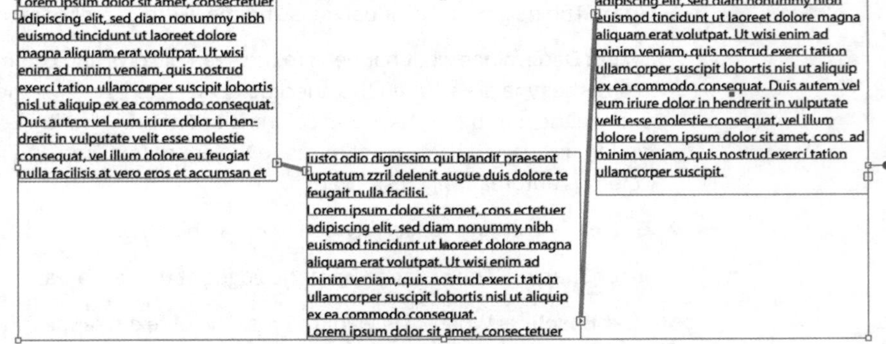

FIGURE 5-5: Threaded text areas flow from one area to another.

If you want to automatically fill with placeholder, right-click (Windows) or Ctrl-click (Mac) and choose Fill with Placeholder Text.

If you no longer want the text to be threaded, choose Type ➪ Threaded Text ➪ Remove Threading, which eliminates all threading from the text shapes. To remove one or more, but not all, shapes from the threading, select the shape you want to remove from the threading and choose Type ➪ Threaded Text ➪ Release Selection.

Wrapping text

Wrapping text isn't quite the same as wrapping a present — it's easier! A *text wrap* forces text to wrap around a graphic, as shown in Figure 5-6. This feature can add a bit of creativity to any piece.

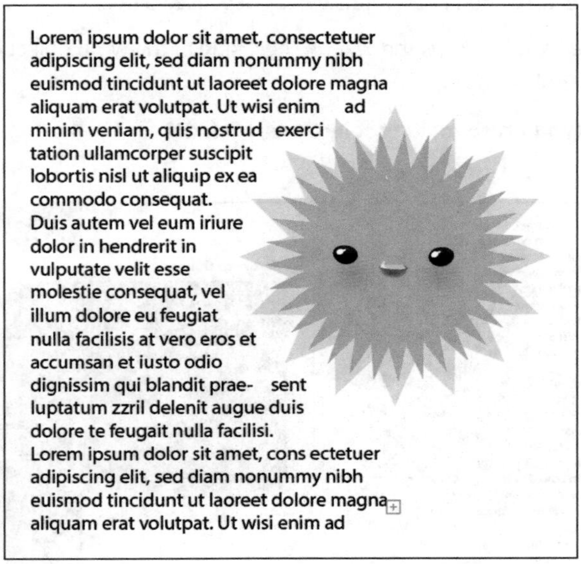

Lorem ipsum dolor sit amet, consectetuer adipiscing elit, sed diam nonummy nibh euismod tincidunt ut laoreet dolore magna aliquam erat volutpat. Ut wisi enim ad minim veniam, quis nostrud exerci tation ullamcorper suscipit lobortis nisl ut aliquip ex ea commodo consequat. Duis autem vel eum iriure dolor in hendrerit in vulputate velit esse molestie consequat, vel illum dolore eu feugiat nulla facilisis at vero eros et accumsan et iusto odio dignissim qui blandit prae- sent luptatum zzril delenit augue duis dolore te feugait nulla facilisi. Lorem ipsum dolor sit amet, cons ectetuer adipiscing elit, sed diam nonummy nibh euismod tincidunt ut laoreet dolore magna aliquam erat volutpat. Ut wisi enim ad

FIGURE 5-6: The graphic is forcing the text to wrap around it.

First, create a text area and either enter text or paste text into it. Then place an image that you can wrap the text around. Feel free to use the Illustrator file named `ToledoSun` that is located in your Book05_Illustrator folder that you downloaded in Book 1, Chapter 1. Follow these steps to wrap text around another object or group of objects:

1. **Select the wrap object.**

   This object is the one you want the text to wrap around.

2. **Make sure that the wrap object is on top of the text you want to wrap around it by choosing Object⇨Arrange⇨Bring to Front.**

   REMEMBER

   If you're working in layers (which we discuss in Chapter 7 of this minibook), make sure that the wrap object is on the top layer.

3. **Choose Object⇨Text Wrap⇨Make.**

   An outline of the wrap area is visible.

4. **Adjust the wrap area by choosing Object⇨Text Wrap⇨Text Wrap Options.**

   The Text Wrap Options dialog box appears, as shown in Figure 5-7.

   You have these options:

   - *Offset:* Specifies the amount of space between the text and the wrap object. You can enter a positive or negative value.
   - *Invert Wrap:* Wraps the text on the inside of the wrap object instead of around it.

5. **When you finish making selections, click OK.**

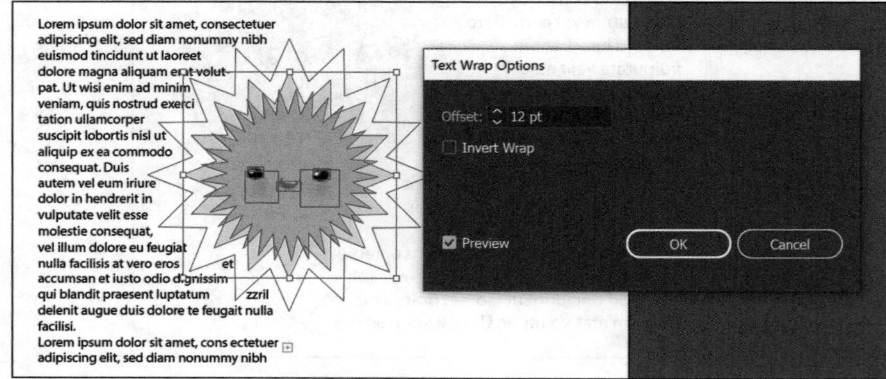

FIGURE 5-7: Adjust the distance of the text wrap from the object.

If you want to change the text wrap at a later point, select the object and choose Object⇨Text Wrap⇨Text Wrap Options. Make your changes and click OK.

If you want to unwrap text from an object, select the wrap object and choose Object⇨Text Wrap⇨Release.

Outlining text

Illustrator gives you the opportunity to change text into outlines or artwork. Basically, you change the text into an object, so you can no longer edit that text by typing. The plus side is that it saves you the trouble of sending fonts to everyone who wants to use the file. Turning text into outlines makes it appear as though your text was created with the Pen tool. You want to use this tool when creating logos that will be used frequently by other people or artwork that you may not have control over.

To turn text into an outline, follow these steps:

1. **Type some text on your page.**

   For this example, just type a word (say, your name) and make sure that the font size is at least 36 points. You can change your font size in the Font Size box in the Control panel at the top. You want to have it large enough to see the effect of outlining it.

2. **Switch to the Selection tool and choose Type⇨Create Outlines.**

   You can also use the keyboard command Ctrl+Shift+O (Windows) or ⌘ +Shift+O (Mac).

   The text is now grouped together in outline form.

3. **If you're being creative, or just particular, and want to move individual letters, use the Group Select tool or choose Object⇨Ungroup to separate the letters, as shown in Figure 5-8.**

FIGURE 5-8: Letters converted to outlines.

WARNING

When you convert type to outlines, the type loses its *hints*, which are the instructions built in to fonts to adjust their shape so that your system displays or prints them in the best way based on their size. Without hints, letters such as lowercase *e* or *a* might fill in as the letter forms are reduced in size. Make sure that the text is the approximate size it might be used at before creating outlines. Because the text loses the hints, try not to create outlines on text smaller than 10 points.

Putting text on a path, in a closed shape, or on the path of a shape

Wow — that's some heading, huh? You've probably seen text following a swirly path or inside a shape. Maybe you think that accomplishing such a task is too intimidating to even attempt. In this section, we show you just how easy these tasks are! Some Type tools are dedicated to putting type on a path or a shape (refer to Figure 5-1), but we think you'll find that the key modifiers we show you in the following sections are easier to use.

Creating text on a path

Follow these steps to put type on a path:

1. **Create a path with the Pen, Line, or Pencil tool.**

   Don't worry if it has a stroke or fill applied.

2. **Select the Type tool and simply cross over the start of the path.**
3. **When an I-bar with a squiggle appears (which indicates that the text will run along the path), click.**

   The stroke and fill of the path immediately change to None.

4. **Start typing, and the text runs on the path.**
5. **Choose Window⇨Type⇨Paragraph and change the alignment in the Paragraph panel to reposition where the text falls on the path.**

   Alternatively, switch to the Selection tool and drag the first of the three I-bars that appear, as shown in Figure 5-9. This allows you to freely move the text on that path. The path in Figure 5-9 was created with the Pen tool. Make sure that you don't cause text to overflow. A red plus sign appears if your text can no longer fit on the path.

Flip the text to the other side of a path by clicking and dragging the middle I-bar under or over the path.

Creating text in a closed shape

Putting text inside a shape can add spunk to a layout. This feature allows you to custom-create a closed shape with the shape tools or the Pen tool and flow text into it. Follow these steps to add text inside a shape:

1. **Create a closed shape — a circle or an oval, for example.**
2. **Select the Type tool and cross over the path of the closed shape.**

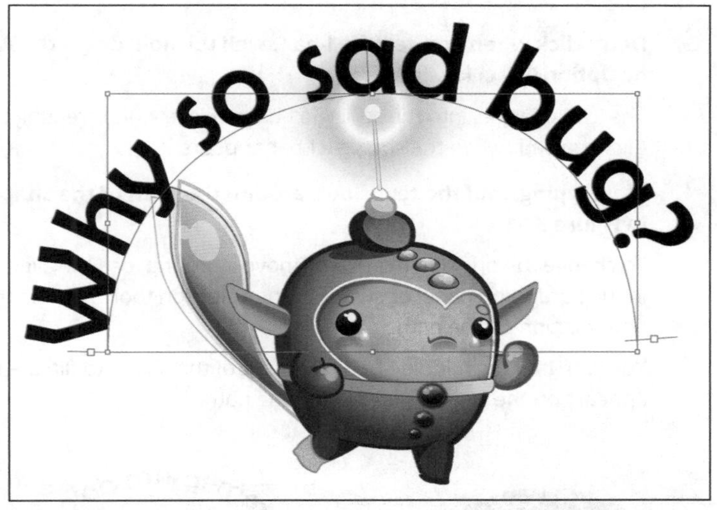

FIGURE 5-9: Use the Selection tool to drag the I-bar to adjust the text.

3. **When you see the I-bar swell or become rounded, click inside the shape, as shown in Figure 5-10.**
4. **Start typing, and the text is contained inside the shape.**

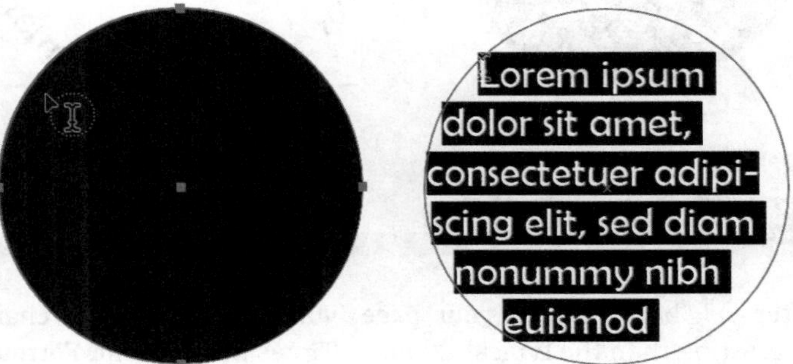

FIGURE 5-10: Position the type cursor over a shape, when you see the bloated cursor, as shown on the left, click to insert text into the shape.

Text on the path of a closed shape

Perhaps you want text to run around the edge of a shape instead of inside it. Follow these steps to have text created on the path of a closed shape:

1. **Create a closed shape, such as a circle.**
2. **Select the Type tool and cross over the path of the circle.**

3. **Don't click when you see the I-bar swell up; hold down the Alt (Windows) or Option (Mac) key instead.**

   The icon changes into the squiggle I-bar you see when creating text on a path. Click the path when the squiggle I-bar appears.

4. **Start typing, and the text flows around the path of the shape, as shown in Figure 5-11.**

   To change the origin of the text or move it around, use the alignment options in the Paragraph panel or switch to the Selection tool and drag the I-bar to a new location on the path.

   You can drag the middle I-bar in and out of the shape to flip the text so that it appears on the outside or inside of the path.

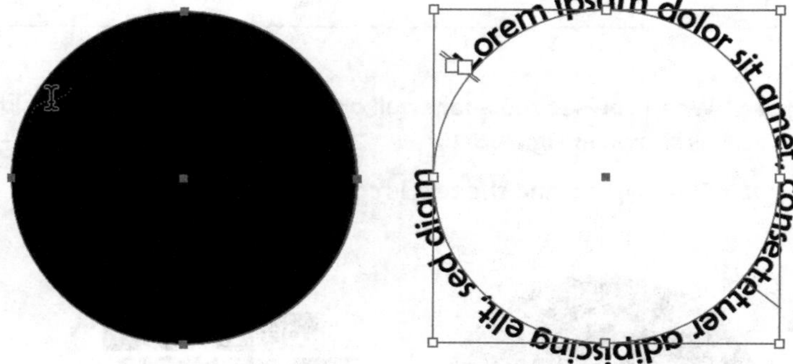

FIGURE 5-11: Holding down the Alt or Option key flows text around a closed shape.

Assigning Font Styles

After you have text on your page, you'll often want to change it to be more interesting than the typical 12-point Times Roman font. Formatting text in Illustrator isn't only simple, but you can also do it multiple ways. In the following list, we name and define some basic type components (see Figure 5-12):

- **Font:** A complete set of characters, letters, and symbols of a particular typeface design.
- **X height:** The height of type, based on the height of the small *x* in that type family.

FIGURE 5-12: Components of type.

- **Kerning:** The space between two letters. Often used for letters in larger type that need to be pulled closer together, such as *W i.* Kern a little to slide the *i* in a little closer to the *W,* maybe even moving into the space occupied by the *W,* as shown in Figure 5-13. Kerning doesn't distort the text; it only increases or decreases the space between two letters.

FIGURE 5-13: Before kerning (left) and after (right).

Wi Wi

- **Leading:** Space between the lines of text.
- **Tracking:** The space between multiple letters. Designers like to use this technique to spread out words by increasing the space between letters. Adjusting the tracking doesn't distort text; it increases or decreases the space between the letters, as shown in Figure 5-14.

FIGURE 5-14: Tracking set at 0 (top) and 300 (bottom).

AGI TRAINING

A G I T R A I N I N G

TIP

Pretty good tracking and kerning have already been determined in most fonts. You don't need to bother with these settings unless you're tweaking text for a more customized look.

- **Baseline:** The line that type sits on. The baseline doesn't include *descenders,* type that extends down, like lowercase *y* and *g.* You adjust the baseline for trademark signs or mathematical formulas, as shown in Figure 5-15.

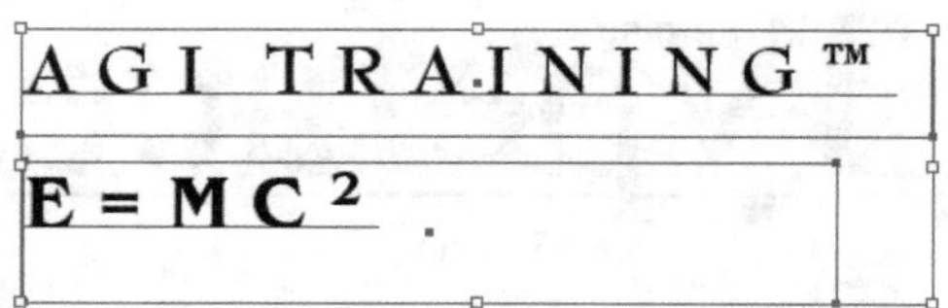

FIGURE 5-15: Adjust the baseline for superscript.

TIP

The keyboard shortcuts for type shown in Table 5-1 work with Adobe Illustrator, InDesign, and Photoshop.

TABLE 5-1 Keyboard Shortcuts for Type

Command	Windows	Mac
Align left, right, or center	Shift+Ctrl+L, R, or C	Shift+⌘ +L, R, or C
Justify	Shift+Ctrl+J	Shift+⌘ +J
Insert soft return	Shift+Enter	Shift+Return
Reset horizontal scale to 100 percent	Shift+Ctrl+X	Shift+⌘ +X
Increase or decrease point size	Shift+Ctrl+> or <	Shift+⌘ +> or <
Increase or decrease leading	Alt+↑ or ↓	Option+↑ or ↓
Set leading to the font size	Double-click the leading icon in the Character panel	Double-click the leading icon in the Character panel
Reset tracking or kerning to 0	Alt+Ctrl+Q	Option+⌘ +Q
Add or remove space *(kerning)* between two characters	Alt+→ or ←	Option+→ or ←
Add or remove space *(kerning)* between two characters by 5 times the increment value	Alt+Ctrl+→ or ←	Option+⌘ +→ or ←
Add or remove space *(kerning)* between selected words; if selected, it changes tracking. Leave cursor in between words.	Alt+← or →	Option+← or →
Increase or decrease baseline shift	Alt+Shift+↑ or ↓	Option+Shift+↑ or ↓

Using the Character Panel

To visualize changes you're making to text and to see characteristics that are already selected, choose Window⇨Type⇨Character or press Ctrl+T (Windows) or ⌘ +T (Mac), which opens the Character panel. Click the menu icon in the upper-right corner to see a panel menu of additional options. Choose Show Options, and additional type attributes appear, such as baseline shift, underline, and strikethrough. Keep in mind that the same options you can see in the Character section in the Properties panel.

TIP

Pressing Ctrl+T (Windows) or ⌘ +T (Mac) is a toggle switch to either show or hide the Character panel. If you don't see the Character panel appear at first, you may have hidden it by pressing the keyboard shortcut. Just try it again.

The following list explains the options in the Character panel (see Figure 5-16):

FIGURE 5-16: The Character panel shows additional options.

» **Font:** Select the font you want to use from this drop-down list.

TIP

You can click and drag across the font name in the Character panel or Control panel, and press the up- or down-arrow key to automatically switch to the next font above or below on the font list. Do this while you have text selected to see the text change live!

» **Set Font Style:** Select the style (for example, Bold, Italic, or Bold Italic) from this drop-down list. The choices here are limited by the fonts you have loaded. In other words, if you have only Times regular loaded in your system, you don't have the choice to bold or italicize it.

» **Type Size:** Set the size of the type in this combo box. Average readable type is 12-point; headlines can vary from 18 points and up.

» **Leading:** Select how much space you want between the lines of text in this combo box. Illustrator uses the professional typesetting method of including the type size in the total leading. In other words, if you have 12-point and want it double-spaced, set the leading at 24 points.

» **Kerning:** To use this combo box, you first need to place the cursor between two letters. Then, to push the letters farther apart from each other, you can click the up arrow or type a value; to decrease the spacing between the letters, type a lower value, even negative numbers, or click the down arrow.

» **Tracking:** Use the Tracking combo box by selecting multiple letters and increasing or decreasing the space between them all at once by clicking the up or down arrows or by typing a positive or negative value.

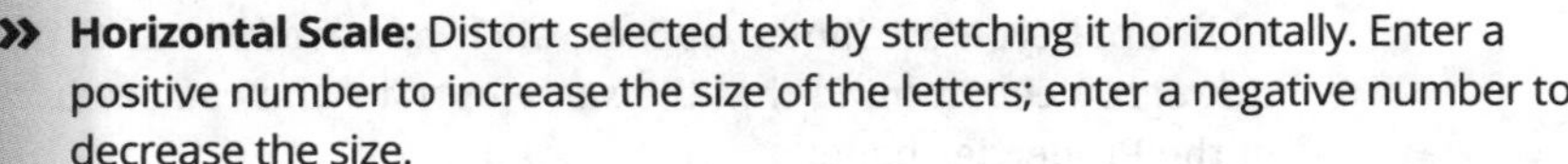

» **Horizontal Scale:** Distort selected text by stretching it horizontally. Enter a positive number to increase the size of the letters; enter a negative number to decrease the size.

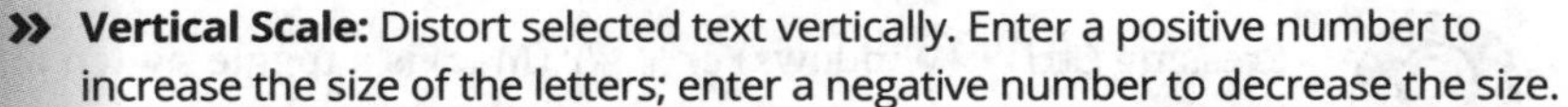

» **Vertical Scale:** Distort selected text vertically. Enter a positive number to increase the size of the letters; enter a negative number to decrease the size.

Using horizontal or vertical scaling to make text look like condensed type often doesn't give good results. When you distort text, the nice thick and thin characteristics of the typeface also become distorted and can produce weird effects.

» **Baseline Shift:** Use baseline shift for trademark signs and mathematical formulas that require selected characters to be moved above or below the baseline.

» **Character Rotation:** Rotate just the selected text by entering an angle in this text field or by clicking the up or down arrows.

» **Underline and Strikethrough:** These simple text attributes underline and strikethrough selected text.

» **Language:** Select a language from this drop-down list. ***Note:*** The language you specify here is used by Illustrator's spell checker and hyphenation feature. We discuss these features in the later section "Text Utilities: Your Key to Efficiency."

Using the Paragraph Panel

Access the Paragraph panel quickly by clicking the Paragraph hyperlink in the Control panel or by choosing Window⇨Type⇨Paragraph. This panel, shown in Figure 5-17, has all the attributes that apply to an entire paragraph (such as alignment and indents, which we discuss in the next two sections, and hyphenation, which we discuss later in this chapter). For example, you can't flush-left one word in a paragraph — when you click the Flush Left button, the entire paragraph flushes left. To see additional options in the Paragraph panel, click the menu icon in the upper-right corner of the panel (the panel menu) and choose Show Options. The attributes for this panel are also available on the Properties panel.

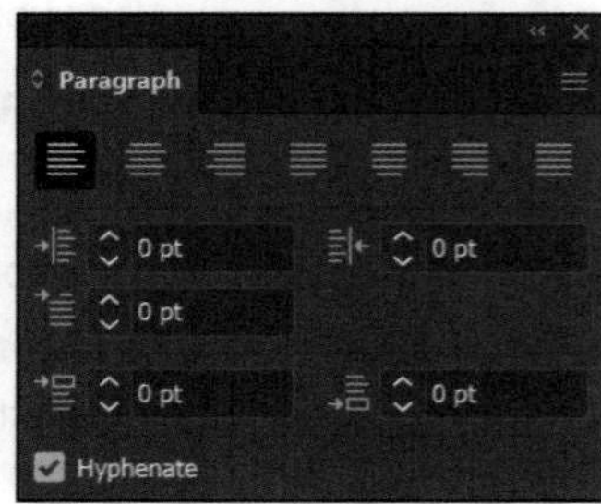

FIGURE 5-17:
Use this panel to open typographic controls that apply to paragraphs.

Alignment

You can choose any of the following alignment methods by clicking the appropriate button on the Paragraph panel:

- **Flush Left:** All text is flush to the left with a ragged edge on the right. This is the most common way to align text.
- **Center:** All text is centered.
- **Flush Right:** All text is flush to the right and ragged on the left.
- **Justify with the Last Line Aligned Left:** Right and left edges are both straight, with the last line left-aligned.
- **Justify with the Last Line Aligned Center:** Right and left edges are both straight, with the last line centered.
- **Justify with the Last Line Aligned Right:** Right and left edges are both straight, with the last line right-aligned.
- **Justify All Lines:** In this *forced justification* method, the last line is stretched the entire column width, no matter how short it is. This alignment is used in many publications, but it can create some awful results.

Indentation

You can choose from the following methods of indentation:

- **First Line Indent:** Indents the first line of every paragraph. In other words, every time you press the Enter (Windows) or Return (Mac) key, this spacing is created.

  To avoid first-line indents and space after from occurring — if you just want to break a line in a specific place, for example — create a line break or a soft return by pressing Shift+Enter (Windows) or Shift+Return (Mac).
- **Right Indent:** Indents from the right side of the column of text.
- **Left Indent:** Indents from the left side of the column of text.

Use the Eyedropper tool to copy the character, paragraph, fill, and stroke attributes. Select the text you want to change, select the Eyedropper tool, and click the text once with the attributes you want to apply to the selected text.

By default, the Eyedropper affects all attributes of a type selection, including appearance attributes. To customize the attributes affected by these tools, double-click the Eyedropper tool to open the Eyedropper dialog box.

Text Utilities: Your Key to Efficiency

After you have text in an Illustrator document, you may need to perform various tasks within that text, such as search for a word to replace with another word, check your spelling and grammar, save and create your own styles, or change the case of a block of text. You're in luck because Illustrator provides various text utilities that enable you to easily and efficiently perform all these otherwise tedious tasks. In the following sections, we give you a quick tour of these utilities.

Find and Replace

Generally, artwork created in Illustrator isn't text heavy, but the fact that Illustrator has a Find and Replace feature can be a huge help. Use the Find and Replace dialog box (choose Edit ➪ Find and Replace) to search for words that need to be changed, such as changing *Smyth* to *Smith*, or to locate items that may be difficult to find otherwise. This feature works much like all other search-and-replace methods.

Spell checker

Can you believe there was a time when Illustrator didn't have a spell checker? Thankfully, it does now — and its simple design makes it easy to use.

To use the spell checker, choose Edit ➪ Spelling ➪ Check Spelling and then click the Start button in the dialog box that appears. The spell checker works much like the spell checker in Microsoft Word and other popular applications: When a misspelled word is found, you're offered a list of replacements. You can choose to fix that instance, fix all instances, ignore the misspelling, or add the word to the dictionary.

If you click the arrow to the left of Options, you can set other specifications, such as whether you want to look for letter case issues or have the spell checker note repeated words.

Note: The spell checker uses whatever language you specify in the Character panel. We discuss this panel in the earlier section "Using the Character Panel."

TIP

If you work in a specialized industry that uses loads of custom words, save yourself time by choosing Edit➪Edit Custom Dictionary and then adding your own words. We recommend that you do so before you're ready to spell check a document so that the spell checker doesn't flag the custom words later (which slows you down).

Don't forget that just because a word is spelled correctly doesn't mean it's the correct word in the context of a sentence. You should still proofread manually your text.

The Hyphenation feature

Nothing is worse than trying to read severely hyphenated copy. Most designers either use hyphenation as little as possible or avoid it altogether by turning off the Hyphenation feature.

Here are a few things you should know about customizing your hyphenation settings if you decide to use this feature:

- **Turning the Hyphenation feature on or off:** Activate or deactivate the feature in the the Properties panel by clicking on the More icon underneath the Paragraph section (see Figure 5-18); If you don't want to use the Hyphenation feature, turn it off by deselecting the Hyphenation checkbox.
- **Setting specifications in the Hyphenation dialog box:** Set specifications in the dialog box that determine the length of words to hyphenate, the number of hyphens to be used in a single document, whether to hyphenate capitalized words, and how words should be hyphenated. The Before Last setting is useful, for example, if you don't want to have a word, such as *liquidated* hyphenated as *liquidat-ed.* Type **3** in the Before Last text field and Illustrator won't hyphenate words if it leaves only two letters on the next line.

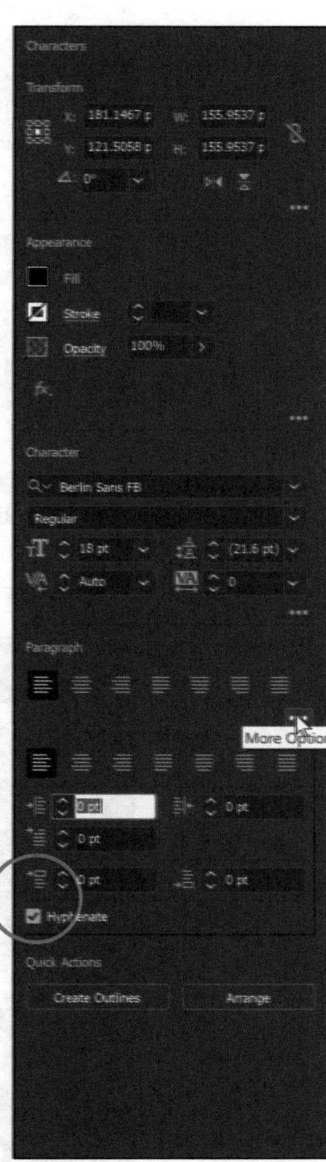

FIGURE 5-18:
Customizing hyphenation settings.

» **Setting the Hyphenation Limit and Hyphenation Zone:** They're not diets or worlds in another dimension — the Hyphenation Limit setting enables you to limit the number of hyphens in a row. For example, type **2** in the Hyphenation Limit text field so that you never see more than two hyphenated words in a row. The Hyphenation Zone text field enables you to set up an area of hyphenation based on a measurement. For example, you can specify **1 inch** to allow for only one hyphenation every inch. You can also use the slider to determine whether you want better spacing or fewer hyphens. This slider works only with the Single-Line Composer (the default).

The Find Font feature

The Find Font feature enables you to list all fonts in a file that contains text and then search for and replace fonts (including the font's type style) by name. Do this by using the Find Font dialog box (see Figure 5-19), opened by choosing Type⇨Find Font. Select the font you want to replace from the Fonts in Document list. Next, select a font from the Replace with Font From list. Click open the Replace with Fonts From drop-down menu to choose the font you want from the Document, Recent, or System. Click the Change button to replace the font (or click the Change All button to replace all instances of the font) and then click OK. That's it!

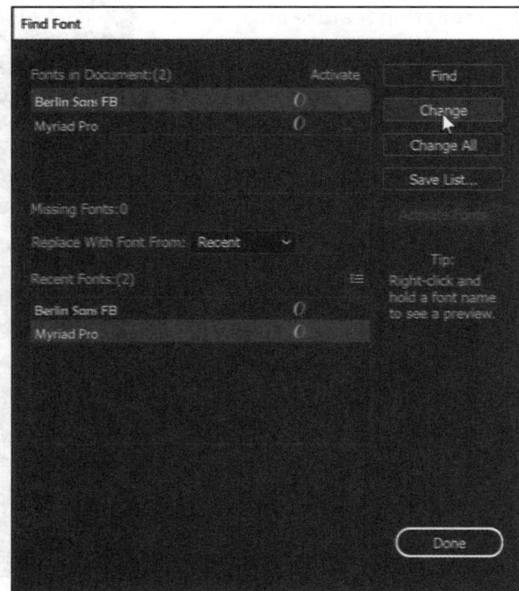

FIGURE 5-19: Use the Find Font dialog box to find and replace typefaces.

The Change Case feature

Doesn't it drive you crazy when you type an entire paragraph before discovering that you somehow pressed the Caps Lock key? Fix it fast by selecting the text, choosing Type⇨Change Case, and then choosing one of these options:

- **Uppercase:** Makes the selected text all uppercase
- **Lowercase:** Makes the selected text all lowercase
- **Title Case:** Capitalizes the first letter in each word
- **Sentence Case:** Capitalizes just the first letter in selected sentences

In Illustrator, you use the same type engine used by InDesign for high-quality text control. You're working, as a default, in what's referred to as Single-Line Composer. Select Single or Every Line composer from the Paragraph panel menu.

The options include the following:

- **Single-Line Composer:** Useful if you prefer to have manual control over how lines break. In fact, this method had been in place in the past. The Single-Line Composer option doesn't take the entire paragraph into consideration when expanding letter space and word spacing, so justified text can sometimes look odd in its entire form. (See Figure 5-20, left.)
- **Every-Line Composer:** A professional way of setting text; many factors are taken into account as far as spacing is concerned, and spacing is based on the entire paragraph. With this method, you see few spacing issues that create strange effects, such as the ones on the left in Figure 5-20, right.

AGI was founded as a training provider and maintains a presence as a resource for companies and individuals looking to become more productive with electronic publishing software. AGI maintains a strong relationship with electronic publishing software companies including Adobe Systems and Quark as a member of their authorized training provider network. AGI is also a private, licensed school in the Commonwealth of Pennsylvania.

AGI was founded as a training provider and maintains a presence as a resource for companies and individuals looking to become more productive with electronic publishing software. AGI maintains a strong relationship with electronic publishing software companies including Adobe Systems and Quark as a member of their authorized training provider network. AGI is also a private, licensed school in the Commonwealth of Pennsylvania.

FIGURE 5-20: Single-Line Composer (left) and Every-Line Composer (right).

Text styles

A *text style* is a saved set of text attributes, such as font and size. Creating text styles keeps you consistent and saves you time by enabling you to efficiently implement changes in one step rather than have to select the text attributes for each instance of that style of text (say, a heading or caption). So when you're finally happy with the way your headlines appear and how the body copy looks or when your boss asks whether the body copy can be a smidgen smaller (hmm, how much is a smidgen?), you can confidently answer, "Sure!"

If you've created styles, changing a text attribute is simple. What's more, the change is applied at once to all text using that style. Otherwise, you would have to make the attribute change to every occurrence of body text, which could take a long time if your text is spread out.

Illustrator offers two types of text styles:

- **Character:** Saves attributes for individual selected text. If you want just the word *New* in a line of text to be red, 20-point Arial, you can save it as a character style. Then, when you apply it, the attribute apply to only the selected text (and not to the entire line or paragraph).
- **Paragraph:** Saves attributes for an entire paragraph. A span of text is considered a paragraph until it reaches a hard return or paragraph break. Note that pressing Shift+Enter (Windows) or Shift+Return (Mac) is considered a soft return, and paragraph styles continue to apply beyond the soft return.

You can create character and paragraph styles in many ways, but we show you the easiest and most direct methods in the following subsections.

Creating character styles

Create a character style when you want individual sections of text to be treated differently from other text in the paragraph. So rather than repeatedly apply a style manually, you create and implement a character style. To do so, open a document containing text and follow these steps:

1. **Set up text with the text attributes you want included in the character style in the Character panel, and then choose Window⇨Type⇨Character Styles.**

   The Character Styles panel opens.

2. **Select the text from Step 1 and Alt-click (Windows) or Option-click (Mac) the New Style button (theplus sign icon) at the bottom of the Character Styles panel.**
3. **In the Character Styles Options dialog box that appears, name your style and click OK.**

   Illustrator records which attributes have been applied already to the selected text and builds a style from them.

4. **Create another text area by choosing Select⇨Deselect and using the Type tool to drag out a new text area.**

   We discuss using the Type tool in the earlier section "Creating text areas."

5. **Change the font and size to dramatically different choices from your saved style and type some text.**

6. **Select some (not all) of the new text and then Alt-click (Windows) or Option-click (Mac) the style name in the Character Styles panel.**

   Alt-click (Windows) or Option-click (Mac) to eliminate any attributes that weren't part of the saved style. The attributes of the saved character style are applied to the selected text.

TIP

When you create a new panel item (any panel) in Adobe Illustrator, InDesign, or Photoshop, we recommend that you get in the habit of Alt-clicking (Windows) or Option-clicking (Mac) the New Style button. This habit allows you to name the item (style, layer, or swatch, for example) while adding it to the panel.

Creating paragraph styles

Paragraph styles include attributes that are applied to an entire paragraph. What constitutes a paragraph is all text that falls before a hard return (you create a hard return when you press Enter in Windows or Return on the Mac), so this could be one line of text for a headline or ten lines in a body text paragraph.

To create a paragraph style, open a document that contains text or open a new document and add text to it; then follow these steps:

1. **Choose Window⇨Type⇨Paragraph Styles to open the Paragraph Styles panel.**

2. **Find a paragraph of text that has the same text attributes throughout it and put your cursor anywhere in that paragraph.**

   You don't even have to select the whole paragraph!

3. **Alt-click (Windows) or Option-click (Mac) the Create New Style button (the plus sign icon at the bottom of the Paragraph panel) to create a new paragraph style; give your new style a name.**

   Your new style now appears in the Paragraph Styles panel list of styles.

4. **Create a paragraph of text elsewhere in your document and make its attributes different from the text in Step 2.**

5. **Put your cursor anywhere in the new paragraph and Alt-click (Windows) or Option-click (Mac) your named style in the Paragraph Styles panel.**

   The attributes from the style are applied to the entire paragraph.

TIP

If style doesn't seem to work, make sure that the Character Style is set to Normal Character Style.

Updating styles

When you use existing text to build styles, reselect the text and assign the style. In other words, if you put the cursor in the original text whose attributes were saved as a style, it doesn't have a style assigned to it in the Styles panel. Assign the style by selecting the text or paragraph and clicking the appropriate style listed in the Styles panel. By doing so, you ensure that any future updates to that style apply to that original text and to all other instances.

Before updating a style, make sure you have nothing selected (Select ⇨ Deselect). To update a style, simply select its name in either the Character or Paragraph Styles panel. Choose either Paragraph Style Options, or Character Style Options from the panel menu, which you access by clicking the Menu icon in the upper-right corner of the panel. In the resulting dialog box (see Figure 5-21), make changes by clicking the main attribute on the left and then updating the choices on the right. After you do so, all tagged styles are updated.

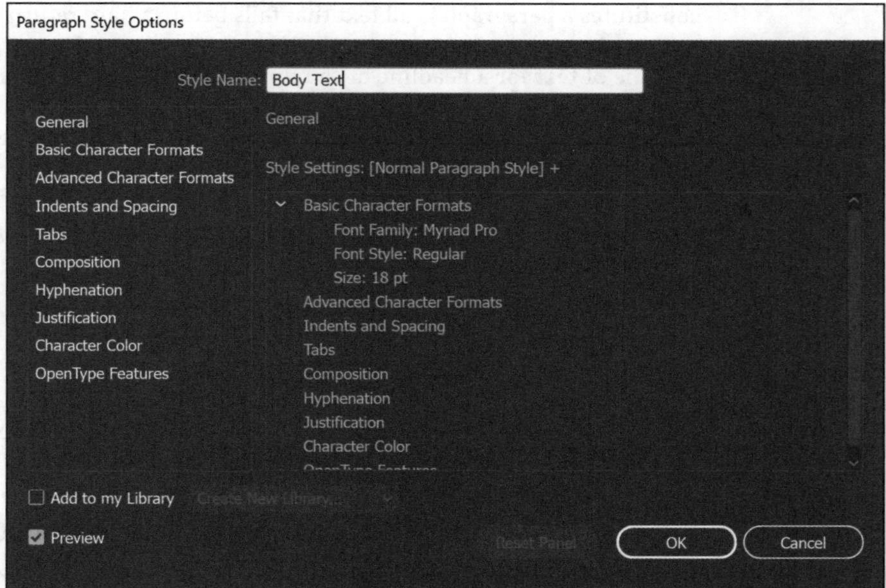

FIGURE 5-21: Updating a paragraph style.

WARNING

Documents created in older versions of Adobe Illustrator (Version 10 or earlier) contain *legacy text*, which is text using the older text engine. When these files are opened, you see a Warning dialog box. If you click the Update button, any text on the document will most likely reflow, causing line breaks, leading, and other types of spacing to change.

Click the OK button to update the file after it's opened to lock down the text. If necessary, you can use the Type tool to click a selected text area to update only the contained text. Another Warning dialog box appears that gives you the opportunity to update selected text, copy the text object, or cancel the text tool selection. This method is the best way to see which changes are occurring so that you can catch any spacing issues right off the bat. See Figure 5-22 for samples of the three options in the Warning dialog box.

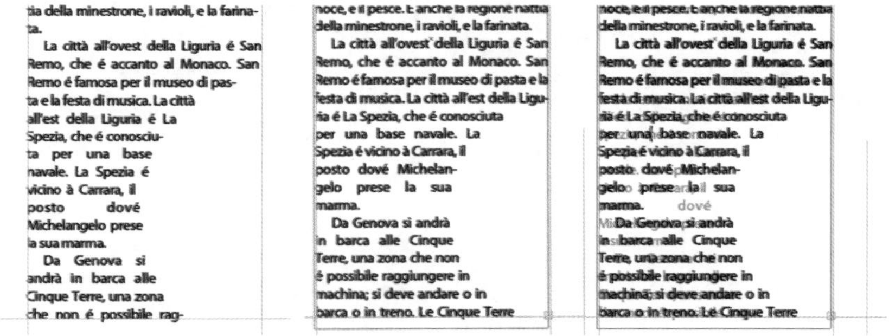

FIGURE 5-22: Original text (left), updated text (center), and text object copied (right).

REMEMBER

If you click Copy the Text Object, you can use the underlying locked copy to adjust the new text flow to match the old. Throw away the legacy text layer by clicking and dragging it to the Trash icon in the Layers panel, or click the visibility eye icon to the left of the Legacy Text layer to hide it when you're finished.

» Working with the Transform panel for placement

» Changing ruler origin

» Rearranging, hiding, and locking objects

» Masking objects

Chapter 6
Organizing Your Illustrations

You can create incredible effects in Illustrator, but if don't keep your content neat and aligned you can create a design nightmare. By learning the tricks to using rulers, guides, and grids, you can also create your artwork faster. In this chapter, you focus on organizational tricks that save you time and make your artwork look professional.

Setting Ruler Increments

Using rulers to help accurately place objects in an illustration is not difficult, but not knowing how to use them effectively in Illustrator can drive you straight over the edge.

To view rulers in Illustrator, choose View⇨Rulers⇨Show Rulers or press Ctrl+R (Windows) or ⌘ +R (Mac). When the rulers appear, their default measurement setting is the point (or whichever measurement increment was last set up in the preferences).

To change the ruler increment to the measurement system you prefer, use one of these methods:

- Create a new document and select a measurement unit in the New Document dialog box.
- Right-click (Windows) or Control-click (Mac) the horizontal or vertical ruler and pick a measurement increment from the contextual menu that appears.
- Choose Edit⇨Preferences⇨Units (Windows) or Illustrator⇨Preferences⇨Units (Mac) to open the Preferences dialog box.

REMEMBER

Change the ruler unit only by using the General drop-down list in the Preferences dialog box. If you change the measurement unit on the Stroke and Type tabs, you can end up with 12-*inch* type rather than that dainty 12-*point* type you were expecting.

REMEMBER

Setting general preferences changes them in all future documents.

- Choose File⇨Document Setup to change the measurement unit for only the document you're working on.

Using Guides

Guides can help you create more accurate illustrations. After a guide is created, you can turn its visibility off or on quickly with the View menu. You can use two kinds of guides in Illustrator:

- **Ruler guides:** You create these straight-line guides by clicking the ruler and dragging out to the artboard.
- **Custom guides:** These guides, created from Illustrator objects such as shapes or paths, are helpful for replicating the exact angle of a path, as shown in Figure 6-1.

Creating a ruler guide

A ruler guide is the easiest guide to create: Click the vertical or horizontal ruler anywhere and drag it to the artboard, as shown in Figure 6-2. By default, the horizontal ruler creates horizontal guides (no kidding), and the vertical ruler creates vertical guides. You can press Alt-drag (Windows) or Option-drag (Mac) to change the orientation of the guide. The vertical ruler then creates a horizontal guide, and the horizontal ruler then creates a vertical guide.

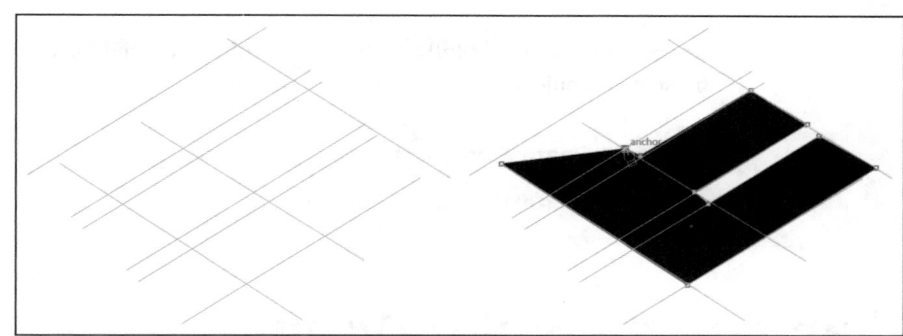

FIGURE 6-1: Turn selected paths and shapes into custom guides. In this example, the Pen tool followed the path to create a slanted letter.

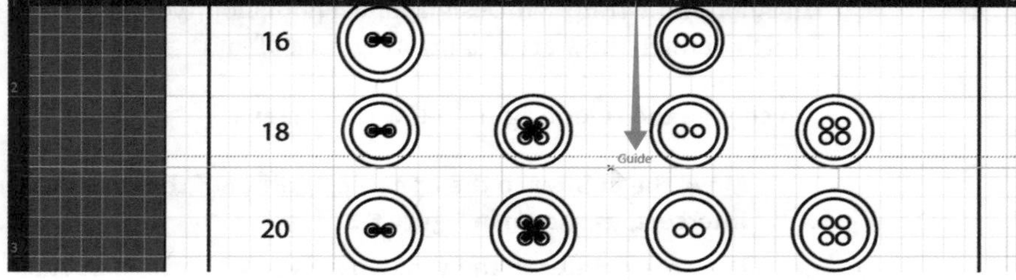

FIGURE 6-2: Click the ruler and drag out a guide.

Creating a custom guide

Create a custom guide by selecting a path or a shape and choosing View⇨Guides⇨Make Guides. The selected object turns into a nonprinting guide. Changing a path into a guide isn't a permanent change. Select the guide and then choose View⇨Guides⇨Release Guides to turn guides back into paths.

Using the Properties Panel for Placement

Placing shapes and paths precisely where you want them can be difficult even if you have steady hands. Save yourself some aggravation by using the Properties panel to achieve this. If you don't see the Properties panel, choose Window⇨Properties. Enter *x*, *y* coordinates in the Properties panel to position objects exactly where you want them.

TIP

In Adobe Illustrator, the Reference Point Indicator icon is on the left side of the Properties panel. Click the handle of the icon to change the point of reference. As a default, the upper-left is where zero starts. To measure from the upper-left corner, click the indicator on the handle there. If you want to know the exact center of an object, click the center point in the indicator. The point of reference is the spot on the object that falls at the *x*, *y* coordinates, which specify the placement of the selected object:

- **»** ***x* coordinate:** From left to right
- **»** ***y* coordinate:** From top to bottom

Changing the Ruler Origin

In Adobe Illustrator, InDesign, and Photoshop, you can change the *ruler origin*, which defines the start of a printing area of an image.

To change the ruler origin, follow these steps:

1. **Move the pointer to the upper-left corner of the rulers where the rulers intersect, as shown in Figure 6-3.**

   Drag the pointer to the spot where you want the new ruler origin.

   While you drag, a crosshair in the window and in the rulers indicates where the new ruler origin will be placed.

TIP

You can restore the original ruler origin by double-clicking the ruler intersection.

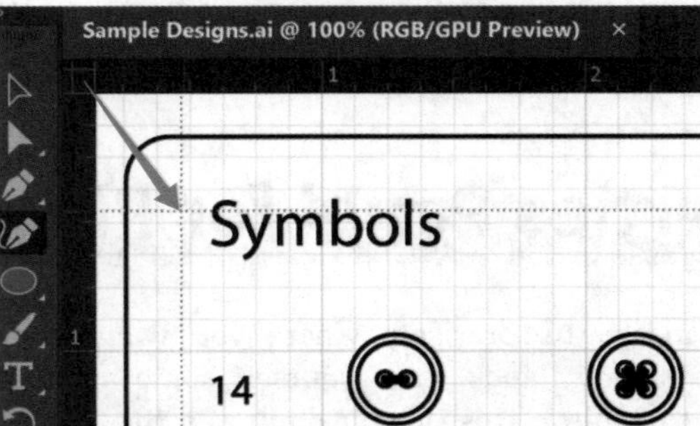

FIGURE 6-3: Changing the ruler's origin.

What You Need to Know about Object Arrangement

Just like the stacks of paper on your desk, new objects in Illustrator are placed on top of existing objects. Change their order by choosing the Object⇨Arrange Menu Options.

The easiest choices are to bring an object to the front or send it to the back. The results of sending forward or backward can be unnerving if you don't know the exact order in which objects were created. The illustration in Figure 6-4 shows three objects that we plan to rearrange using four available choices. Figure 6-5 shows the result of each choice.

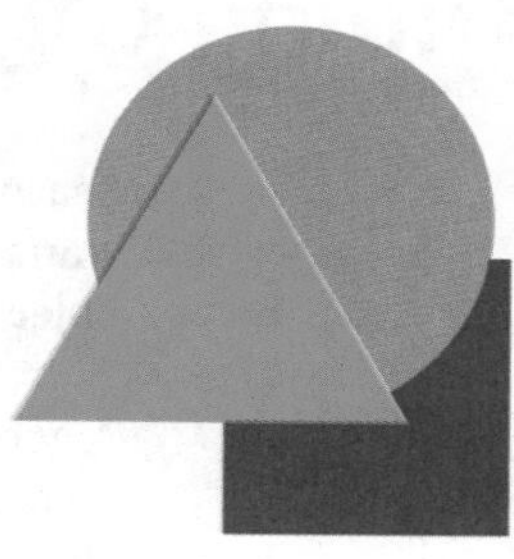

FIGURE 6-4: Objects in their original positions.

To change the stacking order, select the object (or objects) whose placement you want to change and then choose one of these commands:

- **Object⇨Arrange⇨Bring to Front:** Moves the selected object to the top of the painting order. In Figure 6-5, the square is brought in front by using the Bring to Front command.

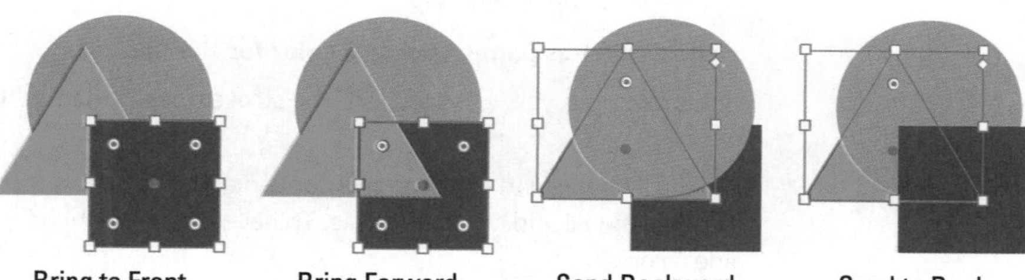

FIGURE 6-5: Rearranging objects; compare to original art.

- **Object⇨Arrange⇨Bring Forward:** Moves a selected object in front of the object created just before it or one level closer to the front. In Figure 6-5, the triangle is moved in front of the square by using the Bring Forward command.
- **Object⇨Arrange⇨Send Backward:** Moves a selected object so that it falls under the object created just before it or one level back. In Figure 6-5, the triangle is sent backward so that it's just under the circle.
- **Object⇨Arrange⇨Send to Back:** Moves a selected object to the bottom of the painting order. In Figure 6-5, the triangle is placed on the bottom by using the Send to Back command.

Hiding Objects

Seasoned Illustrator users love the Hide command. Use it when the object you want to select is stuck behind something else or when you need to select one object and another repeatedly activates instead.

A good opportunity to use the Hide command is when you're creating text inside a shape. In Chapter 5 of this minibook, we show you that as soon as you turn a shape into a text area, the fill and stroke attributes turn into None. Follow these steps to hide a shape:

1. **Create a shape.**

   For this example, we created an ellipse.

2. **Click the Fill color box at the bottom of the Illustrator Tools panel and then choose Window⇨Swatches.**

   The Swatches panel appears.

3. **In the Swatches panel, choose a color for the fill.**

   In this example, yellow is selected. The stroke doesn't matter; this one is set to None.

   Clicking a shape with the Type tool converts the shape to a text area and converts the fill and stroke to None. To have a colored shape remain, you must hide a copy.

4. **After selecting a colored shape, choose Edit⇨Copy; alternatively, you can press Ctrl+C (Windows) or ⌘ +C (Mac).**

   This step makes a copy of your shape.

5. **Choose Edit⇨Paste in Back or press Ctrl+B (Windows) or ⌘ +B (Mac).**

   This step puts a copy of your shape exactly in back of the original.

6. **Choose Object⇨Hide⇨ Selection or press Ctrl+3 (Windows) or ⌘ +3 (Mac).**

   The copy of the shape is now hidden; what you see is your original shape.

7. **Switch to the Type tool by selecting it in the Tools panel or pressing T.**
8. **Use the cursor to cross over the edge of the shape and change it to the Area Type tool.**

   You use the Area Type tool to type text in a shape.

9. **When you see the type insertion cursor swell up, as shown in Figure 6-6, click the edge of the shape.**

   The insertion point is now blinking inside the shape, and the fill and stroke attributes of the shape have been changed to None.

10. **Type some text, as shown in Figure 6-7.**

11. **When you finish entering text, choose Object➪Show All or press Ctrl+Alt+3 (Windows) or ⌘+Option+3 (Mac).**

    The colored shape reappears with the text in the middle of it, as shown in Figure 6-7.

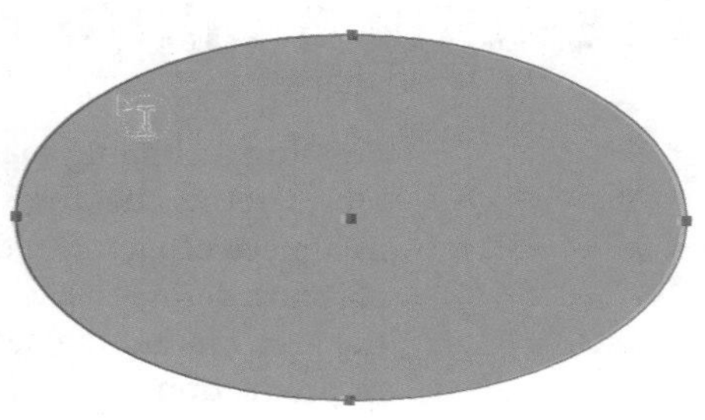

FIGURE 6-6: The type insertion cursor on the edge of a shape.

Use the Hide command anytime you want to tuck away objects for later use. We promise: Nothing hidden in Illustrator will be lost. Just use the Show All command, and any hidden objects are revealed, exactly where you left them. (Too bad the Show All command can't reveal where you left your car keys!)

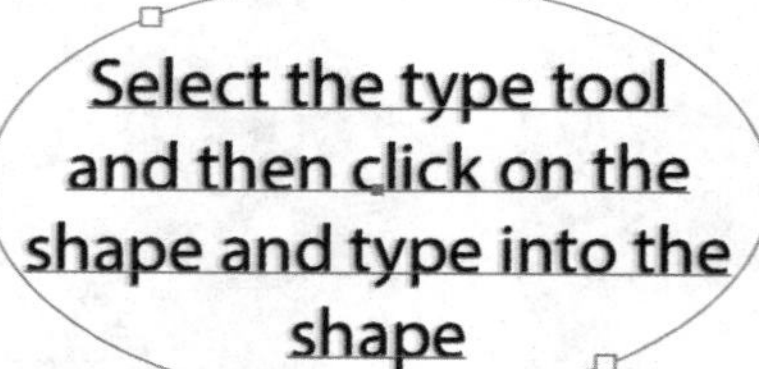

Select the type tool
and then click on the
shape and type into the
shape

FIGURE 6-7: Type directly in the shape and then show the shape you hid earlier.

Locking Objects

Being able to lock items is handy when you're building an illustration. The Lock command not only locks down objects you don't want to change, but also drives anyone crazy who tries to edit your files. In fact, we mention locking mainly to help preserve your sanity. Sometimes you need to make simple adjustments to another designer's artwork and can't, unless the objects are first unlocked. Follow these instructions:

- **To lock an object:** Choose Object➪Lock ➪Selection or press Ctrl+2 (Windows) or ⌘+2 (Mac) to lock an object so that you can't select it, move it, or change its attributes.
- **To unlock an object:** Choose Object➪Unlock All or press Ctrl+Alt+2 (Windows) or ⌘+Option+2 (Mac). Then you can make changes to it.

You can also lock and hide objects with layers. See Chapter 7 in this minibook for more information about using layers.

Creating a Clipping Mask

Creating a clipping mask may sound complex, but it's easy and highlights some topics in this chapter, such as arranging objects. Similar to peering through a hole in a piece of paper to the objects underneath it, a *clipping mask* allows a topmost object to define the selected shapes underneath it; with a clipping mask, however, the area around the defining shape is transparent, as shown in Figure 6-8. Some in production may refer to using a clipping mask as creating a *silhouette* for an image. Read on to see how to create these mask examples.

FIGURE 6-8: A clipping mask around an image and another example that clips other Illustrator objects.

You may recall what a film mask looks like — it's black to block out the picture and clear where you want to view an image, as shown in Figure 6-9.

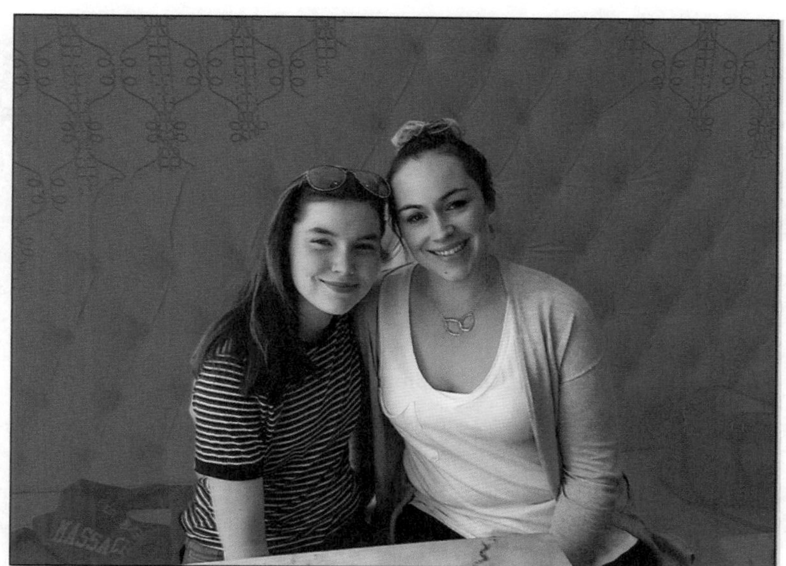

FIGURE 6-9: A conventional film mask.

The clipping mask feature uses the same principle as the conventional film mask. It hides the area outside the mask area. To create a clipping mask, follow these steps:

1. **Choose File⇨Place to place an image.**

   Choose any image that you have handy for this or use one of the image samples, such as Runonbeach–`small.jpg`, that are located in the Images folder inside the DummiesCCFiles you downloaded in Book 1, Chapter 1.

   Masks work with objects created in Illustrator and with objects placed (scanned or otherwise imported) there.

2. **Create the item you want to use as a mask by using the Pen tool to create a shape or a closed path.**

   For example, in Figure 6-9, the custom pen path is the mask. (The photo underneath it is the placed image from Step 1.) The shape is placed where the mask will be created. The shape's color, fill, and stroke values don't matter because they automatically change to None when you create a mask. In this example, you can use a circle if you like, as the example in Figure 6-10 uses.

   Note: When creating a clipping mask, make sure that the object to be used as a mask is a closed shape and is at the top of the stacking order.

3. **Use the Selection tool to select the placed image and the shape.**

   Shift-click to add an object to the selection.

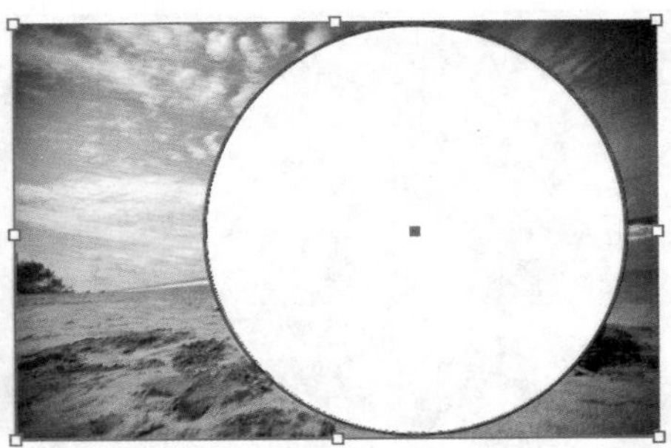

FIGURE 6-10: Position the shape that you want to be the mask on top of the image and then select both before creating a mask.

4. **Choose Object➪Clipping Mask➪Make.**

   Alternatively, you can use the keyboard shortcut Ctrl+7 (Windows) or ⌘+7 (Mac) to create the clipping mask.

   Ta-da! You created the clipping mask. Masked items are grouped, but you can use the Direct Selection tool to move the image or mask individually.

5. **To turn off the clipping mask, choose Object➪Clipping Mask➪Release.**

TIP

You can also use text as a clipping mask: Type a word and ensure that it's positioned over an image or another Illustrator object (or objects). Then select both the text and the object and choose Object➪Clipping Mask➪Make.

Creating a Mask of Illustrator Objects Using the Blend Tool

In this example, you find out how to use the Blend tool in order to create the varied lines and mask them into a circle. Follow these steps if you want to replicate the circle shown in Figure 6-8 in the previous section. You can also access this sample file by opening the file named `MaskExamples.ai` in the Book05_Illustrator folder inside the DummiesCCfiles that you downloaded in Book 1, Chapter 1.

1. **Select the Rectangle tool and click and drag out a thin, long rectangle.**
2. **Using the Fill and Stroke options in the Properties panel assign a black fill and no stroke to the rectangle.**

3. **Switch to the Selection tool. Hold down the Alt/Option key and click and drag the rectangle down about an inch or two from the original rectangle.**

   Remember that holding down Alt/Option key *clones* whatever you have selected.

4. **Using the Direct Selection tool, make the second rectangle thinner than the original. You can do this by switching to the Direct Selection tool and clicking on the Artboard to make sure nothing is selected. Then click on the bottom of the rectangle and click and drag up, as you see in Figure 6-11.**

TIP

   Make sure your bounding box is turned on if you want to easily see transform anchor points. You can find the bounding box option under the View menu. If it states Hide Bounding box, you are all set. If it states Show Bound Box, click on it to more easily make your transformations.

5. **Press W to select the Blend tool and click on the top rectangle, and then click on the second rectangle. The larger rectangle is blended into the smaller rectangle, as shown in Figure 6-11.**

REMEMBER

   If you do not see the tool you need on the Tools panel, click on the Edit Tools button at the bottom of the Tools panel and find the Blend tool and drag it onto your tools panel.

FIGURE 6-11: Create a rectangle and clone it, making the second thinner than the first. Then select both rectangles and use the Blend tool to blend the larger rectangle into the smaller rectangle.

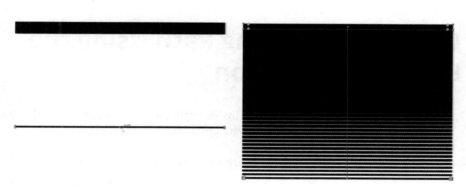

6. **An automatic amount of steps are created to create the blend. Double-click on the Blend tool and choose Specified steps from the Spacing drop-down menu. Enter a number of steps to change the blend. In this example, shown in Figure 6-12, the steps were changed to 20.**

FIGURE 6-12: You can change the effect of the blend by changing the amount of steps.

7. **Now that the blend is finished, create a circle and place it on top of the blend.**
8. **Select both the circle and the Blend and choose Object ➪ Clipping Mask ➪ Make; alternately you could press Control+7 (Windows) or Command+7 (Mac OS) to mask the shapes as well.**

Creating a Clipping Path Using the Draw Inside Button

You can use the Draw Inside button to create a clipping path. The button is at the bottom of the Tools panel. Just follow these steps:

1. **Select your artwork and choose Edit ➪ Copy or Cut.**

   In Figure 6-13, the artwork to the right has been cut to fit into the star image.
2. **Select the artwork that you want to "paste into" existing artwork and then click the Draw Inside button.**
3. **Choose Edit ➪ Paste.**

   The artwork is pasted inside the shape, as shown in Figure 6-13.

TIP

After you have pasted your artwork, you can use the Direct Selection tool to reposition it.

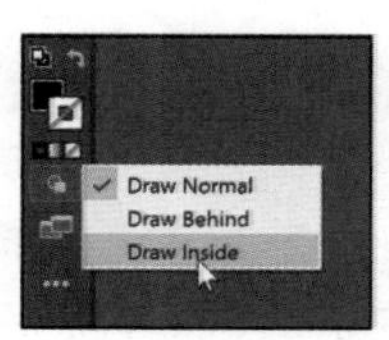

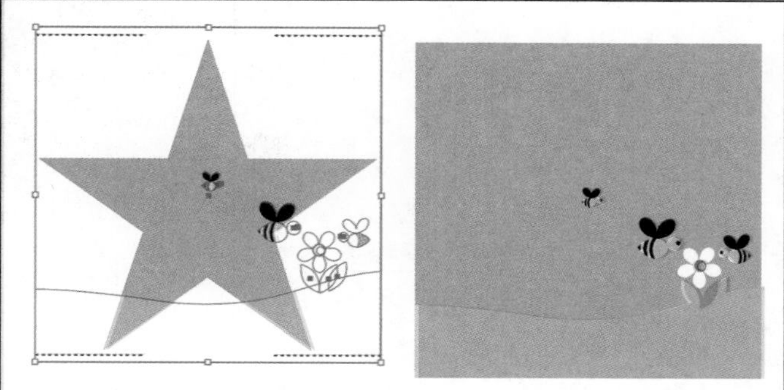

FIGURE 6-13: Select artwork and copy or paste it.

» Using layers for a selection

» Changing the stacking order of a layer

» Moving and cloning objects to another layer

» Hiding and locking layers

Chapter 7
Using Layers

This chapter shows you how helpful layers can be when you're producing complex artwork. Layers are similar to clear films stacked on top of your artwork: You can place content (text, shapes, and other objects) on a layer, lift up a layer, remove a layer, hide and show layers, or lock a layer so that you can't edit its content. Taking advantage of the Layers feature can help you:

- Organize the painting (stacking) order of objects.
- Activate objects that would otherwise be difficult to select by using either the Selection or Direct Selection tool.
- Lock items that you don't want to reposition or change.
- Hide items until you need them.
- Repurpose objects for artwork variations. For example, a company's business cards use the same logo and company address, but the name and contact information change for each person. In this case, placing the logo and company address on one layer and the person's name and contact information on another layer lets you easily create a new business card just by changing the person's name and contact details.

Creating New Layers

When you create a new Illustrator document, you automatically start with a single layer. To understand how layers work, you can create a new file and follow some steps to create new layers and place objects on them:

1. **Choose File⇨New, choose any profile, such as Mobile, Web, Print, and then any size artboard. The size and profile are irrelevant for this exercise.**
2. **Using the Rectangle tool, click and drag anywhere on the artboard to create a rectangle.**

   The size of the rectangle doesn't matter.
3. **If you don't see the Properties panel, choose Window ⇨ Properties. Click the Fill button and select any color for the rectangle from the swatches that appear, as shown in Figure 7-1.**

   This fill color makes the rectangle easier to identify later.

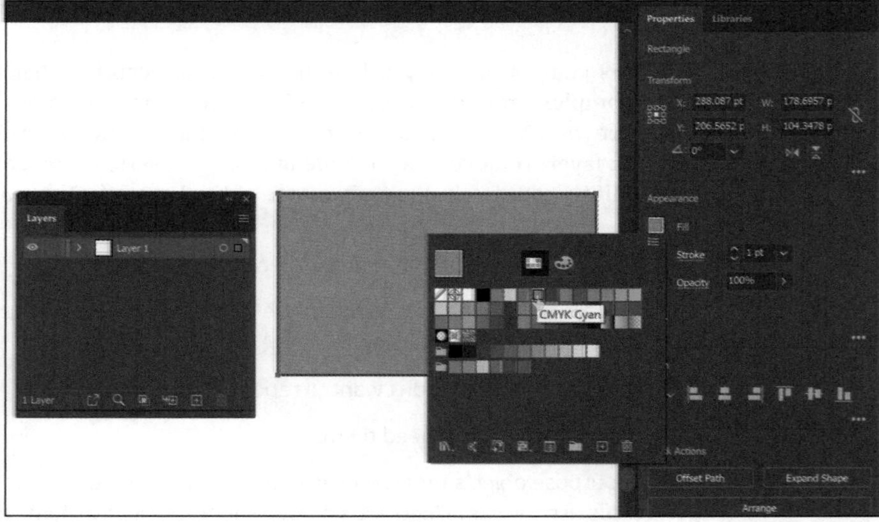

FIGURE 7-1: Create a rectangle to notice how the Layers panel indicates the selection with a blue selection square to the right.

4. **Choose Window⇨Layers to see the Layers panel. If it replaces the Properties panel you can click and drag the tab and drag it away to ungroup it from the Properties panel so that you can see them both at the same time.**

   Notice that as soon as you create any shape, the layer becomes active. As you see in Figure 7-1, the blue handle color that appear on the rectangle match the blue color bar you see on the left side of the layer you are working on. It is the same blue that you also see in the small selection square to the right of the radio button.

TIP

The small selection square on the right disappears if you choose Select⇨Deselect. You can use that square to see which layer a selected when you select an object anywhere on your artboard.

Also notice in Figure 7-2 that since you've added a shape an arrow appears to the left of the layer name. This arrow indicates that you now have a *sublayer,* which shows each object on that layer. Click the arrow to expand the layer and show any sublayers nested underneath it; sublayers are automatically created when you add objects.

It helps to name layers with a unique name so that you can identify them while working.

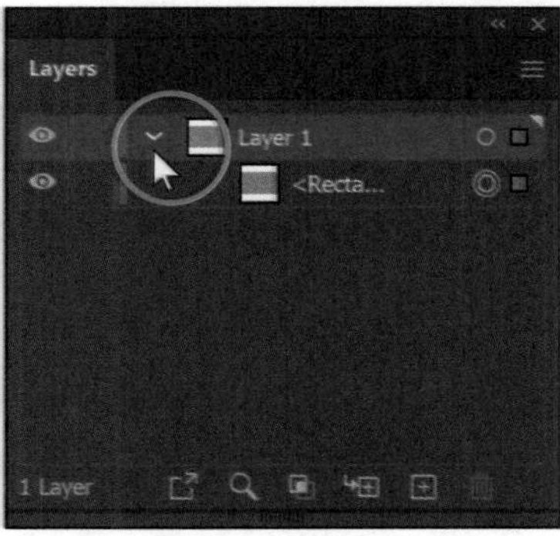

FIGURE 7-2: When you start a document you automatically have one layer. When you add each object you automatically create new sublayers.

5. **Alt-click (Windows) or Option-click (Mac) the Create New Layer button at the bottom of the Layers panel to create a new layer.**

   The Layer Options dialog box appears (see Figure 7-3), and you can use it to name a layer and change the selection color. You don't have to hold down the Alt or Option key when making a new layer, but if you don't, the Layer Options window does not appear.

6. **Enter the name** Circle **for the new layer in the Name text box and click OK.**

   In Figure 7-5, we entered **Circle** because it's the shape we add in Step 7.

   If you want to stay organized, you can name the original layer **Rectangle** by double-clicking Layer 1. Just make sure that you click the circle layer again to make it the active layer.

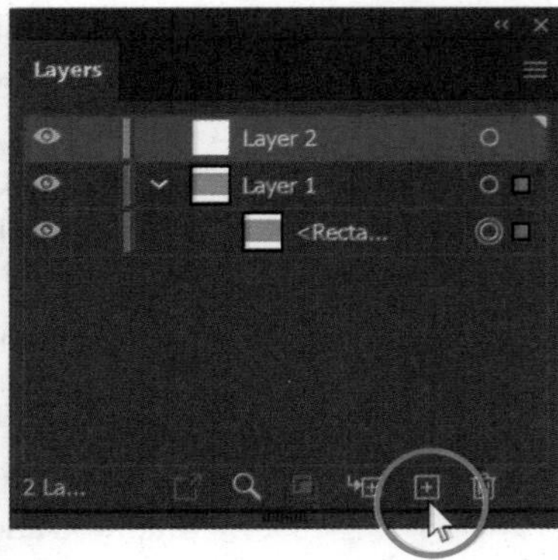

FIGURE 7-3: Alt/Option-click on the New Layer icon at the bottom of the Layers panel in order to see the Layer Options window before a new layer is created.

A new layer is added to the top of the stack in the Layers panel. Look at the selection handles on the circle that you just created: They change to a different color, indicating that you're on a different layer. The different handle colors are for organizational purposes only and aren't visible when imported into another application or printed.

7. **Make a shape on the new layer and overlap the shape you created in Step 2. (See Figure 7-4.)**

8. **Change the fill color for your new shape.**

   Using the Properties panel, click on the Fill and give your circle a different color than the rectangle.

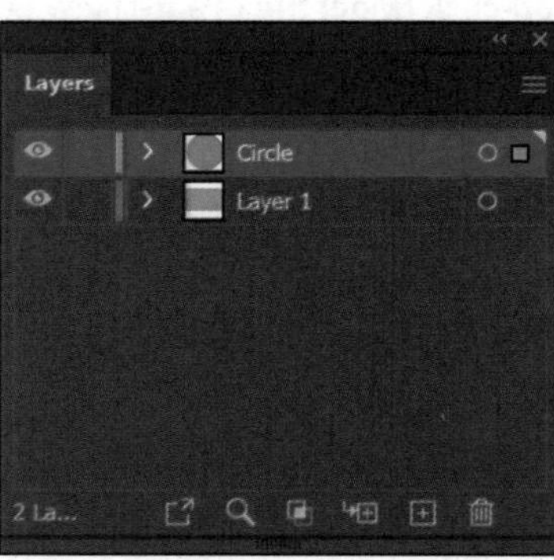

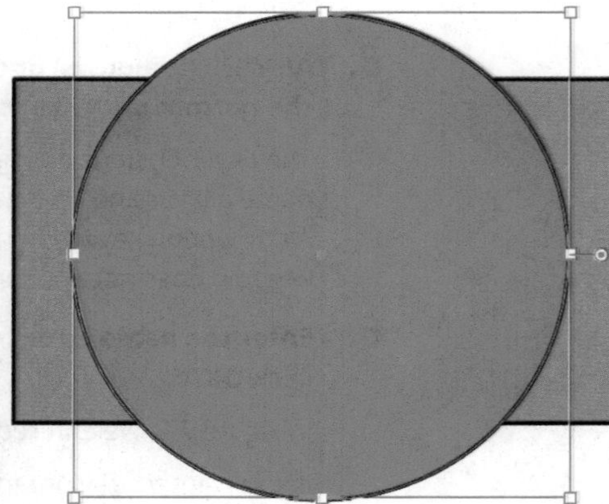

FIGURE 7-4: Create a new circle shape and change the color to be different than what you selected for the rectangle shape.

9. **Try different way of creating a layer this time. Choose New Layer from the panel menu in the upper-right of the Layers panel.**

   The Layer Options dialog box appears.

10. **(Optional) In the Layer Options dialog box, change the color of the selection handles by selecting an option from the Color drop-down list.**

    If you recall, the color you select here is only represented in the selection handles. This can make it easier to locate items on different layers when you create more complex artwork. Notice that you can temporarily hide or lock the contents of the layer using the Layer Options dialog box as well.

11. **Enter the name** Star **in the Name text field, click OK, and then create a star shape on it.**

    For the example shown in Figure 7-6, we entered **Star** into the Name text box and used the Star tool to create a star on the new layer. The Star tool is hidden in the Rectangle shape tool.

    Keep in mind these helpful tips when creating the Star shape:

    - Hold down the Shift key to create a star that is not crooked on your artboard.
    - You can also hold down the Alt/Option key to create a Non-bloated or "puffy" star.

12. **Again, change the fill color of your newest shape so that it's different from the other shapes. Use the Selection tool to move the new shape so that it overlaps the others slightly as you see in Figure 7-5.**

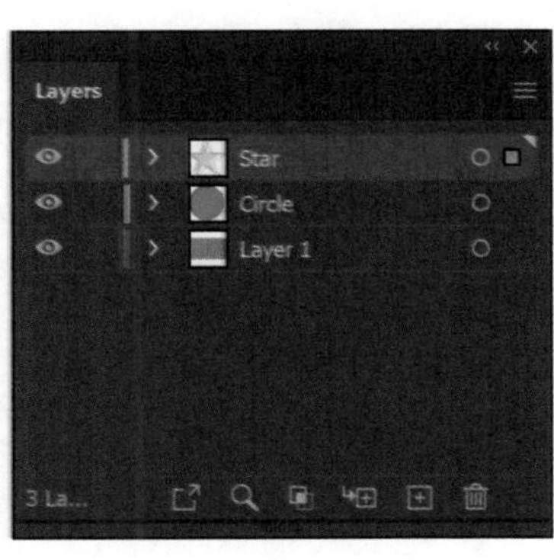

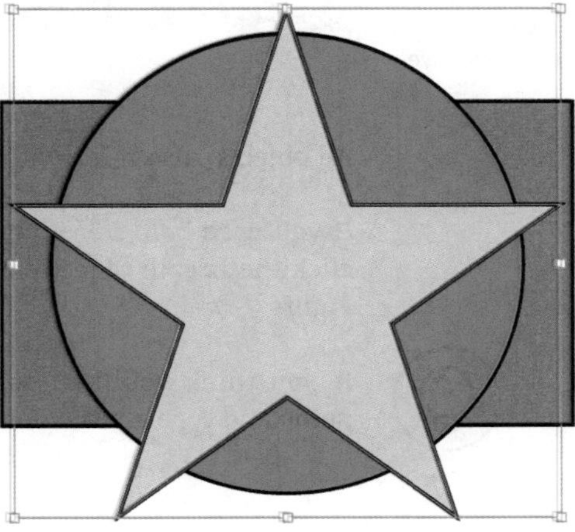

FIGURE 7-5: Create a new layer named Star and add a star shape to it. Then overlap it with the other shapes.

TIP

You can open the Options dialog box for any existing layer by choosing Options for "Layer" (Named Layer) from the panel menu in the Layers panel.

You've created three new layers and now have a file that you can use to practice working with layers.

Using Layers for Selections

As previously mentioned, when you select an object a color selection square appears to the right of the named layer. If you select the radio button directly to the right of the layer's name in the Layers panel, as shown in Figure 7-6, all objects are selected on that layer.

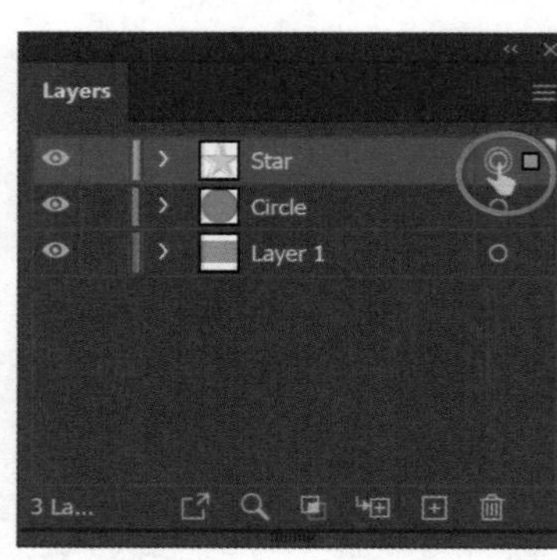

FIGURE 7-6: Click the radio button to the right of the layer's name to select all items on a layer.

All objects, also referred to as *sublayers*, have their own radio buttons.

If you need help selecting a specific graphic's layer, select the graphic, and then click the Locate Object button in the lower left of the Layers panel, as shown in Figure 7-7.

TIP

If you think you'll be selecting sublayers frequently, double-click the default name and type a more descriptive name for that layer.

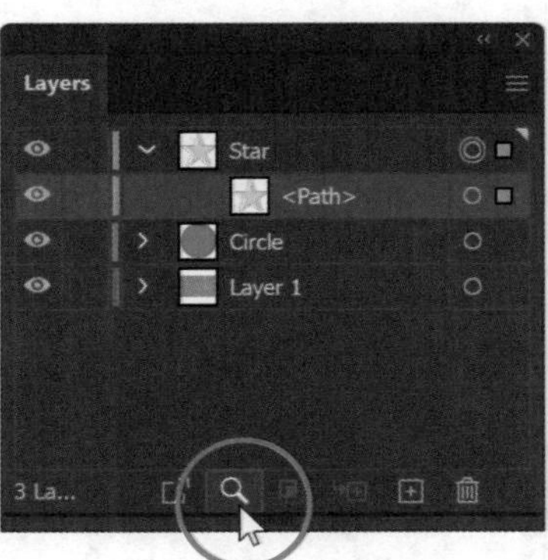

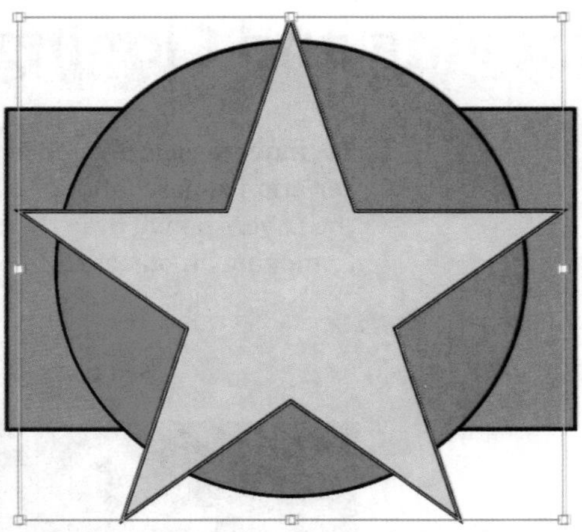

FIGURE 7-7: If you can't find your object in the Layers panel, use the Locate Objects button at the bottom of the Layers panel.

Changing the Layer Stacking Order

In Chapter 6 of this minibook, we tell you about the Object⇨Arrange feature in Illustrator; with layers, this process becomes slightly more complicated. Each layer has its own *painting order,* the order in which you see the layers.

To move a layer (and thereby change the painting or stacking order of the layers), click the Layer in the Layers panel and drag it until you see the blue insertion line where you want the layer to be moved.

As previously mentioned, when you add objects to a layer, a sublayer is automatically created, and it has its own little stacking order that's separate from other layers. In other words, if you choose to send an object to the back and it's on the top layer, it goes only to the back of the objects that are also on that layer, *but* it is still in front of any objects that are on layers beneath it.

REMEMBER

Understanding how the stacking order affects the illustration is probably the most confusing part about layers. Just remember that in order for an object to appear behind everything else, it has to be on the bottom layer (and at the bottom of all objects on that bottom layer); for an object to appear in front of everything else, it has to be on the topmost layer.

Moving and Cloning Objects

To move a selected object from one layer to another, click the small color-selection square (shown in Figure 7-8) to the right of the object's radio button in the Layers panel, drag the object to the target layer, and release. That's all there is to moving an object from one layer to another.

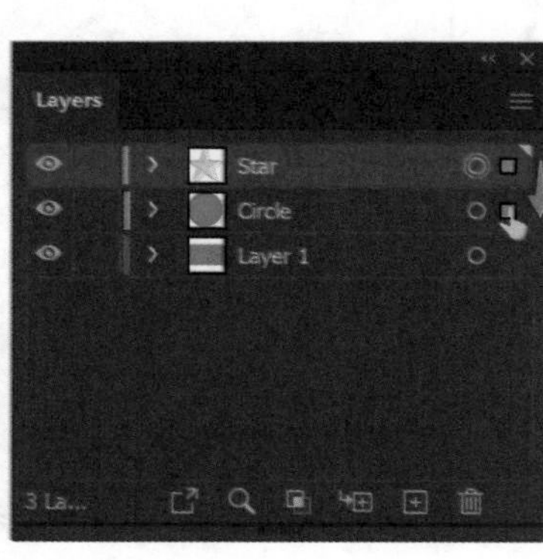

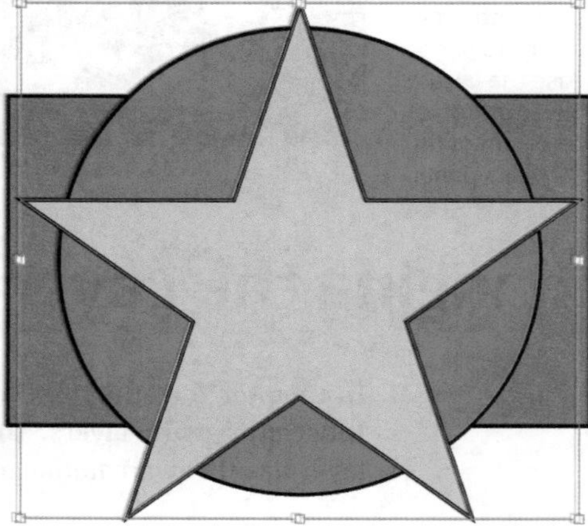

FIGURE 7-8: Drag the small square icon to move the selected object to another layer.

You can also *clone* an item, or make a copy of it while you move the copy to another layer. Clone an object by Alt-dragging (Windows) or Option-dragging (Mac) the color selection square to another layer. A plus sign appears while you drag (so you know that you're making a clone of the object). Release when you reach the cloned object's target layer.

TIP

Choose Paste Remembers Layers from the Layers panel to have Illustrator automatically remember which layer you copied an object from. No matter which layer is active, Illustrator always pastes the object back on the original layer it was copied from.

Hiding Layers

To the left of each layer in the Layers panel is an eye icon — a visibility toggle button. Simply clicking the eye icon hides the layer (the eye disappears, denoting that this layer is hidden). Click the empty square (where the eye icon was) to show the layer again.

Alt-click (Windows) or Option-click (Mac) an eye icon to hide all layers except the one you click; Alt-click (Windows) or Option-click (Mac) the eye icon to show all layers again.

Ctrl-click (Windows) or ⌘-click (Mac) the eye icon to turn just the selected layer into Outline view. In Outline view, all you see are the outlines of the artwork with no stroke widths or fill colors. The rest of your artwork remains in Preview mode, with strokes and fills visible. This tricky technique is helpful when you're looking for stray points or need to close paths. Ctrl-click (Windows) or ⌘-click (Mac) the eye icon again to return the layer to Preview mode.

Locking Layers

Lock layers by clicking the empty square to the right of the Visibility (eye) icon. A padlock icon appears so that you know the layer is now locked. Locking a layer prevents you from making changes to the objects on that layer. Click the padlock to unlock the layer.

- » Using the Swatches and Color panels
- » Working with strokes and fills
- » Changing the stroke width and type
- » Saving and editing colors
- » Discovering patterns
- » Taking advantage of gradients and meshes
- » Copying color attributes
- » Using the Live Paint feature

Chapter 8
Livening Up Illustrations with Color

This chapter is all about making your illustrations come alive with color. You discover how to create new colors, patterns, and gradients as well as how to edit existing colors and save custom colors. We also explain how to apply and edit color attributes to many objects at the same time.

Choosing a Document Profile

Whether you are using the default New dialog box or the legacy dialog box, you are offered the option to choose a color mode. In the default dialog, you can choose from Mobile, Web, Print, Film & Video, or Art & Illustration. If you are using the default New dialog box, the color mode is selected based upon the profile you have selected. (See Figure 8-1.)

FIGURE 8-1: Use the New Document profiles to start in the right color mode, resolution, measurement units, artboard size, and more.

Keep in mind that if you select Mobile, Web, Film & Video, or Art & Illustration, you are defaulted to working in the RGB mode (red, green, blue). If you select Print, your default color mode is CMYK (cyan, magenta, yellow, black).

Along with the color mode, other changes, such as resolution, measurement units, artboard size, and more are also set as a default based upon your selection as well. Read on for more details.

Illustrator CC offers several profiles, and they fall into two basic categories: onscreen and print.

The following New Document profiles are used for onscreen display:

- **Mobile:** This profile ensures a small file size and uses a resolution that is preset for a specific mobile device. You can choose your device from the Size menu.
- **Web:** This profile provides preset options, such as size and resolution, that are optimized for output to the web.
- **Print:** This profile provides preset options for size and resolution defaults that are optimized for print graphics. The colors in the Swatches panel are in CMYK, as well as gradients and patterns.
- **Film & Video:** This profile provides several preset video- and film-specific crop area sizes. Illustrator also creates only square pixel files so that the sizes are interpreted correctly in video applications.
- **Art & Illustration:** This profile creates a document in RGB mode with the Raster Effects Resolution set to 72 ppi.

The Print profile is used for print display. By default, this profile uses a letter-size artboard, but it provides a variety of other preset print sizes to choose from. A document created with this profile is also set to a default resolution of 300 dpi. Use this profile if you plan to send your file to a commercial printer.

TIP

You can change the color mode at any time without losing information by choosing File⇨Document Color Mode.

Using the Swatches Panel

Accessing color from the Properties panel is probably the easiest way to make color choices: You can use the Fill and Stroke controls to quickly access your color, shown in Figure 8-2, and at the same time ensure that the color is applied to either the fill or stroke. This way you don't accidently apply the fill color to the stroke or vice versa. Note that you can also find the Swatches panel by selecting Window⇨Swatches.

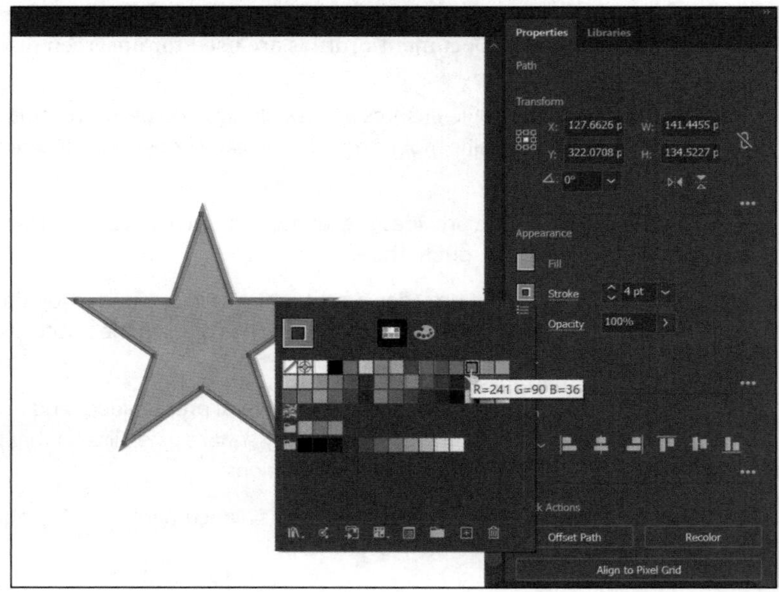

FIGURE 8-2: Use the Fill and Stroke controls in the Properties panel to quickly access color options.

You can also access the Swatches panel, which you open by choosing Window ⇨ Swatches. Although limited in choice, its basic colors, patterns, and gradients are ready to go. You can use the buttons at the bottom of the Swatches panel to quickly open color libraries, as shown in Figure 8-3, select kinds of colors to view, access swatch options, create color groups, add new swatches, and delete selected swatches.

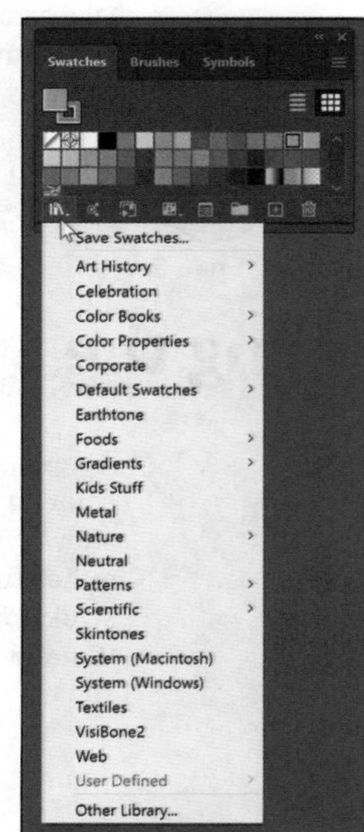

FIGURE 8-3: Additional options for your colors including add a new color library.

You may notice some odd color swatches — for example, the crosshair and the diagonal line.

The crosshair represents the Registration color. Use this swatch only when creating custom crop marks or printer marks. The Registration color looks black, but it's created from 100 percent of all colors. This way, when artwork is separated, the crop mark appears on all color separations.

The diagonal line represents None. Use this option if you want no fill or stroke.

Applying Color to the Fill and Stroke

Illustrator objects are created from *fills* (the inside) and *strokes* (border or path). Look at the bottom of the Tools panel for the Fill and Stroke color boxes. If you're applying color to the fill, the Fill color box must be forward in the Tools panel. If you're applying color to the stroke, the Stroke color box must be forward.

Table 8-1 lists keyboard shortcuts that can be a tremendous help to you when applying colors to fills and strokes.

TABLE 8-1 **Color Keyboard Shortcuts**

Function	Keyboard Shortcut
Switch the Fill or Stroke color box position	X
Inverse the Fill and the Stroke color boxes	Shift+X
Default (black stroke, white fill)	D
None	/
Last color used	<
Last gradient used	>
Color Picker	Double-click the Fill or Stroke color box

TIP

Try this trick: Drag a color from the Swatches panel to the Fill or Stroke color box at the bottom of the toolbar or the top left of the Swatches panel. This action applies the color to the color box you dragged it to. It doesn't matter which is forward!

To apply a fill color to an existing shape, drag the swatch directly to the shape. Select a swatch, hold down Alt+Shift+Ctrl (Windows) or Option+Shift (Mac), and drag a color to a shape to apply that color to the stroke.

Changing the Width and Type of a Stroke

Access the Stroke panel by clicking on Stroke in the Properties panel. In the Stroke options that appear, shown in Figure 8-4, you can choose to change the Width height by clicking and selecting a preset width from the Width drop-down menu, or you can type in a value. You can also customize the *caps* (the end of a line), *joins* (the endpoints of a path or dash), and the *miter limit* (the length of a point). The Stroke panel also enables you to turn a path into a dashed line.

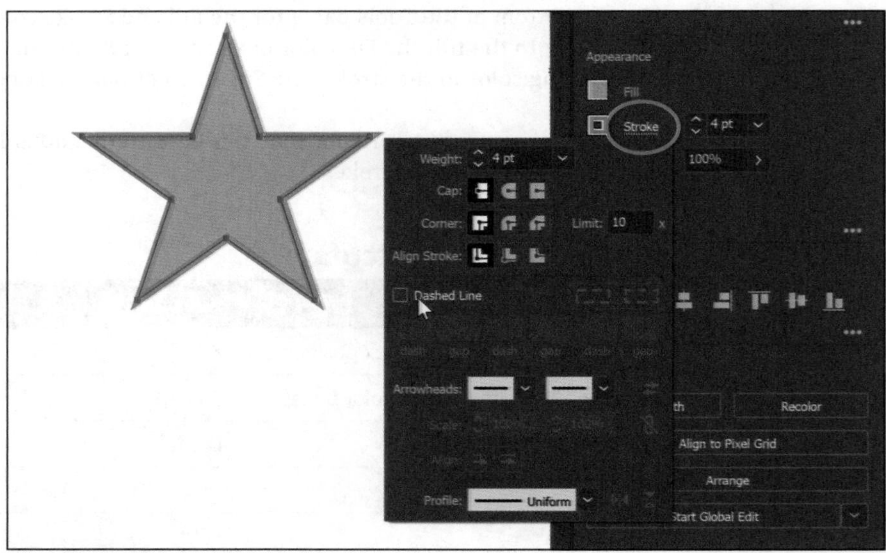

FIGURE 8-4: The Stroke panel options.

If you prefer to work with these options always visible, you can access the Stroke panel from the Windows menu. Click on the panel menu and choose Show options if you want to see all the options that are available to you.

In the Stroke panel options, you can choose to align the stroke on the center (default) of a path, the inside of a path, and the outside of a path as shown in Figure 8-5. Figure 8-6 shows the results. In order to stroke text, as in our example, you need to create text, then change to the Selection tool and choose Type⇨Create outlines. This action converts your text to vector art.

This feature is especially helpful when stroking outlined text. Refer to Figure 8-6 to compare text with the traditional centered stroke, as compared with the options for aligning the stroke inside and outside of a path.

You can also customize the following aspects of a stroke from the Stroke panel by clicking the buttons we describe:

- **Cap Options:** The endpoints of a path or dash
  - *Butt Cap:* Makes the ends of stroked lines square
  - *Round Cap:* Makes the ends of stroked lines semicircular

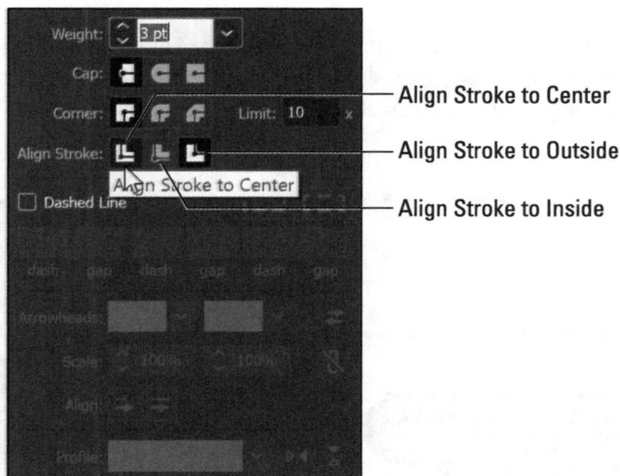

FIGURE 8-5: The Align Stroke options affect the placement of the stroke.

FIGURE 8-6: After converting text to outlines you can move the stroke to the center, inside or outside of the letter shapes.

- *Projecting Cap:* Makes the ends of stroked lines square and extends half the line width beyond the end of the line

» **Join Options:** How corner points appear

- *Miter Join:* Makes stroked lines with pointed corners
- *Round Join:* Makes stroked lines with rounded corners
- *Bevel Join:* Makes stroked lines with squared corners

» **Dashed Lines:** Regularly spaced lines, based on values you set

TIP

To create a dashed line, specify a dash sequence by entering the lengths of dashes and the gaps between them in the Dash Pattern text fields. (See Figure 8-7.) The numbers entered are repeated in sequence so that after you set up the pattern, you don't need to fill in all the text fields. In other words, if you want an evenly spaced dashed stroke, just type the same number in the first and second text fields, and all dashes and spaces will be the same length (say, 12 points). Change that number to **12** in the first text field and **24** in the next to create a larger space between dashes.

TIP

Click the Aligns Dashes button to create improved corners for your dashed object. Keep in mind that Align Dashes can only be used when the default stroke alignment, Align Stroke to Center, is selected.

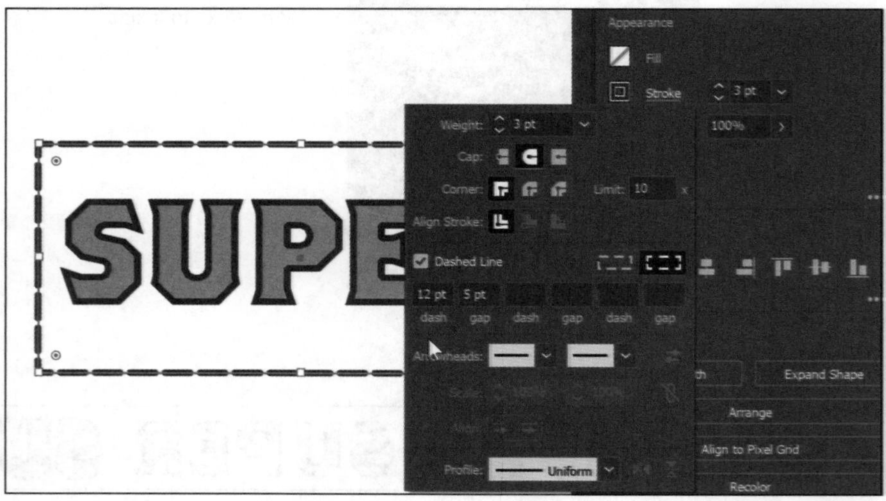

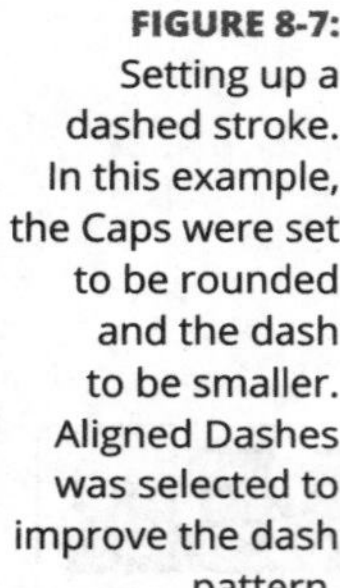

FIGURE 8-7: Setting up a dashed stroke. In this example, the Caps were set to be rounded and the dash to be smaller. Aligned Dashes was selected to improve the dash pattern.

Adding Arrowheads

You can also add arrowheads to one side of a stroke or the other, or even both. This time, you use the Appearance panel to change the way a selected stroke looks. The Appearance panel is helpful in many ways; you will find out more about the Appearance panel in Chapter 11 of this minibook. Follow these steps to use the Appearance panel to change the look of a selected stroke:

1. **Select the Pen tool and click anywhere on your artboard, then move to another location and click again to create a path. Press the Esc key to release you from the path and then return to the Selection tool.**
2. **Choose Window➪Appearance to see the Appearance panel.**
3. **Click on Stroke to see your options, including Arrowheads.**
4. **Using the Arrowhead drop-down list, you can choose to add arrowheads to one or both sides of your path.**

   Keep in mind that if you increase your stroke size, your arrowheads increase as well. This is when the scale option becomes important. As you see in Figure 8-8, the scale of the arrowheads had to be adjusted down to 50%, or they would have been disproportionally large.

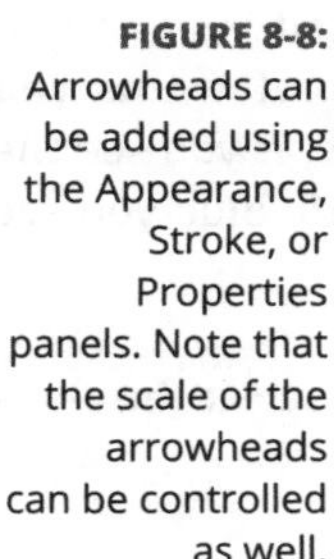

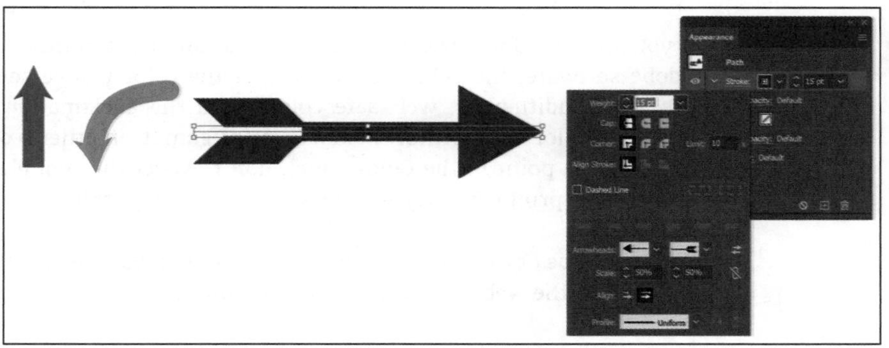

FIGURE 8-8: Arrowheads can be added using the Appearance, Stroke, or Properties panels. Note that the scale of the arrowheads can be controlled as well.

Using the Color Panel

The Color panel (choose Window⇨Color) offers another method for choosing color. You must custom-pick a color using values on the color ramp. You see as a default only the *color ramp* — the large color well spanning the panel. If you don't see all color options, choose Show Options from the Color panel menu. (Click the triangle in the upper-right corner to access the panel menu.)

TIP

If you ever want to create tints of a CMYK color but aren't quite sure how to adjust individual color sliders, just hold down the Shift key while adjusting the color slider of any color. Then watch as all colors move to a relative position at the same time!

As shown in Figure 8-9, the panel menu offers many other choices. Even though you may be in RGB or CMYK color mode, you can still choose to build colors in Grayscale, RGB, HSB (Hue, Saturation, Brightness), CMYK, or web-safe RGB. Choosing Invert or Complement from the panel menu reverses the color of the selected object or changes it to a complementary color, respectively. You can also choose the Fill and Stroke color boxes in the upper-left corner of the Color panel.

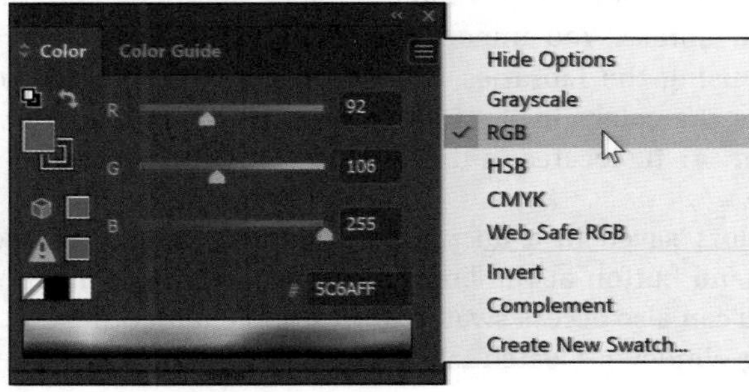

FIGURE 8-9: Different color models are available in the Color panel.

TIP

If you see the infamous cube-and-exclamation-point in the Color panels in most Adobe software, the cube warns you that the color you've selected isn't one of the 216 nondithering, web-safe colors, and the exclamation point warns you that your color isn't within the CMYK print gamut. In other words, if you see the exclamation point in the Color panel, don't expect the cool electric blue you see onscreen to print correctly — it may print as dark purple!

Click the cube or exclamation point symbols when you see them to select the closest color in the web-safe or CMYK color gamut.

Saving Colors

Saving your colors not only keeps you consistent, but also makes future edits easier. Any time you build a color, drag it from the Color panel (or the Tools panel) to the Swatches panel to save it as a color swatch for future use. You can also select an object that uses the color and click the New Swatch button at the bottom of the Swatches panel. (Refer to Figure 8-2 to see this button that looks like a square with a plus sign in the middle of it.)

To save a color and name it at the same time, Control-click (Windows) or Shift-click (Mac) the New Swatch icon. The New Swatch dialog box opens, allowing you to name and edit the color, if you want. By double-clicking a swatch in the Swatches panel, you can access the options at any time.

A color in the Swatches panel is available only in the document in which it was created. Read the next section on custom libraries to see how to import swatches from saved documents.

Building and using custom libraries

When you save a color in the Swatches panel, you're essentially saving it to your own custom library. You import the Swatches panel from one document into another by using the Libraries feature. In order to do this, you must save the file that has the colors in it. If you like, you can try the following steps using `SportsCover.ai` fie located in the Book05_Illustrator folder.

Retrieve colors saved in a document's Swatches panel by clicking the Swatch Libraries menu button at the bottom of the Swatches panel and selecting Other Library. You can also access swatch libraries, including those in other documents, by choosing Window⇨Swatch Libraries⇨Other Library.

Locate the saved document, such as the `SportsCover.ai` file, and click Open. A panel appears with the document name, as shown in Figure 8-10. You can't edit the colors in this panel, but you can use the colors in this panel by double-clicking a swatch (which adds it to the Swatches panel) or dragging it to the current document's Swatches panel where it can be edited.

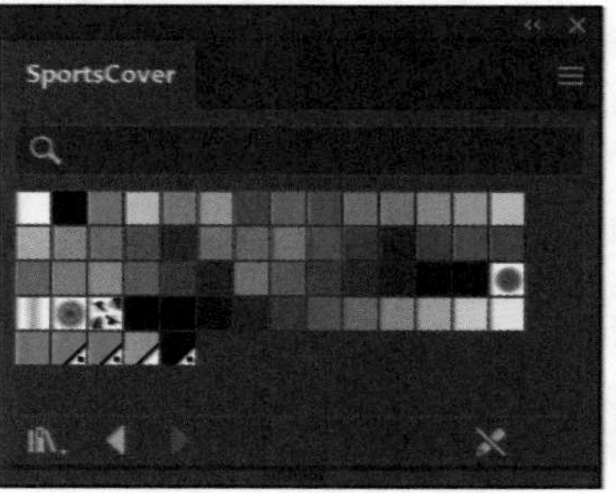

FIGURE 8-10: An imported custom swatch library.

TIP

You can also click the Swatch Libraries button to access color libraries for Pantone colors, web colors, and some neat creative colors, such as jewel tones and metals. Later in this chapter, you find out how you add the Pantone Matching System (PMS) colors to your swatches.

Using the Color Guide and color groups

Perhaps you failed at color selection in art class or just don't feel that you pick colors that work well together. Fortunately, you can use the Color Guide to find colors and save them to organized color groups in your Swatches panel. You can create color schemes based on 23 classic color-harmony rules, such as the Complementary, Analogous, Monochromatic, and Triad options, or you can create custom harmony rules.

Sounds complicated, doesn't it? Fortunately, all you have to do is choose a base color and then see which variations you come up with according to rules you choose. Give it a try:

1. **Choose Window⇨Color Guide. Click on the tab and drag the Color Guide panel away from the docking area so that it does not go away when you select another panel.**

   The Color Guide appears.

2. **Select a color from the Swatches panel. If it isn't visible, choose Window⇨Swatches.**

   Immediately, the Color Guide panel kicks in to provide you with colors related to your original swatch, as you see in Figure 8-11.

3. **You can change the harmony rules quickly by selecting the Harmony Rules drop-down menu, as shown in Figure 8-12.**

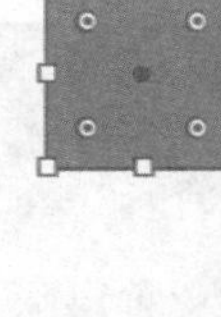

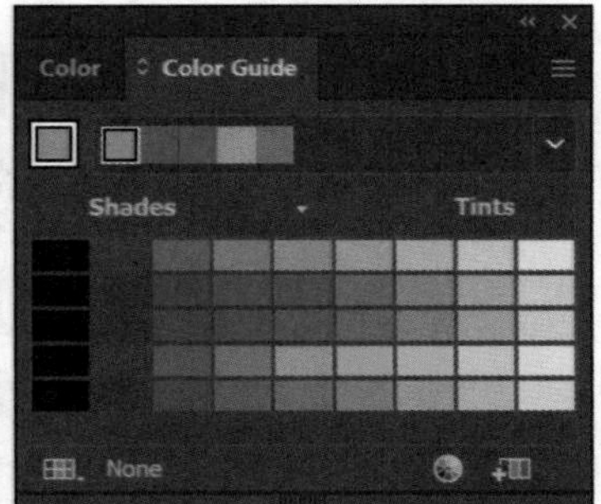

FIGURE 8-11: The Color Guide panel identifies colors that are good combinations for your selected color.

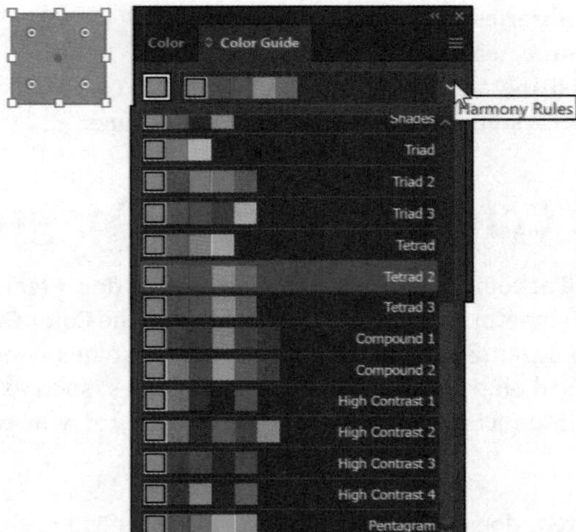

FIGURE 8-12: Make a selection from the Harmony Rules drop-down list.

Editing or Applying Colors

Give the Recolor Artwork feature a go if you have several colors selected and want to quickly try different harmonies to the entire illustration. For this example, you can use the `SportsCover.ai` file from the Book05_Illustrator folder that you downloaded in Book 1, Chapter 1:

1. **Open an illustration with many colors and choose Select⇨All.**
2. **Experiment with different colors by selecting the Edit or Apply colors button at the bottom of the Color Guide panel.**

   The Recolor Artwork dialog box appears.

3. **Change the harmony rules by clicking the Harmony Rules arrow to the right of the color bar.**

   A drop-down list appears with many choices for selecting colors, as shown in Figure 8-13. As you select different colors the entire illustration is changed to match the selected color harmony.

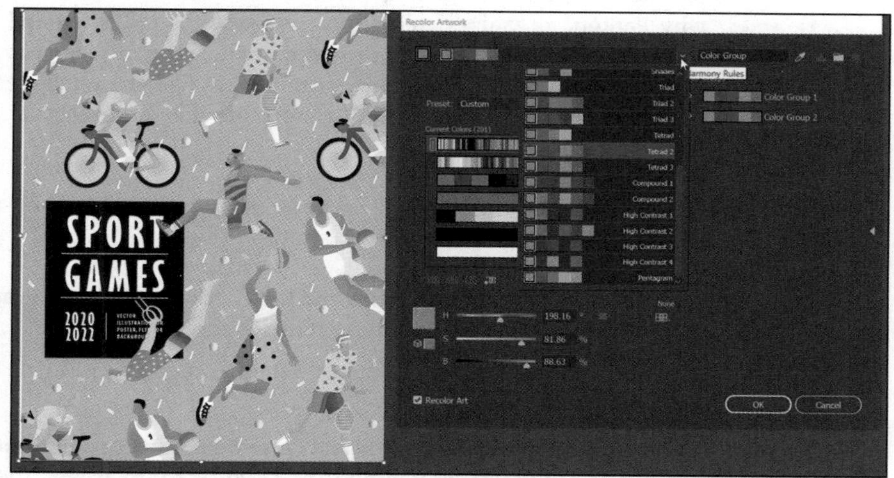

FIGURE 8-13: Choose to recolor artwork all at once using the Edit or Apply colors feature in the Color Guide.

4. **Select a color harmony to see your artwork recolored.**
5. **Save your color selection as a color group by clicking the New Color Group icon, located at the top right of the dialog box.**

   If you like, you can rename the color group by double-clicking the group name in the Color Group text field in the Recolor Artwork dialog box.
6. **Click OK.**

   The color group is added to the Swatches panel.

TIP

You don't have to use the Recolor Artwork dialog box to save a group of colors. You can Ctrl-click (Windows) or ⌘-click (Mac) to select multiple colors and then click the New Color Group button at the bottom of the Swatches panel.

Adding Pantone colors

The Pantone Matching system is popular in the print and fashion industries, but in other industries as well. Pantone organizes nearly 5,000 Pantone Colors into two numbered Systems, one for print and packaging, and the other for product design. You can keep your colors consistent by assigning a Pantone number. An example of a PMS number would be PMS 300, or PMS 220. If you are working in branding you frequently assign a PMS color so that when your logo, or other branded items are produced, they are consistent in color no matter where they are produced.

Note: The appearance of color can change based on the material on which it is produced. In fact, some colors are not achievable at all on a certain material. That's why Pantone organizes colors into two systems – Coated and Uncoated, which indicates whether the color is to be reproduced on a nice coated stock, like you would see in a quality magazine, or uncoated, like you might see in a newspaper, to ensure that the colors included are achievable and reproducible based on the materials used.

If you're looking for the typical swatches numbered in the Pantone Matching System (PMS), follow these steps:

1. **Choose Window ⇨ Swatches and then click the Swatch Libraries menu button at the bottom of the Swatches panel.**

   You can also access the Swatches panel by clicking the Fill in the Properties panel.

2. **From the drop-down list, select Color Books and then Pantone Solid Coated or whatever Pantone library you want to access. The Pantone colors appear in a separate panel.**

As previously mentioned, colors in the Pantone numbering system are often referred to as PMS 485 or PMS 201 or whatever number has been designated. You can locate the numbered swatch by typing the number into the Find text field of the Pantone panel, as shown in Figure 8-14. When that number's corresponding color is highlighted in the panel, click it to add it to your Swatches panel. Many users find it easier to see colored swatches by using List view. Choose Small List View or Large List View from the panel menu.

FIGURE 8-14: Use the Find text box to locate Pantone colors.

Editing Colors

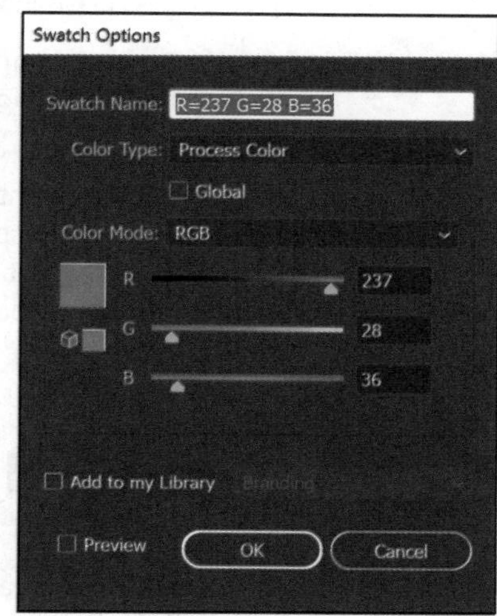

FIGURE 8-15:
Edit a color swatch by double-clicking it.

Edit colors in the Swatches panel by using the Swatch Options dialog box (shown in Figure 8-15), which you access by double-clicking the color or choosing Swatch Options from the Swatches panel menu. Caution! Make sure that you do not have anything selected when you double-click on a color, as it will apply to your selection.

Use the Swatch Options dialog box to:

- **Change color values:** Change the values in a color by using the sliders or typing values in the color text fields. Being able to enter exact color values is especially helpful if you're given a color build to match. Select the Preview checkbox to see results as you make the changes.
- **Use global colors:** If you plan to use a color frequently, select the Global checkbox. If it's selected and you use the swatch throughout the artwork, you have to change the swatch options only one time, and then all instances of that color are updated.

One important option to note in the Swatch Options dialog box is the Color Type drop-down list. You have two choices: spot color and process color. What's the difference?

- **Spot color:** A color that isn't broken down into the CMYK values. Spot colors are used for one- or two-color print runs or when precise color matching is important.

  Suppose that you're printing 20,000 catalogs and decide to run only two colors: red and black. If you pick spot colors, the catalogs have to be run through the press that contains two ink applications: one for black and one for red. If red were a process color, however, it would be created from a combination of cyan, magenta, yellow, and black inks, and the catalogs would need to be run through a press with four or more ink applications in order to build that color. Plus, if you went to a print service and asked for red, what color would you get — fire engine red, maroon, or a light and delicate pinkish red? But if the red you pick is PMS 485, your printer in Kutztown, Pennsylvania, can then print the same color of red on your brochure as the printer making your business cards in Woburn, Massachusetts.

- **Process color:** A color that's built from four colors (cyan, magenta, yellow, and black); used for multicolor jobs.

  For example, you would use process colors to send an ad to a four-color magazine. Its printers certainly want to use the same inks they're already running, and using a spot color would require another ink application in addition to the cyan, magenta, yellow, and black plates. In this case, you convert to process colors any spot colors created in corporate logos or similar projects.

Building and Editing Patterns

Using patterns can be as simple or as complicated as you want. If you become familiar with the basic concepts, you can take off in all sorts of creative directions. To build a simple pattern, start by creating the artwork you want to use as a pattern on your artboard — polka dots, smiley faces, wavy lines, or whatever. Then select all components of the pattern and drag them to the Swatches panel. That's it — you made a pattern! Use the pattern by selecting it as the fill or stroke of an object.

If your pattern doesn't stay in the Swatches panel, it may be that you created a pattern inside one of the shapes of what you want to create as a pattern. It just can't be done using this method, but when you find out about advanced pattern editing, later in this section, you will see how you can complete that task.

You can update patterns you created or patterns that already reside in the Swatches panel. To edit an existing pattern, follow these steps:

1. **Click the pattern swatch in the Swatches panel and drag it to the artboard.**
2. **Deselect the pattern and use the Direct Selection make changes to the pattern.**

   Keep making changes until you're happy with the result.
3. **To update the pattern with your new edited version, use the Selection tool to select all pattern elements and Alt-drag (Windows) or Option-drag (Mac) the new pattern over the existing pattern swatch in the Swatches panel.**

4. When a black border appears around the existing pattern, release the mouse button.

All instances of the pattern in your illustration are updated.

TIP

To add some space between tiles, as shown in Figure 8-16, create a bounding box by drawing a rectangle shape with no fill or stroke (representing the repeat you want to create). Send the rectangle behind the other objects in the pattern and drag all objects, including the bounding box, to the Swatches panel.

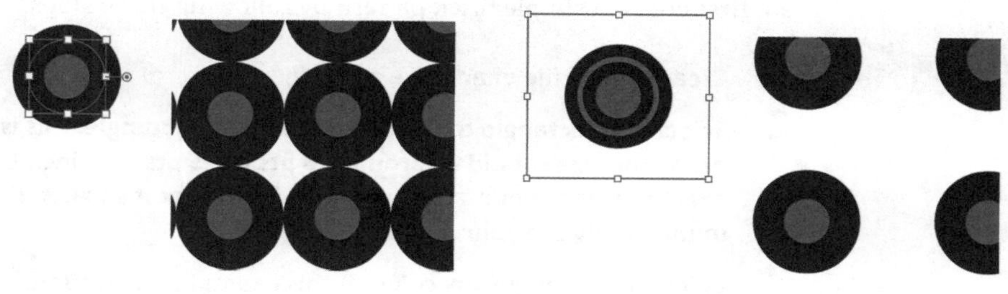

FIGURE 8-16: A pattern created on the left with no bounding box and a pattern on the right created with a transparent bounding box.

Transformations are covered in detail in Chapter 9 of this minibook, but some specific transform features apply to patterns. To scale a pattern, but not the object it's filling, double-click the Scale tool, shown in the margin. In the Scale dialog box that appears, type the value that you want to scale and deselect all options except Transform Patterns, as shown in Figure 8-17. This method works for the Rotate tool as well.

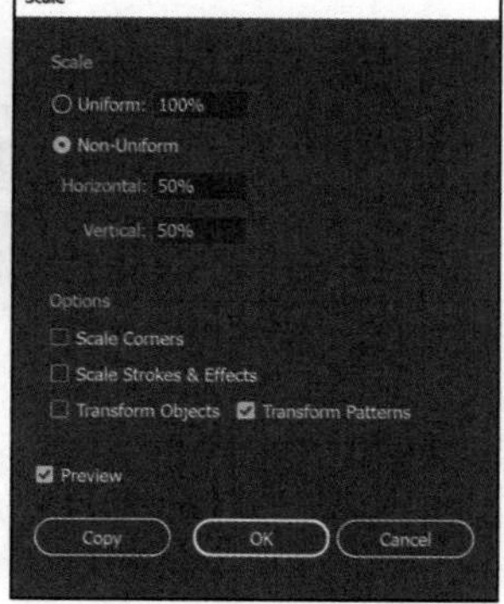

FIGURE 8-17: Choose to scale or rotate only the pattern, not the object that contains the pattern.

Advanced Pattern Editing

In Illustrator CC, you have more control over patterns than ever before. The preceding section tells you how to start your pattern, but what if you want to create the perfect repeat? Or what if you want to align your repeated tile in a different offset — like stacking bricks — instead of just tiling one pattern right next to the other? With Illustrator CC, you can!

These steps show you how to take advantage of the powerful Pattern editor. To do so, first create a simple brick pattern by following these steps.

1. **Create a new file of any size or profile.**
2. **Select the Rectangle tool and make a small rectangle; this is your first brick. The size should be around 75 pts by 25 pts, or 1 inch by .3". An exact size does not matter, you just don't want it so large that you can't immediately see your repeat.**
3. **Color the rectangle any color. In this example, red is used.**
4. **Click and drag the brick you made into the Swatches panel. Your panel is created, as shown in Figure 8-18.**

   This would not be much of a pattern if you left it as is, but in the next few steps you will take edit that pattern in the pattern editing mode.

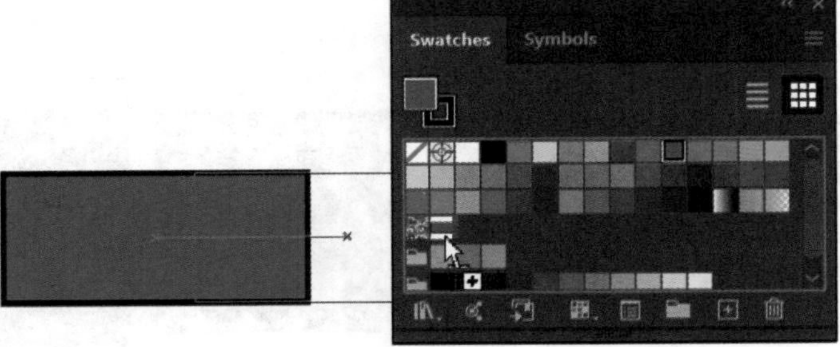

FIGURE 8-18: Click and drag the brick into the Swatches panel to turn it into a pattern.

After you add the pattern to the Swatches panel, double-click the swatch. The Pattern Options panel appears, and the artboard switches to Pattern Editor Preview mode, as shown in Figure 8-19. You can edit your pattern by following these steps:

1. **In the Tile Options panel, open the Tile Type drop-down menu and select Brick by Row, as shown in Figure 8-20.**

   The artboard displays the change in the pattern repeat.

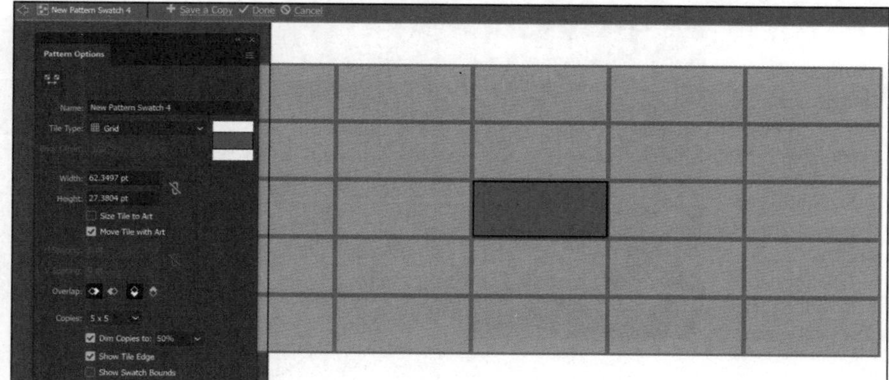

FIGURE 8-19: Pattern options allow you to edit pattern art, change the offset, and more.

FIGURE 8-20: Change the Tile Type repeat to a Brick by Row.

2. **If you want, change the default Brick Offset by opening that drop-down list and selecting another option.**
3. **You can also edit the size of the repeat by manually entering a size in the Width and Height text boxes, as shown in Figure 8-21.**

   If you are a more visual person, click once on the Pattern Tile tool in the upper-left of the Pattern Options panel, and then click and drag the handles on the default bounding box.

   REMEMBER

   Keep in mind that you can return to the original size by selecting the Size Tile to Art checkbox below at any time.
4. **Depending upon your pattern, you might want an overlap. Change the type of overlap — that is, which item is on top of the stacking order — by using the Overlap section in the Pattern Options panel.**

FIGURE 8-21: Change the size of the repeat by entering values into the width and height text boxes, or click the Pattern Tile tool and adjust the bounding box.

5. **(Optional) Make further adjustments to the pattern and edit the artwork if needed.**

   In the lower section of the Pattern Options panel, you can control the visual look of the pattern preview you're seeing, from the amount of copies that you see, to how dimmed the non-original pattern tiles appear.

   You can edit your artwork using the drawing and editing tools, just like you would if you weren't in the Pattern Editor Preview mode. You can see that a pen path is being created in Figure 8-22, and the change is reflected in all the pattern tiles instantaneously.

6. **When you're finished making changes to your pattern, click Done, located to the right of Save a Copy at the top of the editing window.**

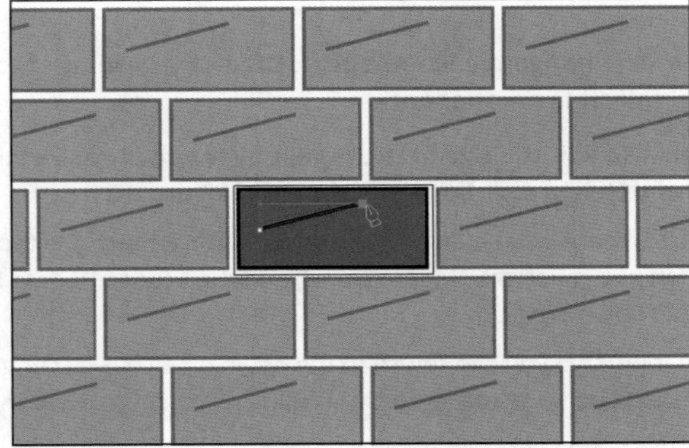

FIGURE 8-22: Add and edit artwork right in the pattern editing preview mode.

Working with Gradients

Create gradients for smooth metallic effects or just to add dimension to illustrations. Gradients are basically just color stops that are created in a ramp. This ramp can transition from one color to another and even transition for various opacities.

Using the Gradient feature you can create a gradient to model your illustration and add dimension, or just use it to add interesting effects to your illustrations. If you want to see default Gradients, choose Window⇨Swatches and then choose Gradient from the Show Swatch Kinds button at the bottom of the Swatches panel. Alternately, you can make your own new gradient by following these steps:

1. **Select an object and fill it with gradient by selecting the Fill in the Properties panel and then clicking one of the default gradients in the Swatches panel. (See Figure 8-23.)**

   You can then access Gradient options in the Properties panel by clicking Fill and then selecting the Gradient Options button. Keep in mind that the Gradient Options button does not appear unless you have filled with a gradient.

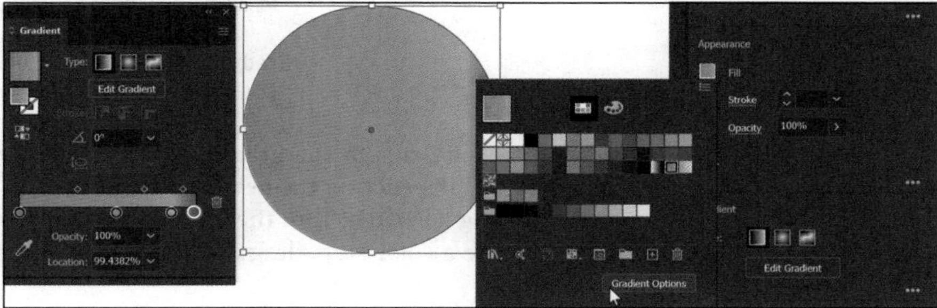

FIGURE 8-23: You can apply a gradient to a selected object.

   You can modify the color, origin, opacity, location, and angle of a gradient from the Gradient tool, Gradient panel, Control panel, and Properties panel. To further edit your gradient, continue with the next steps.

   Notice in Figure 8-24 the Gradient Slider that appears at the bottom of the Gradient panel. On the left and right side of the slider, you see color stops. In this example, you also see additional color stops, along with other labeled parts of this gradient panel that you will have the opportunity to use later in this section.

2. **Click the right color stop to see that you can change the Opacity and the Location using the sliders at the bottom of the Gradient panel.**

   When a color stop is active, the triangle on top turns solid.

3. **Double-click the right color stop to see that a pop-up of the Color panel appears. You can now select a new color for the right color stop.**

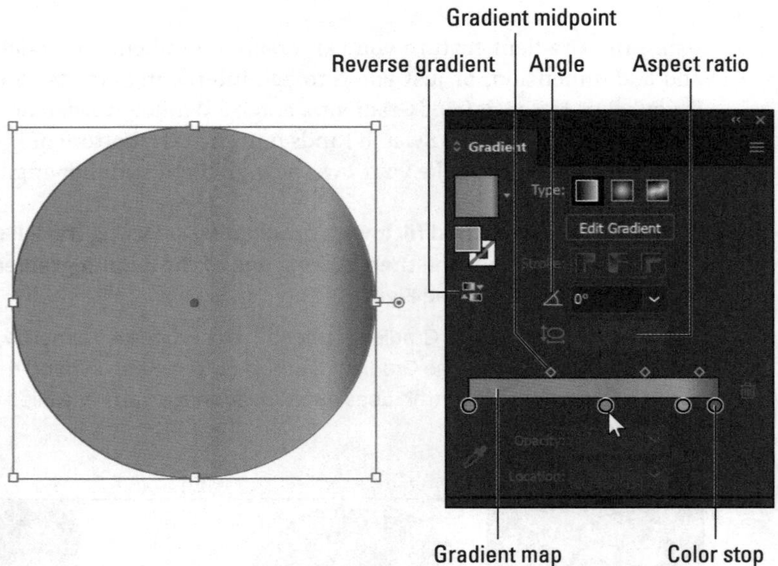

FIGURE 8-24: The Gradient panel offers the ability to create gradients with custom colors and transparencies.

To add color stops, click beneath the gradient ramp, where you want the new color stop, and then double-click on the new stop to choose a color from the Color panel. You can also drag a swatch from the Swatches panel to add a new color to the gradient. To remove a color stop, drag it off the Gradient panel, as you see in Figure 8-25.

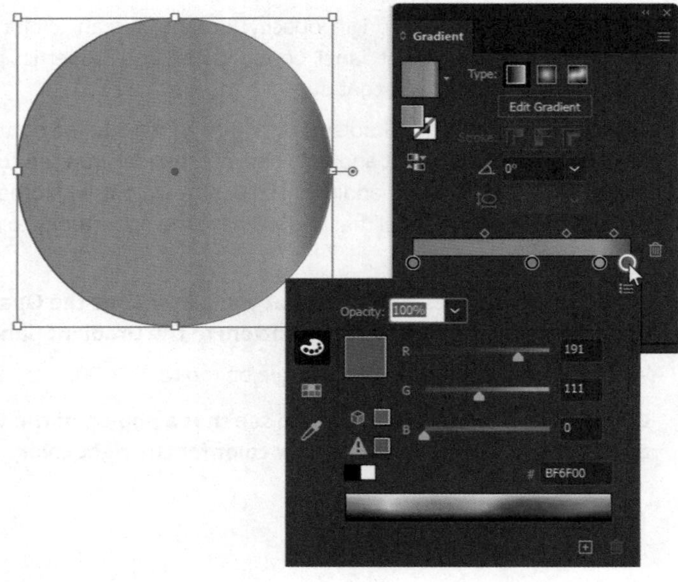

FIGURE 8-25: Add, remove, and reposition colors using the Gradient ramp.

TIP

You can click the gradient ramp to add colors and also to change the opacity of that location of the ramp by entering values in the Opacity text box. This technique is a helpful way to create stripes and other reflective gradients.

Using the Gradient tool

Take advantage of the Gradient tool to adjust the length and direction of the gradient that you apply to objects. To use the Gradient tool, simply select the object that has the gradient applied to it and then click and drag. A color ramp appears, indicating the length and direction of your gradient, as shown in Figure 8-26.

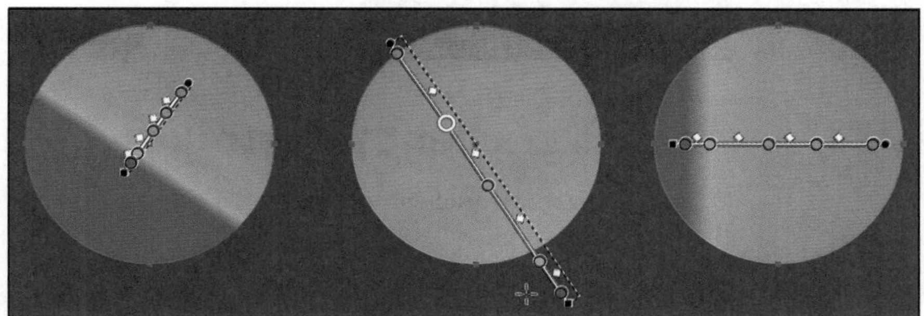

FIGURE 8-26: Use the Gradient tool to change the direction and length of your gradient.

You can also use the Gradient tool to apply and change colors on the color ramp. To use the Gradient tool, follow these steps:

1. **Apply a gradient to an object.**
2. **Select the Gradient tool (or press G) and click and drag to define the length and direction of the gradient.**

   Drag a long path for a smooth, long gradient. Drag a short path for a short, more defined gradient.
3. **Hover over the gradient ramp (on top of the object), as shown in Figure 8-27.**

   The color ramp appears with the color stops visible. This is referred to as a gradient annotator.
4. **Double-click the color stop to open the color panel.**
5. **Select another color and then click outside the Color panel when you're finished editing the gradient ramp.**

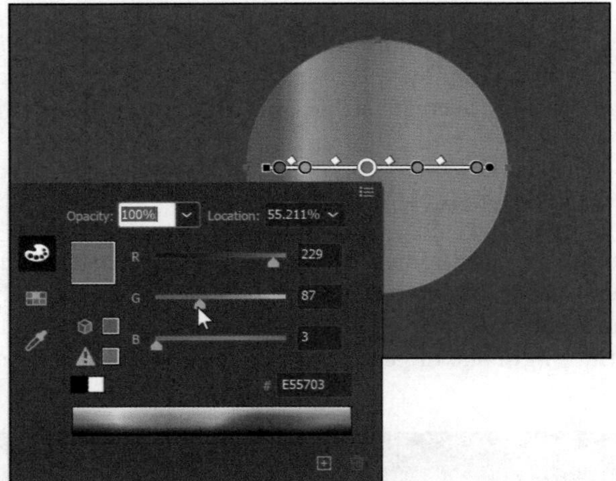

FIGURE 8-27: Use the Gradient tool to change the color stops.

Apply a gradient to a stroke

You can also apply a gradient to a stroke. To apply a gradient to a stroke, follow these steps:

1. **Make sure that you have the stroke of an object selected. You may want to increase the stroke to a larger point size so that you can see the gradient that you will apply a little better.**

   You can do this by bringing forward the stroke at the bottom of the toolbar, or you can click Stroke in the Control panel and then select a gradient from the Swatches menu that appears.

2. **Choose a gradient from the Swatches panel.**

   The gradient is applied.

TIP

You can't change the direction or distance of a gradient applied to a stroke with the Gradient tool. To make these changes, choose Window⇨Gradient and use the angle (for the direction) and the Gradient slider to change the span of the color stops, as you see in Figure 8-28.

Using transparency in gradient meshes

The Gradient Mesh tool helps you to create incredible photo-realistic illustrations using meshes you build on vector objects. If you've already discovered the Gradient Mesh tool, you know how useful this tool is. If you add in the transparency feature, the number of ways you can use it skyrockets.

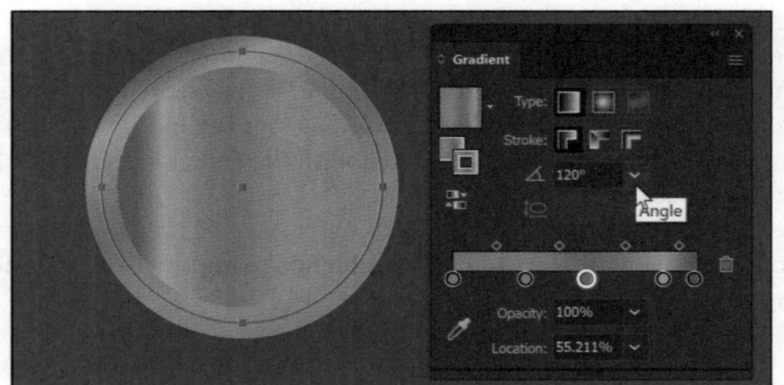

FIGURE 8-28: To edit a gradient on a stroke, you must use the Gradient panel.

Follow these simple steps to try the new transparency feature:

1. **Choose a solid color for your fill and no stroke.**
2. **Create a shape.**

   You can use the `ButterflyOutline.ai` file in the Book05_Illustrator folder.
3. **Choose Object ➪ Create Gradient Mesh and leave the default of 4 Rows and 4 Columns, Appearance, To Center, as shown in Figure 8-29. Click OK.**

   A gradient mesh appears in your shape with multiple mesh nodes.
4. **Using the Direct Selection tool, click an individual node.**
5. **Either choose another color to create a blending effect from one color to another or choose Window ➪ Transparency and decrease the value of the transparency drop-down list.**

   The individual node becomes transparent based on the value you select.

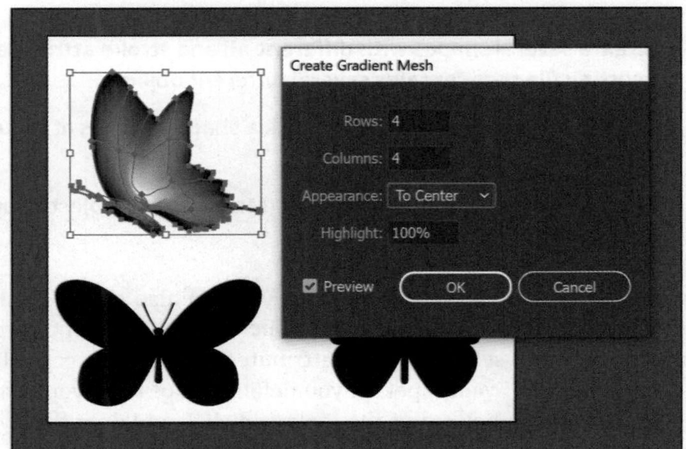

FIGURE 8-29: Choose how many rows and columns you want to start your mesh with.

Keep in mind that if you click on a node, or intersection, and then select a color, the gradient is generated from that node. If you click in the center of a section, the gradient is created from that section, as you see in Figure 8-30. If you no longer want a color node, simply click on it with the Direct Selection tool and press Delete or Backspace.

You can create all sorts of complex blends using the Gradient Mesh without having to create custom gradients.

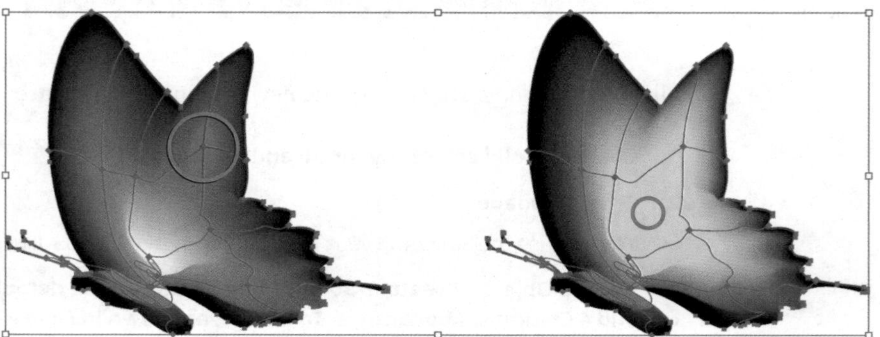

FIGURE 8-30: Using the Direct Selection tool, click on a node to generate the color from that point, or click in the center to add a gradient that spans the section.

Copying Color Attributes

Wouldn't it be helpful if you had tools that could record the fill and stroke attributes and apply them to other shapes? You're in luck — the Eyedropper tool can do just that. Copy the fill and stroke of an object and apply it to another object by using the Eyedropper tool:

1. **Create several shapes with different fill and stroke attributes or open an existing file that contains several different objects.**
2. **Select the Eyedropper tool and click a shape that has attributes you want to copy.**
3. **Alt-click (Windows) or Option-click (Mac) another object to apply those attributes.**

Not only is this technique simple, but you can also change which attributes the Eyedropper applies. Do so by double-clicking the Eyedropper tool; in the dialog box that appears, select only the attributes you want to copy. Remember that if you don't see the Eyedropper on you default set of tools you can click on the Edit Tool button at the bottom of the tools and click and drag the Eyedropper to your tools panel.

Painting Made Easy: The Live Paint Feature

Don't worry about filling closed shapes or letting fills escape from objects with gaps into unwanted areas. Using the Live Paint feature, you can create the image you want and fill in regions with color. The Live Paint bucket automatically detects regions composed of independent intersecting paths and fills them accordingly. The paint within a given region remains live and flows automatically if any paths are moved.

If you want to give it a try, follow these steps to put together an example to experiment with:

1. **Use the Ellipse tool to create a circle on your page.**

   Make the circle large enough to accommodate two or three inner circles. If you would rather, you can use the `LivePaintExercise` file that is in the Book05_Illustrator folder that you downloaded in Book 1, Chapter 1 and skip to Step 7.

2. **Press D (and nothing else).**

   As long as you aren't using the Type tool, you revert to the default colors of a black stroke and a white fill.

3. **Double-click the Scale tool and enter** 75% **in the Uniform Scale text box.**
4. **Press the Copy button and then click OK.**

   You see a smaller circle inside the original.

5. **Press Ctrl+D (Windows) or ⌘+D (Mac) to duplicate the transformation and create another circle inside the last one.**
6. **Choose Select⇨All or press Ctrl+A (Windows) or ⌘+A (Mac) to activate the circles you just created.**
7. **Using the Selection tool, select all the circles. Next, select the Live Paint Bucket tool. The easiest method for finding this tool is to press K, as this will add the Live Paint Bucket to your tools panel.**

   Using the Live Paint Bucket tool click, on your selected objects. This turns the selected objects into a Live Paint Group, as you see in Figure 8-31. Now when you move the Live Paint Bucket tool over them again, the different regions become highlighted, indicating they are ready to paint.

8. **Note the color annotator that appears above the curser. You can press the arrow keys to move in either the left or right direction to access colors in your Swatches panel. Simply position your cursor over the shape area that you wish to fill, then press the arrow keys to find the color you want to use, and then click. If you do not want to use the color annotator you can simply choose your colors from the Swatches panel.**

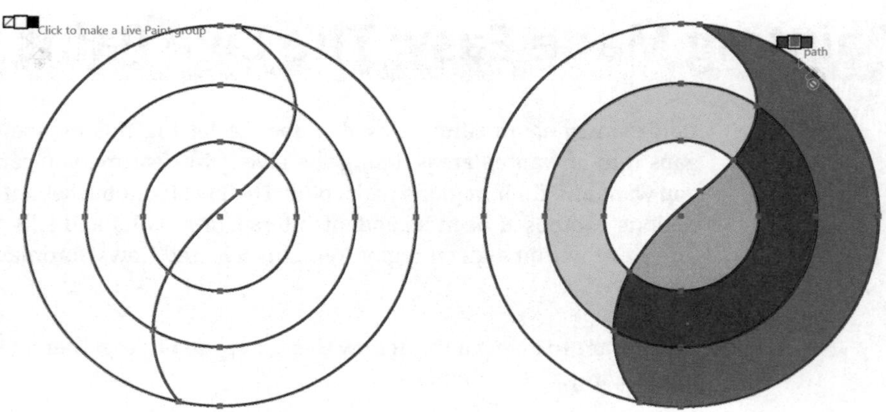

FIGURE 8-31: Using the Live Paint Bucket, click the selected objects, then press the arrow keys to access additional colors in your Swatches panel through the color annotator. Click each section to apply the color.

A companion feature to the Live Paint Bucket is support for gap detection. With this feature in its arsenal, Illustrator automatically and dynamically detects and closes small to large gaps that may be part of the artwork. You can determine whether you want paint to flow across region gap boundaries by using the Gap Options dialog box, accessible by choosing Object ➪ Live Paint ➪ Gap Options.

Before you save a file for an older version of Illustrator that uses the Live Paint feature, first select the occurrences of Live Paint and choose Object ➪ Expand. When the Expand dialog box appears, leave the options at their defaults and click OK. This setting breaks down the Live Paint objects to individual shapes, which older versions of Illustrator can understand.

» Putting the Transform tools to work

» Becoming familiar with the Liquify tools

» Distorting, warping, and otherwise reshaping objects

Chapter 9
Using the Transform and Distortion Tools

You can apply transformations to objects in Illustrator that include scaling, rotating, skewing, and distorting. In this chapter, you discover how to use the general transform tools as well as some of the interesting Liquify and Envelope Distort features available in Illustrator.

Working with Transformations

Using just the Selection tool, you can scale and rotate a selected object. Drag the bounding box handles to resize the object, shown in Figure 9-1, or move outside a handle and then, when the cursor changes to a "flippy" arrow (a curved arrow with arrowheads on both ends), drag to rotate the object.

If you want to scale proportionally, hold down the Shift key while you drag to resize. To rotate an object at 45-degree increments, hold down the Shift key while you're rotating.

TECHNICAL STUFF

When you scale, rotate, or use any other type of transformation in Illustrator, the final location becomes the *zero point.* In other applications, such as InDesign, you can rotate an object by any number of degrees (45 degrees, for example) and later enter **0** for the rotation angle in the Transform panel or in the Rotate dialog box to return the object to its original position. In Illustrator, if you enter **0** for the rotation angle to return a rotated object to its original position, the object doesn't change its position. To return the object to the previous position in Illustrator, you have to enter the negative of the number you originally entered to rotate the object, so you would enter **–45** for the degree of rotation in this example.

FIGURE 9-1: Use the bounding box to resize or rotate a selected object.

Transforming an object

The Rotate, Reflect, Scale, and Shear tools all use the same basic steps to perform transformations. Read on for those basic steps and then follow some individual examples of the most often used Transform tools. The following sections show five ways to transform an object: one for an arbitrary transformation and four others for exact transformations based on a numeric amount you enter. Keep in mind that if you want simple transformations, like flipping and rotating objects on a center axis, you can simply use the controls in the Transform section of your Properties panel.

Arbitrary transformation method

Because this transformation method is arbitrary, you're eyeballing the transformation of an object — in other words, you don't have an exact percentage or angle in mind, and you want to freely transform the object until it looks right. Just follow these steps:

1. **Select an object and then choose a Transform tool (Rotate, Reflect, Scale, or Shear).**
2. **Click the artboard once where you want your axis to start.**

   Be careful where you click because the click determines the point of reference, or *axis point,* for the transformation. In Figure 9-2, the click put the axis point above the right ear. The image is rotated on the axis created at that point.

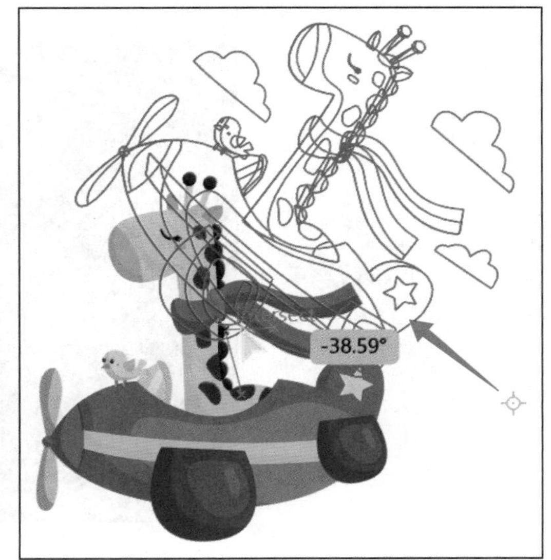

FIGURE 9-2: Click once to create the axis point, and then drag to rotate, reflect, scale, or shear your artwork.

3. **Drag in one smooth movement.**

   Just drag until you create the transformation you want.

TIP

Hold down the Alt (Windows) or Option (Mac) key when dragging to clone a newly transformed item while keeping the original object intact. This trick is especially helpful when you're using the Reflect tool.

Exact transformation methods

In the methods in this section, you discover how to perform transformations using specific numeric information:

Exact transformation method 1 — using the tool's dialog box:

1. **Select an object and then choose the Rotate, Reflect, Scale, or Shear tool.**
2. **Double-click the Transform tool that you selected in the Tools panel.**

   A dialog box specific to your chosen tool appears, as shown in Figure 9-3. In this example, the Scale tool was selected to open the Scale dialog box.
3. **Type an angle, a scaled amount, or a percentage in the appropriate text field.**

   In Figure 9-3, **50%** is entered.
4. **Select the Preview checkbox to see the effect of the transformation before you click OK; click the Copy button instead of OK to keep the original object intact and transform a copy.**

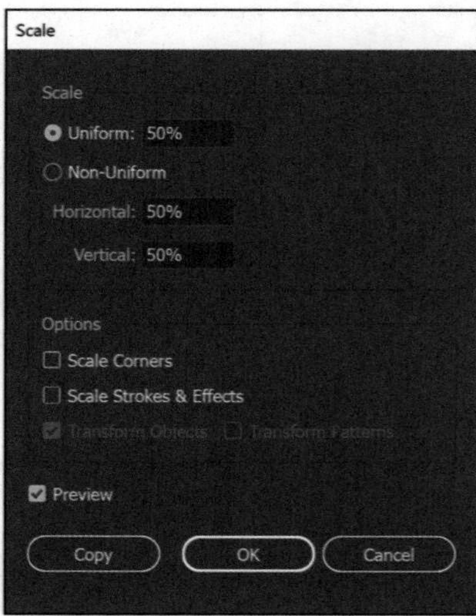

FIGURE 9-3: Double-click a Transform tool to open its dialog box.

Exact transformation method 2 — using the reference point:

1. **Select an object and then choose the Rotate, Reflect, Scale, or Shear tool.**
2. **Alt-click (Windows) or Option-click (Mac) wherever you want to place the reference point.**
3. **In the Transform tool dialog box that appears, enter values and click OK or click the Copy button to apply the transformation.**

   This method is the best one to use if you need to rotate an object an exact amount on a defined axis.

Exact transformation method 3 — using the Transform menu:

1. **Select an object and then choose a transform option from the Object ➪ Transform menu.**

   The appropriate transform dialog box appears.

2. **Enter some values and click OK or the Copy button.**

Exact transformation method 4 — using the Transform section of the Properties panel, or the Transform panel:

1. **Select an object and choose Windows ➪ Properties or Window ➪ Transform to access the Transform panel, both shown in Figure 9-4.**
2. **Enter your values in the text fields for the appropriate transformation.**

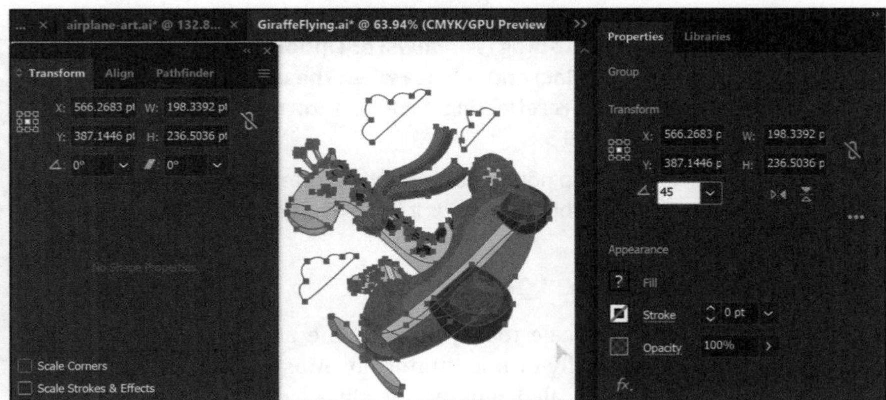

FIGURE 9-4: You can enter exact amounts in the fields in either the Transform panel (left) or Properties panel (right).

Though using the Properties and Transform panels is probably the easiest method when changing values, they doesn't offer you the capability to specify an exact reference point or other options that apply to the individual Transform tools. Keep in mind that by using the square grid on the left side of the Transform panel you can select a point of reference for your object. By using this grid, you can measure from the corners or center point of an object.

Using the Transform tools

In the following sections, you discover how to use some of the most popular Transform tools to create transformations.

The Reflect tool

Nothing in life is perfectly symmetrical, right? Maybe not, but objects not created symmetrically in Illustrator can look off-kilter. Using the Reflect tool, you can reflect an object to create an exact mirrored shape of it; just follow these steps:

1. **Open a new document in Illustrator and type some text or create an object.**

   If you want to reflect text, make sure that you use at least 60-point type so that you can easily see it.

2. **Select the Reflect tool (hidden under the Rotate tool) and click the object; if you're using text, click in the middle of the text baseline.**

   This step sets the reference point for the reflection.

3. **Alt+Shift-drag (Windows) or Option+Shift-drag (Mac) and release when the object or text is reflecting itself, as shown in Figure 9-5.**

   This step not only clones the reflected object or text, but also snaps it to 45-degree angles.

FIGURE 9-5: The completed reflection.

The Scale tool

Using the Scale tool, you can scale an object proportionally or non-uniformly. Most people like to be scaled non-uniformly — maybe a little taller, a little thinner — but on with the topic. Follow these steps to see the Scale tool in action:

1. **Create a shape and give it no fill and a 5-point black stroke.**

   For this example, we created a circle. See Chapter 3 of this minibook if you need a reminder on how to do it.

2. **Select the shape and double-click the Scale tool.**

   The Scale dialog box appears.

3. **Type a number in the Scale text field (in the Uniform section) and click the Copy button.**

   We entered **125** in the Scale text field to increase the size of the object by 125 percent.

4. **Press Ctrl+D (Windows) or ⌘ +D (Mac) to repeat the transformation as many times as you want.**

   Every time you press Ctrl+D (Windows) or ⌘ +D (Mac), the shape is copied and sized by the percentage you entered in the Scale text field. This trick, especially handy with circles, creates an instant bull's-eye!

To experiment with the Scale tool, create different shapes in Step 1 and enter different values in Step 3. Remember that if you type **50%** in the Scale text field, the object is made smaller; surpass 100 percent — say, to 150 percent — to make the object larger. Leaving the Scale text field at 100 percent has no effect on the object.

The Shear tool

The Shear tool lets you shear an object by selecting an axis and dragging to set a shear angle, as shown in Figure 9-6.

FIGURE 9-6: Create perspective with the Shear tool.

The axis is always the center of the object unless you use method 1 or 2 from the earlier section "Exact transformation methods." Use the Shear tool in combination with the Rotate tool to give an object perspective.

The Reshape tool

The Reshape tool lets you select anchor points and sections of paths and adjust them in one direction. You determine that direction by dragging an anchor point with the Reshape tool selected.

The Reshape tool works differently from the other Transform tools. To use it, follow these steps:

1. **Select just the anchor points on the paths that you want to reshape. Deselect any points that you want to remain in place.**
2. **Select the Reshape tool (hidden under the Scale tool) and position the cursor over the anchor point you want to modify; click the anchor point.**

   If you click a path segment, a highlighted anchor point with a square around it is added to the path.
3. **Shift-click more anchor points or path segments to act as selection points.**

   You can highlight an unlimited number of anchor points or path segments.
4. **Drag the highlighted anchor points to adjust the path.**

Using the Transform and Distortion Tools

The Free Transform tool

You use the Free Transform tool in much the same way as you use the bounding box. (See the earlier section "Working with Transformations.") This tool is necessary only if you choose View⇨Hide Bounding Box but want free transform capabilities.

Creating Distortions

You can bend objects — make them wavy, gooey, or spiky — by creating simple to complex distortions with the Liquify tools and the Envelope Distort features.

The Liquify tools

The Liquify tools can accomplish all sorts of creative or wacky (depending on how you look at it) distortions to your objects.

You can choose from eight Liquify tools. You should experiment with these tools to understand their full capabilities. Here are some tips:

- A variety of Liquify tools are available by holding down the mouse button on the default selection, the Width tool. (Table 9-1 shows the Width tool keyboard shortcuts.) If you use the tools frequently, drag to the arrow at the end of the tools. You can then position the tools anywhere in your work area.

TABLE 9-1 Width Tool Keyboard Shortcuts

Width Tool Function	Windows	Mac OS
Create non-uniform widths	Alt-drag	Opt-drag
Create a copy of the width point	Alt-drag the width point	Opt-drag the width point
Copy and move all the points along the path	Alt+Shift+drag	Opt+Shift+drag
Change the position of multiple width points	Shift-drag	Shift-drag
Select multiple width points	Shift-click	Shift-click
Delete selected width points	Delete	Delete
Deselect a width point	Esc	Esc

» Double-click any Liquify tool to open a dialog box specific to the selected tool. The Liquify tools are shown in Table 9-2.

TABLE 9-2 The Liquify Tools

Icon	Tool Name	What It Does to an Object
	Width	Increases the stroke width or height when you click and drag a path.
	Warp	Molds it with the movement of the cursor. (Pretend that you're pushing through dough with this tool.)
	Twirl	Creates swirling distortions within it.
	Pucker	Deflates it.
	Bloat	Inflates it.
	Scallop	Adds curved details to its outline. (Think of a seashell with scalloped edges.)
	Crystallize	Adds many spiked details to the outline of an object, such as crystals on a rock.
	Wrinkle	Adds wrinklelike details to the outline of an object.

» When a Liquify tool is selected, the brush size appears. Adjust the diameter and shape of the Liquify tool by holding down the Alt (Windows) or Option (Mac) key while dragging the brush shape smaller or larger. Press the Shift key to constrain the shape to a circle.

Using the Width tool, cross over a selected path. When a hollow square appears, click and drag outward (or inward), and the stroke width at that location is adjusted. See Figure 9-7.

FIGURE 9-7: Use the Width tool.

If you want a little more accuracy, you can double-click the stroke by using the Width tool and create, modify, or delete the width point by using the Width Point Edit dialog box, shown in Figure 9-8.

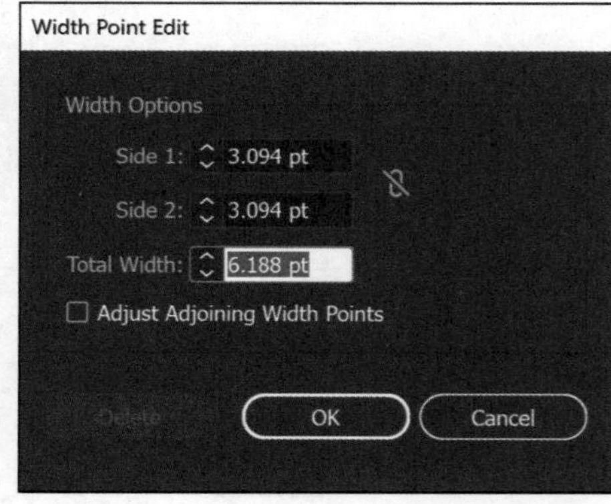

FIGURE 9-8: Double-click the stroke you are working on to open the Width tool options.

TIP

To create a discontinuous width point, just create two width points on a stroke with different stroke widths and then drag one width point on to the other width point to create a discontinuous width point for the stroke.

Using the Envelope Distort command

Use the Envelope Distort command to arch text and apply other creative distortions to an Illustrator object. To use the Envelope Distort command, you can use a preset warp (the easiest method), a grid, or a top object to determine the amount and type of distortion. In this section, we discuss all three methods.

Using the preset warps

Experimenting with warp presets is a little more interesting if you have a word or an object selected before trying out the warp options. To warp an object or some text to a preset style, follow these steps:

1. **Select the text or object that you want to distort and then choose Object ⇨ Envelope Distort ⇨ Make with Warp.**

   The Warp Options dialog box appears.

2. **Select a warp style from the Style drop-down list and then specify any other options you want.**
3. **Click OK to apply the distortion.**

TIP

If you want to experiment with warping but also want to revert to the original at any time, choose Effect ⇨ Warp. You later change or delete the warp effect by double-clicking it in the Appearance panel or by dragging the effect to the trash can in the Appearance panel. Find out more in Chapters 10 and 11 of this minibook about exciting effects you can apply to objects.

Reshaping with a mesh grid

You can assign a grid to an object so that you can drag different points and create your own, custom distortion, as shown in Figure 9-9.

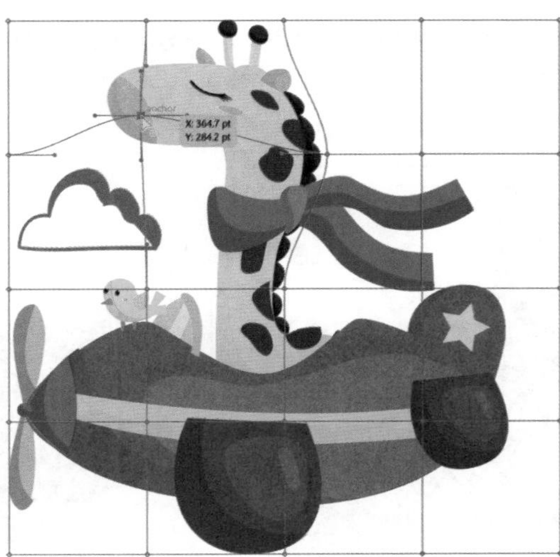

FIGURE 9-9: Custom distortion using a mesh grid.

Follow these steps to apply a mesh grid:

1. **Using the Selection tool, select the text or object that you want to distort and then choose Object ➪ Envelope Distort ➪ Make with Mesh.**

   The Envelope Mesh dialog box appears.

2. **Specify the number of rows and columns you want the mesh to contain and then click OK.**
3. **Drag any anchor point on the mesh grid with the Direct Selection tool to reshape the object.**

TIP

To delete anchor points on the mesh grid, select an anchor point by using the Direct Selection tool and press the Delete key.

You can also use the Mesh tool to edit and delete points when using a mesh grid on objects.

Reshaping an object with a different object

To form letters into the shape of an oval or to distort selected objects into another object, use this technique:

1. **Create text that you want to distort.**
2. **Create the object you want to use as the *envelope* (the object to be used to define the distortion).**
3. **Choose Object ➪ Arrange to ensure that the envelope object is on top, as shown in Figure 9-10.**

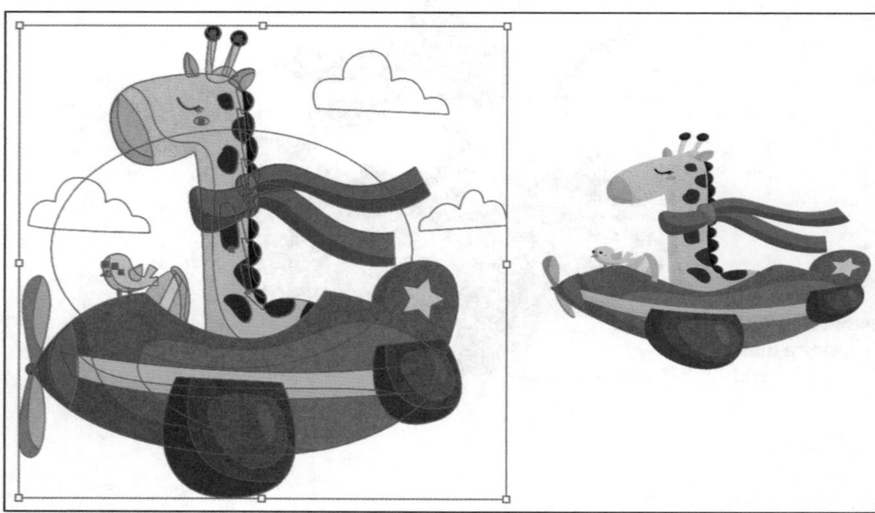

FIGURE 9-10: Position the shape over the text. In this example an oval was used. Then select both and choose Make with Top Object.

4. **Select the text and Shift-click to select the envelope object.**
5. **Choose Object ➪ Envelope Distort ➪ Make with Top Object.**

The underlying object is distorted to fit the shape of the top (envelope) object.

TIP

Choose Effect ➪ Distort and Transform ➪ Free Distort to take advantage of the Free Distort dialog box, shown in Figure 9-11. You can edit or undo effects at any time by clicking or deleting the Free Distort effect from the Appearance menu.

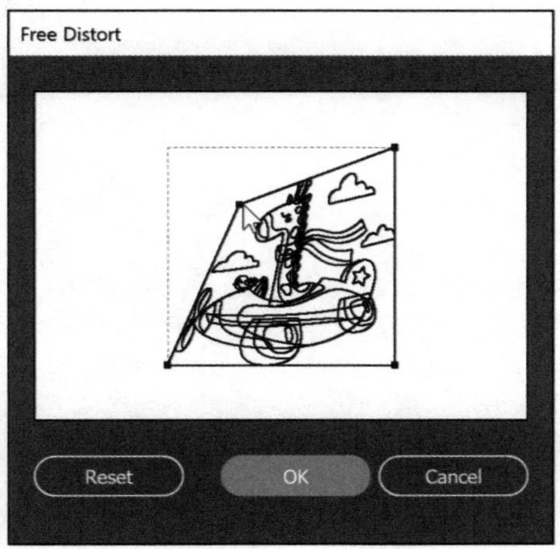

FIGURE 9-11:
Distort an object from the Free Distort dialog box.

Using the Transform and Distortion Tools

» Getting to know the Blend tool

» Using the Symbol Sprayer tool

» Discovering transparency, blend modes, and opacity masks

Chapter 10
Working with Transparency and Special Effects Tools

This chapter is full of neat things you can do using some of the more advanced features in Adobe Illustrator. These special effects tools can help you create art that makes an impact: Discover how to make your art look like a painting with the Gradient Mesh tool, create morphlike blends with the Blend tool, become a graffiti artist by trying out the Symbol Sprayer tool, and see what's underneath objects by using transparency.

The Mesh Tool

If you're creating art in Illustrator that requires solid colors or continuous patterns, you can achieve those results quite easily. But if you're working on an element that requires continuous tones, such as a person's face, you turn to the handy Mesh tool to create smooth tonal variations in your illustration. Choose to blend one color into another and then use the Mesh tool to adjust your blend. You can also apply varying levels of transparency to these mesh points.

The Mesh tool can be as complex or simple as you want. Using the Mesh tool, you can create intense illustrations that look like they were painted with an airbrush or just use the tool to give dimension to an object, as in the illustration shown in Figure 10-1.

FIGURE 10-1: Mesh tool illustrations can be complex or simple.

In this chapter, you discover two ways to create gradient meshes in Illustrator, manually and automatically. By manually selecting where you want mesh points, you gain more freedom. By automatically defining rows and columns, you can be more precise in the placement of your mesh points.

To create a gradient mesh by clicking, follow these steps:

1. **Select an object.**

   If you want to use a sample file, locate the image named `GradientmeshApple.ai` in the Book05_Illustrator folder you downloaded in Book 1, Chapter 1.

2. **Choose Object⇨Create Gradient Mesh.**

   The Create Gradient Mesh dialog box appears.

3. **Set the number of rows and columns of mesh lines to create on the object by entering numbers in the Rows and Columns text fields.**

4. **Choose the direction of the highlight from the Appearance drop-down list.**

   The direction of the highlight determines which way the gradient flows (see Figure 10-2); you have these choices:

- *Flat:* Applies the object's original color evenly across the surface, resulting in no highlight
- *To Center:* Creates a highlight in the center of the object
- *To Edge:* Creates a highlight on the edges of the object

5. **Enter a percentage of white highlight to apply to the mesh object in the Highlight text field.**
6. **Click OK to apply the gradient mesh to the object.**

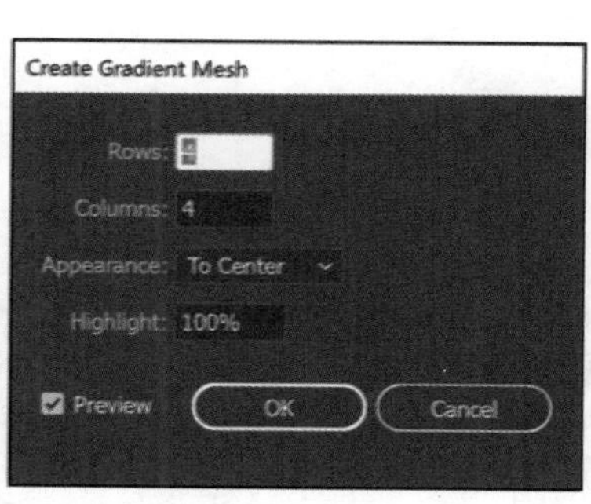

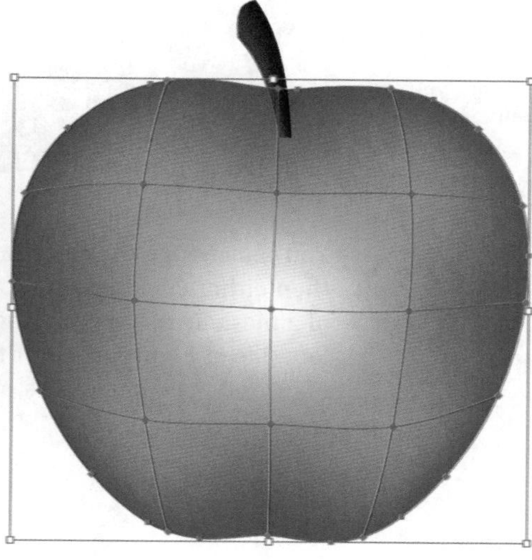

FIGURE 10-2: Set how many rows and columns you want to have in your gradient mesh. In this example, To Center was selected for the Appearance.

You can select individual or multiple mesh points and apply varying levels of transparency to them. Follow these steps to apply transparency to a gradient mesh:

1. **Select a shape tool and click and drag to the artboard to add the shape to your document.**
2. **Give the shape a solid fill (any color) and no stroke.**
3. **Choose the Mesh tool and click anywhere in the object.**

   This step adds a mesh point to your object.
4. **Choose Window⇨Transparency and select a percent from the Opacity drop-down list or type in any other value you want in the Opacity text field.**

   As you can see in Figure 10-3, an illustration was sent behind the object to demonstrate transparency in the object.

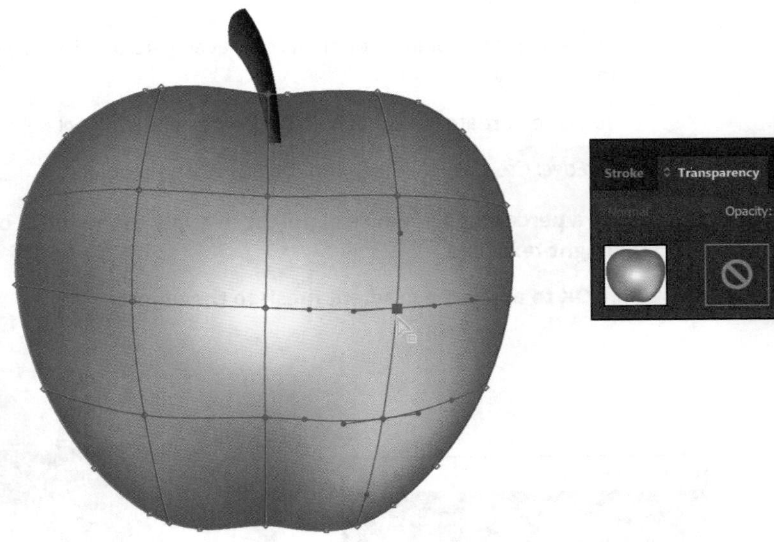

FIGURE 10-3: Using the Direct Select tool, you can define individual transparency values.

TIP

You can change the color in mesh points by choosing the Direct Selection tool and then either clicking a mesh point and picking a fill color or clicking in the center of a mesh area and choosing a fill. Your choice of selecting the mesh point (see the left side of Figure 10-4) or the area between the mesh points (see the right side of Figure 10-4) changes the painting result. To add a mesh point without changing to the current fill color, Shift-click anywhere in a filled vector object.

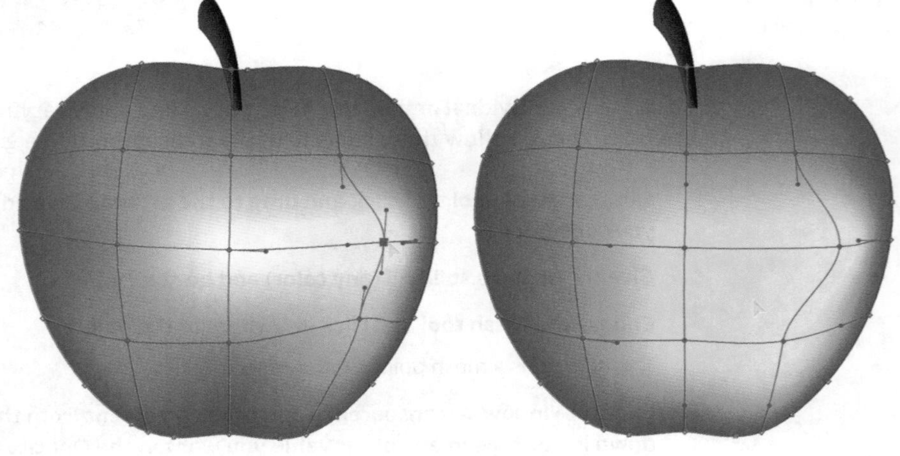

FIGURE 10-4: Using the Direct Selection tool, click and drag points. Also, click a point to change color, or click in-between points to span color further.

The Blend Tool

Use the Blend tool (located in the main Illustrator Tools panel) to transform one object to another to create interesting morphed artwork or to create shaded objects. Using the Blend tool, you can give illustrations a rendered look by blending from one color to another or create an even number of shapes from one point to another. Figure 10-5 shows examples of what you can do with this tool.

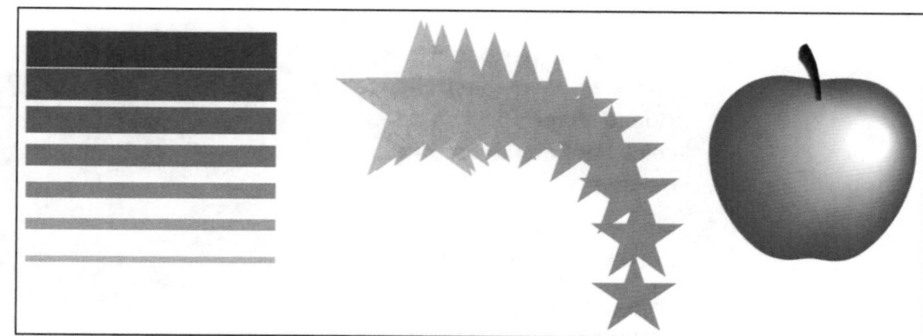

FIGURE 10-5: Some objects using the Blend tool.

Creating a blend

Creating a blend isn't difficult, and as you get used to it you can take the process even further, to create incredibly realistic effects. Follow these steps to create a simple blend from one rectangle size to another, creating an algorithmic stripe pattern (a rectangle of one height blended to a rectangle of another height):

1. **Create a rectangle.**

   Size doesn't matter in this example; just be sure that you can see a difference in shapes when you blend. This example uses a rectangle that measures roughly 4 x 1 inches.

2. **Give your shape a fill and assign None to the stroke.**

   You can use other settings, but we recommend keeping it simple if you're new to working with blends.

3. **Using the Selection tool, click the rectangle and Alt-drag (Windows) or Option-drag (Mac) toward the bottom of the artboard to clone your shape; press the Shift key before you release the mouse button to make sure that the cloned shape stays perfectly aligned with the original shape.**

4. **Reduce the cloned shape to about half its original height by using the Transform panel.**

   If the Transform panel isn't visible, choose Window⇨Transform.

   Alternatively, you can hold down the Shift key and drag the lower-middle bounding box handle, shown in Figure 10-6.

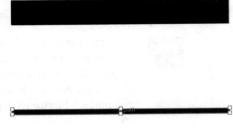

FIGURE 10-6: Reduce the size of the cloned shape.

5. **In the Swatches panel (choose Window⇨Swatches), change the cloned shape's fill to a different color but keep the stroke set to None.**

   Changing the color just helps you see the blend effect a little better.

FIGURE 10-7: Creating a smooth transition between rectangles.

6. **Using the Blend tool, click the original shape and then click the cloned shape.**

   As a default action, the Blend tool creates a smooth blend that transitions from one color to another, as shown in Figure 10-7. To change the blend effect, experiment with the Blend Options dialog box.

Setting Blend options

You can change the way a blend appears by using the Blend Options dialog box: Choose Object⇨Blend⇨Blend Options. From the Spacing drop-down list, change the blend to one of these options:

- **Smooth Color:** Blend steps are calculated to provide the optimum number of steps for a smooth transition.
- **Specified Steps:** Determine the number of steps in a blend by typing a number in the text field to the right of the drop-down list.
- **Specified Distance:** Control the distance between steps in the blend by typing a number in the text field to the right of the drop-down list.

You can also choose between two orientation options:

- **Align to Page:** Orients the blend perpendicular to the x-axis of the page.
- **Align to Path:** Orients the blend perpendicular to the path. You probably won't see a difference when changing orientation unless you've edited the blend path.

TIP

You can easily access the Blend tool options by selecting a blended object and double-clicking the Blend tool in the Tools panel.

If you're feeling adventurous, try changing a smooth blend (such as the one you create in the preceding step list) into a logarithmic blend. In the Blend Options dialog box, select Specified Steps from the Spacing drop-down list and change the value to **5**. This change creates the blend in 5 steps rather than in the more than 200 steps that may have been necessary to create the smooth blend.

Here are a few more tips to help you become more comfortable using blends. You can

- Blend between an unlimited number of objects, colors, opacities, or gradients.
- Edit blends directly with tools, such as Selection, Rotate, or Scale.
- Switch to the Direct Selection tool and edit the blend path by dragging anchor points. A straight path is created between blended objects when the blend is first applied.
- Edit blends that you created by moving, resizing, deleting, or adding objects. After you make editing changes, the artwork is blended again automatically.

The Symbol Sprayer Tool

The super Symbol Sprayer tool is one you must experiment with in order to understand its full potential. In a nutshell, it works like a can of spray paint that sprays, rather than paints, *symbols* — objects that, in Illustrator, can be either vector- or pixel-based. Each symbol is an *instance*.

Exploring the symbol tools

Illustrator comes with a library of symbols ready for use in the Symbols panel. (If the Symbols panel isn't visible, choose Window⇨Symbols.) Use this panel as a storage bin or library to save repeatedly used artwork or to create your own symbols to apply as instances in your artwork, like blades of grass or stars in the sky. You can then use the Symbol tools, described in Table 10-1, to adjust and change the appearance of the symbol instances.

TABLE 10-1 **The Symbol Tools**

Icon	Tool Name	What You Can Do with It
	Symbol Sprayer	Create a set of symbol instances.
	Symbol Shifter	Move symbol instances around; can also change the relative paint order of symbol instances.
	Symbol Scruncher	Pull apart, or put together, symbol instances.
	Symbol Sizer	Increase or decrease the size of symbol instances.
	Symbol Spinner	Orient the symbol instances in a set. Symbol instances located near the cursor spin in the direction you move the cursor.
	Symbol Stainer	Colorize symbol instances.
	Symbol Screener	Increase or decrease the transparency of the symbol instances in a set.
	Symbol Styler	Apply or remove a graphic style from a symbol instance.

TIP

Press the Alt (Windows) or Option (Mac) key to reduce the effect of the Symbol tools. In other words, if you're using the Symbol Sizer tool, you click and hold to make the symbol instances larger; hold down the Alt (Windows) or Option (Mac) key to make the symbol instances smaller.

You can also selectively choose the symbols you want to affect with the Symbol tools by activating them in the Symbols panel. Ctrl-click (Windows) or ⌘-click (Mac) multiple symbols to change them at the same time.

TECHNICAL STUFF

Just about anything can be a symbol, including placed objects and objects with patterns and gradients. If you're going to use placed images as symbols, however, choose File⇨Place and deselect the Link checkbox in the Place dialog box.

Creating and spraying symbols on the artboard

To create a symbol, select the object and drag it to the Symbols panel or click the New Symbol button at the bottom of the Symbols panel. Yes, it's that easy. Then use the Symbol Sprayer tool to apply the symbol instance on the artboard by following these steps:

1. **Select the symbol instance in the Symbols panel.**

   Either create your own symbol or use one of the default symbols supplied in the panel. You can also find additional symbols by selecting the Panel menu in the upper-right of the Symbols panel and selecting Open Symbol Library from the menu that appears. As shown in Figure 10-8, Illustrator has many default libraries that you can choose from; in this example, you'll use the library named Tiki.

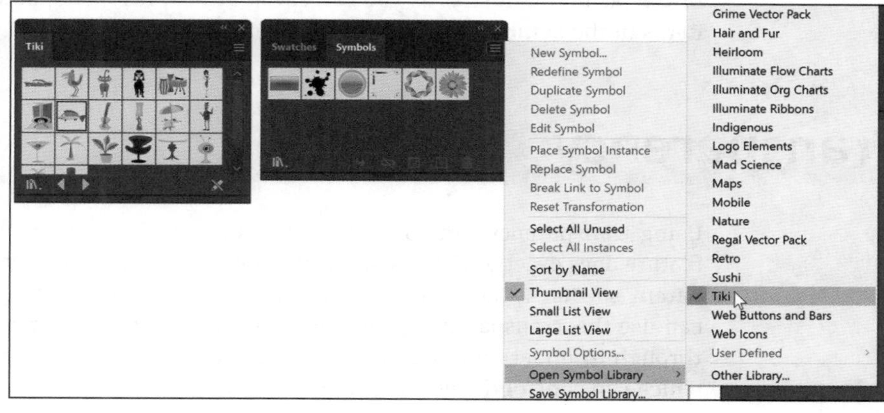

FIGURE 10-8: You can create your own symbol by dragging art into the Symbols panel, or use one of the default libraries.

2. **Click and Drag using the Symbol Sprayer tool, spraying the symbol on the artboard. (See Figure 10-9.)**

   That's it. You can increase or reduce the area affected by the Symbol Sprayer tool by pressing the bracket keys. Press] repeatedly to enlarge the application area for the symbol or press [to make it smaller.

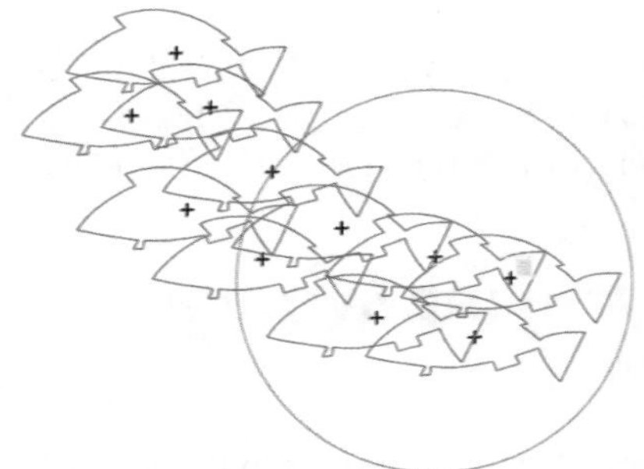

FIGURE 10-9: Using the Symbol Sprayer tool to create fish.

Note that you can access all sorts of symbol libraries from the Symbols panel menu. Find 3D, nature, map, flower, and even hair and fur symbol collections by selecting Open Symbol Library.

TIP

If you want to store artwork that you frequently need to access, simply drag selected objects to the Symbols panel or Alt-click (Windows) or Option-click (Mac) the New Symbol button to name and store the artwork. Retrieve the artwork later by dragging it from the Symbols panel to the artboard. In fact, you can drag any symbol out to your artboard to change or use it in your own artwork. To release the symbol back into its original art elements, choose Object⇨Expand. In the Expand dialog box, click OK to restore the defaults. This is a popular feature in the fashion industry, as you can save buttons, fasteners, and other repeatable items in the Symbols panel.

Transparency

Using transparency can add a new level to your illustrations. The transparency feature does exactly what its name implies — changes an object to make it transparent so that what's underneath that object is visible to varying degrees. You can use the Transparency panel for simple applications of transparency to show through to underlying objects or for more complex artwork using *opacity masks*, which can control the visibility of selected objects.

Choosing Window⇨Transparency opens the Transparency panel, where you can apply different levels of transparency to objects. To do so, create an arrangement of objects that intersect, select the topmost object, and then change the transparency level of the object in the Transparency panel, by either selecting a percentage from the Opacity drop-down list or entering a value of less than 100 in the Opacity text field.

Blend modes

A *blending mode* determines how the resulting transparency will look. To achieve different blending effects, choose different blend modes from the Blend Mode drop-down list in the Transparency panel.

TIP

Truly, the best way to find out what all these modes do is to create two overlapping shapes and start experimenting. Give the shapes differently colored fills (but note that many blending modes don't work with black-and-white fills). Then select the topmost object and change the opacity to anything other than 0% or 100% and change the blending mode by selecting an option from the Blend Mode drop-down list in the Transparency panel. You see all sorts of neat effects and might even pick a few favorites.

We define each blend mode in the following list, but we'll say it again: The best way to see what each mode does is to apply it — so start experimenting.

- **Normal:** Creates no interaction with underlying colors.
- **Darken:** Replaces only the areas that are lighter than the blend color. Areas darker than the blend color don't change.
- **Multiply:** Creates an effect similar to drawing on the page with magic markers, or like the colored film you see on theater lights.
- **Color Burn:** Darkens the base color to reflect the blend color. If you're using white, no change occurs.
- **Lighten:** Replaces only the areas that are darker than the blend color. Areas lighter than the blend color don't change.
- **Screen:** Multiplies the inverse of the underlying colors. The resulting color is always a lighter color.
- **Color Dodge:** Brightens the underlying color to reflect the blend color. If you're using black, there's no change.
- **Overlay:** Multiplies or screens the colors, depending on the base color.
- **Soft Light:** Darkens or lightens colors, depending on the blend color. The effect is similar to shining a diffused spotlight on the artwork.
- **Hard Light:** Multiplies or screens colors, depending on the blend color. The effect is similar to shining a harsh spotlight on the artwork.
- **Difference:** Subtracts either the blend color from the base color or the base color from the blend color, depending on which has the greater brightness value. The effect is similar to a color negative.

- **Exclusion:** Creates an effect similar to, but with less contrast than, Difference mode.
- **Hue:** Applies the *hue* (color) of the blend object to underlying objects but keeps the underlying shading, or luminosity.
- **Saturation:** Applies the saturation of the blend color but uses the luminance and hue of the base color.
- **Color:** Applies the blend object's color to the underlying objects but preserves the gray levels in the artwork; works well for tinting objects or changing their colors.
- **Luminosity:** Creates a resulting color with the hue and saturation of the base color and the luminance of the blend color. This mode is essentially the opposite of Color mode.

Opacity masks

Just like in Photoshop, you can use masks to make more interesting artwork in Illustrator. You can create an opacity mask from the topmost object in a selection or by drawing a mask on a single object. The mask uses the grayscale of the selected object as its opacity mask. Black areas are transparent; shades of gray are semitransparent, depending on the amount of gray; white areas are opaque. Figure 10-10 shows the effect of using an opacity mask.

FIGURE 10-10: An opacity mask converts the topmost object into a mask that then masks out the underlying objects.

To create an opacity mask, follow these steps:

1. **Open the Transparency panel menu.**
2. **Make sure that the Blend Mode drop-down list is set to Normal.**

3. **Create a shape anywhere on the artboard or open a document that has artwork on it.**

   We're using a circle, but the shape doesn't matter. Make sure that the artwork has a fill. A solid color helps you see the effect.

4. **Open the Symbols panel (choose Window➪Symbols Panel) and drag a symbol to the artboard.**

   In this example, we're using the drums symbol from the Tiki Symbol library.

5. **Using the Selection tool, enlarge your symbol so that it fills the shape.**

   See the image on the left in Figure 10-11.

6. **Select both the symbol and the shape and then click the Make Mask button.**

   See the image on the right in Figure 10-11.

   The symbol turns into a mask, showing varying levels of the underlying box of the newly created mask, depending on the original color value.

7. Click the Clip and Invert Mask checkboxes to see how your artwork is affected. To delete an opacity mask, click the Release button.

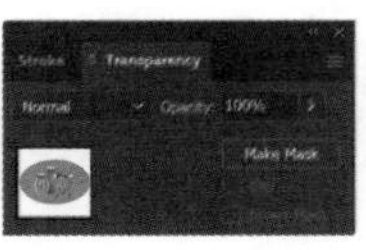

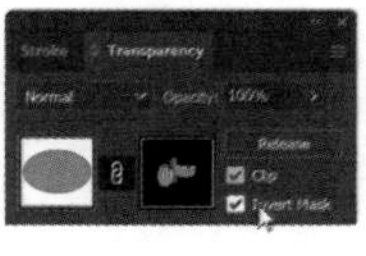

FIGURE 10-11: Creating an opacity mask.

Click the right thumbnail (which represents the mask) in the Transparency panel, and a black border appears around it, indicating that it's active. You can move the items on the mask or even create items to be added to the mask. The mask works just like the regular artboard, except that anything done on the mask side is used only as an opacity mask. To work on the regular artboard, click the left thumbnail.

» Applying effects

» Getting to know the Appearance panel

» Discovering graphic styles

» Making artwork 3D

» Playing with additional fills and strokes

» Creating the illusion of space and distance with the perspective grid

Chapter 11
Using Filters and Effects

Effects give you the opportunity to make jazzy changes, such as drop shadows and zigzagged artwork, to your Illustrator objects. You can even use Photoshop filters directly in Illustrator. In this chapter, you find out how to apply, save, and edit effects; you also take a quick tour of the Appearance panel (your trusty sidekick when performing these tasks). Keep in mind: Just because you can do something doesn't mean you always should! Don't add filters and effects if you believe they will add messiness and distractions to your artwork.

Working with Effects

Adobe Illustrator offers many artistic and creative effects that you can use to add blurs, drop shadows, inner glows, and more. Effects are connected dynamically to the objects they are applied to. You can add, change, and remove effects at any time from the Appearance panel. (Choose Window➪Appearance.)

Understanding the Appearance panel

Using the Appearance panel you can apply multiple effects to one object and even copy them to other objects. Make sure to create a simple object, such as a shape. If the Appearance panel isn't visible, choose Window➪Appearance to open it, as shown in Figure 11-1, alongside an object with several effects applied to it.

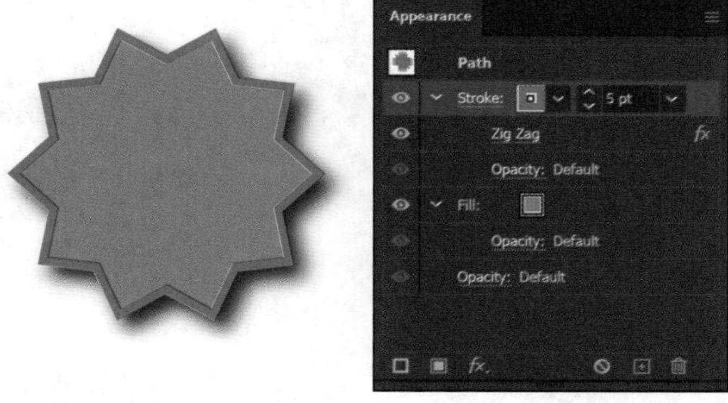

FIGURE 11-1: The Appearance panel not only shows you the properties of an object but also allows you to change them.

If you have no effects applied, you see as a default only a fill and a stroke listed in the Appearance panel. As you create effects, they're added to this list. You can add more strokes and fills to the list; you find out why that capability is so useful in the following sections.

Applying an effect

In this section, you see how to apply an effect. You can choose from many effects, and they're all applied in much the same manner. In this example, you create an interesting effect for a border.

Create a new document using any of the default profiles, and then follow these steps to apply an effect:

1. **Make sure that your Appearance panel is visible. Choose Window➪ Appearance and click and drag the panel out of the dock area so that it doesn't collapse while you're working.**
2. **Select the Rectangle tool and click and drag a rectangle out on the artboard.**

   Any size rectangle is fine for this example.

3. **Click Stroke in the Appearance panel, and change the stroke to** 3 pt.
4. **Using Stroke Color in the Appearance panel, assign any color except the default color of black.**
5. **Notice in the Appearance panel, shown in Figure 11-2, that you can select the entire object or just the Stroke or Fill. By selecting one or the other, you can apply an effect to just the stroke or the fill. In this example, select Stroke.**
6. **Choose Effect⇨Distort and Transform⇨Zig Zag.**

   Choose the settings that work well to make your straight border appear as a zigzag. In Figure 11-3, we selected Smooth to round out the points of the zigzag effect and changed the Ridges per segment to 15.

Keep in mind that you can also use the Properties panel to change the appearance of an object as well. The difference is that once you add multiple effects, you can only access them through the Appearance panel.

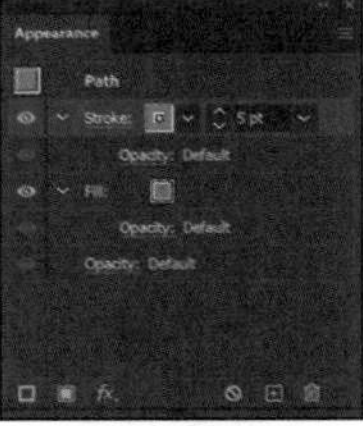

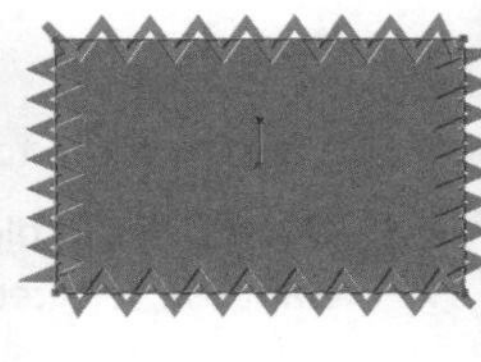

FIGURE 11-2: You can change the stroke and fill attributes right in the Appearance panel. Select only the Stroke, in the Appearance panel and apply a Zig Zag effect.

REMEMBER

Effects are linked dynamically to the object they're applied to. They can be scaled, modified, and even deleted with no harm done to the original object.

To delete an effect, simply select it in the Appearance panel and then click the Trashcan icon at the bottom of the Appearance panel.

Adding a Drop Shadow effect

Creating a drop shadow is a quick and easy way to add dimension and a bit of sophistication to your artwork. The interaction between the object with the drop shadow and the underlying objects can create an interesting look. To add the Drop Shadow effect to an illustration, follow these steps:

1. **Select the object (or objects) to apply the drop shadow to.**

   If you like, you can continue using the rectangle you worked with in the preceding section.

2. **In the Appearance panel, make sure that Stroke is selected.**

3. **Choose Effect⇨Stylize⇨Drop Shadow.**

4. **In the Drop Shadow dialog box that appears, as shown in Figure 11-3, select the Preview checkbox in the lower-left corner.**

   You now see the drop shadow is applied to only the stroke.

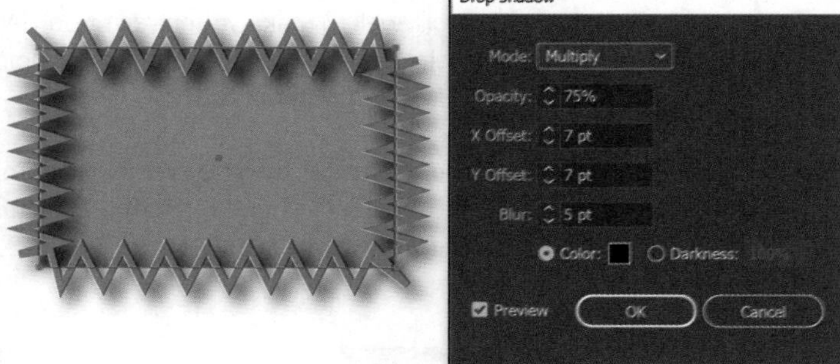

FIGURE 11-3: The Drop Shadow dialog box offers the effect's options and a preview.

5. **Choose from the following options in the dialog box:**

   - *Mode:* Select a blending mode from this drop-down list to choose how you want the selected object to interact with the objects underneath it. The default is Multiply, which works well — the effect is similar to coloring with a magic marker.
   - *Opacity:* Enter a value or use the drop-down list to determine how opaque or transparent the drop shadow should be. If it's too strong, choose a lower amount.
   - *Offset:* Enter a value to determine how close the shadow is to the object. If you're working with text or small artwork, smaller values (and shorter shadows) look best. Otherwise, the drop shadow may look like one big, indefinable glob.

     The X Offset shifts the shadow from left to right, and the Y Offset shifts it up or down. You can enter negative or positive numbers.
   - *Blur:* Use Blur to control how fuzzy the edges of the shadow are. A lower value makes the edge of the shadow more defined.
   - *Color and Darkness:* Select the Color radio button to choose a custom color for the drop shadow. Select the Darkness radio button to add more black to the drop shadow. Zero percent is the lowest amount of black, and 100 percent is the highest.

As a default, the color of the shadow is based on the color of your object, sort of — the Darkness option also has a play in this task. As a default, the shadow is made up of the color in the object if it's solid. Multicolored objects have a gray shadow.

6. **When you're finished, click OK to close the dialog box.**

   Take a look at the Appearance panel, as shown in Figure 11-4, to see that the two Effects that you applied are displayed. Notice that you can turn off and on the visibility of the effect by selecting the Click to Toggle Visibility icon to the left of the effect, and even edit the effect by double-clicking on the effect in the Appearance panel.

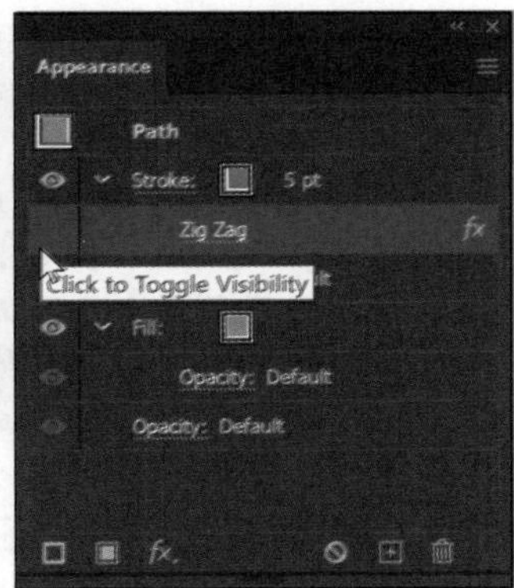

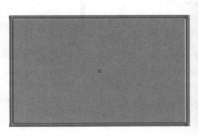

FIGURE 11-4: You can toggle the visibility and edit the effects right from the Appearance panel.

Saving your combination of effects as a graphic style

Graphic styles are an incredible feature that allows you to reuse properties repeatedly on different objects. So, if you always want a .75 dashed line, create it once and then save it as a graphic style. Decide later that you need to change the stroke size to 1 point? No problem. Change it once and it is reflected in all the instances. Read the next section for more details about creating, saving, and editing graphic styles.

1. **Choose Window⇨Graphic Styles, and while you have your rectangle selected, choose New Graphic Style from the Panel menu.**
2. **Name the Style and press OK.**
3. **Now make another rectangle, any size, and click the thumbnail of graphic style that you just saved in the Graphic Styles panel.**

   All the saved attributes are applied to this new shape.

Saving Graphic Styles

A *graphic style* is a combination of all settings you choose for a particular filter or effect in the Appearance panel. By saving this information in a graphic style, you store these attributes so that you can quickly and easily apply them to other objects later. If you have already created an object with effects in a previous section, you will have the opportunity to save those effects as a graphic style later in this section.

To apply a saved a more complex graphic style, follow these steps:

1. **If the Graphic Styles panel is not visible, choose Window⇨Graphic Styles.**

   The Graphic Styles panel appears. Unless you have added your own style, only some default styles appear in this panel.

2. **Click the Library icon in the lower left of the Graphic Styles panel and select the Graphic Style library named Vonster Pattern Styles, as shown in Figure 11-5.**

   A second window appears. It's named Vonster Pattern Styles.

3. **Create a new shape, such as a simple rectangle or an ellipse, and click any graphic style in the Vonster Pattern Styles panel to apply it to an active object.**

   Look at the Appearance panel while you click different styles to see that you're applying combinations of attributes, including effects, fills, and strokes. (See Figure 11-6.)

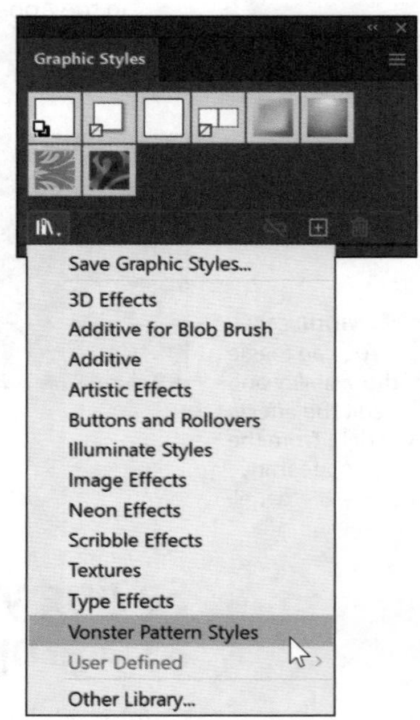

FIGURE 11-5: Access additional graphic styles by selecting the Library icon.

You can store attributes as a graphic style in several ways; we show you two easy methods. If you already applied a combination of attributes to an object, such as the rectangle border that you may have created in the previous section, you can store them by completing one of these tasks:

» With the object selected, Alt-click (Windows) or Option-click (Mac) the New Graphic Style button at the bottom of the Graphic Styles panel. Alt-click (Windows) or Option-click (Mac) allows you to name the style when it's added.

» Drag the selected object directly into the Graphic Styles panel. The panel stores its attributes, but you have to double-click the new style to name it.

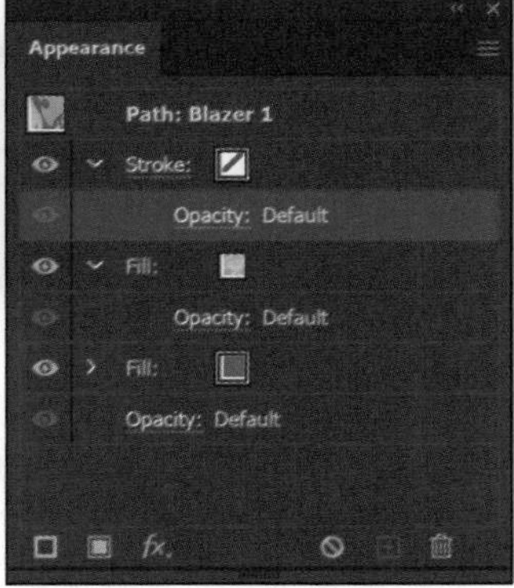

FIGURE 11-6: The Graphic Styles panel stores effects and other attributes.

After you store a graphic style, simply select the object you want to apply the style to and then click the saved style in the Graphic Styles panel.

You can try this by creating another rectangle and then clicking your saved graphic style in the Graphic Styles panel.

Creating 3D Artwork

All Illustrator effects are excellent, but the 3D feature is even better because you can not only add dimension by using the 3D effect, but you can also *map artwork* (wrap artwork around a 3D object) and apply lighting to the 3D object. You can then design a label for a jelly jar, for example, and adhere it to the jar to show a client.

Here are the three choices for the 3D effect:

- **Extrude & Bevel:** Uses the *z*-axis to extrude an object. For example, a square becomes a cube.
- **Revolve:** Uses the *z*-axis and revolves a shape around it. You can use this option to change an arc into a ball.
- **Rotate:** Rotates a 3D object created using the Extrude & Bevel or Revolve effect or rotates a 2D object in 3D space. You can also adjust the perspective of a 3D or 2D object.

To apply a 3D effect, you need to create an object appropriate for the 3D effect. The Extrude & Bevel feature works well with shapes and text. If you want to edit an object that already has a 3D effect applied to it, double-click the 3D effect in the Appearance panel.

To apply a 3D effect, follow these steps:

1. **Select the object you want to apply the 3D effect to. In this example, select a simple square.**
2. **Choose Effect⇨3D⇨Extrude & Bevel.**

   Options for your chosen 3D effect appear. To see the 3D Extrude & Bevel Options dialog box, see Figure 11-7.
3. **Select the Preview checkbox so that you can see results as you experiment with these settings.**

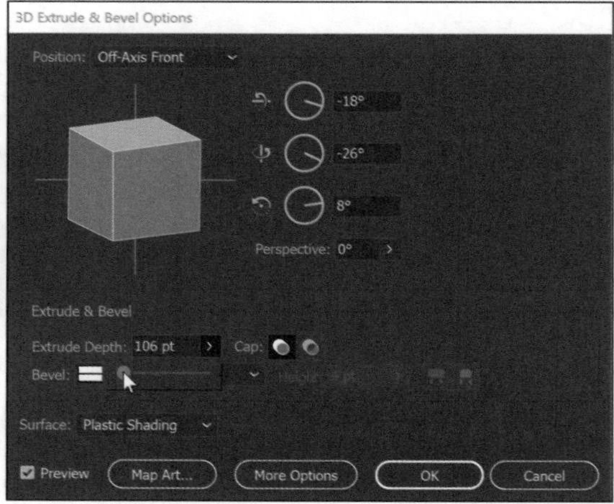

FIGURE 11-7: The Extrude & Bevel Options dialog box.

4. **Click the Preview pane (which shows a cube in Figure 11-8) and drag to rotate the object in space.**

   It makes selecting the proper angle (or *positioning the object in space)* fun to do, or you can choose the angle from the Position drop-down list above the preview.

WARNING

   Never use the Rotate tool to rotate a 3D object, unless you want some funky results; use the Preview pane in the 3D Extrude & Bevel Options dialog box instead.

5. **(Optional) Use the Perspective drop-down list to add perspective to your object.**
6. **In the Extrude & Bevel section of the dialog box, select a depth for your object and a cap.**

   The cap determines whether your shape has a solid cap on it or is hollow, as shown in Figure 11-8.

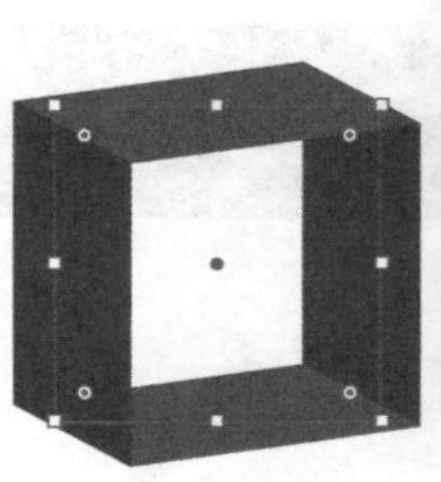

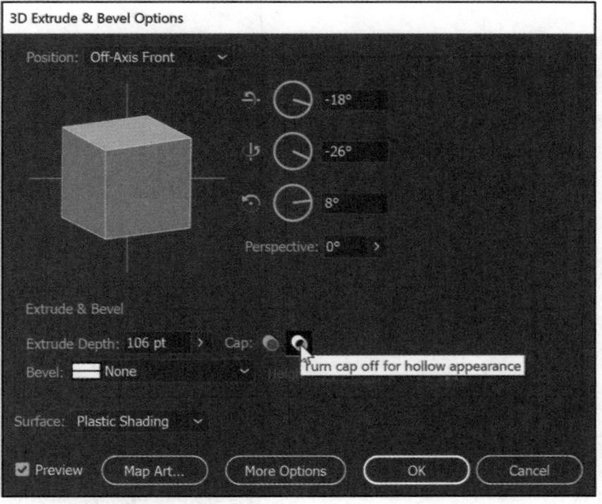

FIGURE 11-8: Turn the cap on or off to make an object appear solid or hollow.

7. **Select a bevel (edge shape) from the Bevel drop-down list and set the height using the Height drop-down list.**

   You have two choices to apply the bevel:

   - *Bevel Extent Out:* The bevel is added to the object.
   - *Bevel Extent In:* The bevel is subtracted from the object.

8. **Select a rendering style from the Surface drop-down list or click the More Options button for in-depth lighting options, such as changing the direction or adding lighting.**
9. **Click the Map Art button.**

   The Map Art dialog box opens, as shown in Figure 11-9. Use this dialog box to apply artwork to a 3D object.

10. **Using the Surface arrow buttons, select the surface you want the artwork applied to and then select a symbol from the Symbol drop-down list.**

    The result is shown on the left in Figure 11-9.

11. **Click OK to close the dialog box.**

REMEMBER

Keep these points in mind when mapping artwork:

- An object must be a symbol to be used as mapped artwork. You simply need to select and drag to the Symbols panel the artwork you want mapped. By doing so, you make it a selectable item in the Map Art dialog box.
- The light gray areas in the Preview pane are the visible areas based on the object's present position. Drag and scale the artwork in this pane to place the artwork where you want it.

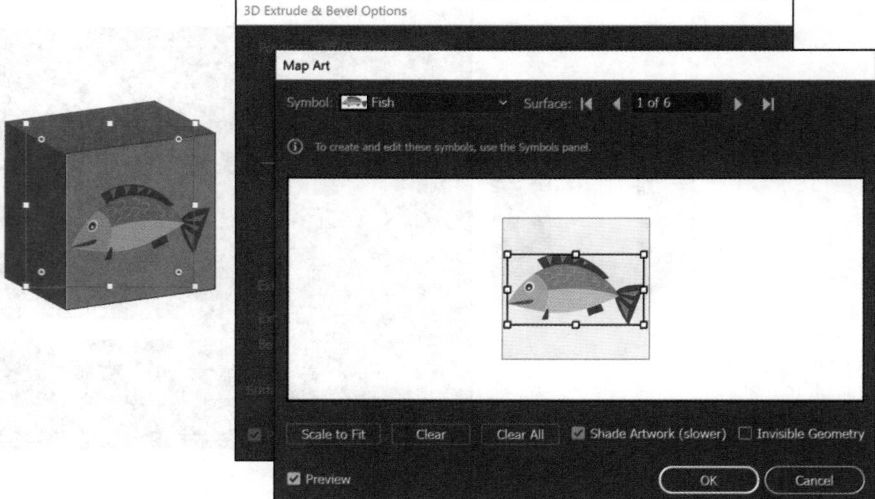

FIGURE 11-9: In the Map Art dialog box, you can select a surface and apply a symbol to it.

- Shaded artwork (enabled by selecting the Shaded Artwork checkbox at the bottom of the Map Art dialog box) looks good but can take a long time to render.

Note: All 3D effects are rendered at 72 dpi (dots per inch; low resolution) so as not to slow down the processing speed. You can determine the resolution by either choosing Effect ➪ Document Raster Effects Settings or saving or exporting the file. You can also select the object and choose Object ➪ Rasterize. After you rasterize the object, you can no longer use it as an Illustrator 3D object, so save the original!

Adding Multiple Fills and Strokes

Using the panel menu in the Appearance panel, you can add more fills and strokes. You can use this feature to put differently colored fills on top of each other and individually apply effects to each one, creating truly interesting and creative results.

Just for fun, follow along to see what you can do to a single object from the Appearance panel:

1. **Create a star shape.**

   Neither the size of the shape nor its number of points matters — just make the shape large enough that it is not too tiny.

2. **Use the Properties panel to fill the shape with yellow and give it a black stroke; make sure the stroke is set to 1 point.**
3. **Choose Window⇨Appearance panel.**

   Notice that the present fill and stroke are listed in the Appearance panel. Even in its simplest form, the Appearance panel helps track basic attributes. You can easily take advantage of the tracking to apply effects to just a fill or a stroke.
4. **Select only Stroke in the Appearance panel.**

   This is so that the next effect applied only applies to the stroke, not the fill.
5. **Choose Effect⇨Path⇨Offset Path.**
6. **In the Offset Path dialog box that appears, enter** –10 pt **in the Offset text box and select the Preview checkbox, as you see in Figure 11-10.**

   Notice that the stroke moves into the fill rather than on the edge.
7. **Change the offset to a number that works with your star shape and click OK.**

   Depending on the size of your star, you may want to adjust the amount of offset up or down.

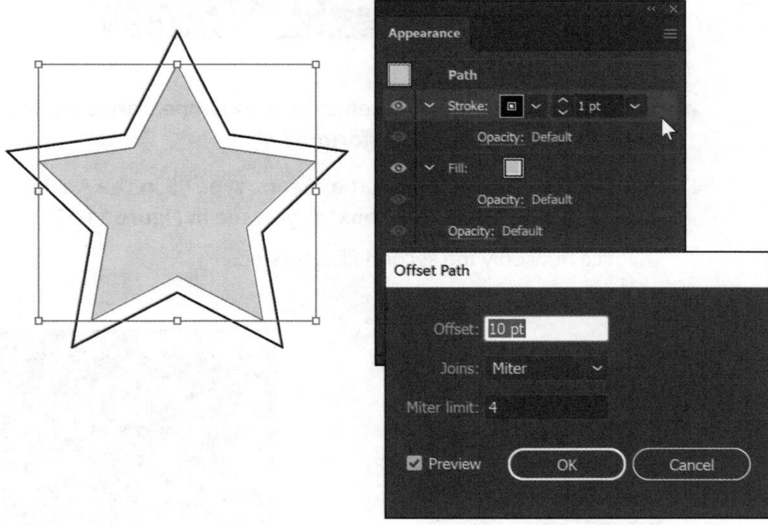

FIGURE 11-10: Select to offset just the stroke of an object.

You can add multiple fills to an object. You would do this if you want to apply a pattern, gradient, or any other fill, and use transparency of blending options in order to see each of the fills in a unique way. This can be lots of fun when trying to create interesting textures.

1. **From the panel menu of the Appearance panel, select Add New Fill to add a new fill to the star shape. Click the arrow to the right of the new fill color swatch to open the Swatches and change the color of the new fill, as shown in Figure 11-11.**

   You will not see this other fill yet, as it is perfectly positioned under the original fill.

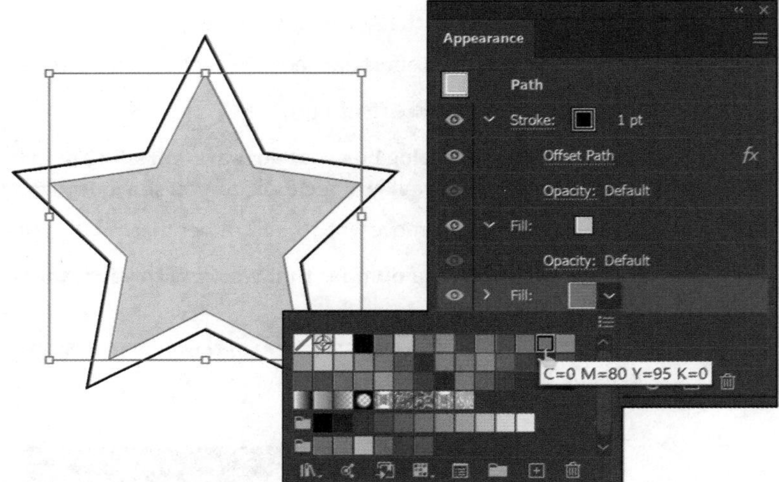

FIGURE 11-11: Change the new fill to a different color right in the Appearance panel.

2. **Make sure the top Fill is selected in the Appearance panel and choose Effect⇨Distort and Transform⇨Twist.**
3. **In the Twist dialog box that appears, type** 45 **in the Angle text field and select the Preview checkbox, as you see in Figure 11-12.**

   Notice how only the second fill is twisted.

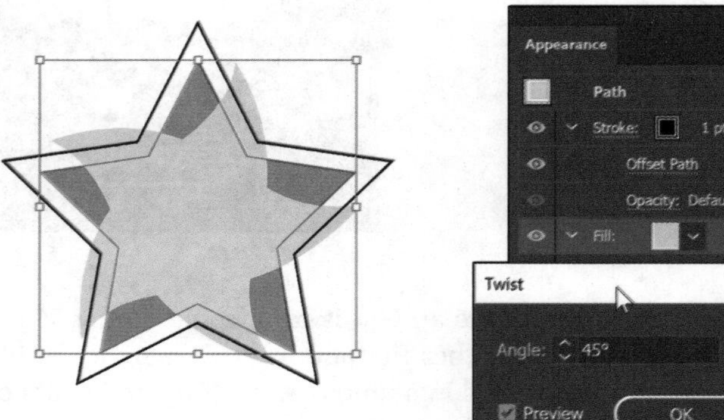

FIGURE 11-12: Choose to twist one of the fills.

4. **Click OK to close the Twist dialog box.**

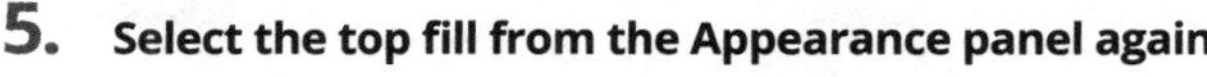

5. **Select the top fill from the Appearance panel again.**

REMEMBER

Always be sure to select the fill or stroke you want before doing anything meant to change just that specific fill or stroke.

6. **In the Transparency panel (choose Window⇨Transparency), select 50% from the Opacity drop-down list or simply type** 50% **in the Opacity text field.**

Now you can see your original shape through the new fill.

7. **With the top fill still selected, change the color or choose a pattern in the Swatches panel for a truly different appearance.**

Continue playing with combinations of fills and strokes for hours if you want. We hope that this process "clicks" and that you can take it further on your own.

Using the Perspective Grid

You can create and edit artwork based on the perspective grid feature, shown in Figure 11-13. The grid is a huge help in creating successful perspective illustrations. To access the Perspective Grid you can click on the Perspective Grid tool, or press Shift+P. Keep in mind that if you do not see the Perspective Grid tool, you can select the Edit toolbar button at the bottom of the Tools panel, then locate the Perspective Grid tool in the list of available tools and drag it to your existing Tools panel.

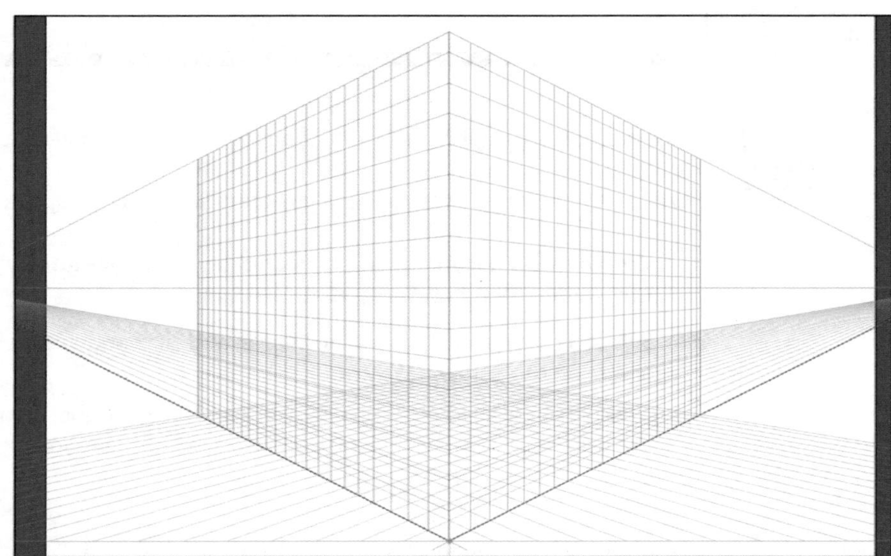

FIGURE 11-13: Build and edit illustrations by using the Perspective Grid tool.

To quickly show or hide the default perspective grid, press Ctrl+Shift+I (Windows) or ⌘+Shift+I (Mac).

To select the perspective plane you wish to draw on, click the Plane Widget that appears in the upper left, as shown in Figure 11-14. To draw with no perspective, click the gray area surrounding the grid, but yet still inside the circle.

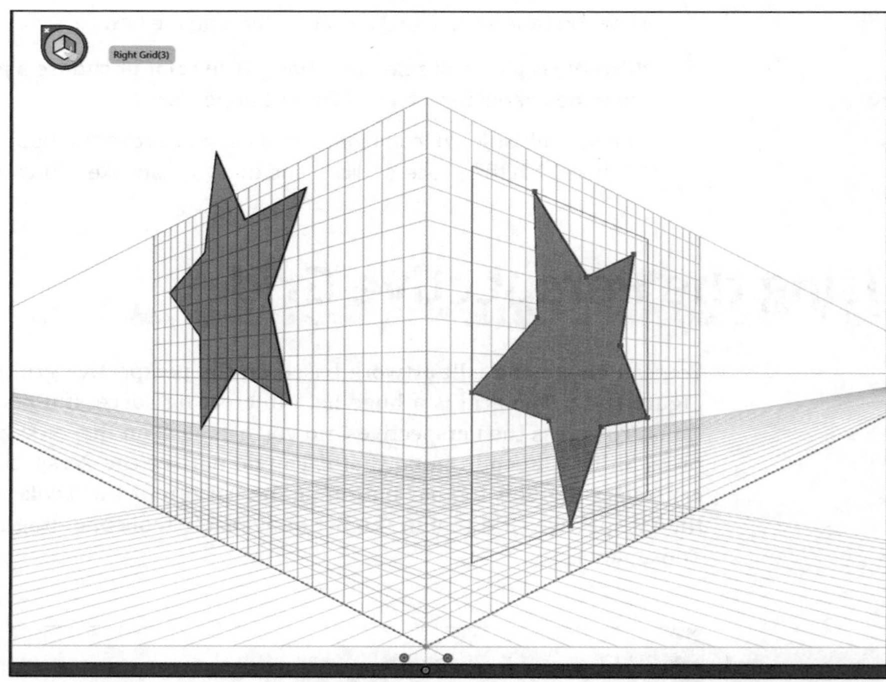

FIGURE 11-14: Click a plane to activate it before adding a shape. In this example, the left star had the left grid active, and the second star had the right grid active.

You can use the Perspective Grid tool on the toolbar to fine-tune the grid.

Here are some simple steps to help you start using the perspective grid:

1. **Create a new document and turn on the perspective grid by pressing Ctrl+Shift+I (Windows) or ⌘ +Shift+I (Mac).**
2. **Select a shape tool, such as the Ellipse, and click and drag it to create the shape on the perspective grid.**

   Notice that the shape's perspective is controlled by the grid. Remember that you can use the circle in the upper left to change which plane you are drawing to.

3. **Select the Perspective Selection tool. If you don't see it, you can press Shift-V to change to that tool.**

4. **Using the Perspective Selection tool, click and drag the shape to see that it's adjusted in position and location using the perspective grid.**

   Figure 11-15 shows the result of dragging an ellipse with the Perspective Selection tool. You can resize and clone in perspective by Alt- or Option-dragging the selected object.

   Note that you can also select and drag the grid itself by using the Perspective Selection tool.

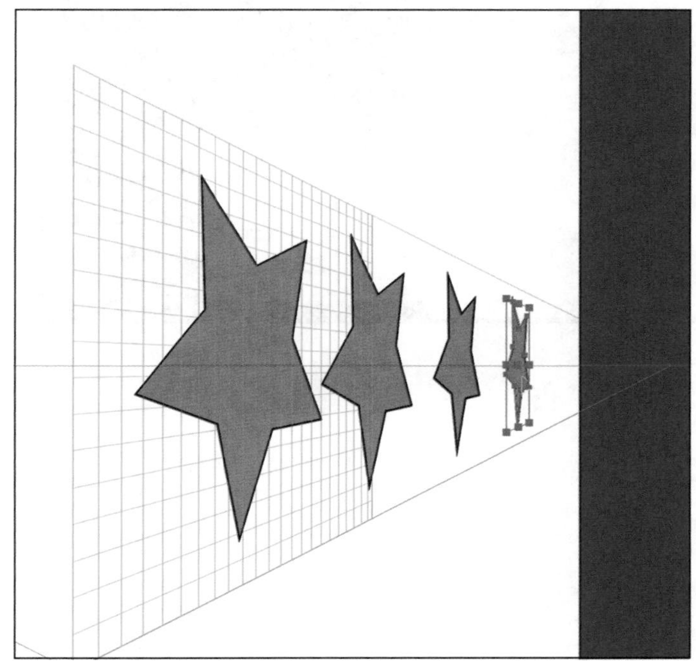

FIGURE 11-15: Click and drag to change the perspective using the Perspective Selection tool. In this example Alt/Option was held down to clone the star in perspective.

5. **Choose View⇨Perspective Grid, as shown in Figure 11-16, and then select the type of perspective to apply.**

   In this step, you further customize your grid and choose other options to make your illustrations more precise.

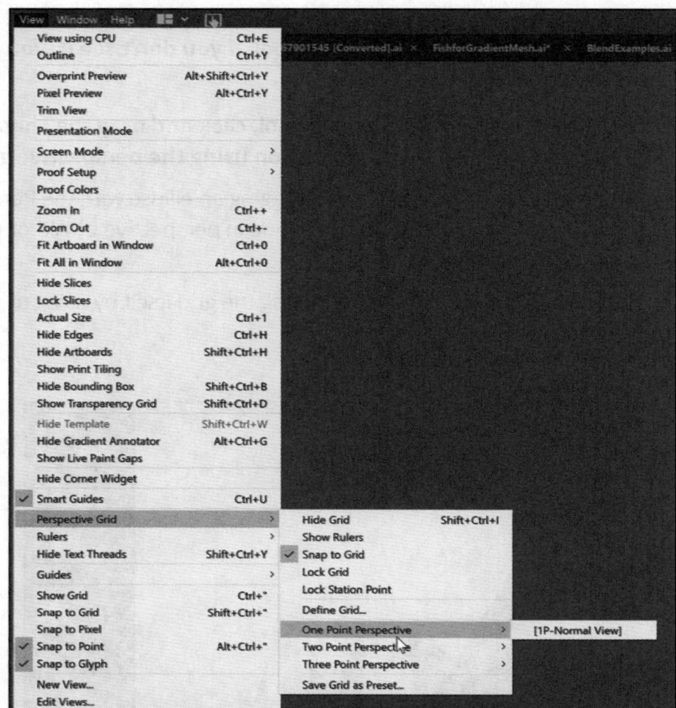

FIGURE 11-16: Customize settings for the perspective grid by using the View menu.

» Exporting files for use in other applications

» Preparing art for the web

» Flattening your transparency

» Printing in Illustrator

Chapter 12
Using Your Illustrator Images

So you have completed a beautiful artwork, but you don't know what to do with it. Should you have a party and invite all interested clients to stand around your monitor and admire it? Should you share or sell your artwork by printing it or posting it on the Internet?

In this chapter, you discover how to use your illustrations in a variety of workflows, from using Illustrator files in page layout programs to exporting files for Photoshop (and other programs) and the web. This chapter helps you use your artwork effectively and to understand the saving and flattening choices available in Adobe Illustrator.

Saving and Exporting Illustrator Files

In this section, you discover how general choices differ in the Save As dialog box (choose File⇨Save As) and describe their benefits.

When you choose File⇨Save or File⇨Save As, you save your file in one of these formats: Adobe Illustrator, EPS, PDF, AIT (Illustrator Template), SVG, or SVG Compressed. You discover information about these formats throughout this chapter.

TIP

If you need a file format not listed in the regular Save As dialog box, choose File➪Export As to see additional choices. Using the File Export command, you can choose to save your files in any format listed in Table 12-1.

TABLE 12-1 **Available File Formats via Export**

File Format	Extension
AutoCAD Drawing	.dwg
AutoCAD Interchange File	.dxf
BMP	.bmp
CSS	.css
Enhanced Metafile	.emf
Flash	.swf
JPEG	.jpg
Macintosh PICT	.pct
Photoshop	.psd
PNG	.png
Targa	.tga
Text Format	.txt
TIFF	.tif
Windows Metafile	.wmf

WARNING

Many formats *rasterize* artwork, so they no longer maintain vector paths and the benefits of being vector. Scalability isn't unlimited, for example. If you think that you may want to edit your image again later, be sure to save a copy of the file and keep the original in the `.ai` format.

The native Adobe Illustrator file format

If you're working with the programs in the Creative Cloud, the best way to save a file is as a native Illustrator `.ai` file. For instance, the `.ai` format works with Adobe applications such as Adobe InDesign for page layout, Adobe Dreamweaver for web page creation, Adobe Photoshop for photo retouching, and Adobe Acrobat for cross-platform documents.

Understanding when it's best to use the .ai format is important. Saving your illustration as an .ai file ensures that it's editable; it also ensures that any transparency is retained, even if you use the file in another application.

To save and use a file in native Illustrator format, follow these steps:

1. **Make an illustration that has some transparency applied to one of the objects.**

   You can also use the file named `BabikBagels.ai` located in your Book05_Illustrator folder that you downloaded in Book 1, Chapter 1.

2. **Choose File ➪ Save As.**

   Select Adobe Illustrator Document (`.ai`) from the Save As Type drop-down list, name the file, and click Save.

3. **Leave the Illustrator Native options at their defaults and click OK.**

TIP

After you follow the preceding step list to prepare your Illustrator file, you can use the illustration in other Adobe applications:

- **Adobe Acrobat:** Open the Acrobat application and choose File ➪ Open. Locate the `.ai` file. The native Illustrator file opens within the Acrobat application.
- **Adobe InDesign:** Choose File ➪ Place. This method supports transparency created in Adobe Illustrator, as shown in Figure 12-1. (However, copying and pasting from Illustrator to InDesign does *not* support transparency.)

FIGURE 12-1: InDesign supports the transparency created in your Illustrator file, even over text.

- **Adobe Photoshop:** Choose File ➪ Place. By placing an Illustrator file into Adobe Photoshop, you automatically create a Photoshop smart object. You can scale, rotate, and even apply effects to the Illustrator file and return to the original illustration at any time. Read more about smart objects in Photoshop in Book 3, Chapter 9.

If you want to take an Illustrator file into Photoshop for further editing, choose File ➪ Export and select the Photoshop (.psd) format from the Save As Type drop-down list. Choose a resolution from the Options window. If you used layers, leave the Write Layers option selected.

In Photoshop, choose File ➪ Open, select the file you just saved in Illustrator in .psd format, and click Open. The file opens in Photoshop with its layers intact.

» **Adobe XD:** Use the integration features built in to Adobe Illustrator to cut and paste directly into Adobe XD. If you choose Edit ➪ Copy from Adobe Illustrator, you can then switch to Adobe XD and choose Edit ➪ Paste. The Paste dialog box appears.

Saving Illustrator files back to previous versions

When saving an .ai or .eps file, you can choose File ➪ Save As, choose an Illustrator format, and then click OK.

When the Illustrator Options dialog box appears, choose a version from the Version drop-down list. Keep in mind that any features specific to newer versions of Illustrator aren't supported in older file formats, so make sure that you save a copy and keep the original file intact. Adobe helps you understand the risk of saving to older versions by putting a warning sign next to the Version drop-down list and showing you specific issues with the version you selected in the Warnings window.

The PDF file format

If you want to save your file in a format that supports more than a dozen platforms and requires only Acrobat Reader, available as a free download at www.adobe.com, choose to save your file as a PDF (Portable Document Format) file.

If you can open an Illustrator file in Acrobat, why would you need to save a file as a PDF? For one thing, you can compress a PDF to a smaller size; also, the receiver can double-click the file and then Acrobat or Acrobat Reader launches automatically.

Depending on how you save the PDF, you can allow some level of editability in Adobe Illustrator. To save a file as a PDF, follow these steps:

1. **Choose File ➪ Save As, select Illustrator PDF (.pdf) from the Save As Type drop-down list, and then click Save.**

2. **In the Adobe PDF Options dialog box that appears, choose one of these options from the Preset drop-down list:**

   - *Illustrator Default:* Creates a PDF file in which all Illustrator data is preserved. PDF files created using this preset can be reopened in Illustrator with no loss of data.
   - *High Quality Print:* Creates PDF files for desktop printers and proofers.
   - *PDF/X-1a:2001:* The least flexible but still quite powerful delivery method for PDF content data; requires that the color of all objects be CMYK (cyan, magenta, yellow, black) or spot colors. Elements in RGB (red, green, blue) or Lab color spaces or tagged with International Color Consortium (ICC) profiles are prohibited. All fonts used in the job must be embedded in the supplied PDF file.
   - *PDF/X-3:2002:* Has slightly more flexibility than the X-1a:2001 method (see preceding bullet) in that color managed workflows are supported elements in Lab and attached ICC source profiles may also be used.
   - *PDF/X-4:2008:* Based on PDF 1.4, which includes support for live transparency and has the same color management and ICC color specifications as PDF/X-3. You can open PDF files created for PDF/X-4 compliance in Acrobat 7.0 and Reader 7.0 and later.
   - *Press Quality:* Creates a PDF file that can be printed to a high-resolution output device. The file will be large but maintain all information that a commercial printer or service provider needs in order to print files correctly. This option automatically converts the color mode to CMYK, embeds all fonts used in the file, prints at a higher resolution, and uses other settings to preserve the maximum amount of information contained in the original document.

TIP

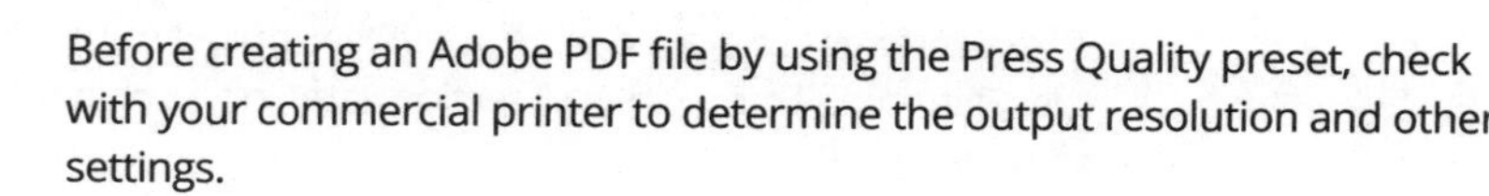

Before creating an Adobe PDF file by using the Press Quality preset, check with your commercial printer to determine the output resolution and other settings.

3. **Click Save PDF to save your file in PDF format.**

REMEMBER

If you want to be able to reopen the PDF file and edit it in Illustrator, make sure that you leave the Preserve Illustrator Editing Capabilities checkbox selected in the Adobe PDF Options dialog box.

In the Adobe PDF Options dialog box, to the left of the preset choices, are options you can change to customize your settings. Skim the options to see how to change resolution settings and even add printer's marks. Take a look at Book 6 to find out more about additional Acrobat PDF options.

TIP

Want a press-quality PDF but don't want to convert all your colors to CMYK? Choose the Press setting and then click the Output options. In the Color Conversion drop-down list, select No Conversion.

The EPS file format

Encapsulated PostScript File (EPS) is the file format that most text editing and page layout applications accept; EPS supports vector data and is completely scalable. Because the Illustrator .eps format is based on PostScript, you can reopen an EPS file and edit it in Illustrator at any time.

To save a file in Illustrator as an EPS, follow these steps:

1. **Choose File ⇨ Save As and select EPS from the Save As Type drop-down list.**
2. **From the Version drop-down list, select the Illustrator version you're saving to.**
3. **In the EPS Options dialog box that appears (shown in Figure 12-2), select the preview from the Format drop-down list:**
   - *8-Bit Color:* A color preview for either Mac or PC
   - *Black & White:* A low-resolution, black-and-white preview
4. **Select either the Transparent or Opaque radio button, depending on whether you want the non-image areas in your artwork to be transparent or opaque.**
5. **Specify your transparency settings.**

   These settings are grayed out if you haven't used transparency in the file. (See the "Flattening Transparency" section, later in this chapter, for more about this setting.)
6. **Leave the Embed Fonts (for other applications) checkbox selected to leave fonts you used embedded in the EPS file format.**
7. **In the Options section, leave the Include CMYK PostScript in RGB Files checkbox selected.**

   If you don't know which Adobe PostScript level you want to save to, leave it at the default.
8. **Click OK to save your file in EPS format.**

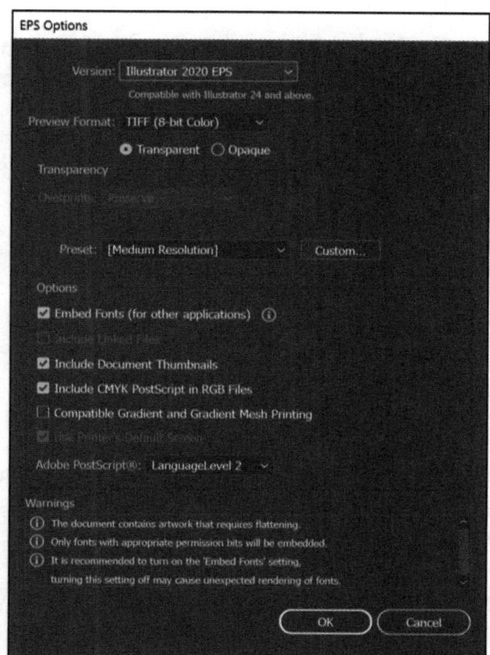

FIGURE 12-2: Select a preview and other important settings in the EPS Options dialog box.

The SVG file format

SVG (Scalable Vector Graphic) is a vector format that describes images as shapes, paths, text, and filter effects. This is huge for those in web and application design as you can save icons and other vector art in the SVG format and not worry about saving multiple sizes. Users can enlarge their view of an SVG image without pixelization, or loss of detail. The SVG format also provides support for text and colors so that web and app users see images as they appear on your Illustrator artboard.

TECHNICAL STUFF

The SVG format is XML-based and offers many advantages to developers and users alike. With SVG, you can use XML and JavaScript to create graphics that respond to user actions such as highlighting, tooltips, audio, and animation.

To save an in the SVG format, follow these steps:

1. **Have an illustration open, something that is built as a vector image is best. Use the `BabikBagel.ai` image again, if you like.**
2. **Choose File⇨ Save As and select SVG from the Save as type menu. The SVG Options window appears. In this window, you have the choice of subsetting fonts, creating a link to a placed image, embedding it in your SVG file, or preserving the editing capabilities in Adobe Illustrator.**

3. If you are looking for additional developer options, select the More Options button at the bottom, or copy the code that is generated by selecting the SVG Code button.

Keep in mind that SVG files can even be used in PowerPoint. When placed, you can change the size, color, and even access the vector shape for editing it by Ungrouping it.

Saving Your Artwork for the Web

If you need to save individual illustrations for the web, you can use the Save for Web feature. This dialog box opens a preview pane where you can test different file formats before you save the file. There are more efficient methods for saving multiple files at the same time that are discussed later in this section, but this initial method is helpful as you can preview results of your settings before saving.

TIP

Before going through the steps to save a graphic for the web, make sure that the artboard is resized to the bounds of the graphic.

To save an Illustrator file that you intend to use in a web page, just follow these steps:

1. Choose File ⇨ Export ⇨ Save for Web (Legacy).

The Save for Web dialog box appears, showing your artwork on the Optimized tab. (See Figure 12-3.)

FIGURE 12-3: Use Save for Web to optimize and preview your images for the web. In this example, the original is on the left, and the JPEG preview is on the right.

2. **Select a tabbed view: Original, Optimized, or 2-Up.**

   You see as a default the artwork in Optimized view, which previews the artwork as it will appear based on the settings to the right. The best choice is 2-Up view because it shows your original image versus the optimized version.

3. **Select a setting for your file from the options on the right.**

   If you want to make it easy on yourself, select a preset from the Preset drop-down list. Keep these points in mind:

   - *Graphics Interchange Format (GIF)* is generally used for artwork with spans of solid color. GIF isn't a lossy format. You can make your artwork smaller by reducing the number of colors in the image — hence the choices, such as GIF 64 No Dither (64 colors). The lower the number of colors, the smaller the file size. You can also increase or decrease the number of colors in the file by changing the preset values in the Color text field or by clicking the arrows to the left of the Color text field.
   - *Dithering* tries to make your artwork look like it has more colors by creating a pattern in the colors, as can be seen in Figure 12-4. It looks like a checkerboard pattern up close and even far away. It also makes a larger file size, so why use it? Many designers don't like the effect and choose the No Dither option.

FIGURE 12-4: The GIF format has an option for dithering, shown on the right.

   - *Joint Photographic Experts Group (JPEG)* is used for artwork that has subtle gradations from one shade to another. Photographs are often saved in this format. If you have drop shadows or blends in your artwork, select this format. JPEG is a *lossy* file format — it reduces an image to a lesser quality and can create odd artifacts in your artwork. You have choices, such as High, Medium, and Low, in the Settings drop-down list — choose wisely. You can also use the Quality slider to tweak the compression.

- *PNG-8* is quite similar to the GIF file format. Unless you have a certain reason for saving as PNG-8, stick with the GIF file format.
- *PNG-24* supports the best of two formats (GIF and JPEG). The Portable Network Graphics (PNG) format supports not only those nice gradients from one tonal value to another (such as JPEGs), but also transparency (such as GIFs). With the Transparency box checked, the PNG format allows for incremental transparency; the GIF format, on the other hand, is either fully transparent in areas or not.

4. **When you're satisfied with your chosen settings, click Save.**

TIP

When saving illustrations for the web, keep these points in mind — they make the whole process much easier for you and anyone who uses your illustrations:

» **Keep file sizes small.** Don't forget that if you're saving illustrations for a web page, many other elements will also be on that page. Try to conserve on file size to make downloading the page quicker for viewers with dial-up connections. Visitors typically don't wait more than ten seconds for a page to download before giving up and moving to another website.

When you make your choices, keep an eye on the file size and the optimized artwork in the lower-left corner of the preview window. A GIF should be around 10K on average, and a JPEG around 15K. (Though these rules aren't written in stone, *please* don't try to slap a 100K JPEG on a web page.)

TIP

You can change the download time by clicking the panel menu in the upper-right corner of the Save for Web & Devices dialog box and choosing Optimize to File Size. Then you can enter a final file size and have Illustrator create your settings in the Save for Web dialog box.

» **Preview the file before saving it.** If you want to see the artwork in a web browser before saving it, click the Preview button at the bottom-left corner of the Save for Web dialog box. The browser of choice appears with your artwork in the quality level and size in which it will appear. If no browser is selected, click the Select Browser Menu icon (to the right of the Preview button) to choose Other and then browse to locate a browser you want to use for previewing. Close the browser to return to the Save for Web dialog box.

» **Change the size.** Many misconceptions abound about the size of web artwork. Most people generally view their browser windows in an area measuring approximately 1366×768 pixels. Even viewers with large, high-resolution monitors often don't want to have their entire screens covered by a browser window. When choosing a size for your artwork, choose one with proportions similar to these. For example, if you want an illustration to occupy about a third of the browser window's width, make your image about 455 pixels wide

(1366 ÷ 3 = 455). If you notice that the height of your image is more than 800 pixels, whittle the height in size as well or else viewers have to scroll to see the whole image (and it will probably take too long to download).

Use the Image Size tab to enter new sizes. As long as the Constrain Proportions checkbox is selected, both the height and width of the image change proportionally. Click the Apply button to change the size, but don't close the Save for Web dialog box.

» **Finish the save.** If you aren't finished with the artwork but want to save your settings, hold down the Alt (Windows) or Option (Mac) key and click the Remember button. (When you aren't holding down the Alt or Option key, the Remember button is the Done button.) When you're finished, click the Save button and save your file in the appropriate location.

Saving Multiple Assets out of Illustrator

Perhaps you have created artwork for an interactive design such as a mobile application. Using Illustrator's new Assets Export feature, you can easily collect the artwork that you need for development at one time.

To take advantage of this feature, either use the asset.export file available in the DummiesCCFiles folder located at `www.agitraining.com/dummies` or use your own file. Then follow these steps:

1. **Choose Window ⇨ Assets Export to open the Asset Export panel.**
2. **At this point, simply click and drag artwork from the artboard to the Assets Export panel, as shown in Figure 12-5.**

   Note that if your artwork contains many objects, and is not grouped together, you need to hold down the Alt/Option key while dragging it into the Asset Export panel. Otherwise, each object becomes its own asset.
3. **Continue adding other artwork to this panel.**

   Note at the bottom of the panel that you have a 1X scale selected as a default. Essentially, this means that you are set up to save every one of your assets in the panel at 100 percent. Depending upon the platform your assets will be used on, you may need to supply each piece of artwork in not only 1X (100%) size, but also 2X (200%) and 3X (300%). This can be set up easily in the Asset Export panel.

FIGURE 12-5: Click and drag artwork that you want to export into the Asset Export panel.

4. **To add additional saved artwork in other scales, press the + Add Scale button at the bottom of the Asset Export panel. The default 2X and 3X scales are added, but you can choose another size or enter your own custom size by clicking the arrow indicated in Figure 12-6.**
5. **Choose a file format for each scale by selecting a file type from the Format drop-down menu.**
6. **Name your exported files by double-clicking on the default name in the Asset Export panel. You can add a suffix in the scale attributes section as well.**
7. **Select the Files you want exported. Press Export when you are ready to export the files that you loaded into the Asset panel. A Pick Location window appears that contains an appropriate folder in which to save your exported files. Either create a new folder, or select a folder and choose Select Folder (Windows) or Choose (Mac).**

   Your files have been exported.

FIGURE 12-6: Working with the Add Scale button in the Asset Export panel.

Note: If you would like additional options, select the Launch the Export for Screens button to the left of Export. In this dialog box, you can choose to export only some or all of your artwork or select to export your artboards as individual assets.

Flattening Transparency

You may find that all those cool effects you added to your illustration don't print correctly. When you print a file that has effects, such as drop shadows, cool gradient blends, and feathering, Illustrator *flattens* them by turning into pixels any transparent areas that overlap other objects and leaving what it can as vectors.

To understand flattening, look at Figure 12-7 to see the difference between the original artwork (on the left) and the flattened artwork (on the right). Notice that in the figure, when the artwork was flattened, some areas turned into pixels. But at what resolution? Flattening helps you determine the quality of art — before receiving an unpleasant surprise at the outcome. If you want to try this on your own, open the file named `BabikBagelsFlattening.ai` in the Book05_Illustrator folder.

FIGURE 12-7: Artwork before and after flattening is applied.

Keep in mind that you do not have to flatten most artwork. If it is staying on-screen, for an app or website, flattening is not necessary. On the other hand, if you are trying to print a complex illustration, or getting unexpected results when printed, flattening offers the option of breaking down the illustration into a more manageable print file.

Flattening a file

If you've taken advantage of transparency or effects using transparency (which we discuss in Chapter 10 of this minibook), follow these steps to produce the highest-quality artwork in your file:

1. **Make sure that the artwork you created is in CMYK mode.**

   You can change the document's color mode by choosing File ➪ Document Color Mode. ➪ CMYK Color.

2. **Choose Effect ➪ Document Raster Effects Settings.**

   The Document Raster Effects Settings dialog box appears, as shown in Figure 12-8. Image should have CMYK as the color mode.

3. **Choose the resolution you want to use by selecting an option in the Resolution area.**

   Select the Screen (72 ppi) option for web graphics, Medium (150 ppi) for desktop printers and copiers, and High (300 ppi) for graphics to be printed on a press.

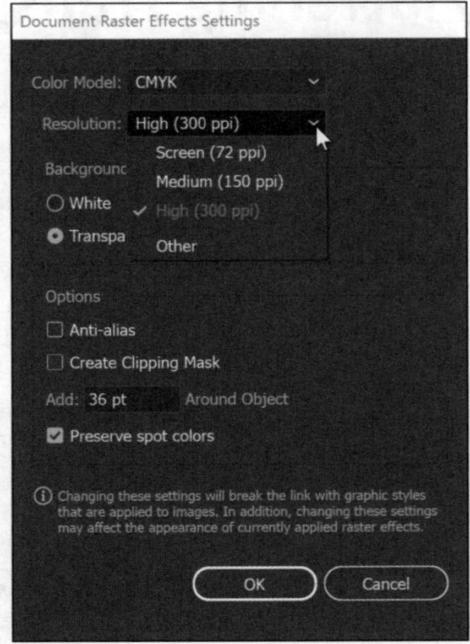

FIGURE 12-8: Choosing the quality of rasterized artwork.

4. **Choose whether you want a white or transparent background.**

   If you select the Transparent option, you create an alpha channel that's retained if the artwork is exported into Photoshop.

5. **You can generally leave the items in the Options section deselected:**

   - *The Anti-Alias checkbox* applies antialiasing to reduce the appearance of jagged edges in the rasterized image. Deselect this option to maintain the crispness of fine lines and small text.
   - *The Create Clipping Mask checkbox* creates a mask that makes the background of the rasterized image appear transparent. You don't need to create a clipping mask if you select the Transparent option for your background.
   - *The Add __ Around Object text field* adds the specified number of pixels around the rasterized image.
   - *The Preserve Spot Colors checkbox* keeps any spot colors you have defined intact. Leave this selected.

6. **Click OK.**

   The next step is to set the transparency options in the Document Setup dialog box.

7. **Choose File ⇨ Document Setup.**

   From the Transparency section in the middle of the dialog box, click the Preset drop-down list and select the Low, Medium, High, or Custom option. Select the Low option for onscreen viewing, the Medium option for printers and copiers, or the High option for press quality. To control more settings, click the Custom button to the right of the drop-down list.

8. **Click OK.**

TIP

If you customize settings regularly, choose Edit ➪ Transparency Flattener Presets to create and store your own presets.

You can apply the flattening in several ways. Here are three simple methods:

- **Select the objects that require flattening and choose Object ➪ Flatten Transparency.** Select a default setting or a custom preset (that you created) from the Preset drop-down list and click OK.
- **Choose File ➪ Print and select Advanced from the list of print options on the left.** Select a preset from the Overprint and Transparency Flattener options. If you used the Attributes panel to create overprints (for trapping used in high-end printing), make sure to preserve the overprints.

REMEMBER

  Overprints aren't preserved in areas that use transparency.

- **Choose File ➪ Save As and choose Illustrator EPS.** In the Transparency section of the EPS Options dialog box, choose a flattening setting from the Preset drop-down list. If your transparency options are grayed out, your file has no transparency.

Using the Flattener Preview panel

If you want to preview your flattening, open the Flattener Preview panel by choosing Window ➪ Flattener Preview.

The Flattener Preview panel doesn't apply flattening, but it shows you a preview based on your settings. Click the Refresh button and choose Show Options from the panel menu. Test various settings without flattening the file. Experiment with different settings and then save your presets by selecting Save Transparency Flattener Preset from the panel menu. You can access the saved settings in the Preset drop-down list in the Options dialog boxes that appear when you save a file in EPS format or in the Document Setup dialog box.

To update the preview, click the Refresh button after making changes.

TIP

Zoom in on artwork by clicking in the Preview pane. Scroll the artwork in the Preview pane by holding down the spacebar and dragging. Zoom out by Alt-clicking (Windows) or Option-clicking (Mac).

Printing from Illustrator

Printing from Illustrator gives you lots of capabilities, such as printing composites to separations and adding printer's marks.

To print your illustration, follow these steps:

1. **Choose File ⇨ Print.**
2. **In the Print dialog box that appears, select a printer if one isn't already selected.**
3. **If the PPD isn't selected, select one from the PPD (PostScript Printer Description) drop-down list.**

   A *PPD* is a printer description file. Illustrator needs one in order to determine the specifics of the PostScript printer you're sending your file to. This setting lets Illustrator know whether the printer can print in color, the paper size it can handle, and its resolution, as well as many other important details. If your printer is not a PostScript printer, this area will be grayed out.
4. **Choose from other options.**

   For example, use the General options section to pick pages to print. In the Media area, select the size of the media you're printing to. In the Options area, choose whether you want layers to print and any options specific to printing layers.
5. **Click the Print button to print your illustration.**

That's it. Although printing illustrations can be quite simple, the following list highlights some basic points to keep in mind as you prepare one for printing:

- **Print a composite:** A *composite* is a full-color image, where all inks are applied to the page (and not separated onto individual pages — one apiece for cyan, magenta, yellow, and black). To ensure that your settings are correct, click Output in the Print Options pane on the left side of the Print dialog box and select Composite from the Mode drop-down list.
- **Print separations:** To separate colors, click Output in the Print Options pane on the left side of the Print dialog box; from the Mode drop-down list, choose the Separations (Host-Based) option. Select the In-RIP Separations option only if your service provider or printer asks you to. Other options to select from are described in this list:

- The resolution is determined by your PPD, based on the dots per inch (dpi) in the printer description. You may have only one option available in the Printer Resolution drop-down list.
- Select the Convert Spot Colors to Process checkbox to make your file four-color.
- Click the printer icons to the left of the listed colors to turn off or on the colors you want to print.

» **Printer's marks and bleeds:** Click Marks and Bleeds in the Print Options pane on the left side of the Print dialog box to turn on all printer's marks, or just select the ones that you want to appear.

Specify a bleed area if you're extending images beyond the trim area of a page. If you don't specify a bleed, the artwork stops at the edge of the page and doesn't leave a trim area for the printer.

TIP

After you create a good set of options specific to your needs, click the Save Preset button (which is the disk icon to the right of the Print Preset drop-down list). Name your presets appropriately; when you want to use a particular preset, select it from the Print Preset drop-down list at the top of the Print dialog box for future print jobs.

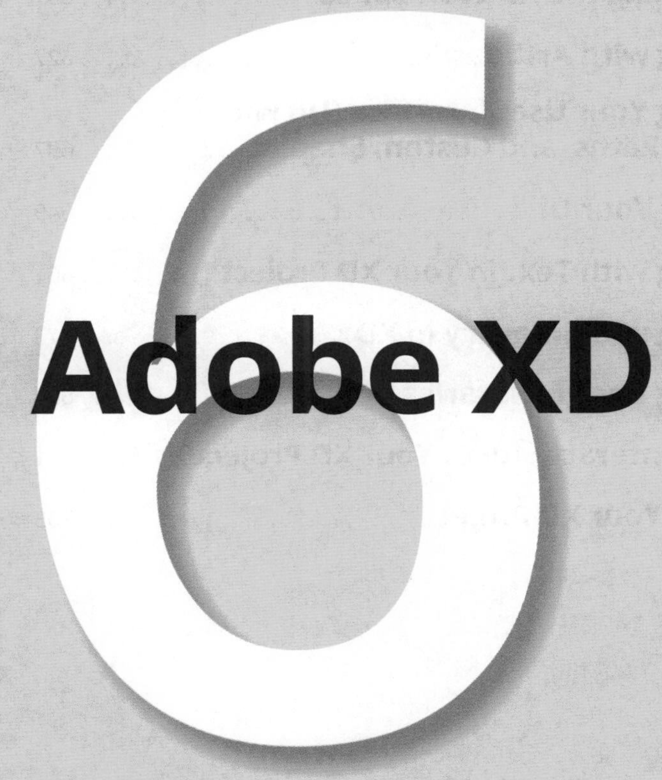

Adobe XD

Contents at a Glance

» Creating your first screen

» Comparing Design versus Prototype mode

Chapter 1
Introducing the XD Workspace

Adobe Experience Design (XD) is a high-fidelity prototyping tool used to demonstrate apps and website designs. In this chapter, you find out how to use XD's design tools to create your UI (user interface) quickly using features such as shapes, text, grids, reusable components, layers, and more. Keep in mind that Adobe XD is updated at a rather fast pace so there may be slight differences in the feature set from the time of writing. Another note: Although XD works on both the Windows and Mac platforms, you may see slight differences in the UI. Those differences will be addressed when appropriate.

Adobe Experience Design is included in your Creative Cloud subscription. If you do not have it already installed, launch your Creative Cloud application and install it before trying to follow along with this minibook.

Understanding the Modern User Interface

XD's clean and simple UI has been created from scratch with the user's experience in mind. After you discover the primary tools, you will easily be able to integrate XD into your production workflow. XD is like Adobe Illustrator; in fact, many

Illustrator users prefer the simplicity of XD's interface. To discover these tools, you can build a new document and follow along with the exercises in this chapter. Throughout the chapter you will take steps toward building a complete, working prototype with art, images, text, and hyperlinks. You can then dig into the details of those steps in the following chapters.

Creating a new artboard

In this chapter, you can follow along to learn while using some of the basic features. As with most of the other Adobe applications, you will want to define an artboard size before working in Adobe XD. You can choose from artboards that are standard screen sizes, or create a custom-sized artboard. The size of the artboard can be changed at any time. To create a new artboard, follow these steps:

1. **Launch Adobe XD.**

   A start screen appears that introduces you to the XD application and offers you a choice of screen sizes. Your selection creates an artboard of that specified size. Figure 1-1 shows the drop-down menu for the iPhone 11 Pro Max artboard. By opening other drop-down menus, you can access other mobile sizes.

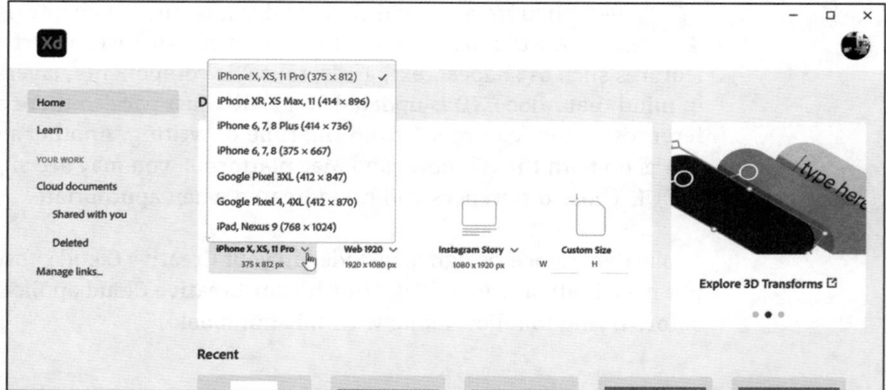

FIGURE 1-1: Selecting the size of the artboard.

2. **Select an artboard size from the drop-down menu and press the Return or Enter key.**

   The artboard appears in the XD workspace.

   The workspace includes a work area in the middle, a main menu across the top (Mac only), an application menu above the work area, a toolbar to the left, and a Property Inspector on the right; see Figure 1-2.

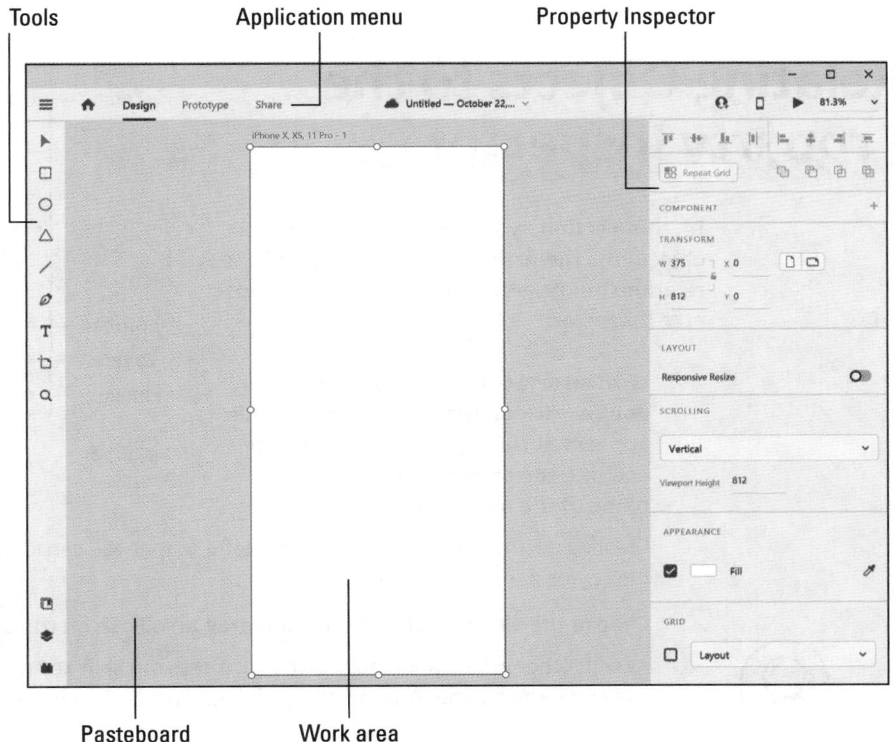

FIGURE 1-2: In the Windows XD workspace, the menu is accessed in the top-left corner. On the Mac, the Main menu appears at the top of the workspace.

Using the tools

In XD, primitive shapes, as well as more advanced drawing tools, are available to you on the toolbar. In addition, some other tools, such as the Select and the Artboard tools, might be familiar to you if you use Adobe Illustrator. See all tools in Figure 1-3.

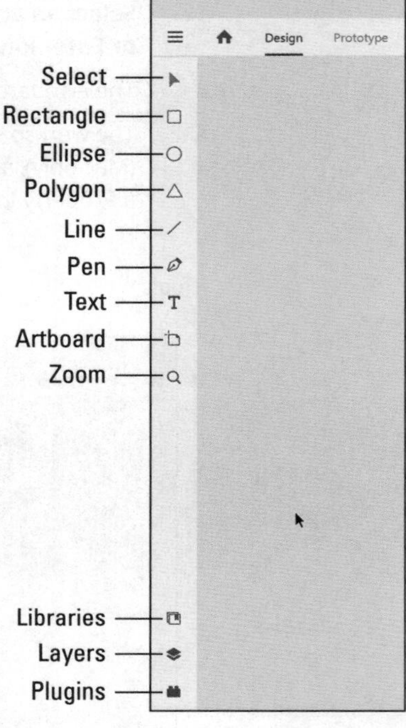

FIGURE 1-3: XD tools.

Relating Objects to the Property Inspector

In this section, you use the tools to place objects on the artboard so you can see the relationship between objects and the Property Inspector:

1. **With an artboard of any size open, double-click on the artboard name that appears at the top of the artboard to activate the textbox and change the name of the artboard to "home."**

   Creating a name for your artboard is helpful so that you can identify it easier in the Layers panel or other menus.

2. **Choose the Ellipse tool and click and drag an ellipse on the screen.**

   TIP

   As with other Adobe applications, you can press the Shift key while dragging to create a perfect circle. You can also press Alt/Option+Shift while dragging to create a circle generated from a center point.

   When the ellipse is complete, its properties now appear in the Properties Inspector on the right, as shown in Figure 1-4.

3. **The Ellipse tool is still selected. To avoid unexpectedly creating another circle, click the Select tool.**

Centering an object on the screen

The Align options at the top of the Properties Inspector are helpful when you want to align the circle with the center of the artboard, horizontally or vertically (or both). The Distribute options are not relevant unless you have more than one object selected. Further down the Properties panel are other items, such as Repeat Grid and Component. These features are discussed in context later in this book.

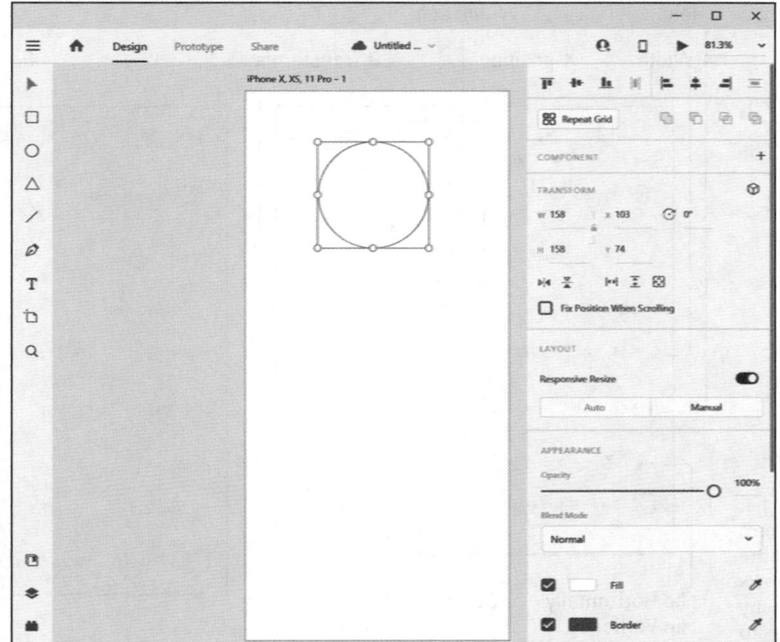

FIGURE 1-4: When an object is selected, its properties appear in the Properties panel.

Using transform features

In the Transform section, shown in Figure 1-5, you can adjust the size of your selected object using the Width and Height textboxes and change the position of your selected object by changing the *x*, *y* coordinates. In this section, you work with these and also flip your object vertically and or horizontally.

Some advance features such as scroll groups allow you to define areas of content that can be scrolled horizontally, vertically, or in all directions. This is perfect when you are trying to replicate an experience of panning a map or when you have multiple, scrollable, panels on one screen. You will find out more about scroll groups in Chapter 2, "Working with Artboards."

Click the 3D Transform button in the upper-right to access 3D transformation tools such as *x* and *y* rotation in a 3D space. The 3D transform mode the rotate icons appear. To use these, choose whether you want to rotate on the *x*-, *y*-, or *z*-axes.

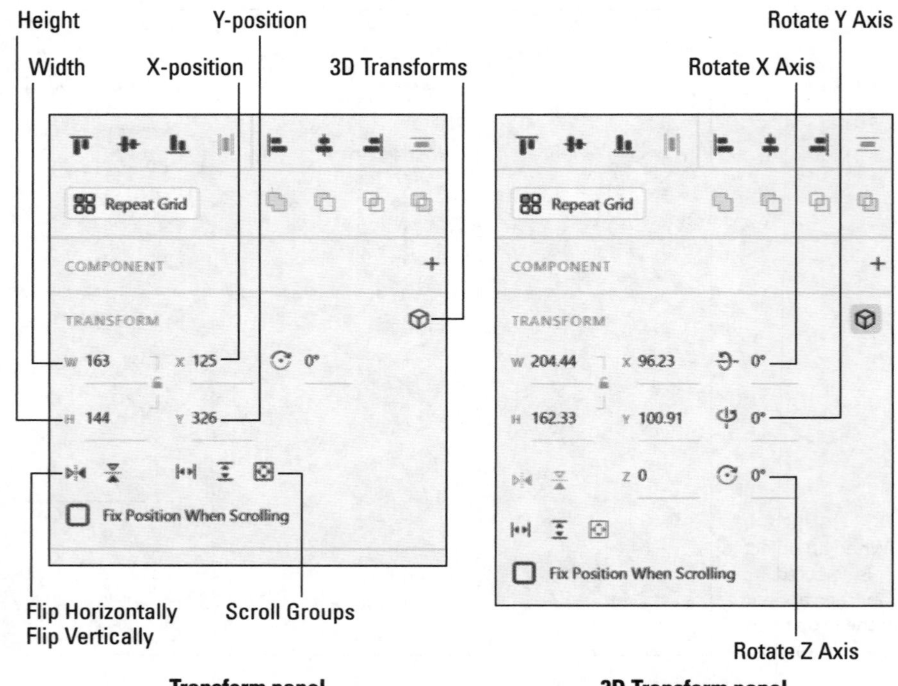

FIGURE 1-5: The transform section of the Properties panels offers many opportunities to resize and shape your object(s).

Fixing position when scrolling

Click the Fix Position When Scrolling checkbox to keep the selected object from scrolling with the rest of the content. This is a great feature that allows you to build a navbar that stays put while the rest of the content on a screen scrolls. This will be demonstrated when you preview the screen, and it is discussed in more detail in the next chapter.

Laying it all out

In this section, two simple choices appear under Responsive Resize: Auto and Manual. You can choose to have your selected object automatically resize, based upon the artboard, or stay fixed in size. You could use this feature when designing multiple views of your website or app. Many designers clone an artboard by using Alt/Opt and dragging an artboard to a new location and then resizing it to show a view of their design for phone, web, or tablet.

To try these features out, follow these steps:

Auto resizing

If you simply want imagery and other content to shrink and grow when an artboard is expanded or reduced in size, you can take advantage of the Auto Resize feature. Try these simple steps to see how it works:

1. **If you do not already have an artboard, use the Artboard tool to create one in any size.**
2. **While the artboard is selected, turn on Responsive Resize in the Layout section of the Properties Inspector.**
3. **Select any shape and click and drag to add it to the artboard. In this example, select an ellipse.**
4. **Click the Select tool.**
5. **Hold down the Ctrl (Windows) or Command (Mac) key and click on the artboard to activate it.**
6. **Click and drag one of the side handles to widen the artboard. Note that the ellipse automatically resizes with the artboard, as shown in Figure 1-6.**

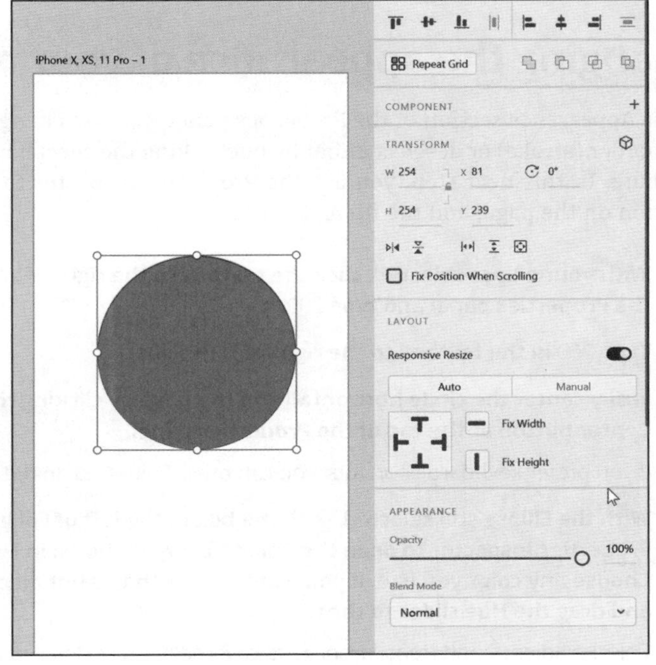

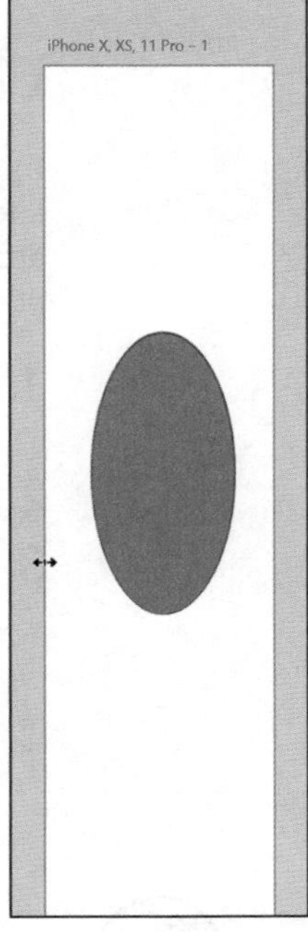

FIGURE 1-6: You can set-up your objects to resize automatically when the artboard is resized. See results of the resize on the right.

Introducing the XD Workspace

Manual resizing

If you don't want all the content on your artboard to automatically resize, you can select some of your objects and keep their size fixed. This basically freezes them so that they do not expand or shrink when the artboard size is adjusted. Here's how:

1. **If you do not already have a shape on an artboard, create one now.**
2. **Click the Responsive Resize switch in the Layout section of the Properties panel.**
3. **Click the Manual button and click Fix Width and Fix Height to turn off those settings.**

   The lines in the UI are blue if the feature is set on, gray if they are turned off.
4. **Select the artboard by clicking on the Artboard name, or Ctrl/Command-clicking on any empty space in the artboard.**
5. **Grab one of the side handles and click and drag to resize the artboard. Notice that the object resizes with the artboard.**
6. **Press Ctrl/Command-Z to undo and try fixing the size of the object by selecting it again and clicking back on the Fixed Width and Height controls. Now when you resize, it stays the same size.**

Changing the appearance of your object

In the Appearance section of the Properties panel, you can change the opacity, fill and border (stroke) or deselect either by unchecking the checkbox to the left of the attribute. In this next step, you use the Properties Inspector to change the size, position on the page, and the fill and border colors:

1. **With your ellipse selected, click the textbox to the right of W (Width) in the Properties panel and type** 200**.**
2. **Type** 200 **in the textbox to the right of H (Height).**
3. **Easily center the circle horizontally on the page by clicking the Align Center button at the top of the Properties panel.**

   If you prefer keyboard shortcuts, you can press Shift+C to do the same thing.
4. **With the Ellipse still selected, click the box to the left of Fill in the Properties Inspector to open the Color Picker, as shown in Figure 1-7. Choose any color you like. If you want a color that is not displayed, click and drag the Hue slider to the right.**

REMEMBER

Color, gradients, and transparency are covered in more detail in Chapter 3 of this minibook.

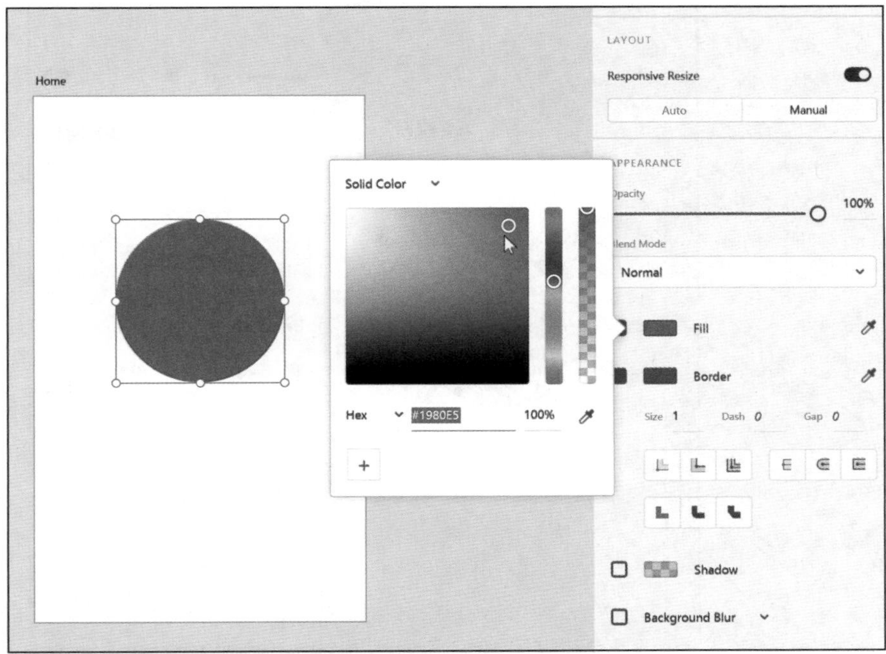

FIGURE 1-7: Choose a color for your fill from the color picker.

5. **Click the box to the left of Border to select any stroke color.**

6. **Select the width value below the Border and type** 10 **so that you can see a 10 pt border on your ellipse.**

   You have many options to select when you want to use a border, as you see in Figure 1-8. These options allow you to change the corner points, add dashed lines, and more. You will find out more about borders on Chapter 3, "Creating Your User Interface (UI) with Shapes, Paths, and Custom Shapes."

TIP

   You can use sliders to change values in the Properties pane — like for your border, for example: Just click and drag up or down on the selected value.

7. **As a bonus, add a drop shadow using the Properties Inspector: With your ellipse still selected, check Shadow. The values that appear beneath are the X and Y coordinates and the blur.**

   If you have been following along, choose to close your file by clicking on the Close box in the upper-right corner. You do not need to save this file; you will be starting a new example in the next section.

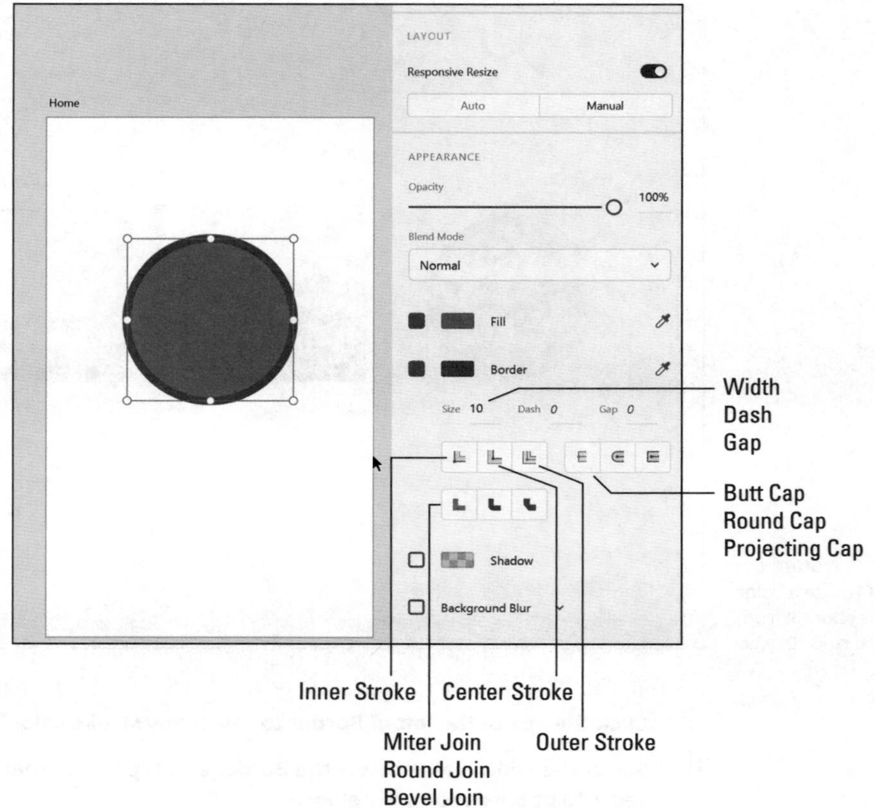

FIGURE 1-8: Check the border option to see additional properties.

A Little Bit about Artboards

In the previous section, you became familiar with ways to create objects and change their properties. In this section, you create a set of three artboards in order to investigate the properties associated with artboards:

1. **Create a new XD document by pressing Ctrl+N (Windows) or Command+ N (Mac OS).**
2. **For this example, choose any of the default size artboards displayed in the New dialog box. In this example, the iPhone artboard was selected.**

   After you click on the artboard size, a new document appears with the first artboard ready to go.

Saving your file

This is a good time to save this file. In Adobe XD, you can save either locally or to the Adobe Cloud. Why two methods? It depends on the amount of collaboration you are looking to enable. By saving your file as a cloud document, you can access it across multiple platforms and devices. This can be helpful when you want to share your document for comments and or development. You find out more about saving as a cloud document in Chapter 9, "Sharing your XD Project." Now, however, you will save the file locally, which means to the hard drive on your current device:

1. **Click and hold the burger menu in the upper-left and then choose Save as a Local Document (Windows) or File ➪ Save as a Local Document (Mac).**
2. **In the File Name section, name this** practice **and browse to a location where you want to save this file. Keep the Save as Type as XD File and press Save. Keep this file open.**
3. **Double-click on the artboard name, at the top of the artboard, and change the name to** home**. When you start creating multiple artboards it is good practice to provide each with a descriptive name.**

   Now add a simple graphic element to the home screen.
4. **Using the Ellipse tool, click and drag the artboard to create an ellipse. It can be any size or color, but position it somewhere near the top of the screen.**
5. **With the ellipse selected, click the Align Center icon at the top of the Properties Inspector.**

   Your screen should look similar to the one shown in Figure 1-9.

   Now add some text to this screen.
6. **Select the Type tool and then click and release to create a text insertion point.**

   If you click and drag, you will create a text area that is not as dynamic when you add additional characters to your text. More details about controlling text are covered in Chapter 5, "Working with Text in Your XD Project."

home

FIGURE 1-9: Create a screen and add an ellipse with any color fill.

Introducing the XD Workspace

7. **Type the word** Welcome **in the Properties Inspector and choose any size type in Text section — for this example, choose 50 pts. Also, choose Center to center the text. To make sure the text object is in the center of the artboard, click the Align Center (Horizontally) button at the top of the Properties Inspector.**

   Your screen should look similar to the example in Figure 1-10. For more type features, see Chapter 5, "Working with Text in Your XD Project."

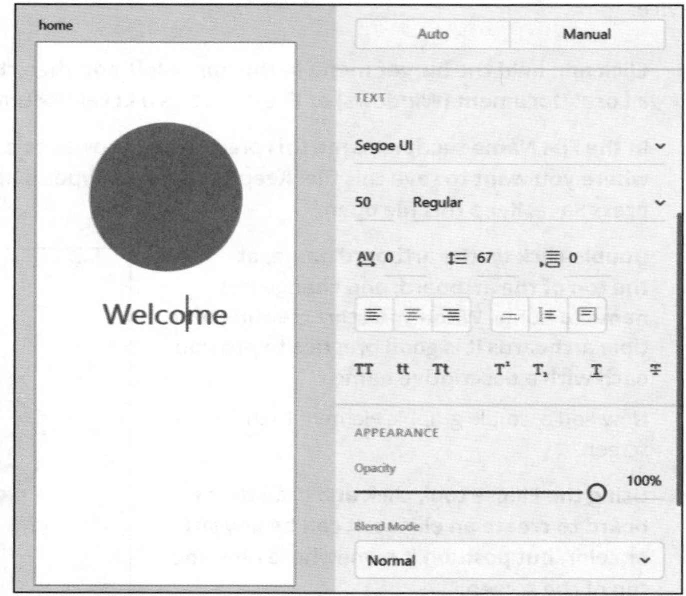

FIGURE 1-10: The screen now has a graphic element and a text object.

Cloning your text objects to the first artboard

These steps add additional text elements to the first page. Later, you will create links from these text objects to other artboards:

1. **Using the Select tool, click the *Welcome* text that you just created in the previous section.**
2. **Hold down Alt/Option and drag a clone of the text down and away from the first text.**

3. **Double-click on the new text field and type** See dog.
4. **Repeat the cloning step by selecting the textbox that states "See dog" and then Alt/Opt-dragging it down underneath the original text.**
5. **Double-click the text and change it to** See cat.

   The result should look like Figure 1-11.
6. **If you want to evenly distribute the lines of text, select all three and then click the Distribute Vertically icon at the top of the Properties Inspector. (See Figure 1-11.)**

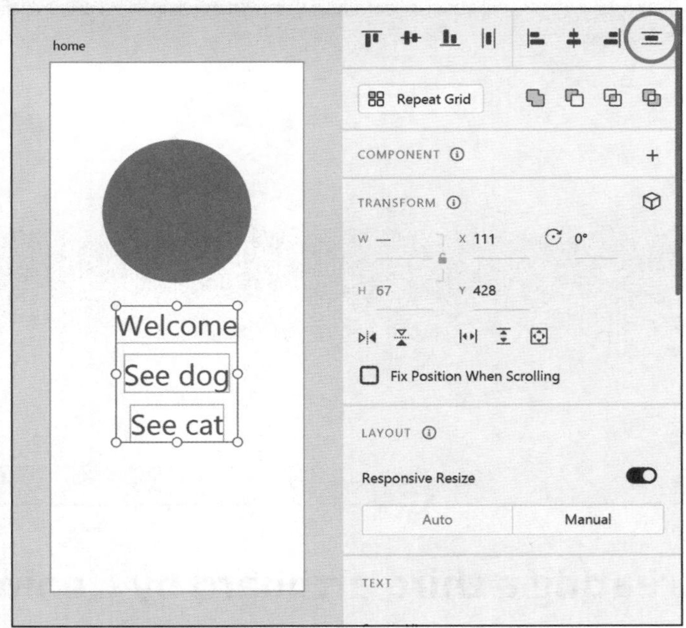

FIGURE 1-11: Clone your original text object by holding down the Alt/Option key and dragging it to another location.

Duplicating your artboard

There may be instances when you want save steps by using an existing artboard to build other screens. Follow these steps to duplicate the artboard you named home in the previous section.

1. **Click on the name of your layer, home, at the top of the artboard to select it.**

2. **In the Layers panel on the left, look for your artboard name, right-click (Windows) or Control-click (Mac) it, and select Duplicate from the contextual menu, as shown in Figure 1-12.**

A new artboard appears above home named home-1.

3. **Double-click the home-1 name in the Layers panel, and change it to** dog.

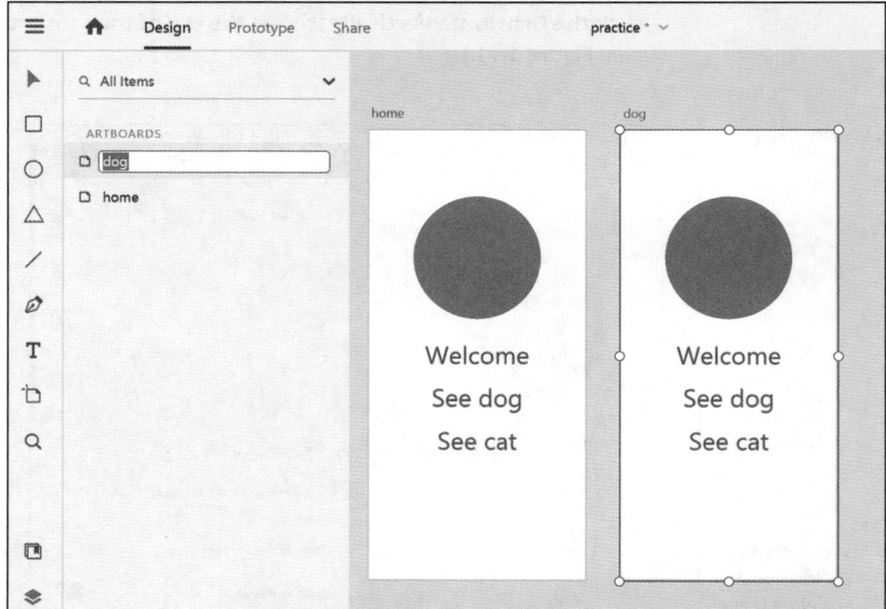

FIGURE 1-12: Duplicate and rename your new layer.

Creating a third artboard by cloning

You can also clone an existing artboard to start building from existing elements on a screen. You clone objects in XD much like you would in other applications, by using the Alt or Option key. Follow these steps to create a third artboard:

1. **Make sure you have the Select tool active.**

2. **Click the dog name at the top of the dog artboard you just created to select the artboard.**

3. **Hold down the Alt (Windows) or Option (Mac OS) key and click and drag a third artboard to another location in your XD document.**

4. **Double-click the name at the top of the new artboard and change it from** dog-1 **to** cat.

Editing your text to fit the screens

In this section, you change the text on the dog and cat pages so that you can link to them later.

1. **On the dog page, double-click the text object that states *See dog* and change it to** This is a dog.
2. **On the cat page, double-click the text object that states *See cat* and change it to** This is a cat.
3. **Save the file by pressing Ctrl+S (Windows) or Command+S (Mac).**

Experimenting with artboard properties

You now have three artboards: home, dog, and cat. In this section, you change the background color of your artboards and turn off and on precision layout and grid tools that help you keep your designs neat:

1. **With the any artboard selected, click the box to the left of Fill, in the Appearance section of the Properties Inspector, and select any color of background for your new artboard. For this example, select a light tint so that text in the design is still visible.**
2. **Check the box to the left of Layout in the Grid section of the Properties Inspector in order to display a layout grid that is visible to you only while working in XD. Change it to a square grid by clicking and holding down Layout and selecting Square.**
3. **Change the size to 20 pixels, as shown in Figure 1-13. Keep in mind that you can hide this grid at any time by unchecking the Grid checkbox.**

Adding an image

To make our app just a little more exciting, you can add some images to the cat and dog pages. This won't produce a beautiful app, maybe, but it will certainly create a functional one:

1. **Remove the circle graphics on all three screens by selecting them and pressing Delete or Backspace.**
2. **Select the dog artboard and then choose Burger menu⇨Import (Windows) or File⇨Import (Mac).**

   An Open dialog box appears.

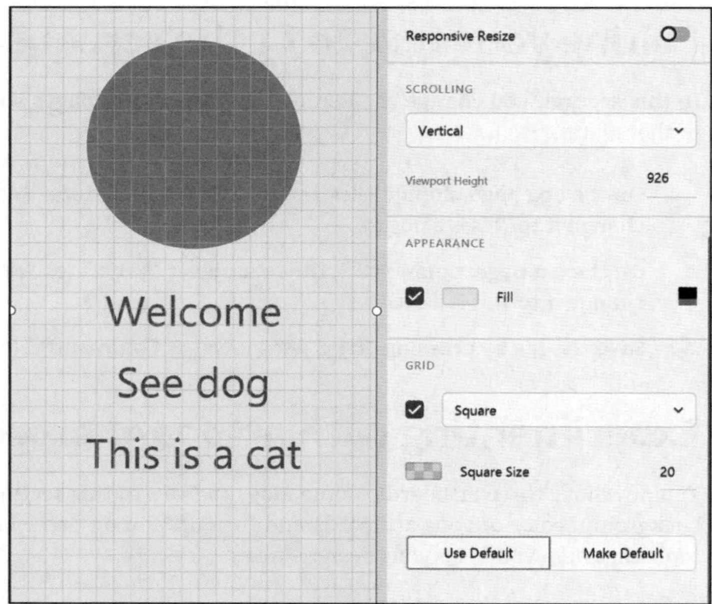

FIGURE 1-13: Change the background color and add a grid in the Properties Inspector.

You can import any images that you like, but if you downloaded the DummiesCCFiles folder, noted in Book 1, Chapter 1, you can use the prepared exercise files for XD.

3. **Navigate to the DummiesCCFiles folder and to the folder named Book07_XD and select the image named `BrownWhitePup`. Hold down Ctrl (Windows) or Command (Mac) and click the `home-icon.svg` file to add the second image to your selection, and then press Import.**

   The files are added to the dog artboard.

4. **Arrange the dog image and the home button on the page. If you want to horizontally align them on the center of the page, press the Align Center button at the top of the Properties panel or press Shift+C.**

5. **Now select the cat artboard and repeat the procedure, choosing the Burger menu⇨Import (Windows) or File⇨Import (Mac). In this example, the `GrayKitty,jpg` image and the `home-icon.svg` image are used.**

6. **If you want to add an image to the home page, use the image named `babyanimals01`.**

After you have arranged the images on the artboards, the result may look similar to Figure 1-14. Note that in this example the square grid lines have been turned off so that you can see the screens better.

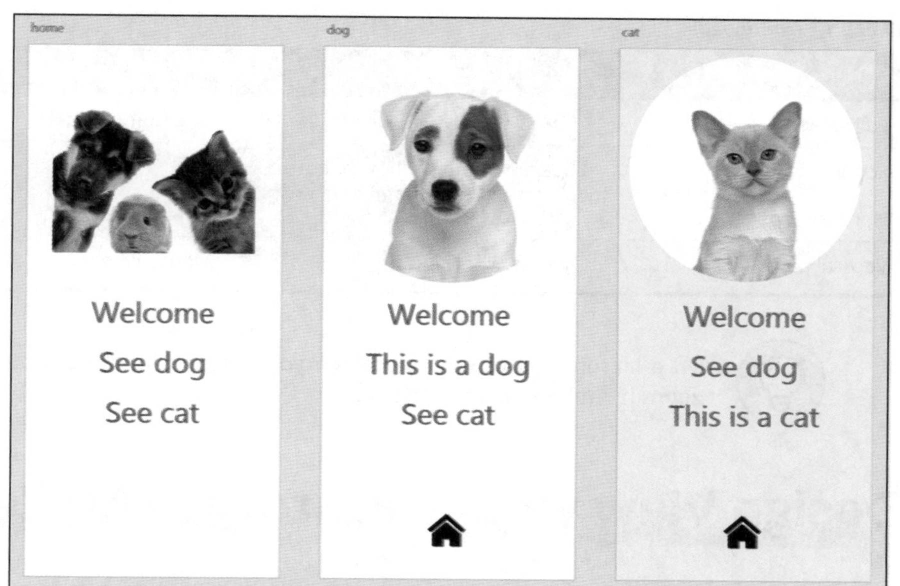

FIGURE 1-14: Import images into your document.

Keyboard Shortcuts for Navigating the Workspace

Now that you have several artboards, you can practice navigating your workspace. Practice using the keyboard shortcuts provided in Table 1-1.

TABLE 1-1 **Keyboard Shortcuts for Navigating the Workspace**

Function	Windows Shortcut	Mac Shortcut
Fit all artboards in window	Ctrl+0	Command+0
Zoom in	Ctrl++ (plus sign) or Alt+mouse wheel	Command++ (plus sign) or Option+mouse wheel
Zoom out	Ctrl+– (minus sign)	Command+– (minus sign)
Go to 100%	Ctrl+1	Command+1
Go to 200%	Ctrl+2	Command+2
Zoom to selection	Ctrl+3	Command+3

(continued)

TABLE 1-1 ***(continued)***

Function	Windows Shortcut	Mac Shortcut
Full screen	N/A	Ctrl+Command+F
Scroll	Spacebar and push with mouse	
Horizontal pan	Ctrl+mouse wheel	Ctrl+mouse wheel
Vertical pan	Mouse wheel	Mouse wheel

TIP

On a laptop, you can simply pinch your fingers in and out on your trackpad to zoom in and out of your artboard.

Design View versus Prototype Mode

Up to this point, you have been working in the default Design mode. This is the view that you work in when creating artwork. Notice at the top of the XD workspace that you have three different modes in which you can work: Design, Prototype, and Share. You learn more advanced features about prototyping in the following chapters of this minibook, but think of this as a quick run-through:

1. **Press Crtl+0 (Windows) or Command+0 (Mac) to make sure that you are viewing all three of your artboards.**
2. **Click Prototype in the Application menu.**

   Now you define an artboard to be your default screen; this is typically your home screen. You'll discover later how to create more advanced interactions, but for right now, start an initial flow. The home screen that was created in this exercise is the first screen of the app. To default to that screen showing up first, define it as the home screen.
3. **Click the home artboard. A gray Home icon appears in the upper-left corner. Click the Home icon. It turns blue, indicating that the artboard is now successfully set as the start screen for your interactive prototype, as seen in Figure 1-15.**

Creating your links

Creating an interactive link is done by creating a wire from one screen to another. You can now link your screens together. To do so, just follow these steps:

FIGURE 1-15: Click Prototype to enter the mode where you can define the start of your flow and create links to other screens.

1. **Click the *See dog* text object on your home artboard. Notice that a blue arrow appears to the right of the object. Click that arrow and drag a wire from the dog text object to the dog artboard.**

   When you release, the target for your link is displayed. There are many options in the Prototype mode, including the ability to link to other pages, create animations, rollovers, show overlays, and more. These are listed as options in the Type drop-down menu of the Properties menu. In this exercise, keep the Type set to Transition.

2. **Click the cat text object on the home page, click the blue connecting arrow, and drag the wire to the cat artboard.**
3. **Allow users to return home by selecting the `home-icon.svg` icons on both screens and clicking and dragging from the blue connecting arrow back to the home artboard, as you see in Figure 1-16.**

Testing your prototype

You can test your prototype by using the Desktop Preview feature:

1. **Select the home artboard.**
2. **Click the Desktop Preview icon in the upper right of the XD workspace, as shown in Figure 1-17.**

3. **Click on the objects that you defined as links, such as See dog, and See cat, and the home icons.**

   You can either allow the Desktop Preview window to stay open while you work — it will show live updates — or you can close it to keep it out of view.

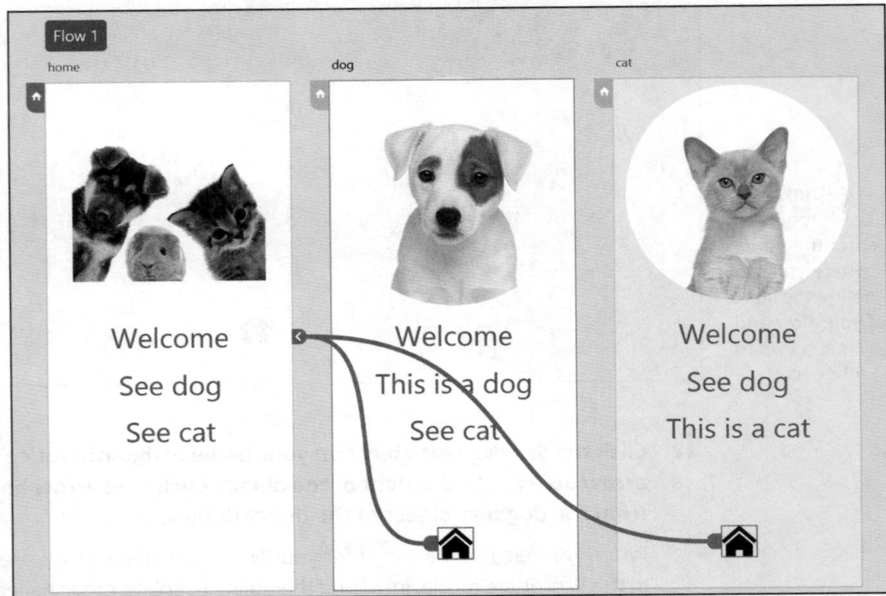

FIGURE 1-16: Click and drag a wire from the object's blue connecting arrow to the artboard that you want to link it to.

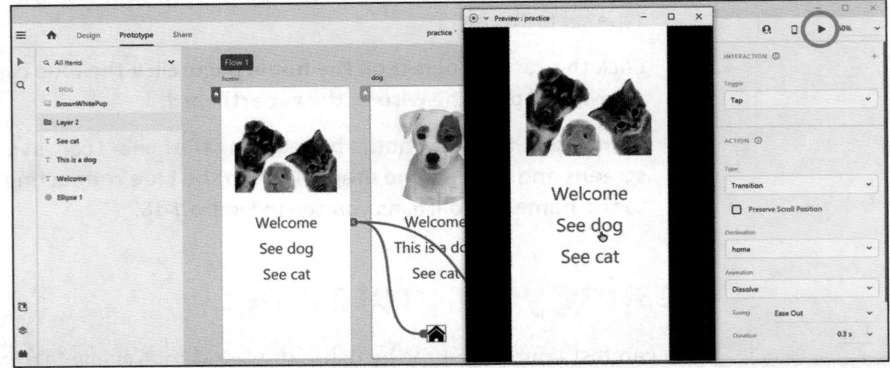

FIGURE 1-17: Preview your prototype by clicking the Desktop Preview button.

» Understanding the relationship with layers

» Adding objects to your artboards

» Adding the scroll feature

» Using artboard guides

Chapter 2
Working with Artboards

Artboards is a term used in many of Adobe's applications, but in Adobe Experience Design, artboards represent screens for your application or pages in your website. Without screens, you have little interactivity, so it is important that you know how to add, delete, edit the size, and organize these artboards. You spent a little time on artboards in the previous chapter, but now you have the opportunity to dive further into the details.

Starting Your Document and Adding an Artboard

Throughout this chapter, lessons take you step-by-step through the process of working with artboards in XD. You can follow along, or you can just read so that you understand the important points you should know about artboards.

When you first launch XD, a Start screen appears that allows you to start your XD document with an artboard sized for your needs. You can see additional related sizes by clicking on a category such as IPhone, Web, Instagram Story, and Custom. You can also enter your own size by typing in values underneath the Custom selection, as seen in Figure 2-1.

To follow along, choose the IPhone X. XS, 11 Pro (375 x 812) option underneath the IPhone category.

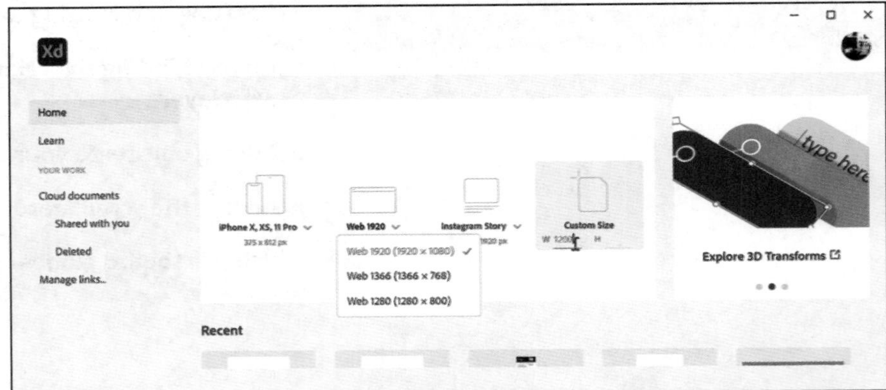

FIGURE 2-1: In the Start menu, choose from preset artboard sizes or type in custom values.

Your document appears with an artboard ready so that you can start creating.

Creating New Artboards

In this example, several artboards with content are created so that you can practice adding, duplicating, and deleting your own artboards.

As a default, artboards are named with the template that you selected followed by a slash and a number in sequential order. In this next step, you type in a custom name:

1. **Double-click the artboard name that appears on top of the artboard in the work area and change the name to Home.**

   Note: To change the name, you can also select the Layers icon at the bottom of the toolbar and double-click on the artboard name that appears in the Layers panel.

2. **Using the Text tool, click and drag to create a text area anywhere on the initial artboard that appears. Type** This is my home screen **— as shown in Figure 2-2 — so that you can visually identify it while working. You can use the Property Inspector to format it anyway that you want.**

FIGURE 2-2: Change the artboard name and add some text to the screen.

Resizing an Artboard

After you have created an artboard, you can change its size, either by dragging handles surrounding the artboard or by entering values in the Property Inspector.

TIP

If your artboard becomes unselected, you can reselect it by choosing the Select tool and then holding down the Ctrl (Windows) or Command (Mac) key and clicking on the artboard.

Adding a New Artboard

You can easily add artboards as you work in Adobe XD. These are typically the same size throughout an application, but you can change properties such as the orientation (portrait or landscape). Here's how:

1. **Add an artboard of the same size by selecting the Artboard tool and clicking in the pasteboard area of your document.**
2. **Change the orientation for this new artboard by clicking the Landscape icon in the Properties Inspector, as you see in Figure 2-3.**
3. **Double-click the default name above the artboard and change the name to** Home Landscape**.**

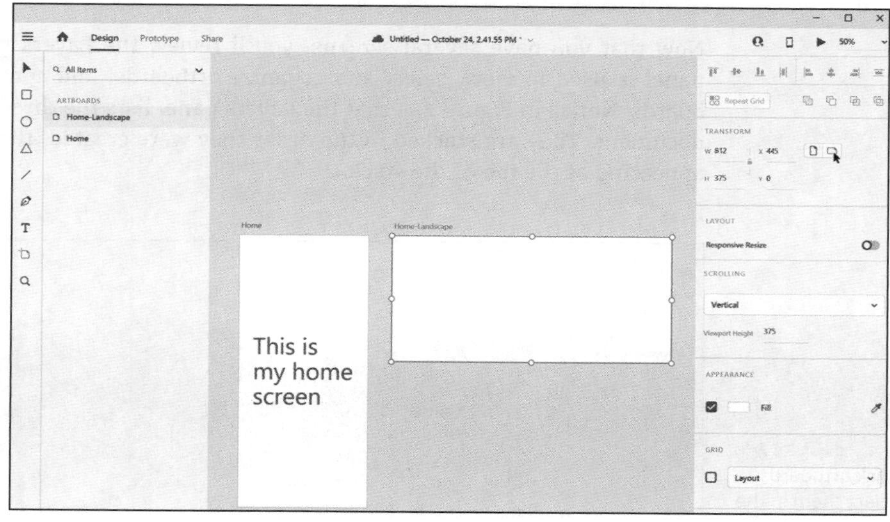

FIGURE 2-3: Change the orientation of your artboard in the Properties panel.

4. **(Optional) Add a screen from other templates using your Property Inspector. To do so, make sure the Artboard tool is still selected and scroll down in the Properties panel until you see a list of available templates. In Figure 2-4, the iPhone X, XS, 11 Pro option was selected, and a new artboard was created to that size.**
5. **Double-click the name of this artboard and change it to** Gallery.

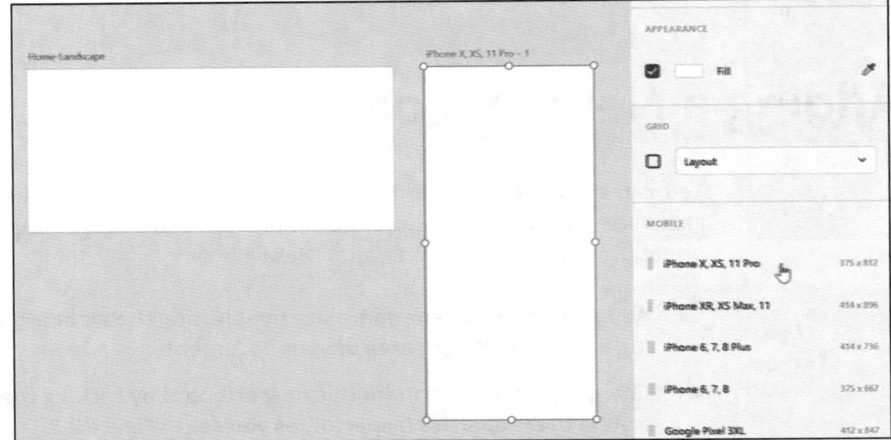

FIGURE 2-4: Add a new artboard by selecting a template in the Properties panel.

Using the Layers Panel

Now that you have several screens, you'll review the Layers panel. The Layers panel is used to view, select and organize artboards, and content on your artboards. Notice in Figure 2-5 that the Layers panel lists the three artboards in this document. They are stacked in the order they were created, the newest artboard appearing at the top of the stack.

FIGURE 2-5: Click artboards to select them in the Layers panel.

In the following steps, you work with the Layers panel. Keep in mind that layers in XD work a little differently than they do in programs such as Adobe Photoshop and Illustrator. In XD, each artboard, object, group of objects automatically are included as objects in your layers panel.

Adding Objects to Your Artboards

Follow this example to see how XD organizes objects within the appropriate artboard:

1. **Select any artboard by clicking it with the Select tool.**

   In this example, the center, Home-Landscape artboard is selected.

2. **Select any shape from the Tools panel and click and drag to add an object to that artboard.**

3. **Repeat Step 2 so that you have two objects on the artboard.**

   Notice in Figure 2-6 that the Layers panel has changed to show you only that active artboard with the objects on the artboard listed beneath it.

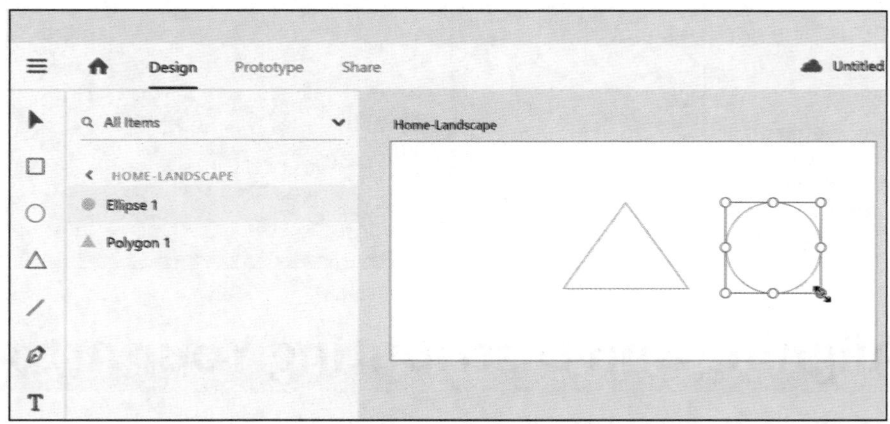

FIGURE 2-6: Objects are listed underneath the layer that they are associated with.

4. **Click the arrow to the left of the artboard name to return to the view where all artboards are listed.**

   This artboard and object layer organization helps to make it easier to find items on specific artboards.

Arranging Artboards in Your Pasteboard Area

You can reposition an artboard within the pasteboard area by clicking and dragging on the name of the artboard. If you would like to select multiple artboards, use any of these methods:

- Shift+click (while holding down the Control key on a Mac) to add artboards to your selection.
- Click and drag a marquee around the entire artboard(s) that you want to select, as shown in Figure 2-7.
- Click and drag through the artboard names in the Layers panel.
- Shift-click to add artboards to the selection in the Layers panel.

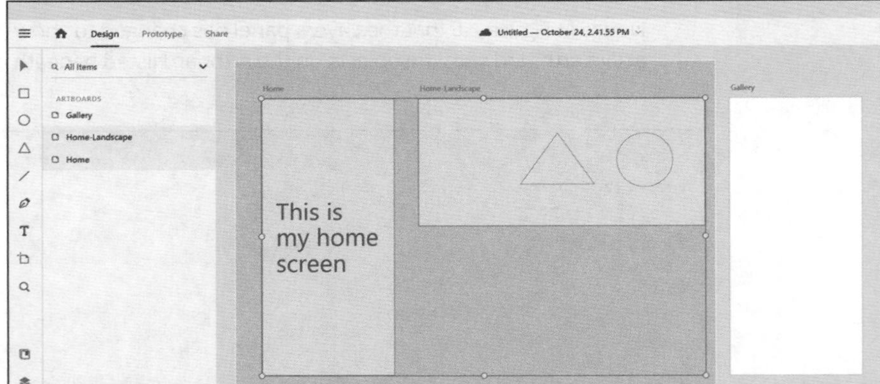

FIGURE 2-7: Click and drag to surround the entire artboard(s) that you want to select.

Aligning and Distributing Your Artboards

If you find your pasteboard is becoming messy and unorganized, you can align and distribute your artboards using the Properties panel. To do this, follow these steps:

1. **Select the artboards that you wish to reposition using the Align and Distribute features.**

2. **Locate the Align and Distribute icons, shown in Figure 2-8, and click any Align option and any Distribute option.**

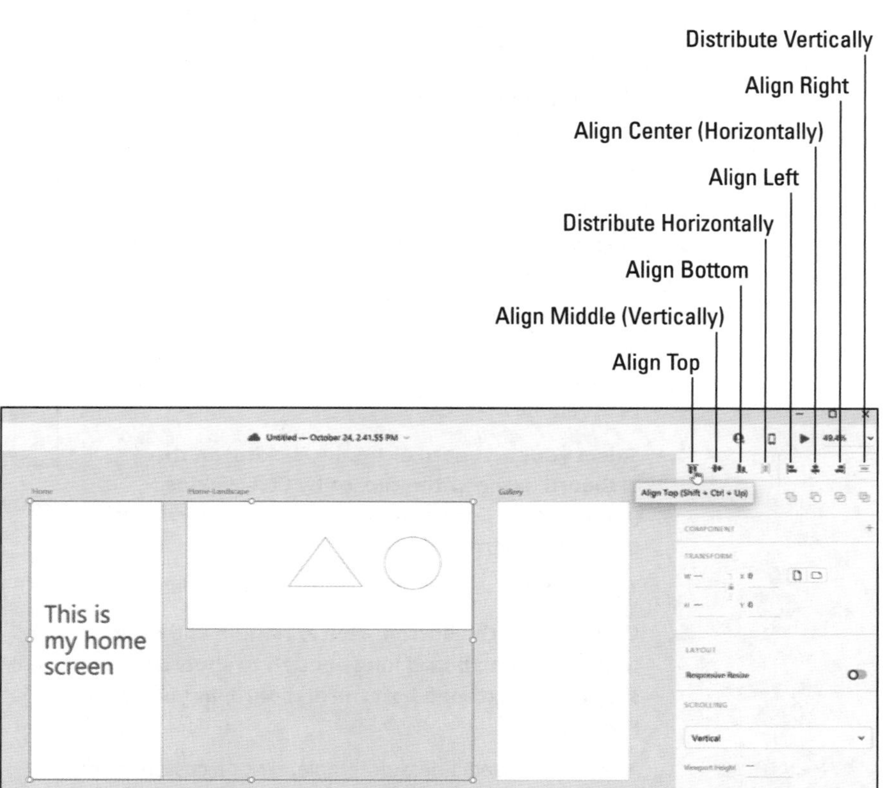

FIGURE 2-8: The Align and Distribute icons.

Adding Scrolling Capability to an Artboard

Often you want to show items scrolling in your prototype. You can simulate scrolling by changing the size of the artboard while keeping something called the View Point at the original artboard size. If you were following along earlier, you can use any of the artboards you have created, or create a new file with an artboard of any size.

You first need to create content that extends beyond your screen area. For this, we add some easy shapes:

1. **Select the Ellipse tool and click and drag to add a circle to your artboard.**
2. **Click Repeat Grid in the Properties panel.**

   This is an incredible feature that you find out more about in Chapter 5, "Working with Text in Your XD Project."
3. **Click and drag the bottom handle to repeat the circle shape, as shown in Figure 2-9.**

   Make sure that you drag beyond the bottom of the artboard to see the scrolling in action. You can adjust the spacing of the shapes by clicking and dragging in between the objects.
4. **Select your artboard either by clicking on the artboard name at the top, or by Ctrl-clicking (Windows) or Command-clicking (Mac) anywhere on the artboard.**

   You know it is selected when the handles appear.
5. **Click only on the bottom-center handle and drag to make the artboard long enough to show the shapes you extended off the artboard, as you see in Figure 2-10.**

   Note: If you used a default template (not a custom-sized artboard), a dotted guide indicates the height of the original artboard. Any area beyond this Viewport is in the scrolling area.

   Note: If you are using a default template, the Viewport Height value will not change when you click and drag to make your artboard longer. If Viewport Height does change, you more than likely created your artboard with custom values. No worries; just select Vertical from the Scrolling drop-down menu and enter the original height of your artboard into the Viewport text field after you extend your artboard.
6. **Now you can test the scrolling by clicking the Desktop Preview button in the upper right of your workspace. Note that when the preview appears you can scroll through your content.**

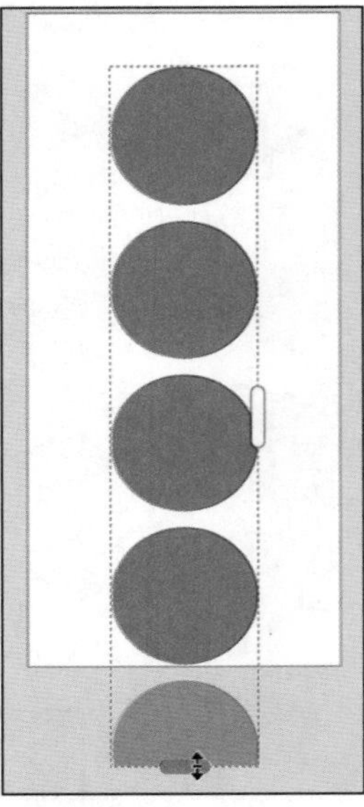

FIGURE 2-9: Click and drag the bottom handle to repeat the shape.

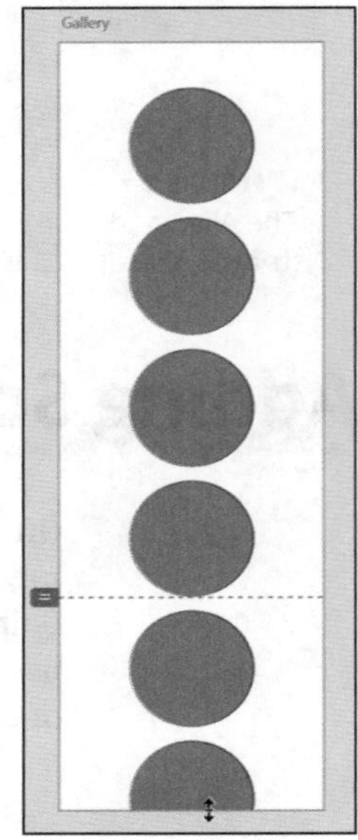

FIGURE 2-10: Click on the center bottom handle of the artboard and extend it.

Adding Guides to Your Artboard

XD offers a handy guide feature to help you align objects on your artboard. By clicking outside the left or top side of your artboard and dragging into your artboard you can create a guide, as shown in Figure 2-11. This guide is not visible in the preview mode and can be repositioned as many times as you want. If you don't need it anymore, simply click and drag it back off your artboard.

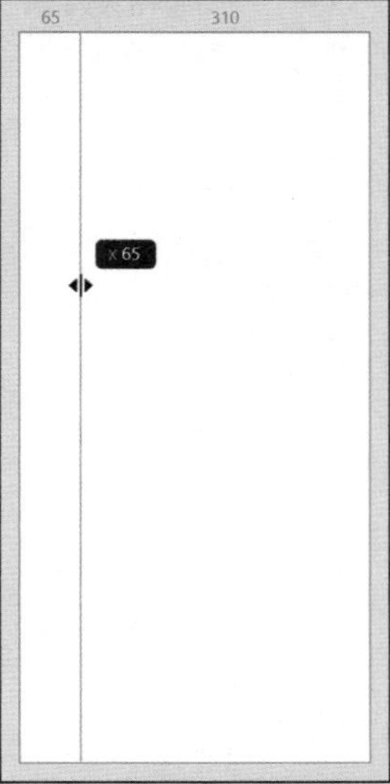

FIGURE 2-11: Drag a guide in from the side or top of your artboard.

TIP

If you don't see the guides, right-click or Control-click and select Guides⇨Show All Guides.

If you enjoying using the guide feature, here are some helpful keyboard shortcuts. Keep in mind that you must have selected the guides in order to use the Show and Hide keyboard shortcut.

Function	Mac shortcut	Windows shortcut
Show/ Hide artboard guides	⌘ + ;	Ctrl + ;
Lock/Unlock artboard guides	Shift + ⌘ + ;	Shift + Ctrl + ;

Now that you are familiar with the basics of creating and organizing artboards, you can start discovering more about how you can organize your content on those artboards.

» Applying fills

» Applying borders

» Customizing and saving shapes

» Using the Pen tool

» Discovering Pen tool tips and tricks

Chapter 3 Creating Your User Interface (UI) with Shapes, Paths, and Custom Shapes

You can create your User Interface using objects such as squares, circles, custom shapes, and paths right in Adobe XD. If you are familiar with Adobe Illustrator, the tools in XD are similar.

Working with the Basic Shapes

In previous chapters, you were introduced to the steps to create simple shapes and change their properties; in this chapter, you discover how to create these shapes and turn them into the UI items you need.

What's included in XD

There are three main shapes to choose from, Rectangle, Ellipse, and Polygon, but you can customize these foundational shapes to fit any need. Combining them with the Boolean feature allows you to add and delete shapes from each other, and using the Pen tool you can pretty much create any UI elements you need.

If you prefer keyboard shortcuts, you will enjoy the list in the following table. You don't have to press Ctrl/Command to use these shortcuts; simply press R and click and drag to create a rectangle, E for an ellipse, Y for a polygon, and so on.

Tool	Windows	Mac
Select	V	V
Rectangle	R	R
Ellipse	E	E
Polygon	Y	Y
Line	L	L
Pen	P	P
Text	T	T

Determining the size

You might want to create a new document and add an artboard so that you can follow along in this chapter. The Artboard tool has a keyboard shortcut too; it is the letter A. Simply press A and then select your artboard size from the Properties panel.

Once you have an artboard go ahead and create a shape, any shape by selecting the Rectangle (R), Ellipse (E), or Polygon (Y) tool, and then clicking and dragging the shape on your screen.

After you have created the square or circle, you can check and change the size by using the Properties panel. In Figure 3-1, you see an active rectangle and its W (Width) and H (Height), as well as its positioning, which is represented in the X and Y coordinates. In most cases, you don't have the luxury of randomly create shapes of various sizes, so the Properties panel helps you pay attention to those details.

Note: The shape tools are sticky, and the tool will remain selected until you switch to another tool such as the Select tool. Keep this in mind because if you try to move your shape, you may just accidently create another shape.

Using a Shape to Create a Button

In this section, you will create a simple button with a solid fill and some text. To follow along, have an artboard of any size ready to go:

1. **Select the Rectangle tool and click and drag to create a rectangle of any size on an artboard.**
2. **With the rectangle still active, type** 200 **into the Width textbox and** 70 **into the Height textbox. Press Tab after you type each to confirm the value.**

   Now add a solid fill color to your shape.
3. **Make sure your rectangle is still selected. Click the white box to the left of Fill in the Properties panel to open the color picker.**
4. **Scroll through the colors by clicking and dragging on the Hue slider. After you have entered a hue color that you like, click a color in the color panel to apply it to your rectangle.**

TIP

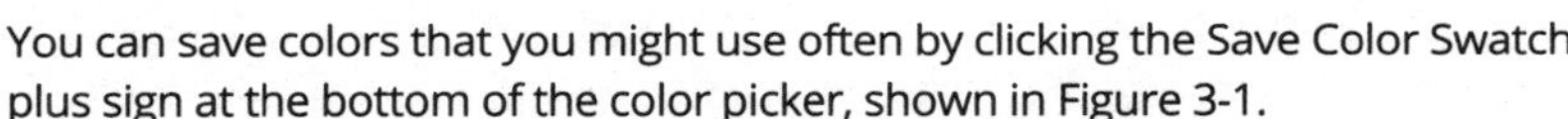

You can save colors that you might use often by clicking the Save Color Swatch plus sign at the bottom of the color picker, shown in Figure 3-1.

TIP

You can apply transparency to your fill by using the slider on the right side of the color picker window.

FIGURE 3-1: Save a color by clicking the Save Color Swatch plus sign.

You can leave the corners square or you can round the corners of the shape by either of the following options:

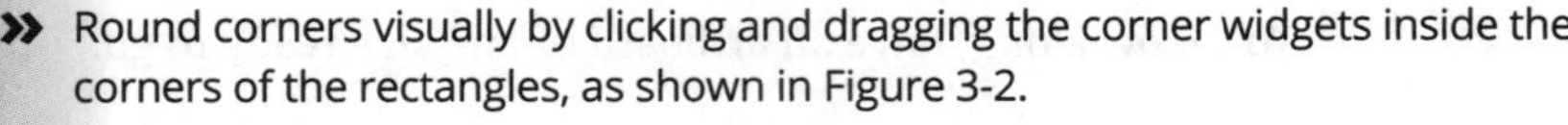

TIP

- Round corners visually by clicking and dragging the corner widgets inside the corners of the rectangles, as shown in Figure 3-2.

  You can hold down the Alt (Windows) or Option (Mac OS) key and drag one corner radius to make it an independent value.

- Enter values for your corner radius in the Corner Radius textboxes in the Properties panel. Notice in Figure 3-2 that you can enter different values for different corners.

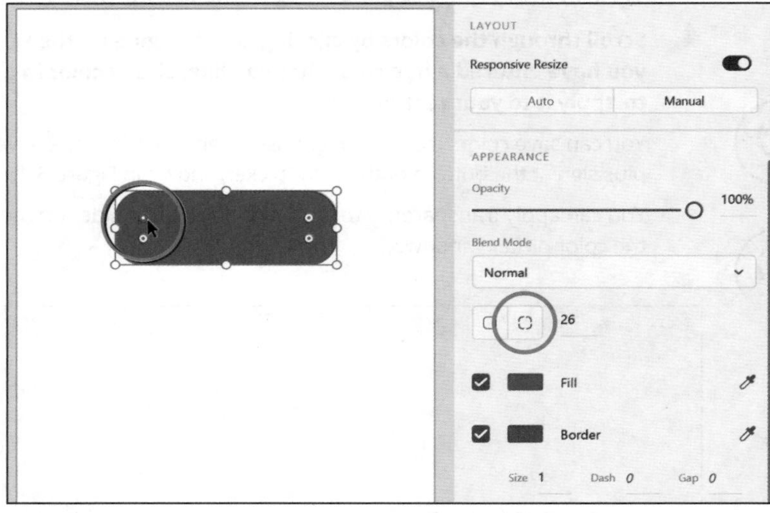

FIGURE 3-2: Click and drag the corner widgets to round your corners, or enter values into the Properties panel to add varying sizes in the corner radius.

Creating a border

A *border*, also known as a *stroke*, can be applied to any of your shapes and paths. When you create a border, you select both the width and color.

1. **Select the rectangle you created for the button in the previous section.**
2. **Make sure the Border in the Properties panel is checked; then click the box to the right of the checkbox to open the color picker and select a color that you would like to use for your border.**
3. **To select a width, type a value into the Width textbox.**

As mentioned in Chapter 1 of this minibook, you can change other properties of your border by adding or changing the values for dashes, gaps, stroke alignments, and more. (See Figure 3-3.) To see how some of these properties can affect your border, open the file named `Examples of borders` located in the Book07_XD folder that you downloaded with the rest of the class files in Book 1, Chapter 1.

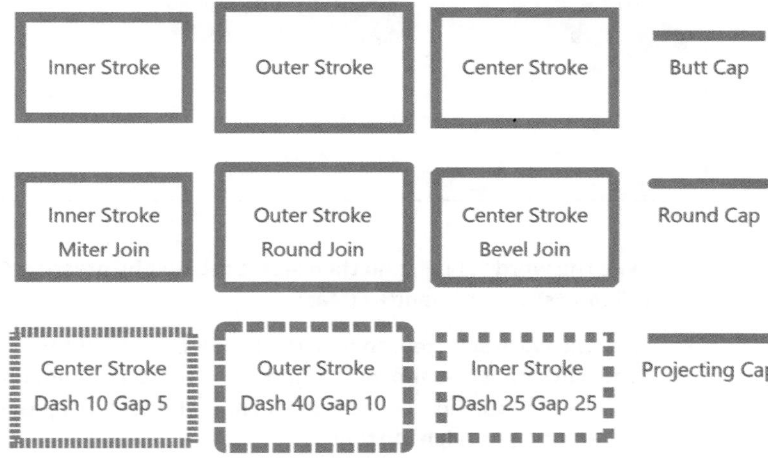

FIGURE 3-3: Examples of how the properties for a border look when applied to an object. This example has a 25 pt. stroke.

Sampling a color using the eyedropper

Both the Fill and the Border have an eyedropper icon positioned off to the right. You can select the Eyedropper tool in the Properties panel and then click any color located within your current project. This can be helpful when you have imported images that you want to sample colors from, as shown in Figure 3-4, or if you are trying to match controls that were created in a different application.

Adding text to your button

Now add the text to the button:

1. **Select the Type tool from the Tools panel and then click inside your button.**

   A text cursor appears. Don't worry about the exact placement of the textbox because this can be moved later.

TIP

   If you click and release to create your text, you enter text at a point. When you type text by clicking on a point on your canvas, you get a horizontal line of text that begins where you click and expands as you enter characters. The line expands or shrinks as you add or delete text, but it won't wrap to the next line. This is just an easy way of adding small amounts of text to your artboard.

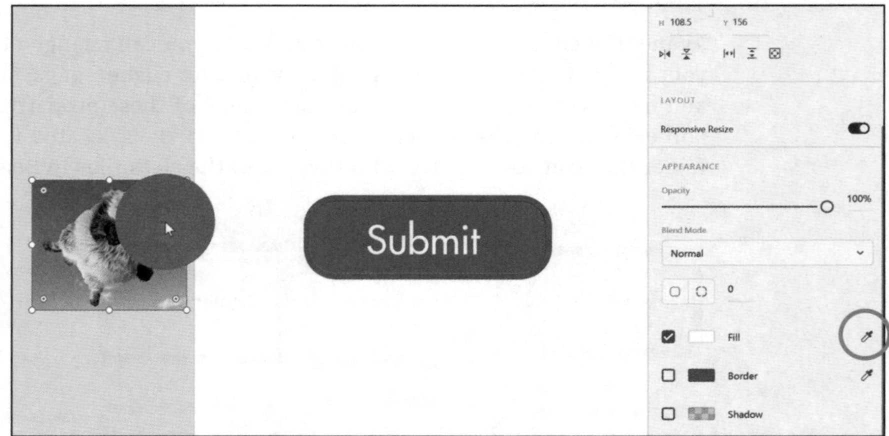

FIGURE 3-4: Use the eyedropper to sample colors from your project.

2. **Type the word** Submit**, and then select the text by pressing Ctrl+A (Windows) or Command+A (Mac).**

3. **With the word selected, change the properties in the Text area of the Properties panel, as shown in Figure 3-5, to select a font and a type size you like. You can also change the alignment by selecting the Left Align, Center, or Right Align buttons.**

TIP

You can select your current font size in the Font size textbox and press the arrow up or down to increase or decrease the font size.

4. **Click the Fill box to change the color of your text.**

Creating a gradient

Adobe XD includes advanced tools that make it simple to create gradients. To create and use a gradient in your project, follow these steps:

1. **Create or select any shape.**

   At the time of writing, a gradient can only be applied to a fill.

2. **Click the box to the left of Fill in the Properties panel to open the color picker.**

3. **Click Solid Fill at the top of the picker and from the drop-down menu select Linear or Radial Gradient, as shown in Figure 3-6.**

   A gradient ramp appears with default colors.

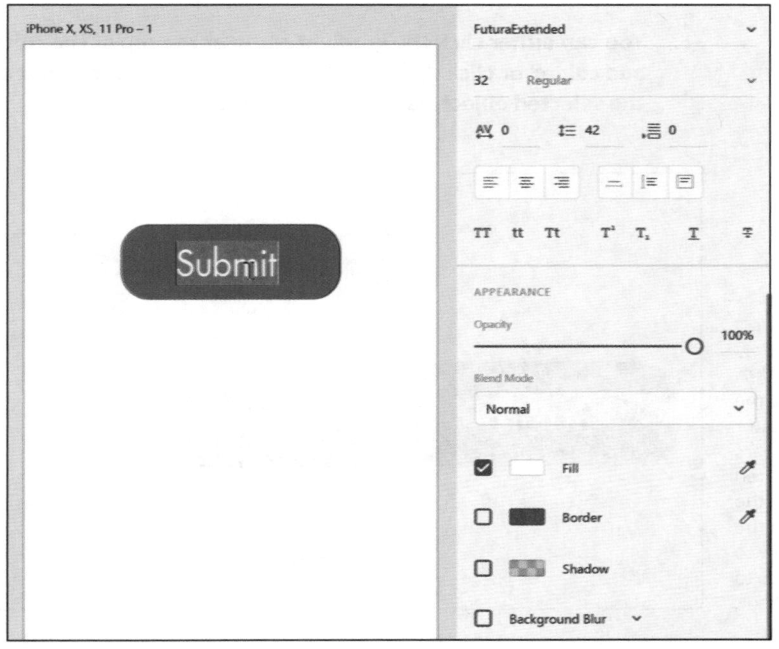

FIGURE 3-5: Select your type properties using the Text section in the Properties panel.

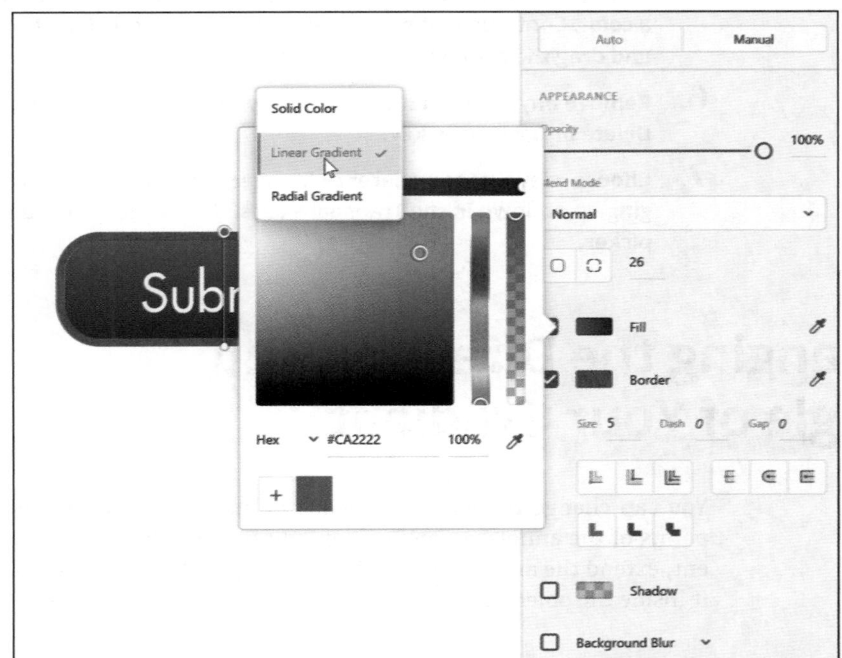

FIGURE 3-6: Select Gradient from the color picker.

4. **You can either click the gradient ramp at the top of the color picker to add colors, or click directly on the gradient annotator that appears inside the selected object, as you see in Figure 3-7.**

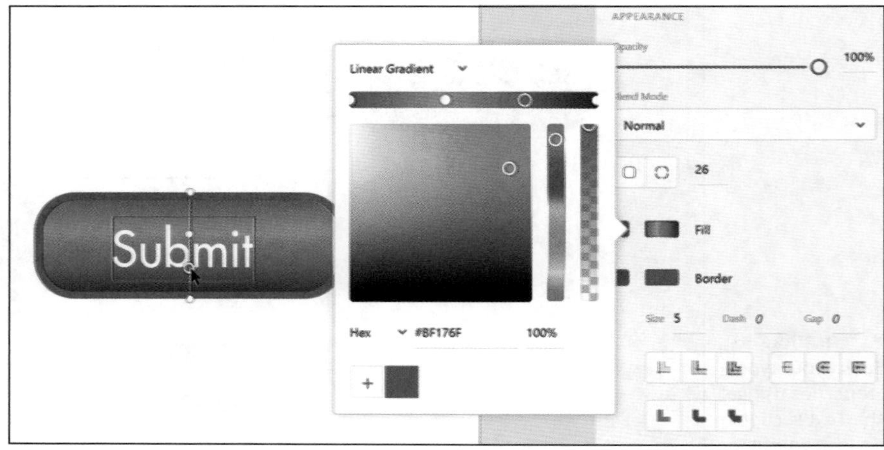

FIGURE 3-7: You can click the color annotator that appears on the selected object, or on the color ramp in the Fill window to add color stops to your gradient.

5. **After you click to add a color stop, you can then use the picker to select a color. Continue adding color stops with various colors if you like. Click and drag the color stops to reposition them.**
6. **Remove any unwanted color stops by selecting one and then pressing the Delete or Backspace key.**
7. **Choose a level of transparency by selecting a color stop and then dragging up or down in the Transparency slider on the right side of the color picker.**

Changing the Direction and Angle of Your Gradient

You can change the length or radius of a gradient quickly by grabbing the end points of the annotator, and repositioning them. To create a more gradual gradient, extend the annotator outside of the object; to make it more compressed, drag it inside the object, as shown in Figure 3-8.

FIGURE 3-8: Click and drag the gradient annotator in order to change the direction and distance of a gradient.

TIP

If you spent a great amount of time creating just the perfect fill and stroke, you can copy that appearance and apply it to other objects. Simply right-click on the object whose properties you want to copy, and choose Copy from the contextual menu. Then right-click on another option and choose Paste Appearance.

Using the Pen Tool

Using the Pen tool in Adobe XD is very similar to using the Pen tool in other Adobe applications. See Chapter 4 in Book 5 for more in-depth features that you can take advantage of with the Pen tool in all the Adobe applications. In this section, you are provided with simple tips to help you create your own custom shapes and paths:

- Create a straight path by clicking a start point and then Shift-clicking to create an end point. This creates a path that is either perfectly straight, 45 degrees, or 90 degrees.
- If you want to stop the Pen tool without closing a shape — in order to create a straight line, for instance — simply press the Esc key.
- When you make a mistake, press Ctrl+Z (Windows) or Command+Z (Mac) in order to undo the step; keep the Pen tool active so you can continue creating your path.
- Create a curved, Bézier path by clicking and dragging, as shown in Figure 3-9.

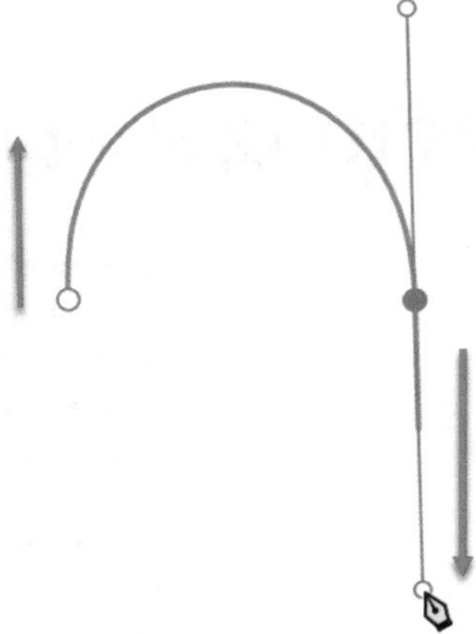

FIGURE 3-9: Click and drag with the Pen tool to create a custom Bézier path.

- Change the direction of a path by Alt- (Windows) or Option- (Mac) clicking and dragging on the directional line. (See Figure 3-10.)
- Create a closed shape by clicking and connecting from one point to another. Closed shapes can be filled to create UI elements and masks, as shown in Figure 3-11. More about images and masks is covered in Chapter 6 of this minibook.

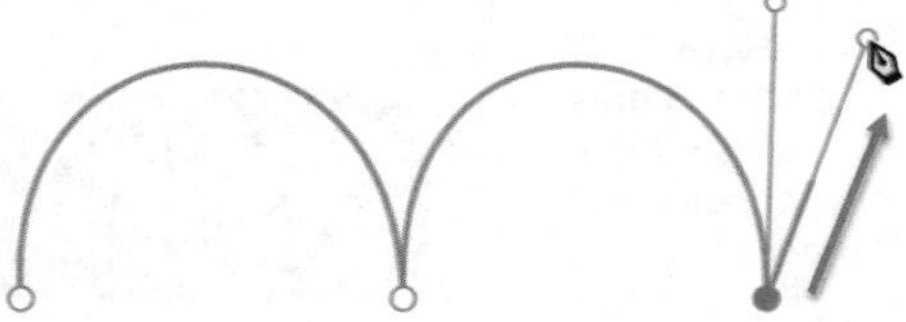

FIGURE 3-10:
Hold down the Alt or Option key to change the direction of your curved Bézier path.

FIGURE 3-11:
The Pen tool can be used to create custom shapes in XD.

Editing shapes

When working with objects in XD you create shapes and paths in the Design mode or double-click on an object to enter the edit mode. In the edit mode you can customize the shape by selecting and moving individual anchor points, adding new anchor points or deleting anchor points, as you see in Figure 3-12.

FIGURE 3-12:
Double-click to edit you shape. You can then move, add, or delete anchor points.

To release the edit mode, simply click off the object or press the Esc key.

Creating Custom Shapes

If your goal is to create a custom shape without using the Pen tool, you can take advantage of the following options in the Property Inspector:

- **Add:** Choose two or more shapes that you would like to unite together and choose Add.
- **Subtract:** Select two or more shapes and then click Subtract. All of the shapes arranged on top are subtracted from the bottom shape.
- **Intersect:** Select two or more shapes and click Intersect. Only the portion of the shapes that overlapped remain.
- **Exclude Overlap:** Select two or more shapes and click Exclude Overlap. Any areas where shapes are overlapped are removed.

See Figure 3-13 for examples of examples of these path selections.

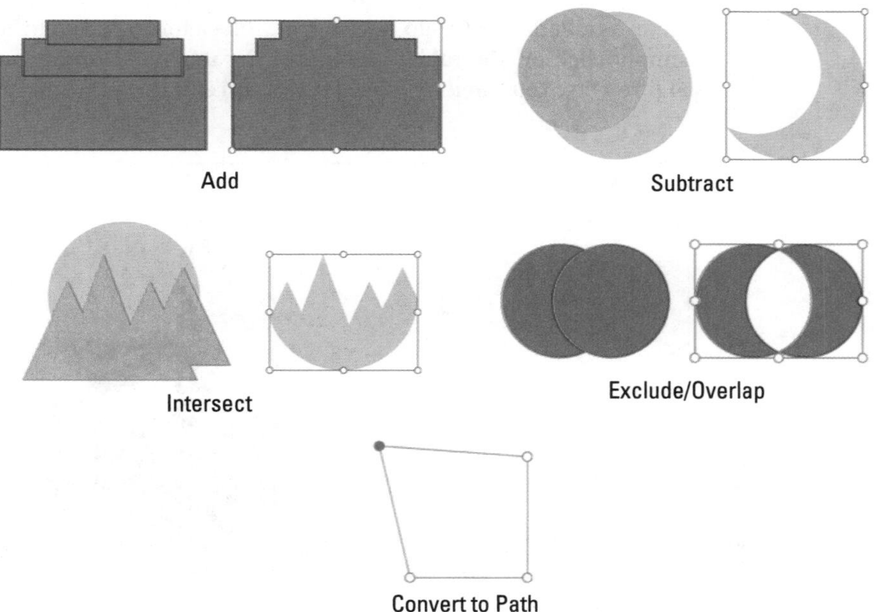

FIGURE 3-13: Examples of the Path tools in the Path menu item.

Adding lines

Using the line tool you can create a rule in your UI. When the line has been created, you can use the same options that are available for borders. Change the width, add dashes, or even control the end points to be rounded using Round caps, or to be squared off using the Butt cap.

To create a straight line that falls on a 90 or 45 degree angle, simply select the Line tool and then click and drag while holding down the Shift key, as shown in Figure 3-14.

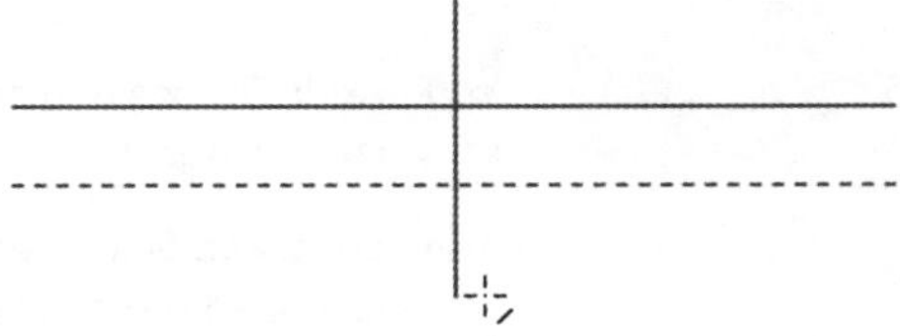

FIGURE 3-14: Click and drag while holding down the Shift key to create straight horizontal and vertical lines.

Importing from Illustrator

Keep in mind that you can cut and paste shapes and other art directly from Illustrator into XD. If you are copying and pasting a vector object you can continue editing it right in XD. Objects that are copied and pasted into XD land in the layers panel as a Path or a Group, which looks like a folder. To edit the individual objects simply click on the folder and select the individual components, as you see in Figure 3-15. You can then resize, color, and edit the points of the pasted vector art.

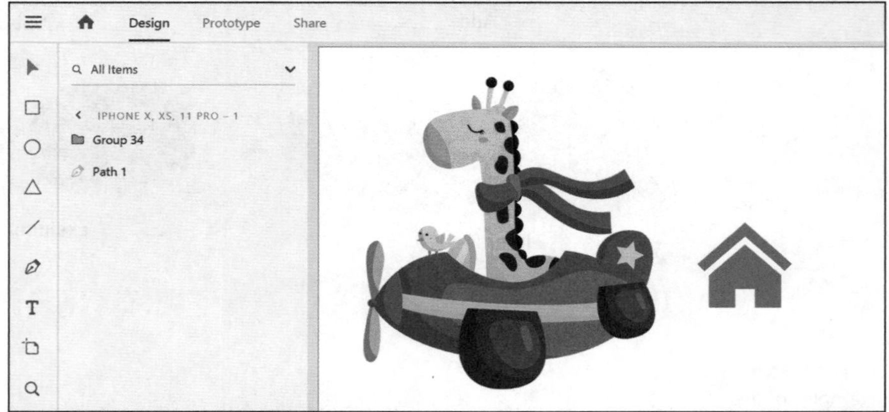

FIGURE 3-15: You can edit vector artwork that was copied and pasted from Illustrator.

» Grouping and ungrouping

» Cloning and duplicating objects

» Transforming objects

» Aligning and distributing

Chapter 4
Building Your UI

Now that you have experience working with the tools and the Properties panel, you can put it all together and start designing your UI.

Keep in mind that the terms *layer* and *object* are used interchangeably in this XD minibook, as each object automatically appears as layer in the Layers panel. Even if an item is grouped, it still exists as its own layer within the larger group.

Throughout this chapter, you discover how to organize, select, and group your objects in Adobe XD. You also find out how to scale and rotate your objects, as well as align and distribute them evenly.

Layers and the Objects They Contain

As previously mentioned, each object that you add to an XD project becomes its own layer. As you add new objects, they stack on top of the existing layers, not only visually, but in the Layers panel as well. The newest object is always at the top of the stack.

Fortunately, XD helps you more readily locate these objects by keeping them organized by artboard. You can narrow down the objects that you see in the Layers panel by selecting an artboard.

You are going to experiment with the layers panel in this chapter. You'll discover a wide amount of helpful features like how to use layers to select, lock, duplicate, group, and align objects.

Artboards and layers

The best way to explain layers is for you to give them a try yourself. Follow these steps to start your own document:

1. **Create a new file and select any size template that you wish.**

   In this example, we selected the iPhone X, XS, 11 Pro template.

2. **Click the Layers icon in the lower-left of the window, as shown in Figure 4-1, to see that the artboard, which was created by default, is shown in the panel labeled Artboards.**

   You can also use the keyboard shortcut Ctrl+Y (Windows) or Command+Y (Mac) to toggle the view of this panel on and off.

   At first, you might be wondering why an artboard is listed in the Layers panel, but don't worry — it will make sense after you start adding objects to this artboard. Note that the artboard has the default name based upon the template you selected and the screen number that is associated with this document.

FIGURE 4-1: Click the Layers icon to reveal a panel that displays your initial artboard.

3. **Select the Artboard tool from the Tools Panel and click anywhere in the work area to add a second artboard.**

   Note that you now have two artboards listed, with the newest artboard appearing at the top of the list and with a -2 following the template name, as shown in Figure 4-2.

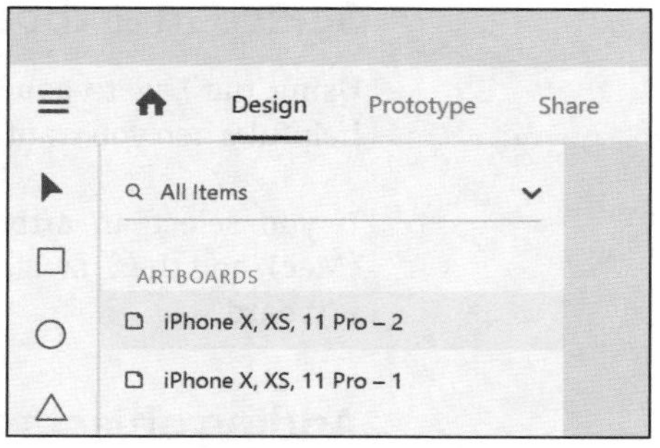

FIGURE 4-2: As you add artboards, they appear at the top of the stacking order. The last one you added is on top.

4. **Keep this file so that you can experiment with other layer specific features discussed throughout this chapter.**

What can you do with your artboards inside the Layers panel?

Using the Layers panel that appears when you click on the Layers icon, you can rearrange the stacking order, rename, duplicate, delete, and select artboards.

Rearrange the order of the artboards

Click and drag up or down an artboard in the Layers panel in order to move it up or down in the stacking order. If you like working with contextual menus, you can right-click any artboard listed in the Layers panel to see a list of features you can use, including Copy, Paste, Duplicate, Delete, Rename, and Export. Following are other methods you can use to take advantage of these same features.

Rename your artboards

As you add additional artboards, it is important to name them correctly. This is a major part of the structure of your UI, as you will not only link from one artboard to another, you will also drop UI elements on various artboards throughout the process. It is important that you always find what you are looking for! You can rename your artboards using one of these methods:

- Double-click the artboard name that appears directly above the artboard, and then type a new name.
- Double-click the artboard name that appears in the Layers panel, and type a new name.

Select an artboard in which to add an object

Using the Layers panel, you can select a specific artboard to activate it; this is helpful when you want to import an image and have it land on the correct artboard.

If you select an artboard, choose Menu⇨Import, (Windows) or File⇨Import (Mac), and then navigate to select an image, it automatically lands on the selected artboard.

Adding objects and how they appear in the Layers panel

Hopefully, you are starting to understand that the Layers panel shows you every item that is included in your project. For this next section, it would be helpful if you work with an example of at least two artboards with several shapes or other objects on them. If you want to use our exercise file, look for the `LayerExercise` file inside the Book07_XD folder that you downloaded with the rest of the DummiesCCFiles in Book 1, Chapter 1.

When you have the `LayerExercise` file open and when nothing is selected, note the three artboards in the Layers panel. If you click on an object such as the image on the Snacks artboard, you now see only the items on that artboard in the Layers panel, as shown in Figure 4-3.

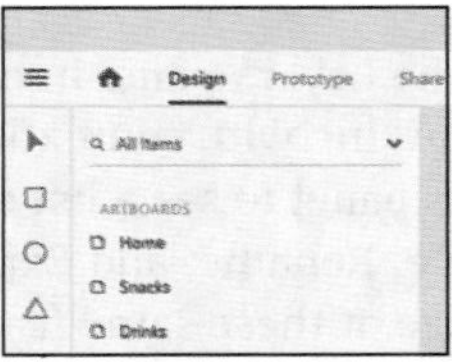

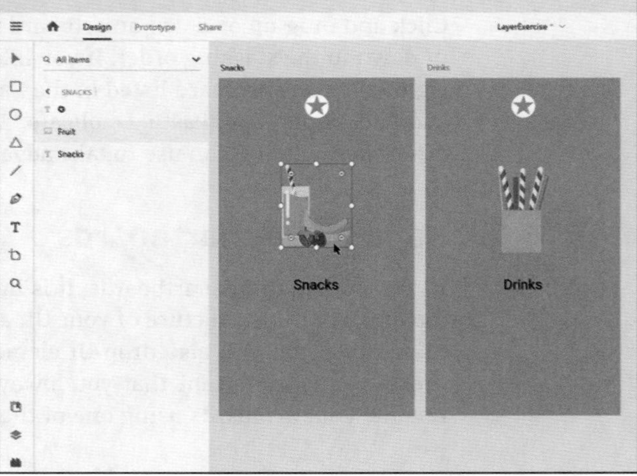

FIGURE 4-3: On the left, you see all artboards; on the right, you see one artboard as an object on that artboard is selected.

Note: You only see the active artboard and any objects that are on that artboard. If you want to see another artboard, you can either press the back arrow to the left of the artboard name in the Layers panel or simply select an object on a different artboard in your work area. This either takes you back to the entire list of artboards, or directly to the artboard with the selected object.

Keep this file open if you want to experiment with the other layer and object features discussed throughout this chapter.

Finding artboards and objects

Speaking of keeping artboards organized so that you can find the objects you need, note that there is also a Search feature built right into the top of the Layers panel. If you name an Artboard "Snacks," for instance, you can easily filter out the other artboards in the Layers panel by typing in "Snacks" into the Search feature. See Figure 4-4.

This Search feature also works to help you find individual objects, across all artboards. In Figure 4-4, note that three objects with *blue* in their names are found in two separate artboards; you can find these by typing *blue* into the Search feature. When done searching, make sure to click the X to the right of your search term to see your other objects.

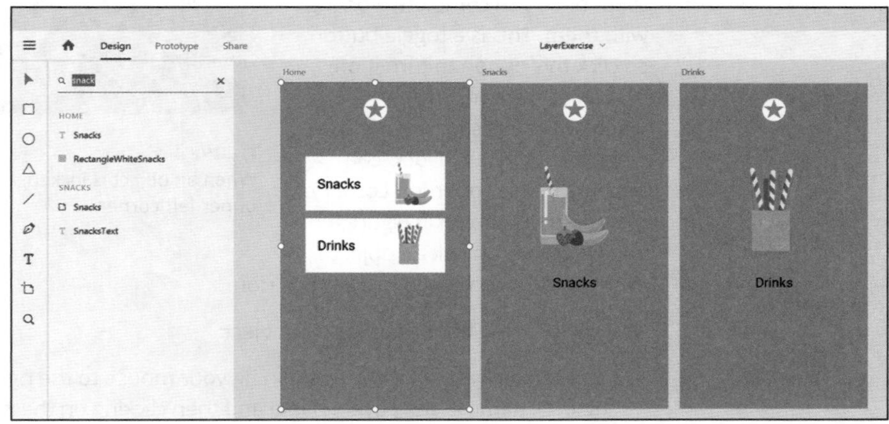

FIGURE 4-4: Use the Find feature in the Layers panel to find artboards and objects by name.

Exporting, hiding, and showing your layers

When you position your cursor over a layer, you see related options appear on the right of the layer name, as shown in Figure 4-5. These options may be difficult to find at first but are very useful. Objects have three options, artboards have the Export option.

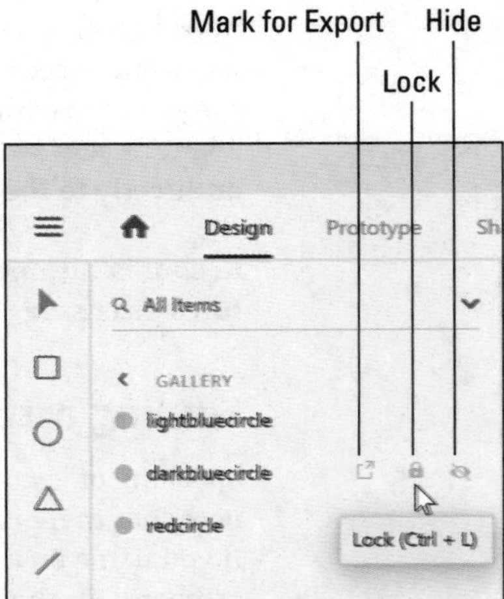

FIGURE 4-5:
Layer options appear to the right when you position your cursor over the layer.

- **Mark for Export:** This is how you tag something that you want exported when saving your file for development. You'll find out more about marking for export in Chapter 9, "Sharing your XD Project." Essentially, marking for export is how you will save artwork out in a selected format such as .svg, .png, or .jpg to use in development. You determine the format if you choose Export from the menu in the upper-left (Windows) or File (Mac).
- **Lock:** To keep other objects in position, you may want to lock them once you complete working with them. This is a toggle button so click this option to turn it off and on as needed. You can tell when an object is locked because when it is selected, a lock icon appears in the upper-left corner of the object, as shown in Figure 4-6. You can easily unlock an object right on the artboard by clicking this icon.

FIGURE 4-6:
When an object is locked, a lock appears in the upper-left corner.

  Use any of these options to lock an object:

  - You can lock an object by positioning your mouse to the right of the object's name in the Layers panel and then clicking on the padlock icon that appears. You can also click in that same location to unlock an object.
  - You can lock an item by right-clicking (Windows) or Control-clicking (Mac) and selecting Lock from the contextual menu. After an item is locked, you can choose Unlock from the same menu.

- In order to lock an item while you are working on the artboard, you can press the keyboard shortcut Ctrl+L (Windows) or Command+L (Mac). The keyboard shortcut is a toggle key, meaning that you can select the locked object and press Ctrl/Command+L again to unlock it.

» **Hide:** You can hide an object to see what is behind it without moving it, or to try various options in your design. To hide an object, click the eye icon that appears when you cross over the area to the right of the named object in the Layers panel. If you like keyboard shortcuts, press Ctrl/Command + , (comma). At the time of this writing there is a feature request for an Unhide All option.

Deleting an artboard

Be careful with deleting artboards. It is very easy to do, because you only have to select an artboard in the Layers panel and press the Delete or Backspace key. Fortunately, this step can be undone using Ctrl+Z (Windows) or Command+Z (Mac OS).

Renaming objects

To keep your files organized and make associating interactions easier, consider naming your objects with a logical naming structure. Keep your file names in lowercase or with just an initial capital (no spacing) to keep it uniform with the typical standard for naming files used in development. Most importantly . . . name it something you will recognize if you try to select it from a list.

Examples might include *header_nav* or *Submit-button*.

Selecting objects

As with selecting artboards, you can switch to the Select tool and then click on an object to select it, or you can follow these options for selecting additional objects:

» Using the Select tool, click an object listed in the Layers panel and then Shift-click on any other. Using the Shift key to add to a selection activates any other objects that are in between in the stacking order displayed in the Layers panel.

» Ctrl-click (Windows) or ⌘-click (Mac OS) to select noncontiguous items in the Layers panel.

» Click one object in the artboard and then Shift-click to add additional objects into the selection. When selecting via the artboard, using the Shift key does not select objects in between, as it does in the Layers panel.

» Using the Select tool, click and drag a marquee selection across the items that you want to select, as shown in Figure 4-7. You do not have to fully encompass the object with the marquee . . . just touching it adds it to the selection.

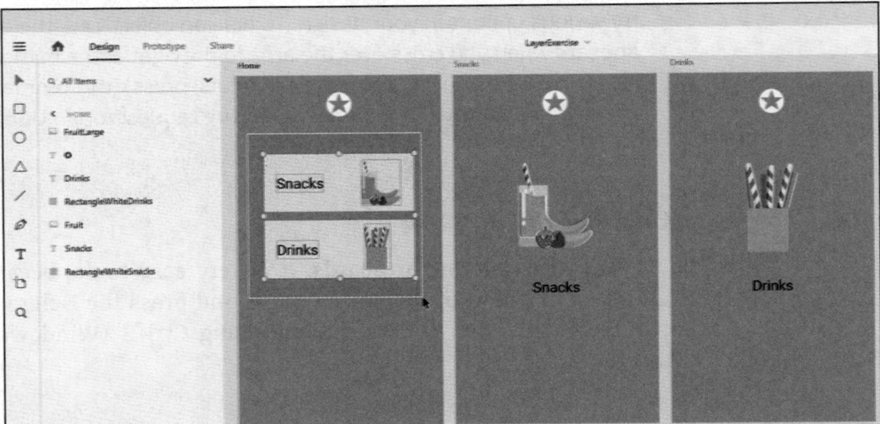

FIGURE 4-7: Click and drag using the Select tool to activate multiple objects.

Grouping and ungrouping objects

Now that you know how to select multiple objects, it seems a logical time to explain grouping. By grouping object, you can keep them together so that selecting them simultaneously is easier. This is especially important if you are building UI made of several parts, such as a button that consists of a shape, icon, and text.

You can group objects by selecting them in the Layers panel and then right-clicking to select Group. You can also group by selecting the items on the artboard and then pressing Ctrl+G (Windows) or Command+G (Mac). In Figure 4-8, you see a group created using the `LayerExercise` file.

Notice that when you first group items they appear as Group in the Layers panel. You can, and should, rename the group with a more logical name so you can identify the object(s) as you work. Do this by double-clicking the word *Group* in the Layers panel and then typing a new name in the activated textbox. Remember to keep your names in order, just like when you name objects.

Copying and pasting

Use these features if you want to copy something from one artboard to another or from one project to another. Use any of these options to cut and paste:

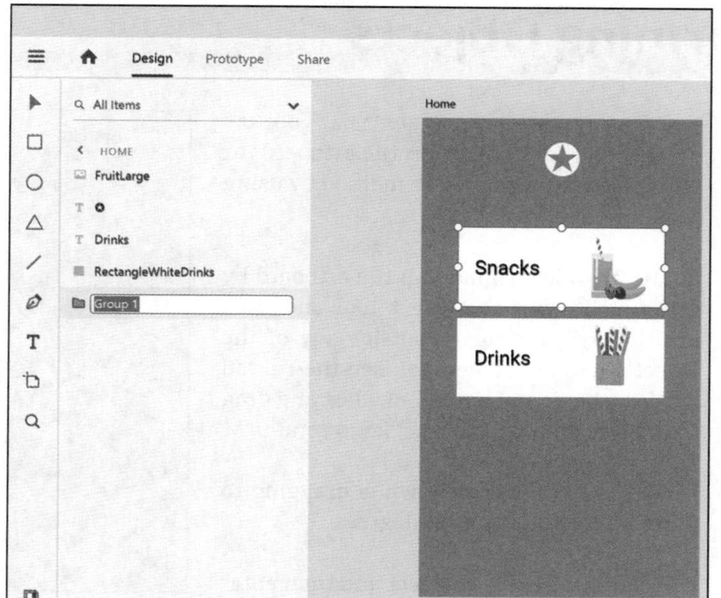

FIGURE 4-8: You should select UI elements that are meant to stay together and group them.

- Select an object right in the Layers panel and right-click it. In the contextual menu that appears, select Copy, and then right-click it again to select Paste. The object is duplicated in the same location right on the artboard. If you prefer, you can also use the typical keyboard shorts for copy and paste, which are Ctrl+C for copy and Ctrl+V, for paste (Windows), or Command+C and Command+V (Mac).
- As with artboards, you can duplicate an object by right-clicking it in the Layers panel and selecting Duplicate from the contextual menu that appears.
- Right-click a selected object and choose Copy from the contextual menu; then, go to another location in your project, right-click, and select Paste from the contextual menu in order to duplicate the object. When you copy and paste in XD, the object is pasted exactly on top of itself if you stay on the same artboard. Otherwise, it recalls the exact position and uses those same coordinates if you activate another artboard before pasting.
- You can hold down the Alt (Windows) or Option (Mac) key and drag a selected object in order to clone an object right on the artboard.

Transforming Objects

You can scale and rotate your objects quickly and easily right on the artboard, or you can use a more precise method by using the Properties panel.

Rotate an object right from the artboard by selecting it with the Select tool and then positioning your cursor outside any of the object's handles. When you see the curved arrows, shown in Figure 4-9, click and drag in the direction you would like to rotate.

TIP

Hold down the Shift key while dragging to jump in increments of 15 degrees.

You can also select an object and then enter a value in the Rotation textbox in the Properties panel.

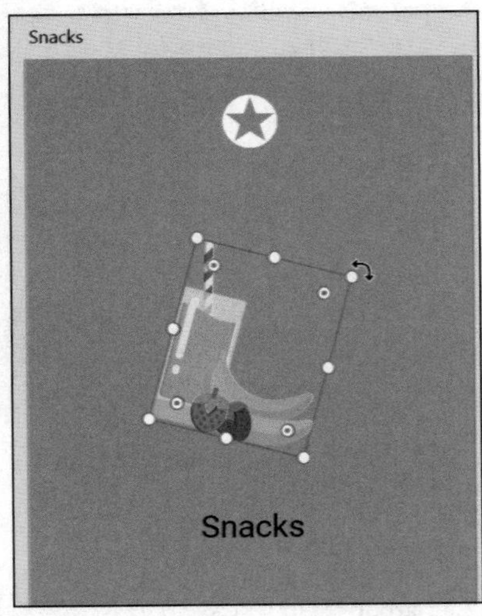

FIGURE 4-9: Click and drag outside an object's handle in order to rotate it.

Aligning and Distributing

After you start creating many objects, you need to consider the position of those objects and how they relate to each other. In order to lessen noise in your design, you want to keep icons and text aligned and distributed in a clean manner. You can check your alignment in several ways, including using a grid or using guides.

Be smart about your guides

In Chapter 2, you found out how to drag horizontal and vertical guides onto you artboard; however, XD provides you with smart guides that are visible as needed. These are the guides that you see when dragging and positioning objects on your artboard.

To use the smart guides, simply pay attention to the lines and shaded areas that appear as you work. Note in Figure 4-10 that when the one diamond is dragged, a guide appears indicating that it is aligned with the diamond on the left. The shaded area also indicates that the space between the second and third diamond is the same as between the first and second.

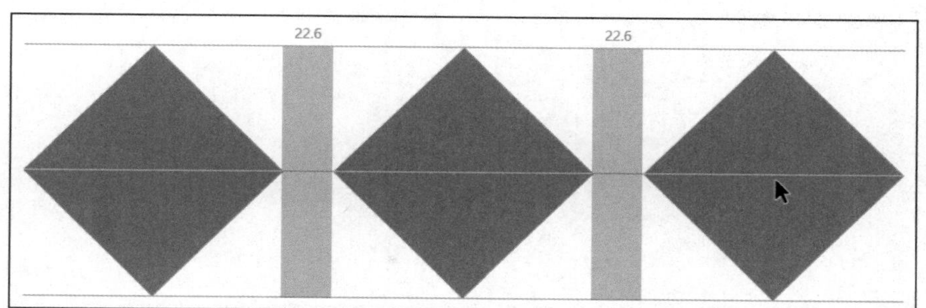

FIGURE 4-10: You can use Smart Guides to align and distribute objects.

Using the Align and Distribute feature

No worries if you already placed your objects on the artboard and then want to align or distribute them. You can use the Align and Distribute feature available at the top of the Properties panel, as shown in Figure 4-11. Note that the Align features appear only if you have one or two objects selected. Distribute functions need to have three or more objects to apply.

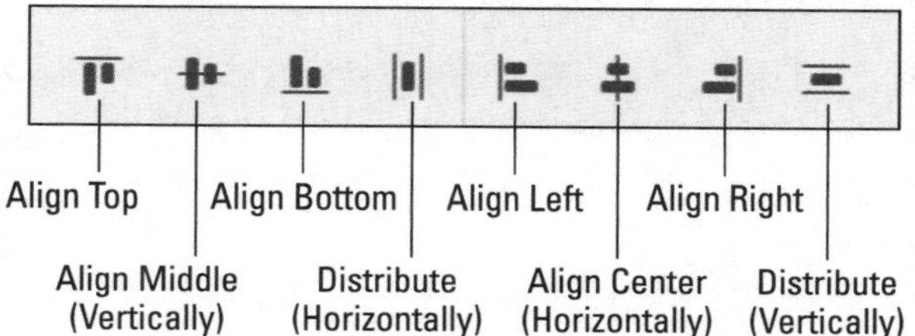

FIGURE 4-11: The Align and Distribute feature at the top of the Properties panel.

If you tend to get horizontal and vertical mixed up, keep this in mind.

Keep in mind the trick to determining horizontal versus vertical. The horizon is horizontal. Anything going from left to right on your page should be distributed horizontally. Any objects positioned from the top to the bottom should be distributed vertically.

» Adding and stylizing your text

» Importing text

» Using the Repeat Grid feature

» Importing updated text into your repeated grid

Chapter 5
Working with Text in Your XD Project

Of course, text is used throughout user interfaces (UI). Text in XD is used for elements such as buttons, navbars, as well as standard copy with links to other content, and for general content such as descriptions, directions, lists, and more. In this chapter, you find out how to add text to your UI and control the look using the Properties panel. You also discover how to import text from other sources and create a repeated list of items using the Repeat Grid feature. Keep in mind that the Adobe Experience (XD) application is continually changing through its development, so you may see slight changes in UI.

Using the Text Tool

Adding text is just a matter of selecting the Text tool from the Tools panel and then indicating where you want it placed on your artboard by either clicking or clicking and dragging.

When you click to create a text insertion point, the text width area is undefined and text runs on until you press the Return key. If you click and drag to create a text area you create a text area that forces the text to be contained within that area. The text area can be adjusted to be larger or smaller using the handles.

In Figure 5-1, the text on the left was created by clicking and releasing. Text continues until you press the Return key. On the right of Figure 5-1, you see a text area that was created by clicking and dragging a text area on the artboard.

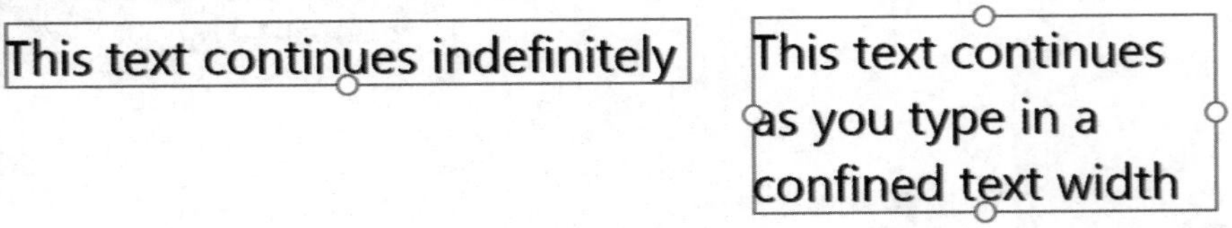

This text continues indefinitely

This text continues as you type in a confined text width

FIGURE 5-1: A text insertion created by simply clicking and releasing (left); a text area created by clicking and dragging (right).

There are benefits to both. If just creating a word or two, it is easier to click and let the text expand as needed. You can also click and drag on the resize handle at the bottom to dynamically change the text size, as shown in Figure 5-2.

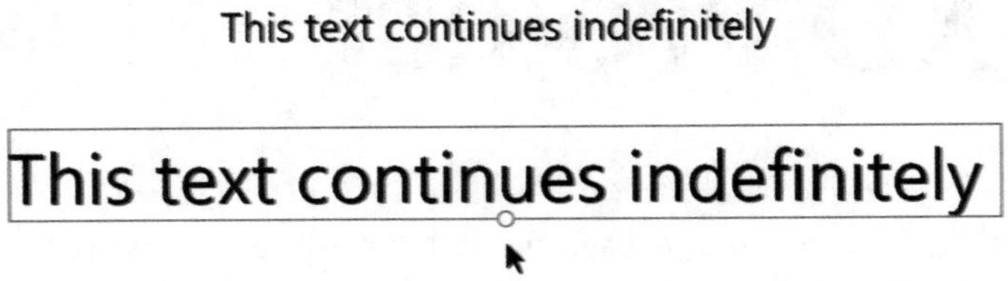

This text continues indefinitely

This text continues indefinitely

FIGURE 5-2: Click and drag on the bottom handle in order to resize the text.

Adjusting Text Properties

After you have text added, you can start changing the properties. Text properties include Font Family, Font Size, Font Weight, Alignment, Character Spacing (kerning and tracking), Line Spacing (leading), Fill, Border (stroke), Shadow (drop shadow), and Background Blur.

Before we dig into the properties, the next couple of sections outline some important points you should understand about text properties and how they affect the readability of your text on the screen.

About readability and font selections

The selection of a font is of critical importance. You may have many font choices in your system, but only a select few will cut it as a successful choice for onscreen reading. Some characteristics to look for in your font selection include the following:

- **Serif or non-serif?** Figure 5-3 shows a font family in the serif style on the bottom and one in sans serif on the top. Serifs have a slight projection that finishes off a stroke of a letter. Sans serif, on the other hand, does not have that slight projection. Sans serif typefaces tend to look more modern and are a popular choice for interactive applications and websites. In the past, it was believed that font families with serifs were easier to read when there was a large amount of text. Tests performed with eye-tracking devices have proven that there is no major difference in readability between serif or sans serif, so feel free to pick either style as long as you pay attention to some of the other font characteristics that follow.

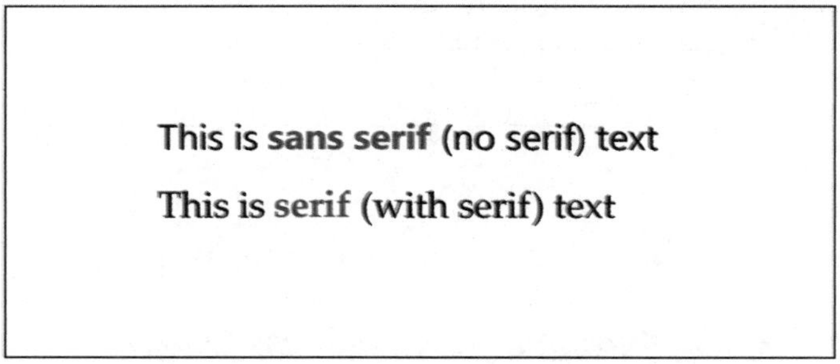

This is **sans serif** (no serif) text

This is serif (with serif) text

FIGURE 5-3: A sans serif typeface on the top, serif on the bottom.

- **Straight, even line widths:** Because your type is going to be created from pixels, it is best not to have lots of variation in the width of your text. In Figure 5-4, you can see the difference, in a pixel preview, between a font choice with variation in the width (top) and a font that is consistent in its width.
- **Strong counters:** *Counters* are the holes that you see in letters like "O," "B," and "R." Pay attention to how large those counters are because they can also cause readability issues if they close up when displayed as pixels on a screen. A pixel view of two different font families appears in Figure 5-5; see how the counters look in each in comparison to each other.
- **Descenders and cap height:** Short descenders and low cap heights are important for readability on the screen. At the top of Figure 5-6 you see a font family with short descenders and low cap heights; at the bottom, a font with long descenders and high cap heights.

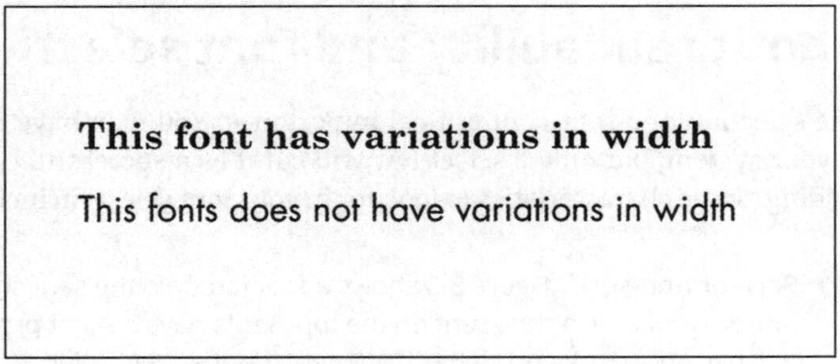

This font has variations in width

This fonts does not have variations in width

FIGURE 5-4: The top font has variation in the thick and thin lines that create the font; the bottom font is consistent and easier to read onscreen.

Minion Pro: abdopqr

Futura: abdopqr

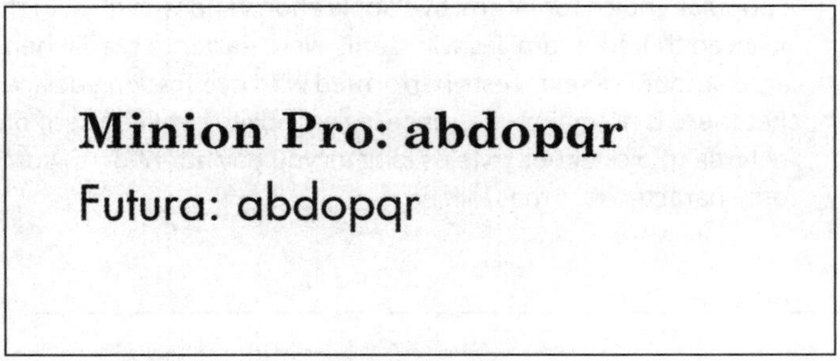

FIGURE 5-5: Look for strong open counters in your font family selection.

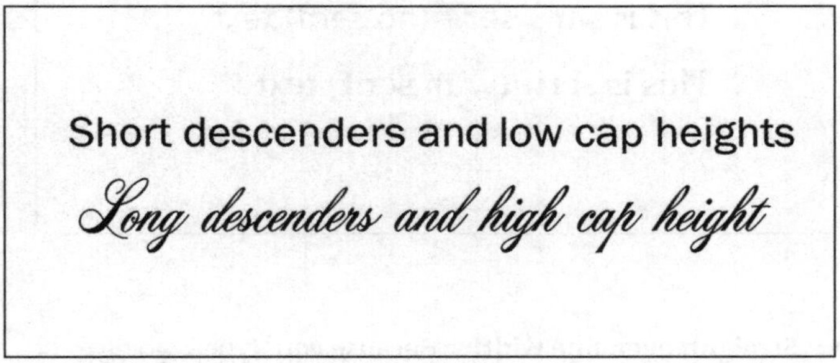

Short descenders and low cap heights

Long descenders and high cap height

FIGURE 5-6: A font family with short descenders and a low cap height is typically easier to read onscreen.

Selecting your font family

Now that you know what makes a better font selection for readability, use the Properties panel to select one. Things to look for when making your selection include the following:

» **Font family:** After you have entered some type on your screen, you can select it and choose the Font Family drop-down menu in the Text section of the Properties panel. If you know the font you want, simply start typing the font's name into the Font Family textbox, and if it is in your system, it will be found in your list.

TIP

Don't forget that you have fonts available to you through Adobe Fonts. See Book 1, Chapter 3 for specifics of how to use this great extra feature that comes with your subscription to the Adobe Creative Cloud.

- **Font weight:** After you select a font family, you can choose a weight from the Font Weight drop-down menu. Depending upon the fonts that you have in your system, you will have different font weight choices; see Figure 5-7.

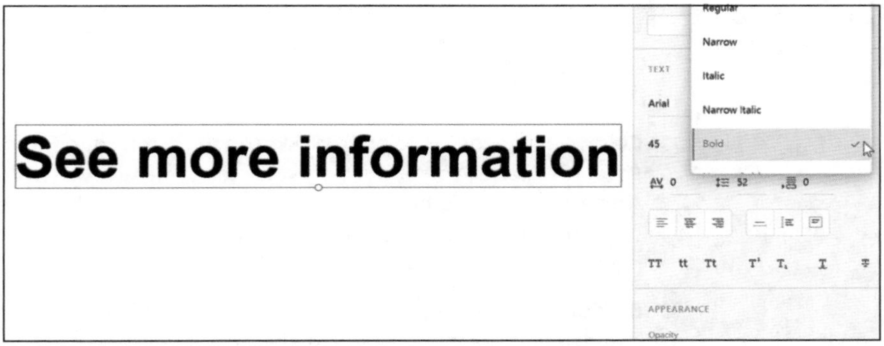

FIGURE 5-7: Selecting a font family and font weight.

- **Font size:** You can simply type a font size into the Font Size textbox, but if you are not sure what size works best use the same option for resizing that is consistent with other Adobe apps, simply select the text and press Ctrl+Shift+ < or > (greater than or lesser than) to enlarger or shrink your selected text. If you are on a Mac it is Command+Shift+ < or > (greater than or lesser than).
- **Alignment:** You can change your selected text to Align Left, Center, or Right using the Alignment buttons in the Text section of the Properties panel.
- **Character spacing:** *Character spacing* refers to the space between characters. This is known as *tracking*. In Figure 5-8, you see the result of increasing the spacing between multiple selected characters. In the bottom example you see kerning, which adjusts the space between just selected characters. You can adjust the space between two letters by selecting the first one and then changing the character spacing. In the example shown, the spacing was changed to –100 in order to bring the *o* and *n* closer together.
- **Line height:** *Line height* is *leading*, the space between the lines of text. This can play an important role in information design, so make sure that you are using this consistently throughout your design. You can also position your cursor below the Line Height textbox and click and drag up or down to decrease and increase your line height.

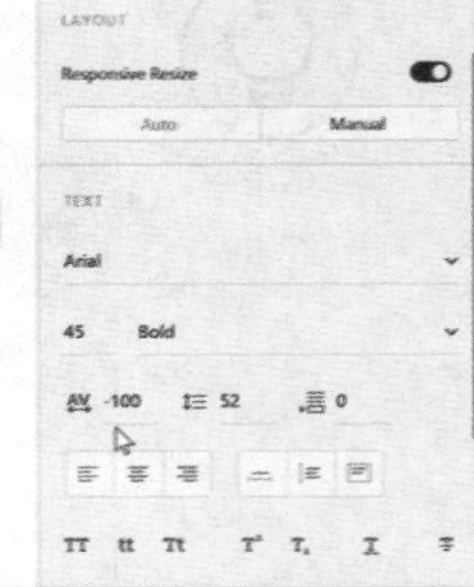

FIGURE 5-8: Character spacing at the top set for 100, making the space larger between each character. Character spacing for just the letter *o* in the bottom example is set to –100, bringing the *o* and *n* closer together.

- **Color:** Change the Fill color by clicking on the box next to Fill color in the Properties panel. You perform this task the same way you would with any shape. Select the Fill box and select a color from the Color Picker. You can also select transparency for your text from the Transparency slide available on the right of the Color Picker.
- **Border:** Applying a border puts a stroke around your text. This is not recommended because it may cause readability issues. Use it sparingly and perhaps only for special effects, or icon fonts. Figure 5-9 shows an example of a larger typeface that has a thin stroke. In a situation like this, a border could work, but not with smaller text.

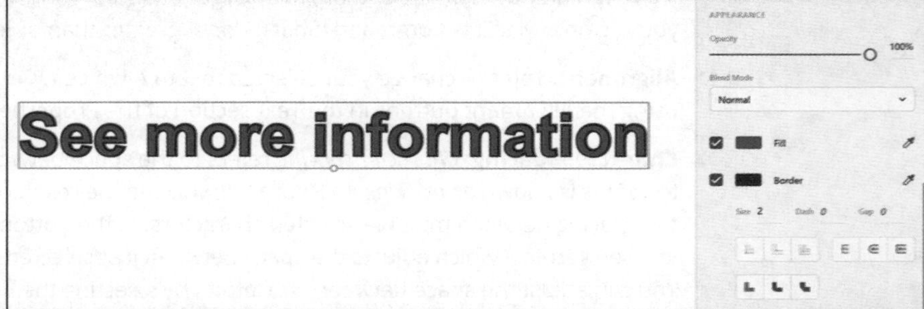

FIGURE 5-9: A small border with a red fill on a dark background.

Saving Your Properties as an Asset

As you build more content in XD, you will likely want to reuse your text styles. You can save selected properties as an asset that you can then apply to other selections at any time. This is pretty similar to saving character styles if you are familiar with that feature that exists in other Adobe products such as InDesign, Illustrator, and Photoshop.

In order to save your character's style follow these steps:

1. **Select text with properties applied that you wish to save.**
2. **Click the Libraries button at the bottom left of the XD workspace.**

   The option for creating three different assets, Colors, Character Styles, and Components, appears in the panel that replaces the Layers panel.

3. **With the text selected, click + Character Style to save it as a style, as shown in Figure 5-10.**

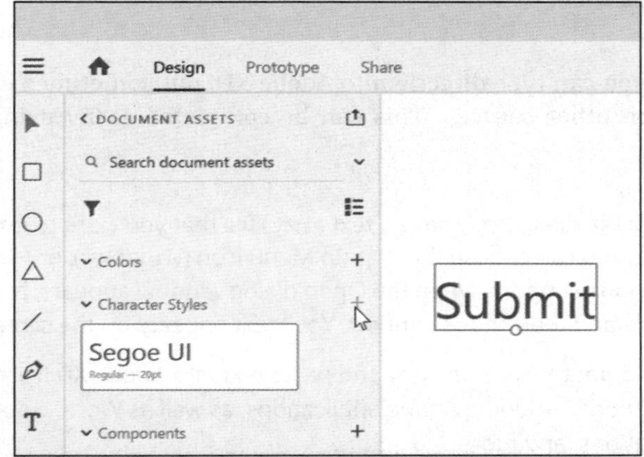

FIGURE 5-10: Save frequently used text properties as a character style.

4. **Click a character style if you wish to rename it. Many designers name these styles appropriately with the CSS name the developer will be using, in other words, H1, H2, or H3.**

Now you can select text on the artboard, and then click on your saved style to apply the same properties.

You can change all instances of a saved character style at one time in your project. Do this by right-clicking on the style and selecting Edit and then make your changes.

If you like this feature, you will certainly like the fact that you can share these styles across multiple XD projects by using libraries. Find out more about libraries in Chapter 7, "Creating Design Libraries."

Using the Spellchecker

We cannot all be perfect, so the dynamic spellchecker will immediately notify you if you are spelling some of your words incorrectly. By default, XD will highlight incorrect words, but if you prefer not to see this feature simply choose the Pop-up menu icon in the upper-right (Windows) of File (Mac) and choose Turn Spellchecker Off.

Importing Text

Of course, you can type directly into Adobe XD, but sometimes it is easier to drop in text from other sources. This can be completed easily using three different methods:

- **Import txt files:** If you have a text (`.txt`) file that you want to import into your project, you can choose the Pop-up Menu icon (Windows), or File (Mac) and then choose Import. When the Open dialog window appears, browse to locate your file and then choose Import. Your text appears on the screen.
- **Cut and paste:** You can copy and paste text into Adobe XD from most other applications, including Adobe applications, as well as Word, Google Docs, text applications, and more.
- **Drag and drop text:** You can also drag and drop `.txt` files right from folders into XD. Simply select the `.txt` file and drag and drop it right onto an artboard in XD. The text appears in a new textbox. In the next section, you see how to drag a `.txt` file right on top of a repeated grid and have your text replaced instantly.

Using the Repeat Grid Feature

Repeat Grid is one of the top timesaving features in Adobe XD. The Repeat Grid feature allows you to create a set of repeated elements with a simple click and drag. When repeated, you can tweak one set of repeated elements and have those changes reflected instantly throughout the rest of the elements. The steps in this section help you experience this great feature.

It is best to follow along with this section if you can. You can find a sample file that has an image and some text in it in the DummiesCCfiles folder at `www.agitraining.com/dummies`. Open the XD folder and choose the `Repeat Grid-Sample` file. Then follow these steps:

1. **Open a file that has elements that you want to repeat vertically (down) the page.**

   As previously mentioned, you can create your own or use the `RepeatGrid-Sample` mentioned in the previous paragraph.

2. **Select all the items that are to be repeated, and then press the Repeat Grid button in the Properties panel. You could also use the keyboard shortcut, Ctrl+R (Windows) or ⌘ +R (Mac).**

   Handles appear on the right and bottom sides of your elements, as shown in Figure 5-11.

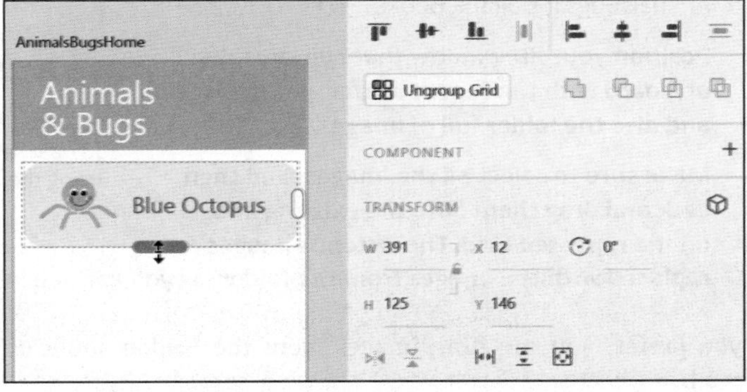

FIGURE 5-11: Select the items that you would like to repeat and then press the Repeat Grid button in the Properties panel.

3. **Click and drag down the bottom handle to see that a repeat is created automatically as you drag, as you see in Figure 5-12. Drag beyond the template to extend the artwork outside the bottom.**

4. **Go back to your original art and make a simple change. Perhaps change the type size of the title, or scale the image a bit. Notice that as you make changes in this first item the changes are reflected throughout the repeat.**

   This can be a huge timesaver, but follow along through the next steps to see how you can easily replace the repeated image and text.

Importing updated images into your repeated grid

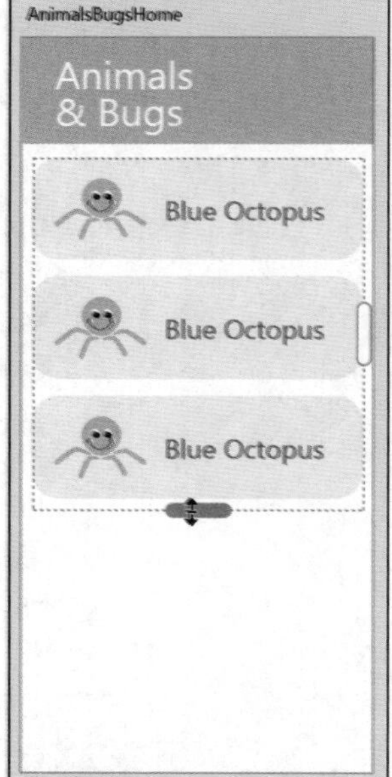

FIGURE 5-12: Repeat selected items easily using the Repeat Grid feature.

After you have repeated items, you may want to demonstrate how your application is going to work with more realistic content. The repeated element does not provide an opportunity to show the various data that your app may be offering. In this next section, you take a folder of images and drop them on top of your repeated grid in order to replace your original image. Just follow these steps:

1. **Locate and open a folder of images in a format that XD supports, such as JPEG, PNG, SVG, or GIF.**

   If you are accessing the DummiesCCfiles folder mentioned earlier, you can find a folder named AnimalsBugs inside the Book07_XD folder.

2. **Position your XD window that contains the artboard with the repeat so that you can see it and also the folder full of images.**

3. **Make sure to select all the images, and then click and drag them onto the existing image on the repeated Grid. The Octopus images are replaced with the images from the folder, as you can see in Figure 5-13.**

TIP

If you prefer, you can drag images from the Bridge application into Adobe XD as well.

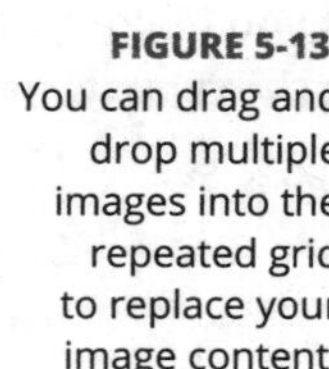

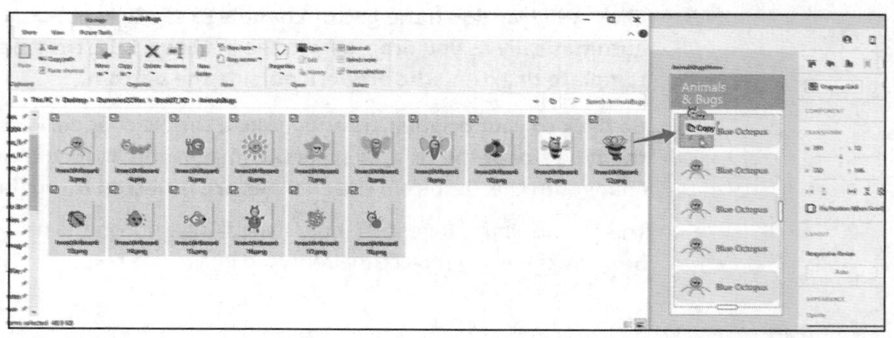

FIGURE 5-13: You can drag and drop multiple images into the repeated grid to replace your image content.

Updating your text in your repeated grid

In this next section, you discover how to update your text in a repeated grid. This is completed in much the same way as with images, but the `.txt` format is vital to having this process work:

1. **Create text in any application that can save the file in a Plain Text format — that is, with an extension of `.txt`.**

   The text can be generated from Notepad, Word, downloaded from Google Docs as plain text, or exported from many other applications. For this example, you can use the sample text file named `Bugtext.txt` in the XD-Files folder on our site.

2. **Position the folder that contains the text file so that you can see the file and the repeated grid.**

3. **As with the process of replacing the repeated image, click and drag the `.txt` file over your repeated grid text. The text is replaced with text from your file.**

   You can see the example as it appears once the imagery and text have been replaced in Figure 5-14.

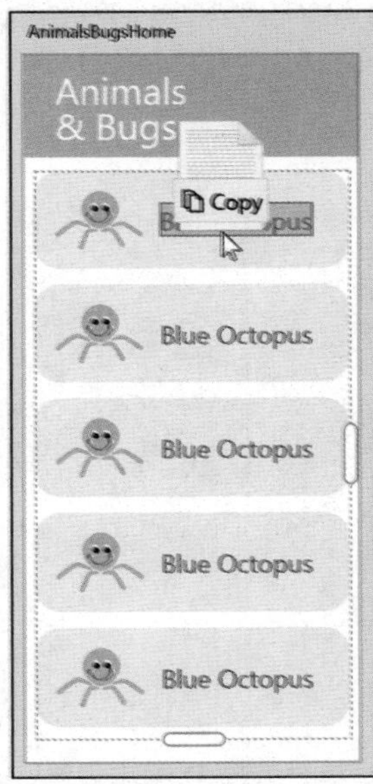

FIGURE 5-14: Drag the text file on top of the text in the Repeated Grid to replace the text.

Working with Text in Your XD Project

4. **If you extended beyond the viewing area, make sure to click the artboard name above the artboard to select it.**
5. **Drag the bottom handle to extend to the length of your repeated grid, as shown in Figure 5-15.**

   This allows the screen to be scrolled in the preview. The dotted line represents the area that previews before the scroll.
6. **Click the Desktop Preview button and scroll to see your repeated grid in action.**

FIGURE 5-15: Click and drag to extend the viewing area of your repeated grid.

» Importing vector graphics

» Working with Illustrator files

» Using custom pen paths for masks

» Exporting artwork and artboards

Chapter 6
Working with Imagery in XD

Using Adobe Experience Design (XD) to create your UI elements for your prototypes is great, but to be able to drop in artwork created in Adobe Illustrator, Photoshop, or other sources is even better.

Importing Artwork into Your Prototype

By this point, you have discovered how to create your own UI using tools such as Rectangle and Ellipse. You even were introduced to the import feature when building a simple prototype in Chapter 1 of this minibook. In this chapter, you spend more time digging down into the features you can take advantage of when importing artwork from other applications and sources. If you do not have images that you can easily access, you can obtain sample images at `www.agitraining.com/dummies`.

In order to follow some of the examples in this section, you should have an empty artboard ready.

There are several methods that you can use in order to bring imagery into your XD prototype:

- **Browsing your file directory:** You can import the following image formats into XD: JPEG, PNG, GIF using your file directory. To import an image using the Pop-up (Windows) or File menu (Mac), choose Pop-up or File⇨Import and locate an image. The keyboard shortcut for this method is Ctrl+Shift+I (Windows) or Command+Shift+I (Mac).
- **Drag and drop:** You can drag and drop images from your file folders, web pages, and even other applications. Simply click and drag the imagery from the source and release your mouse when your cursor is over the artboard. In Figure 6-1, a group of images was all dragged in at once onto an artboard from a file folder.
- **Importing with cut and paste:** You can copy and paste imagery from sources, such as other websites and applications, but transparency is not supported when images are imported in this manner.

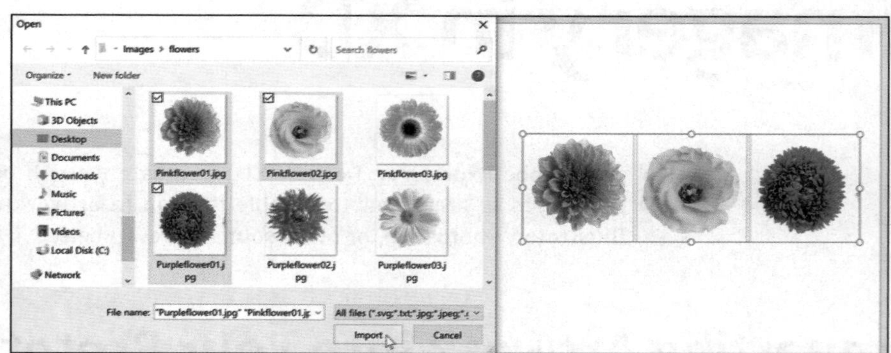

FIGURE 6-1: You can hold down either the Shift or Ctrl/Command key to select multiple files in a folder and drop them all into your project at once.

Working with Images in XD

After you have imported your images you can resize and rotate images just like you would any other shape. For a refresher on how to scale and rotate objects, review Chapter 4 of this minibook.

TIP

If you want to scale an image without losing the center position, hold down the Alt (Windows) or Option (Mac) key while Shift-dragging any corner anchor point. The image scales, larger or smaller, depending upon whether you drag outward or inward, but it stays in position.

You can also easily round the corners of an imported image using the corner widgets, as shown in Figure 6-2. Hold down your Alt (Windows) or Option (Mac) key and drag an individual corner to make one independently rounded.

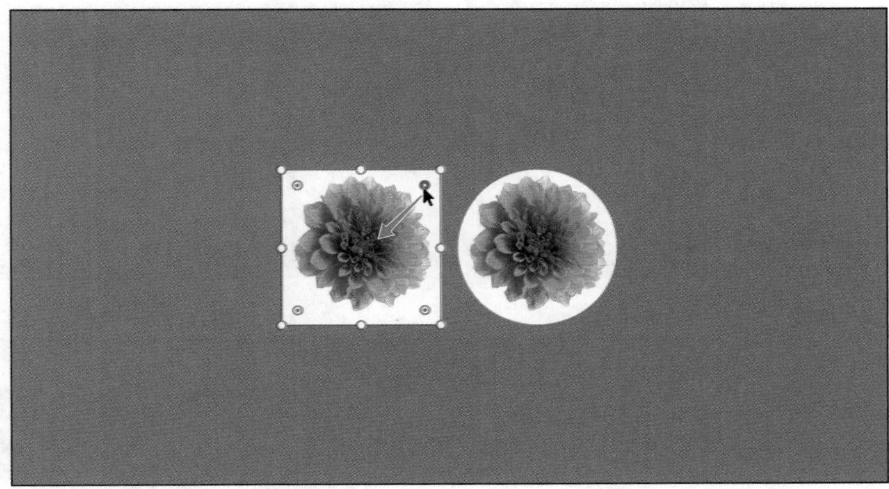

FIGURE 6-2: Click and drag a corner widget in order to round the edges of an image.

Masking your images

By defining a closed shape with a shape tool or pen tool, you can define a mask. A *mask* offers you the ability to put an image into a custom shape. It also offers you the ability to crop out the parts of an image that you do not want to be visible. Follow these steps in order to create a mask:

1. **Import an image on to an artboard.**
2. **Create a shape and size and position it over the image.**

   In this example, we used a triangle shape. We changed the properties of the triangle to 10 points and a 50% Star Ratio, as shown in Figure 6-3. The placement over the image determines what part of the image will show.
3. **Using the Select tool, select both the shape and the image. It is important that the shape be on top of the image; if it is not, right-click (Windows) or Control-click (Mac) and select Bring to Front.**
4. **With both objects selected, right-click (Windows) or Control-click (Mac) and select Mask with Shape.**

   The shape now contains the image, as shown in Figure 6-4. You also see that a Mask Group has been automatically created from the mask and the image in the Layers panel.

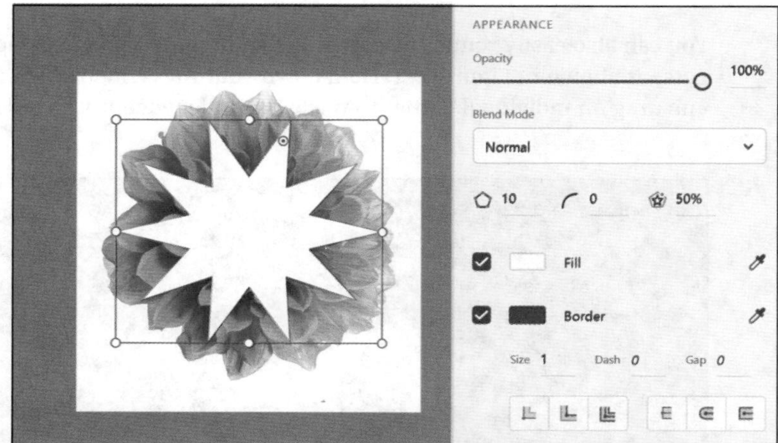

FIGURE 6-3: Position a shape over an image to determine what part of the image you want to show.

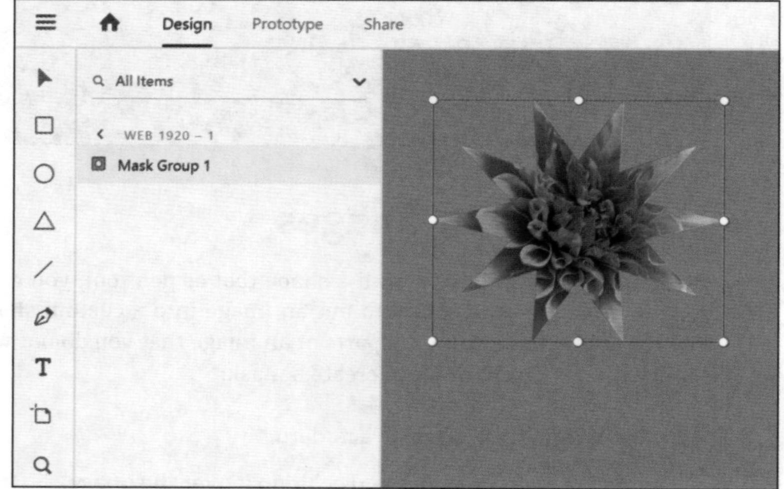

FIGURE 6-4: An image masked with a shape. Masked items are automatically grouped in a Mask Group in the Layers panel.

Note: If you would like to edit the positioning of the shape over the image, simply double-click the masked object using the Select tool. When you are finished repositioning your image and shape, click anywhere else on your artboard. You are also able to double-click on the Mask Group in the Layers panel and select the mask or image independently of each other.

Masking multiple objects

You can mask any object(s) with a shape. Perhaps you have several items that you want to keep together in a circular shape, or you want to crop a series of repeated images. Both of these tasks can be done by following these steps:

1. **After creating your artwork, add a shape on the top, or bring a shape that you wish to mask to the front by right-clicking (Windows) or Control-clicking (Mac) and selecting Bring to Front.**
2. **Select all the objects that are to be masked in addition to the topmost object, which is the mask, and then right-click (Windows) or Control-click (Mac) and choose Mask with Shape. (See Figure 6-5.)**

   If you need to edit objects in a mask independently, you can either right-click (Windows) or Control-click (Mac) and select Ungroup Mask, or open the Layers panel and double-click on the icon to the left of the object named Mask Group. When you double-click, it allows you to select individual objects.

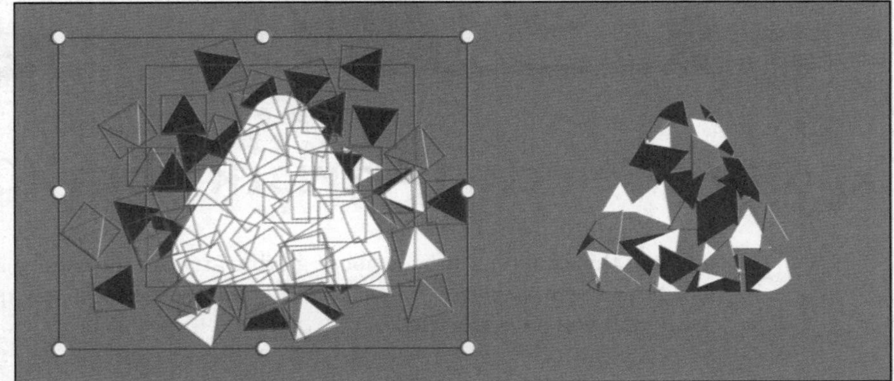

FIGURE 6-5: You can select multiple objects and mask them into one shape.

Importing Vector Images into XD

In addition to importing images, you can also import vector images from several sources, including the Clipboard, Illustrator, SVG (scalable vector graphic), and Sketch (another popular prototyping tool).

Here are two methods that you can use to import editable vector objects into Adobe XD:

» **Copy and paste:** Select your vector object(s) in applications such as Adobe Illustrator or Sketch and copy them. Return to XD and paste. The vector graphics are placed and retain their editability.

» **Choose File ⇨ Import:** If you use File ⇨ Save to save a file from Adobe Illustrator in the SVG format, you can use the import feature in XD to bring in an editable copy of your artwork.

After you have imported your vector artwork into XD, you can continue working with the files and borders using the features available to you in the Properties panel. Keep in mind that vector artwork, though editable typically lands in XD as a group. The group will have to either be ungrouped, or individual parts selected in the imported group within the layers panel, as shown in Figure 6-6.

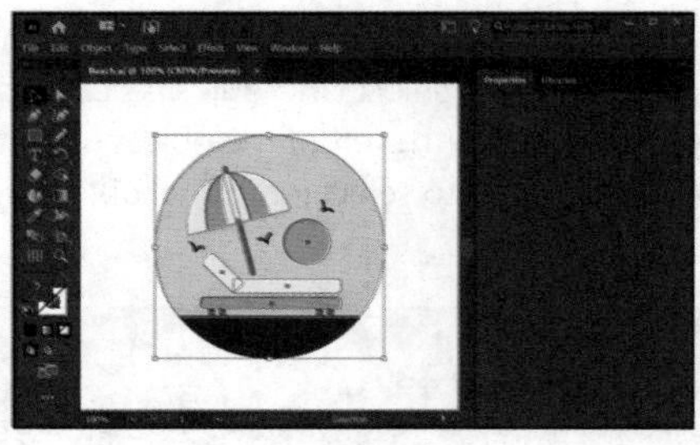

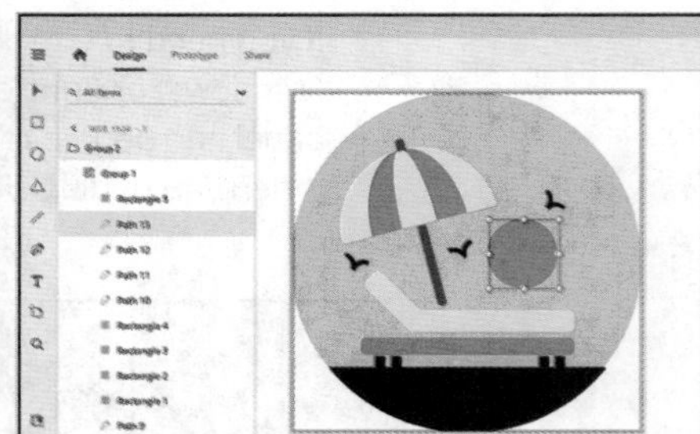

FIGURE 6-6: Artwork copied from Illustrator (left) and pasted into XD (Right).

Exporting Selected Artwork from XD

Most of this chapter has been about getting artwork into Adobe XD; now you find out how to get your artwork *out* of XD. Follow these steps to export just the artwork that you have selected:

1. **Export only selected objects from XD.** Select the Pop-up menu (Windows) or File (Mac and choose Export ⇨Selected. You can also use the keyboard shortcut Control+ E (Windows) or Command+ E (Mac).

   An Export Assets dialog window appears.

2. **Select either the PNG, SVG, or PDF format from the Format drop-down menu.**

   Based upon the format you choose you will be offered additional options. Note that if you choose a vector format, such as SVG, no additional options related to size are offered, but if you choose a format such as JPG or PNG you have options offered that allow you to save the same file in multiple sizes.

   If you choose PNG as your format, you have additional platform options, as shown in Figure 6-7.

   Note: If you are saving as a PNG and want to export one copy, leave the Export For set to Design. The other options, Web, iOS, and Android offer the capability of saving out multiple files at different formats.

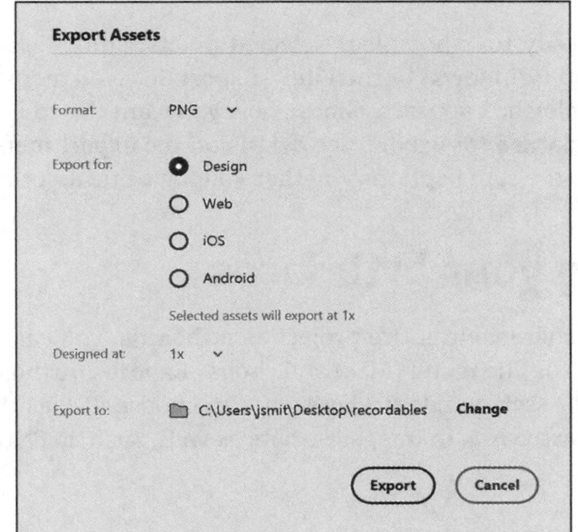

FIGURE 6-7: Depending upon the format you select you have different output options.

- *Web:* Assets are exported at 1x and 2x resolutions.
- *iOS:* Assets are exported at 1x, 2x, and 3x resolutions.
- *Android:* Assets are optimized and exported for a variety of Android screen densities.

3. **Click Change (above the Cancel button) and browse to the location you want to save your exported artwork.**
4. **Press Export.**

Exporting multiple items

The most efficient way to transfer your UI artwork is to mark it for export and then collect it all at once when ready. You can do this using three methods.

- Select the object you wish to export and then click on the Mark for Export icon to the right of the selected object in the Layers Panel.
- Right-click on a selected object and choose Mark for Export from the contextual menu.
- Use the Keyboard shortcut by pressing Shift+E.

When you are ready to export your selected assets, simply choose Export from the pop-up menu (Windows) or the File ⇨ Export Selected menu (Mac); when the Export Assets dialog box appears, choose how you want to export all of the images that have been marked for export. See details on the export menu in the previous section; they are the same options whether you have one asset or many.

Exporting your artboards

If you prefer to share your entire project as artboards, you can open the pop-up menu (Windows) or File menu (Mac) and choose Export ⇨ Artboards. You then see the same Export Assets window where you can choose PDF as the format, or you can export your artboards in image formats as well, such as PNG, SVG, and JPG.

» Saving text styles and colors to a library

» Sharing a library

» Editing a library

» Using UI kits

Chapter 7 Creating Design Libraries

This chapter is all about saving time and keeping your artwork consistent. By creating components, you can save an object in a form that can be dragged and dropped right unto your artboard and then dynamically updated as needed. In this chapter, you also find out how to use existing UI kits that make it easy to locate design patterns for popular platforms such as iOS, Android, and Windows.

Saving Reusable Objects as Components

If you find that you are copying and pasting your UI objects, you should consider converting them to components. *Components* are linked objects that can be reused across all artboards in a document, or even multiple documents if saved in a library. Because they are dynamically linked, changes to one reflect across all instances.

In this next set of steps, we use the icon artwork from the DummiesCCfiles folder at `www.agitraining.com/dummies`. You can use our sample files or one of your own. Whether you choose to use our example file or your own, keep the same file open throughout the chapter and save frequently. The ultimate goal will be to save your file as a Library that allows you to share components, colors, and type styles across multiple XD projects. Read on to see how to prepare your file to be a library.

To convert an object into a component, follow these steps:

1. **Open a document that contains at least one artboard, an icon, and some related text.**

   In this example, a file from the DummiesCCfiles Book07_XD folder named `delivery-symbol-sample` is opened in Adobe XD.

2. **Select an object and some text that you want to reuse. In this example, one of the grouped Submit buttons is selected.**

3. **Right-click (Windows) or Control-click (Mac) and select Make Component from the contextual menu.**

TIP

   If you prefer keyboard shortcuts, you can select an item and press Ctrl+K (Windows) or ⌘+K (Mac) to convert it into a component.

When you've created a component, you see the item listed both in the Layers panel as a layer with a diamond to the left of the name, and in the Libraries panel as a thumbnail of the artwork. You also see a solid diamond in the upper left of the artwork located on the artboard, as shown in Figure 7-1. This diamond indicates that it is the original from which the component was created. Any changes to this artwork automatically updates any other components generated from it.

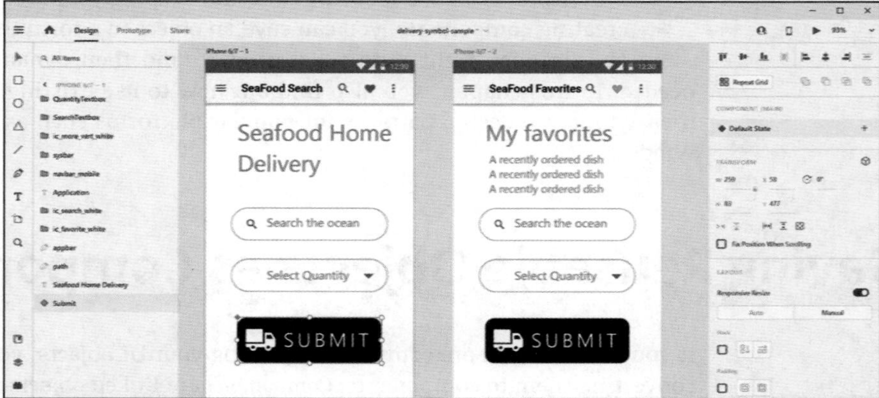

FIGURE 7-1: Make artwork that you wish to reuse frequently as a component.

To use this component, drag a copy out of the Libraries panel, copy and paste the original, or hold down the Alt/Option key and drag the original component in your document window to another location to make a clone. Note in Figure 7-2 that copies of the component do not have the solid green diamond in the upper-left corner of the artwork. In fact, the hollow green diamond they bear indicates that they are a component, not the original. Keep in mind that any changes you want reflected throughout all instances must be made on the original.

FIGURE 7-2: The original component is on the left, a copy on the right.

Making edits to your component

In this section, you see how edits to one instance of your component instantaneously update any other instances. To see how this works, follow these steps:

1. **Select the original artwork that was used to create your component (indicated by a solid diamond in the upper-left corner when selected).**

   If you can't locate the original component, right-click on any instance and choose Edit Main Component from the contextual menu.

2. **In the Properties panel, change the fill to a different color, perhaps red. Note how this change instantly occurs in both instances of the component. This is the case with any property.**

   Keep in mind that you can overwrite the color by selecting any of the copies of the component. This is helpful if you want one different-colored shape. Changing the color in the original does not affect the component that you changed independently.

3. **Now double-click the text in one of the copies of your component to edit it.**

   Remember that changes to the original will affect all components, so you want to pick a component that has a hollow diamond in the upper-left corner.

4. **Change the word from *Submit* to *Send*. Notice that this text change is not reflected in the other instance, as shown in Figure 7-3.**

   This ability to override the component's text makes a great solution when you want consistency in buttons but don't want the same text in each component instance.

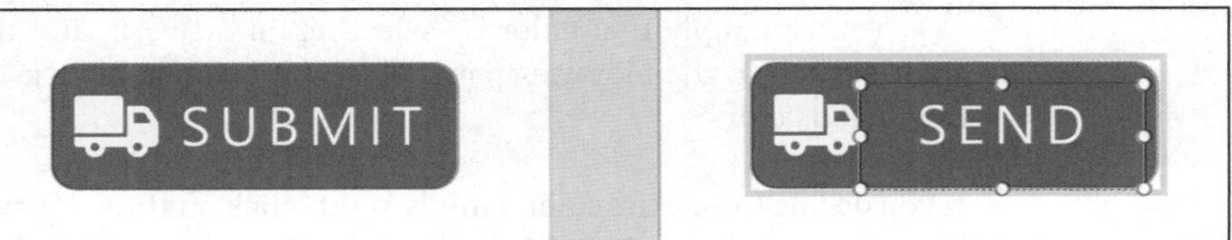

FIGURE 7-3: A change in color is reflected in all instances but a change in the text is not reflected in all instances.

Unlinking a component

If you do not want the changes in your main component to affect all instances, right-click on an instance and choose Ungroup Component from the contextual menu. Now the diamond disappears because this is no longer a component.

If you want to eliminate the component completely, right-click on the thumbnail of the component in the Libraries panel and select Delete from the contextual menu. The instances of that component are not deleted, but they are no longer linked to a component and can be edited independently.

Saving colors in your libraries

If it is important that you use consistent colors throughout your project, you will want to start saving colors in the library as well. This not only saves you time when applying colors, but also helps you to maintain brand consistency.

Saving colors is a simple process; all you need to do is create an object and fill it with a color that you create using the Properties panel. Follow these steps if you want to give it a try:

1. **Have any XD project open.**
2. **Create a rectangle, or any shape.**
3. **In the Properties panel, use the Fill options to build a solid color that you wish to reuse.**
4. **Make sure Border is unchecked because when you save colors it will save both Fill and border colors.**
5. **Choose the Libraries panel from the bottom left of the XD workspace.** **You can also use the keyboard shortcut Ctrl+Shift+ Y.**
6. **Click the plus sign to the right of Colors. The color is added in the Libraries panel, as shown in Figure 7-4.**

Now you can apply that color by selecting an artwork, and then clicking on the color swatch in the Libraries panel. Each time you apply the color it is tagged to this saved color.

If you decide to edit the color, simply right-click on the color swatch in the Libraries panel and choose Edit. When the Color Picker appears, select a new color. It is instantly applied to all instances, as shown in Figure 7-5.

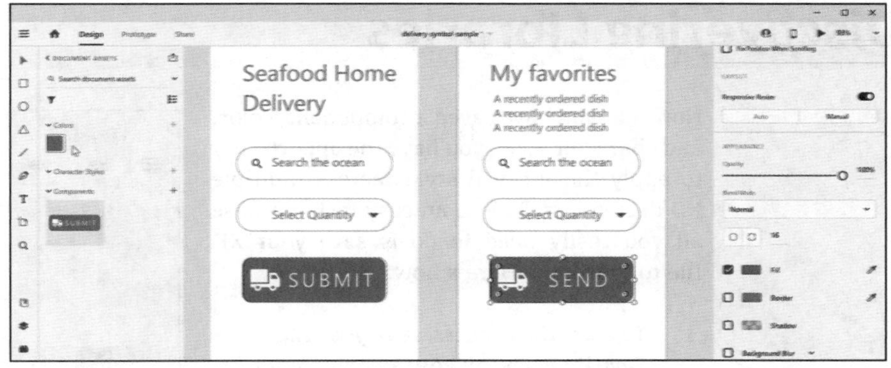

FIGURE 7-4: Save frequently used colors in the Libraries panel.

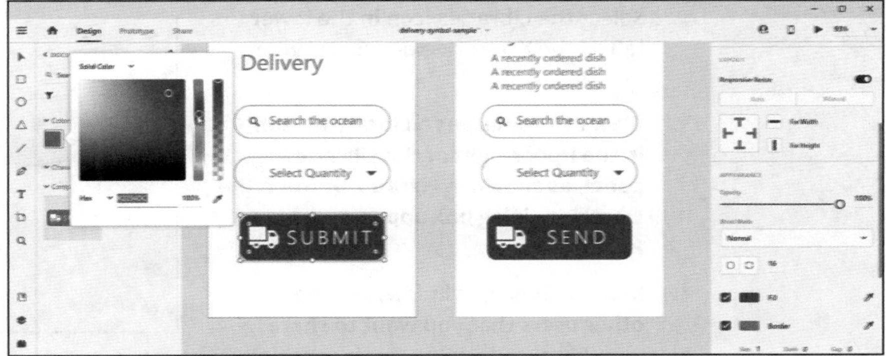

FIGURE 7-5: Update all color instances automatically by saving the color in the library.

Saving Character Styles

Saving character styles is covered in Chapter 5, "Working with Text in Your XD Project." If you read that chapter first, you discovered that you save character styles in your Libraries panel. In this chapter, the concept of saving character styles is taken a step further. Follow these steps to save a character style in your open document. In the next section, you see how you can share components, colors, and your character styles across multiple files.

1. **Select any text in your document, preferably text with its properties set to a style you would want to reuse throughout your project.**
2. **Click Libraries in the lower left of the XD workspace and click the plus sign (+) to the right of Character Styles. Your style has been saved.**

Discovering Libraries

Now that you have saved a component, color, and character style, you have the opportunity to apply those saved styles across multiple XD documents. This is an easy task, because all you really need to do is save your XD file to the cloud. Here's how:

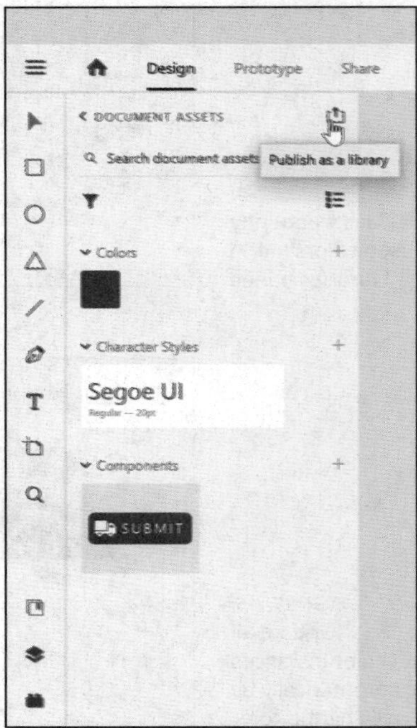

FIGURE 7-6: Publish your XD file as a library to share its assets with others.

1. **To start the process, save your file. Save it locally, on your system, or on the cloud.**
2. **Select the Libraries icon in the lower left of the XD workspace to see your assets.**
3. **Click the Publish as a Library button in the upper right of the Libraries panel, as shown in Figure 7-6. When the Libraries dialog box appears, choose Publish.**
4. **In the following window, invite any other users that you want to share your library file with. This is especially useful if one person or team is managing all the branding assets and wants to quickly share them with others. Do not worry if you don't have the list of people you want to share with right away. You can click the Share Library button that now appears in the top right of your Libraries panel at any time to add others, as shown in Figure 7-7.**

   Now you are ready to use your file as a library.

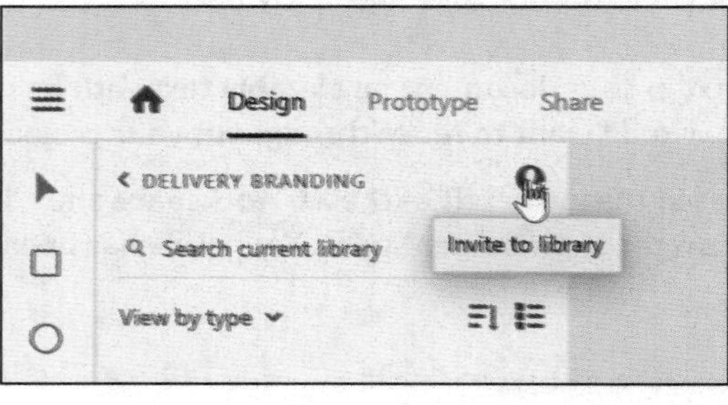

FIGURE 7-7: Share your library at any time by selecting the Share Library icon in the Libraries panel.

Using shared libraries

Now that you have published your file to be used as a library, create a new file so that you can link to your library and take advantage of using the same assets:

1. **In the new XD document, click on the Libraries panel and then click the back arrow to the left of Document Assets. In that view, you see all libraries that are shared with you. You can select a published library or return to the assets in this document by clicking Document Assets. Click the library you just saved. (See Figure 7-8.)**
2. **After selecting your library, you are immediately shown the assets that you are now linked to. Keep in mind that you can add more than one library to an XD document by selecting the arrow to the left of library named at the top of the Libraries panel.**

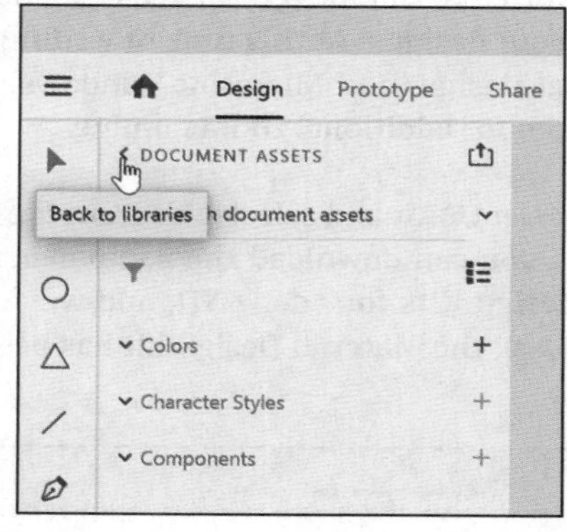

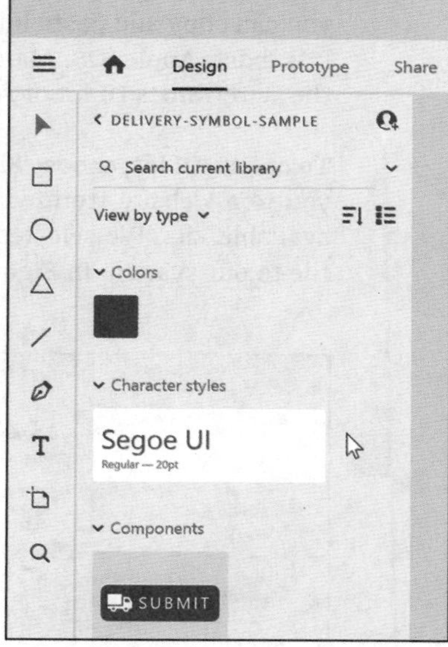

FIGURE 7-8: Click the arrow to the left of Document Assets to find your published library. When it's selected, you can use the same assets.

Updating a published library

After you start using your library in various XD projects, you may discover that a logo needs to be added, or colors need to be revised. To make many revisions, open that original document and be careful with revisions that you make. This is because these revisions could impact any file that is linked. You can also start the revision process right in your open document.

Editing a component

If you are working on an XD project and decide that a component saved in the original library needs to be edited, right-click the component in the Libraries panel and choose Edit in Main Source. The linked library file is opened and your component is selected.

Make the changes you need and then save the document. When XD documents linked to this project are opened, an indicator will alert the user that an update is available.

Taking Advantage of UI Kits

In this section, you are introduced to UI kits. A *UI kit* is a collection of UI elements you can copy and paste into your designs. At this time of writing, three default UI kits exist: Apple iOS, Material Design, and Microsoft Windows. You can also use the Wireframes UI kit or search for additional UI kits online.

To open a UI kit, choose File⇨Get UI Kit and select a kit from the menu. This takes you to a website from which you can download the components for each of the available kits. We selected Design Kits for Adobe XD, and we downloaded an XD file to our system. In Figure 7-9, the Material Design kit has been opened.

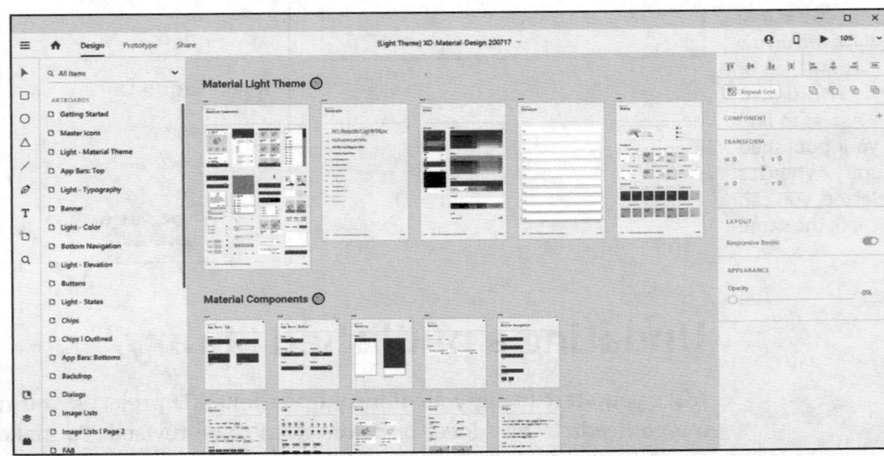

FIGURE 7-9: The Material Design kit provides patterns and other UI elements.

To use these elements, you can either select the items(s) you want from the Layers panel or use your Select tool and then choose Edit⇨Copy. If you prefer, you can use the keyboard shortcut Ctrl+C (Windows) or ⌘ +C (Mac).

Locate an artboard in a different project in which you would like to use the items and choose Edit⇨Paste. The keyboard shortcut is Ctrl+V (Windows) or ⌘ +V (Mac).

Many people refer to these UI kits as *sticker sheets,* because that is almost exactly what they are. You find UI elements you like and you cut and paste them to your own project. Add these to your library to make it easy to drag and drop the UI elements you need quickly and easily.

» Creating transitions between screens

» Previewing your prototype

» Adding animations

» Adding rollovers

» Using plug-ins

Chapter 8
Adding Interactivity to Your XD Project

Now that you have some practice creating UI elements in XD, you can start adding interactivity in the form of links from one UI element to a selected artboard. This interactivity is created in the Prototype mode in Adobe XD. You can use the links that you make in the Prototype mode to demonstrate your information architecture, navigational functions, or changes to objects when users click on them, such as a change in a button or the visibility of a menu. You can even use the Prototype mode to create rollovers, overlays, and animations.

Entering Prototype Mode

If you have been working in XD, you might have a file ready to start working on. If not, you can find a file to use for this chapter at `www.agitraining.com/dummies`. Look for the XD-Files folder inside the DummiesCCfiles folder. In this example, the file named `BabyAnimalsApp` can be used. To enter Prototype mode, follow these steps:

1. **Open a file that has multiple artboards or open the `BabyAnimalsApp` file that was created for this section.**

   This sample prototype is a simple three-screen project that offers you the opportunity to create links from one screen to another.

2. **Select Prototype from the top of the Adobe XD workspace.**

   TIP

   You can also enter the Prototype mode by using the keyboard shortcut Alt+2. Return back to Design mode by pressing Alt+1.

   Start by defining your start page for any clickable flows you want to demonstrate.

3. **Click the gray home button to the left of the Home artboard. It turns blue. Defining this artboard as the Home screen indicates that this artboard is the first screen of your app or website. Your users begin to navigate the app or the website from the Home screen. By default, your Home screen is also set to the first artboard you add a wire to. If you do not see the home screen icon, you may need to zoom in closer to your artboard.**

FIGURE 8-1: Define the first page as your Home. It turns blue, indicating that the artboard is now set as the default start screen of your application.

4. **Click the PuppyButton on the Home artboard; a blue connector arrow appears on the right side of the artwork.**

5. **Click and drag the arrow to the Puppy artboard, as you see in Figure 8-2. In preview mode, this button will create a link, referred to as *wiring*, to the Puppy artboard, essentially working much like a link in an actual application.**

   Test this simple interaction before exploring all the other opportunities you have for interaction.

6. **Click the Preview arrow in the upper-right corner to see that your Home screen appears in a preview mode, as shown in Figure 8-3. Click the PuppyButton to see that you are navigated to the Puppy screen.**

7. **No other interactions have been set at this time, so close the preview window to return back to the XD document.**

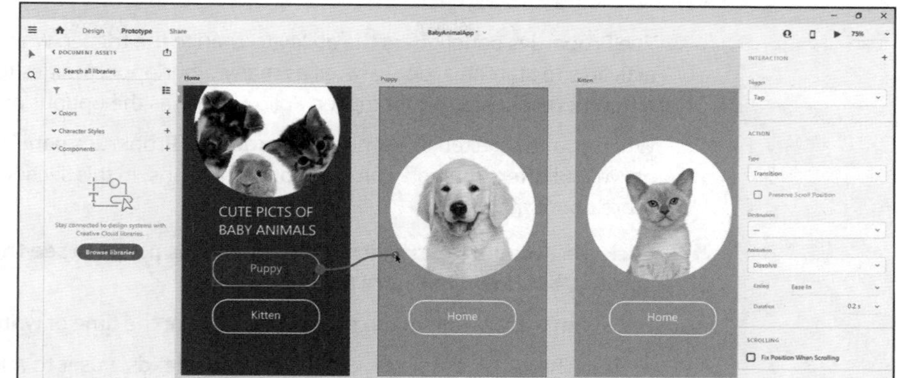

FIGURE 8-2: Click and drag from the connector to wire the button object to the puppy artboard.

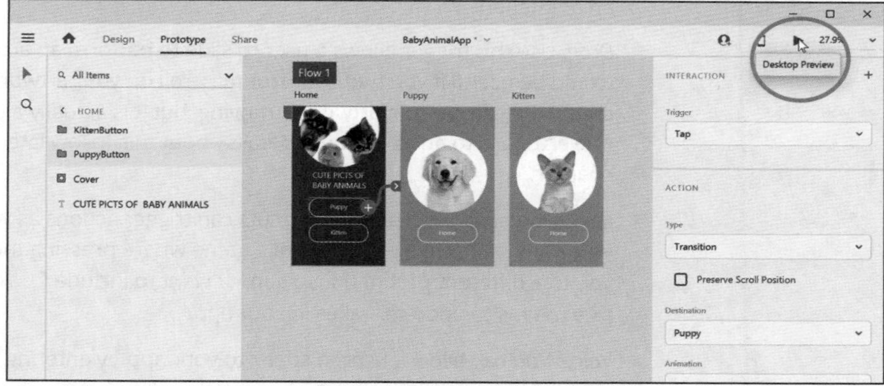

FIGURE 8-3: You can preview interactions using the Preview button in XD.

Discovering the many ways to link your artboards

In this section, you explore the other opportunities you have to create links from an object to another artboard. In the last example, the default trigger of Tap was used, but there are many other triggers that a user can perform, such as Drag, Keys & Gamepad, and Voice. Read on to find out more about these triggers.

In order to see the available options, select a different button to create an interaction with.

1. **Make sure that you are still in the Prototype mode and then click the KittenButton on the Home artboard.**
2. **Click the blue connector button on the right side of the KittenButton and take a look at your Properties panel.**

The Properties panel offers the ability to change the trigger, which the user activates to start an action, as well as many other opportunities for creating different responses to that trigger. Follow along as the options are reviewed.

When you've selected the connecter arrow, the Properties panel becomes active and offers multiple combinations of options. In this section, you learn more about triggers.

3. **Click the default Tap trigger in the Properties panel to see the available options, as shown in Figure 8-4.**

   Here is an explanation of the triggers available at the time of writing:

   - *Tap:* This is the simple click of a button that leads a user to another artboard or location on the same artboard. Tap can also trigger other actions such as showing another artboard, triggering a rollover, or starting an animation.
   - *Drag:* Use this trigger allows a user to slide to perform an action. This would be useful if you had a horizontal scroll on your artwork. You can create the illusion that artwork is moving, but it is actually a transition from one artboard to another. More details about this follow later in this chapter.
   - *Keys & Gamepad:* Keyboard shortcuts can trigger actions as well. Perhaps you want to create a prototype of a game where pressing the arrow moves you to a different part of the screen, or want to include Ctrl/Command+Z as a means to undo mistakes in your app.
   - *Voice:* You can allow a user to speak to your app by entering words. In this feature, your prototype can trigger actions when someone repeats the word you entered. This can be helpful if accessibility is important for your app/website.

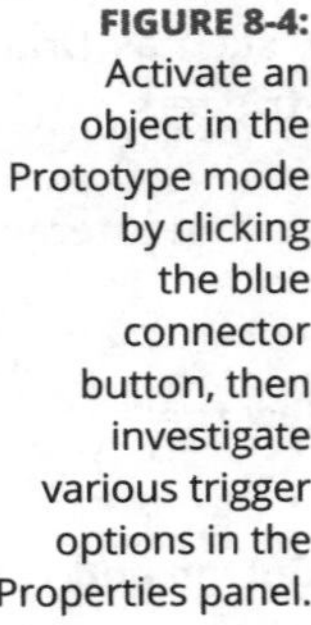

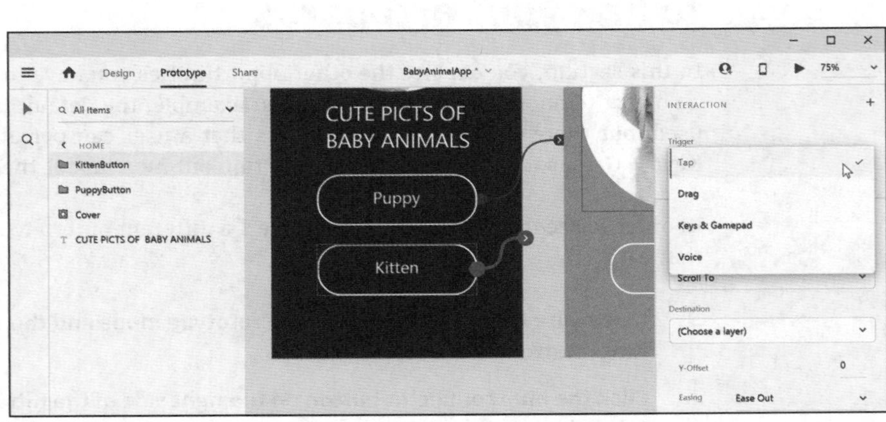

FIGURE 8-4: Activate an object in the Prototype mode by clicking the blue connector button, then investigate various trigger options in the Properties panel.

4. **For this example, choose Tap.**
5. **Scroll down to the Action section and choose Transition from the Type drop-down menu.**

   See descriptions of actions below. Keep in mind that you can apply multiple actions to one object:

   - *Transition:* This provides a simple fade from one artboard to another. Similar to what you would see when you navigate from one screen to another in an app or website.
   - *Auto-Animate:* This is a very cool action that basically morphs what you see on one artboard to another. Using this action, you can make it look like charts are growing or you can bounce and slide menus into place.
   - *Overlay:* Using this action, you can display a selected artboard on top of another. For instance, if you want to create a pop-up menu and have it appear on top of the current screen.
   - *Scroll to:* This action allows you to produce a click that takes the user to another section of the same artboard. This is helpful if you are trying to demonstrate anchors in a screen.
   - *Previous Artboard:* This action is extremely beneficial as you can use it for back buttons or any other control where you want the user to simply return to the last artboard they were viewing.
   - *Audio Playback:* If you want to liven you interactions up with a click or swoosh, you can use this action. This can also be used to play a music track when triggered.
   - *Speech Playback:* This allows you produce voice instructions when triggered.

   For this example, you selected Tap for the trigger and Transition for the action. Now you need to let XD know where you want to transition to. This is the part where you can wire the button to another artboard.

6. **Select the connecter arrow and click and drag the wire to the Kitten artboard.**

   Now you can choose the animation that will occur during the transition from the Home artboard to the Kitten artboard.

7. **Select Dissolve from the Animation drop-down menu. This option fades from one artboard to another.**

   Other options include None, Slide Left, Slide Right, Slide Up, Slide Down, Push Left, Push Right, and Push Up.

TIP

The Slide options slide the new screen over the top of the current content. Push pushes the current content out of the way as the new content "slides" in.

To edit or delete a wire, click to select the wire to edit it or press the Delete key to remove the wire.

To hide the wires temporarily in the Prototype mode, hold down Option (Mac) or Alt (Windows) key.

You created a simple link from a button on one screen to another. Read on if you want to discover how to take advantage of some of the other interactive options such as auto animate.

Creating animations

Creating animations in XD is relatively easy. Essentially, you create an animated transition from one artboard to another. In this example, you create a navigation menu that slides into place. To keep this simple, you create a rectangle that represents a navigation menu on a screen. Keep in mind that you can create an entire navigation menu, group it together, and then follow these same steps to hide and show your navigation menu.

1. **Create a new document (any size will do). Press Alt/Option+1 to make sure you are in the Design mode.**
2. **Double-click on the name at the top of the default artboard to rename the initial artboard** Home-Menu.
3. **Create a rectangle that represents a menu you want to slide into place when triggered. In our example, shown in Figure 8-5, the menu is at the bottom because it is a mobile app. If you like, choose a fill color.**
4. **To keep objects organized, find your rectangle in the layers panel and rename it** NavBar.
5. **Now select the artboard by clicking on its name at the top of the artboard, or by Ctrl/Command-clicking on an empty space on the artboard.**
6. **When selected, hold down the Alt/Option key and drag a cloned copy from the artboard to another location. You now have two identical artboards.**
7. **Change the name of the second artboard to** Home-NoMenu.

   Here comes the easy part. Leave the first artboard the same. Using thre Select tool, click on the NavBar rectangle in the second screen and drag it down and slightly out of the artboard, as shown in Figure 8-6. Leave a little of the rectangle visible because this is what the user will click in order to see the Navbar pop-up.

   Now you create the animation.

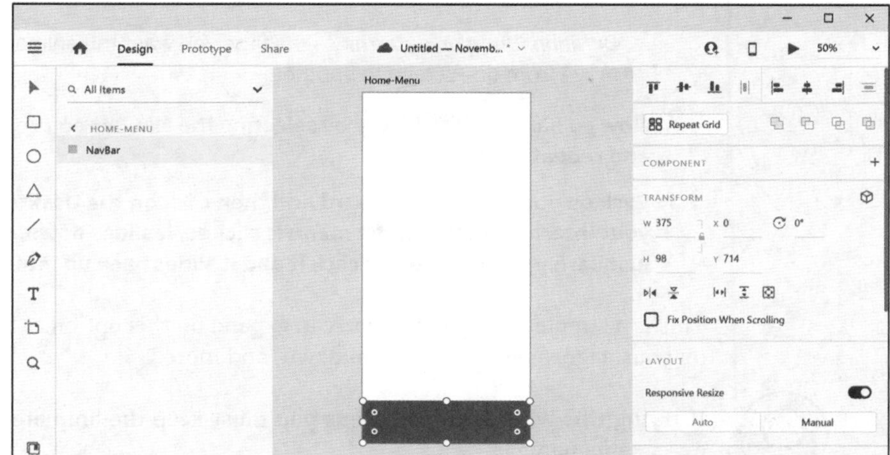

FIGURE 8-5: Create a rectangle that will represent a navbar. Make sure to name it in the Layers panel.

FIGURE 8-6: Clone the artboards and drag one rectangle down, but not totally outside the artboard.

8. **Press Alt/Option + 2 to enter the Prototype mode. Select the rectangle in the first screen.**
9. **Click the blue connecting arrow and drag to the second artboard.**
10. **In the Property Inspector, make sure the trigger is set to Tap, and then change the action to Auto-Animate.**
11. **In Easing, choose Bounce. This adds a fun little bounce to the appearance. If you prefer, you can choose another type of easing.**

    Easing refers to the way in which a motion tween proceeds, the transition from one element to another. You can think of easing as acceleration or deceleration:

    - *Easing:* Easing is an animation technique used to provide a more realistic method of accelerating or slowing down an animation. *Ease in* is when the motion begins slowly before gaining speed. *Ease out* is when an object in motion slows down before coming to a stop.

- *Duration: Duration* determines the time you want the animation to take to move from one screen to another.

12. **Now go back the other way by selecting the NavBar on the second screen and repeating the steps.**

13. **Click on your original artboard and then click on the Desktop Preview to test your interaction. When the menu is clicked, it slides down. When the menu is just barely visible, you can click it and it slides back up into place.**

This is a simple example, but you can expand on this option to have items bounce on your page, arrows flip up and down, and more.

TIP

If trying this with your own items you must keep the animated objects with the exact same name.

Creating a Drag Trigger

In this section, you find out how to use the Drag trigger combined with Auto-Animate in order to create a gallery feature where you can scroll from one image to another. You can also use the Drag trigger to simulate scrollbars and more.

To best utilize the Drag feature, set up three artboards that you can drag to. If you want to start with an example file, open the file named `DragExample`.

1. **Press Alt/Option+2 to make sure that you are in the Prototype mode.**

2. **Using the `DragExample` file, click on the partially exposed cat on the right side of the Screen01 and then click the connector arrow.**

3. **In the Property Inspector, choose Drag for the trigger, Auto-Animate for type, and Screen02 for the designation, as shown in Figure 8-7.**

   What you instructed XD to do is to animate to screen02 when the cat is clicked on.

   Now you apply this same drag feature to a few other images in order to replicate what the drag feature might look like in action.

4. **Select the golden retriever image that is hanging off the left side of Screen02 and then click the connector arrow. Choose Drag as the trigger, and Auto-Animate as the type, but this time choose the destination of Screen01.**

   As you can see, you are providing instructions so that when you drag on the left image you will be transitioned to the left artboard, and when you click on the right artboard you will be transitioned to the right artboard.

5. **Repeat the step above with the other images that are extending off the screen of Screen02 and on the left side of Screen03, as shown in Figure 8-8.**

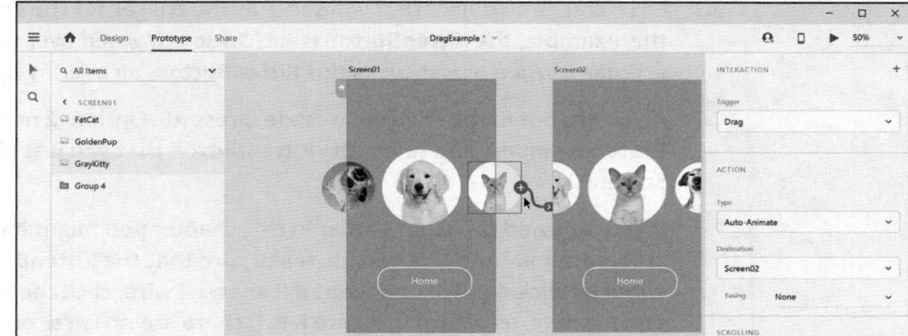

FIGURE 8-7: Implementing the drag feature with animation.

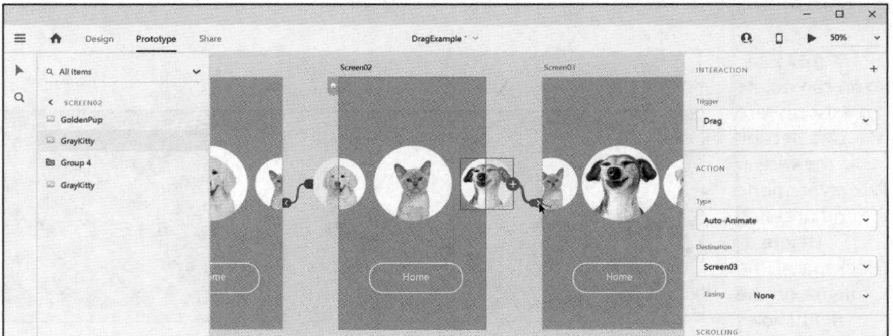

FIGURE 8-8: Apply the Drag trigger to only the small images on either side of the large image to replicate dragging in a gallery.

If you want to test, choose the Desktop Preview button and click and drag on the images on the side to see the animation and transition occur. You can find a completed example file named `DragExampleDone` in the Book07_XD files if you want to investigate.

Creating an Overlay

Creating a pop-up window in XD is a simple process. Follow the example using simple UI objects. Keep in mind that you can create something as complicated as you want as all that is required is that your pop-up UI is placed on its own artboard.

1. **You can create your own file or use the `BabyAnimalApp` file located in the Book07_XD folder that you downloaded in Book 1, Chapter 1. Just keep in mind that you need to have a dedicated artboard that creates your overlay. Typically, this is smaller than the artboard it appears on top of.**
2. **First, create an object that will serve as the trigger for the overlay. In this example, the KittenButton is Alt/Option-dragged over to the Puppy artboard. When a user clicks this KittenButton, an overlay appears.**
3. **If you are not in the Prototype mode, press Alt/Option+2 now to enter the Prototype mode, and select the KittenButton that you just cloned on to Screen02.**
4. **If you followed the steps earlier in this chapter you might have already created a link from this button. Make sure that the KittenButton is not wired by clicking the connector. If it shows a wire, click the arrow and press Delete, as shown in Figure 8-9. If there are no wires on an object, the arrow appears directly to the right of the selected object as shown on the right in Figure 8-9.**

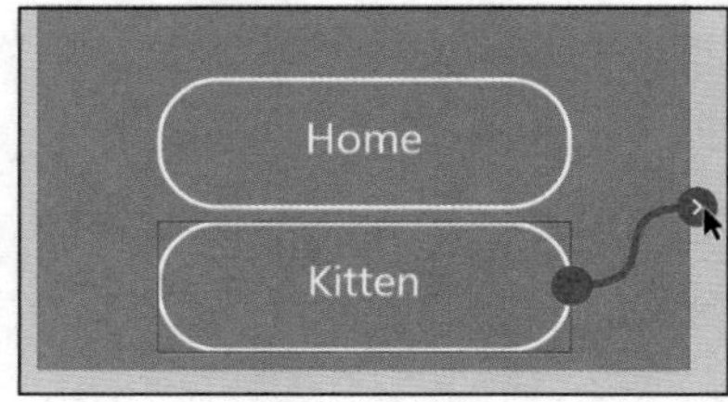

FIGURE 8-9: You can delete an existing wire by selecting the wire in Prototype mode and pressing Delete or Backspace. The figure on the right shows the button with no active wires.

5. **Select the KittenButton and then click the connector arrow to the right. In the Properties panel, choose the following:**
   - *Trigger:* Tap
   - *Type:* Overlay
   - *Artboard:* Popup. This is a named artboard that is in this project.
   - *Animation:* Dissolve
   - *Easing:* Ease In
   - *Duration:* 0.2 seconds

   Note that when you choose Overlay, a green guide appears over the screen indicating where the overlay will appear, as shown in Figure 8-10. You can drag the guide to any location you like, or leave it in the center.

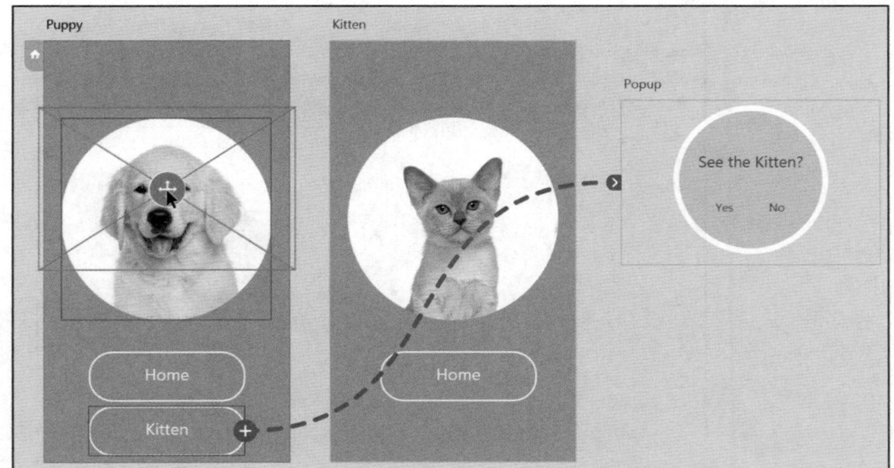

FIGURE 8-10: Click and drag the guide to reposition where the overlay appears on your artboard.

6. **Test your overlay by selecting the Desktop Preview button in the upper-right of the XD workspace. When you click on the KittenButton, the overlay artboard appears. Click the overlay and the window goes away.**

Adding States and Rollovers

Creating additional states and rollovers can add interesting interaction to your prototype, but they can also help inform a user that an action is linked to that object. States can be created that change a property of an object when rollover (Hover State) is clicked. In fact, you can apply any trigger you like to change a state. In this section, you discover how to apply a rollover (Hover State), and a second trigger that will change the object once again.

Follow these steps to create a simple exercise.

1. **Create a new XD document of any size.**
2. **Create a shape that will be used as a button. In the example shown in Figure 8-11, a circle is used. Fill and color it any way you wish.**
3. **In the Layers panel, change the name of your shape to** `rollover`.

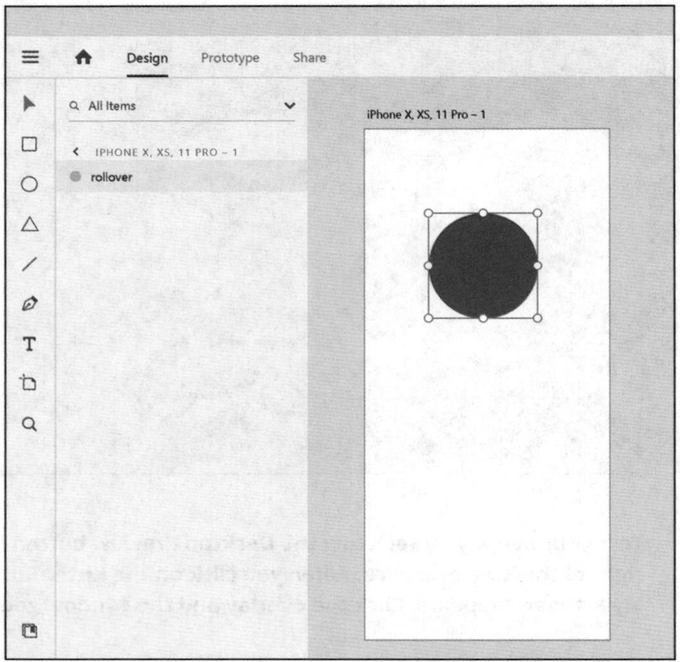

FIGURE 8-11: Create and name a shape object that will be used for a rollover.

4. **In order to create additional states of an object, it must be saved as a component. Make sure the object is selected and then right-click and choose Make Component, or press Ctrl+ K (Windows) or Command+ K (Mac).**

   The shape becomes highlighted in green with a diamond in the upper-left corner, indicating that it is now a saved component.

   Notice that the Properties panel now offers an additional option named States.

5. **Select the plus sign to the right of Default State and choose Hover State, as shown in Figure 8-12.**

6. **While in the Hover State, apply a change to the shape, such as its color.**

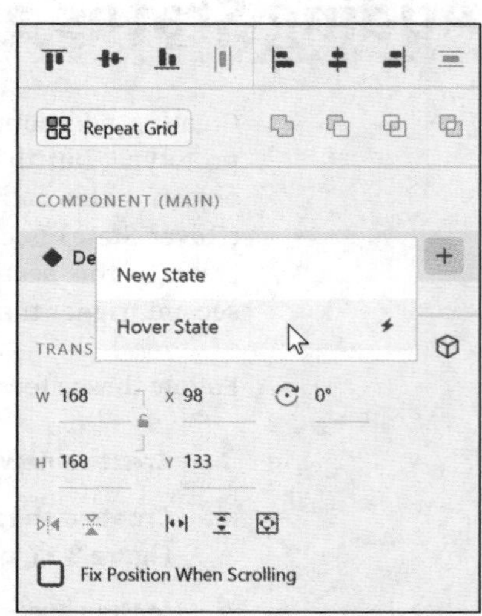

FIGURE 8-12: After an object has been converted to a component you can add additional states.

7. **After changing the color, click back on Default State to return it to its original color.**
8. **Click on the Desktop Preview and hover over the shape. The state changes, and the new color appears; when you are no longer hovering, the shape returns to its default state.**
9. **Return to the XD document and add another state. This time, use the plus sign to the right of Default State and select State 3. Double-click on the name State 3 and enter your own state name; in this example, the name `Resize` is used because the button will become smaller when tapped on.**

TIP

If you accidently add a state, you can delete it by right-clicking on it and selecting Delete from the contextual menu.

10. **While in the Resize state, hold down the Alt+Shift (Windows) or Option+Shift (Mac) keys, click on a corner of your shape, and either enlarge it or shrink it in size. Holding down the Alt/Option and Shift keys keeps the point of reference in the center of the object.**
11. **Click Default State to return to the original object.**
12. **Enter the Prototype mode. Notice that the connector now has a plus sign. This is because you can add additional triggers to one object.**
13. **Click the plus sign in the connector and choose the Tap trigger.**
14. **Choose Auto-Animate for the type, and then select your Resize state as the Destination, as shown in Figure 8-13.**
15. **Click the Desktop Preview to see your rollover work as you cross over and out for the object, and then when you click it enters your Resize state.**

    The connector now indicates that there are two triggers applied to this object. You could apply other triggers if you like, such as Drag, Keys & Gamepad, or Voice.

Keep in mind that this is a powerful feature that you have just been introduced to. You can take this further by creating your own states that show tooltips, resize a button, and more.

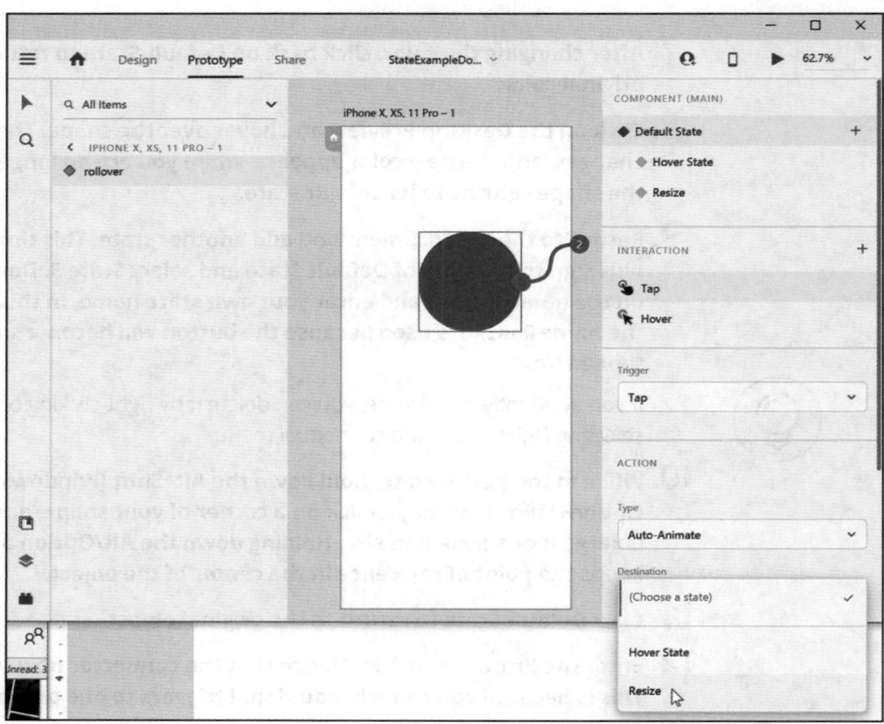

FIGURE 8-13: Create your own state and have it triggered with a Tap.

» Creating and sharing user flows

» Recording your prototype interactions

» Sharing online

» Embedding prototype into Behance

Chapter 9
Sharing Your XD Project

Your prototype exists so that you can test your product before you make a huge investment of time and money. With XD, you can view your preview on your desktop, as you did in the previous chapter; on a device, such as a phone; with the XD app; or on the cloud. In this chapter, you find out about opportunities you have in XD to share to others. You also find out how to share your specs and assets, such as images and code, with your development team.

Viewing Your App/Website on Your Desktop

This section will be a review for those of you who followed XD chapters in order, but even so, keep reading. The Desktop Preview not only offers the ability to test and preview your app on your desktop, but it also allows you to create a video of you using your app in the preview mode. This is helpful when you want to show the typical user flow, or when you want to have a demonstration that goes along well with a user story. Some designers even include these videos in their portfolios.

For the following exercise, you can use your own file or you can use the completed file named `BabyAnimalAppOverlayUserFlow.xd` that is available at `www.agitraining.com/dummies` in the XD-Files folder inside of the DummiesCCfiles folder. Just follow these steps:

1. **Open a file that has already had links applied from one artboard to another using the Prototype mode discussed in Chapter 8 of this minibook.**

   You can use the file named `BabyAnimalAppOverlayUserFlow.xd` inside the Book07_XD folder, inside the DummiesCCFiles folder that you downloaded in Book 1, Chapter 1.

2. **After the file is open, click the Desktop Preview button in the upper right of the work area to see a preview appear. The active artboard is the first to appear in the preview.**

3. **You can control the default screen that appears here by designating it as a user flow. Close the Desktop Preview window and return to the XD document. Click the gray Home icon on the left side of the Adult-Home artboard.**

   This is now your default start screen, demonstrating the first screen an adult might see. You can also name the flow if you like, for instance, Adult-Flow.

   A flow is defined from the home artboard and connects all other artboards that have been linked from the home in the Prototype mode. Multiple flows may share artboards within their flow, but must all have their own unique home artboards to define the start of the flow. Figure 9-1 shows two defined flows, one for a child and one for an adult. Later you will see how you can share these flows with others, such as reviewers, users, and developers.

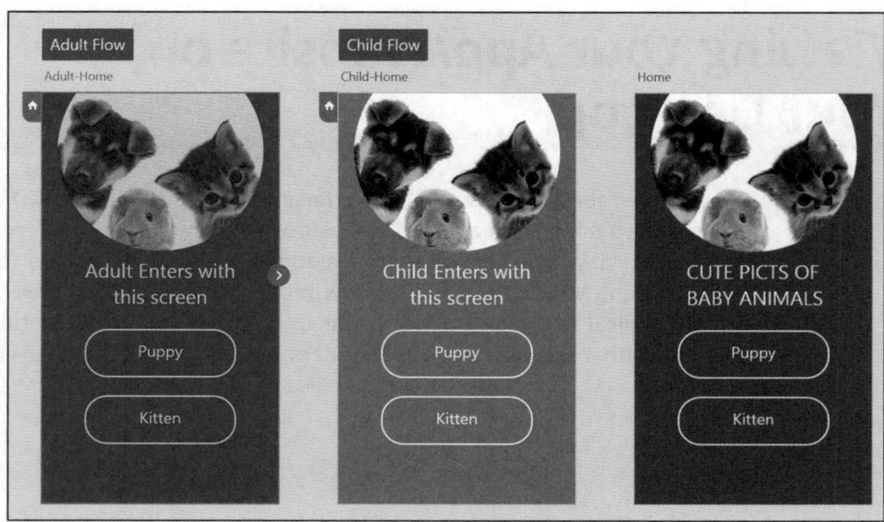

FIGURE 9-1: Create multiple flows from different start pages to demonstrate a particular user group's flow.

4. **If you are using our sample file, press the Desktop Preview button and test the links that were created.**

Creating a Movie for Your User Story

In this section, you record your interactions in this prototype in a movie. You can use this movie to demonstrate a user story, post in your portfolio, or share your ideas with others on your team.

To record a video, simply click the timer in the upper-right corner of the work area, as shown in Figure 9-2. The Record button blinks, indicating that your mouse movements and clicks are being recorded. You also see the timer start.

It is best to practice your movements prior to recording so that there are no unexpected movements or delays.

When you have finished demonstrating your prototype, click on the timer to stop recording. At this point, you can save your movie.

Note: Capturing your movie is much easier on a Mac. Although you can record your video on a PC, it requires that you enter the Windows Game Recording tool by first pressing the Win+G keys.

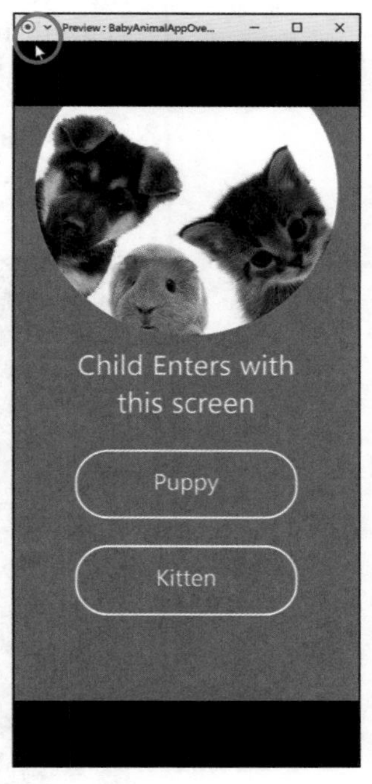

FIGURE 9-2: You can record your interactions in a movie using the Desktop Preview feature.

Using the Device Preview Feature

Adobe XD allows you to create a clickable prototype right on your phone or other mobile device. To do this, you need to do three things:

- **Save your file to the cloud by choosing the pop-up menu⇨Save (Windows) or File⇨Save (Mac).** The default save automatically directs you to the cloud. If you have already saved it locally, like our sample files, then you choose Save As.

» **Download and install the XD mobile app onto your iOS or Android device.** You can find the app at `xd.adobe.com/apps` or at Google Play or the App Store.

» If you want to preview local files, you can **hardwire your device using your USB port and connector on your computer.**

After you have completed these steps, launch the XD app on your device, locate your file in the Cloud directory, and click to activate it. The working prototype is launched in the XD app, as shown in Figure 9-3.

When you are finished reviewing your app, you can return to the main menu by triple-clicking the screen and selecting Exit Prototype, as you see in Figure 9-4.

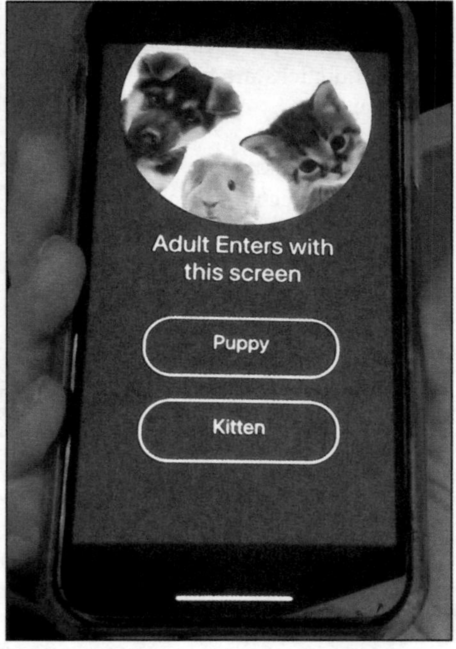

FIGURE 9-3:
Preview your app right on your device by using the XD app.

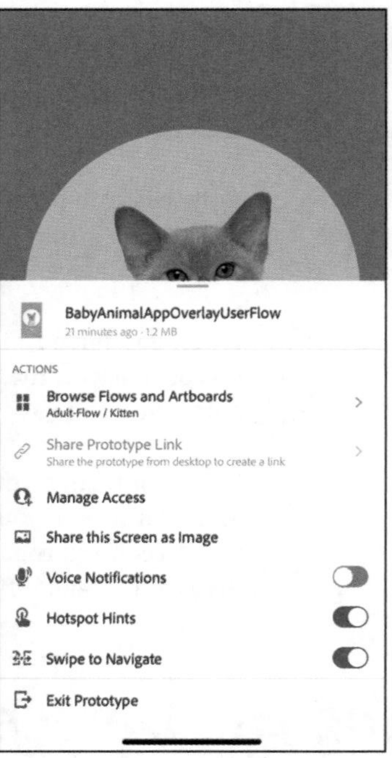

FIGURE 9-4:
Access your main menu in the XD mobile app by triple-clicking on your prototype.

Sharing Your XD File with Others

One of the many benefits of using Adobe XD is the ability to share your prototype with others. By sharing your prototype, you can offer access via a link to your client or remote members of your team and even your development team. In this section you find out how to share your prototype to users, for testing, to your team, for review, and to your developers, so that they can extract valuable content that will save them time.

In order to start the process you must enter the Share Mode: You can do this either by clicking on Share at the top of the XD desktop workspace or by pressing Alt/ Option+ 3. Notice, in Figure 9-5, that the Property Inspector has now changed to show you relevant information for sharing your prototype.

FIGURE 9-5: Switch to the Share mode to share your prototype, via the cloud, with users, reviewers, and developers.

The following sections describe some of the items you can control in this share mode.

Sharing a specific user flow

Note that you can choose a specific user flow using the Link drop-down menu. This is a topic covered in the previous chapter. If you have not defined multiple user flows, the flow starts with the screen that you defined as the Home screen. As a reminder, you define a screen as being the default screen by entering the Prototype mode and selecting the gray home icon to the left of the screen that you want to start with, as shown in Figure 9-6.

Selecting a View setting

Using the View setting you can choose how the prototype is presented to the people you share with. The possibilities are as follows:

- Design Review
- Development
- Presentation
- User Testing
- Custom

The Design Review view

Use this view to provide a clickable prototype that reviewers can navigate. In this view, the reviewers can add comments that will be added to the review document. Test yourself with an open document by selecting the Design Review setting and then pressing Create Link. When the link is created, click on the URL to see the review yourself, or click the Chain icon to copy the link and send it to others. (See Figure 9-6.)

Note: When sharing a prototype, the default is that you are sharing it with anybody who has a link. If security is a concern, you can change the Link Access only to people whom you have invited.

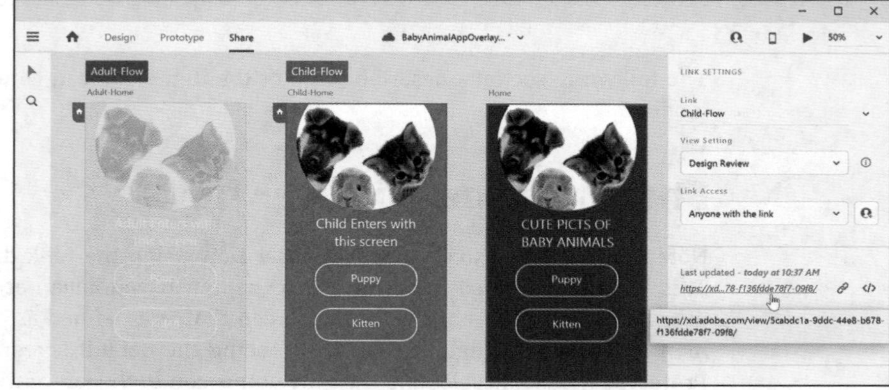

FIGURE 9-6: Create a design review that can be accessed with a URL.

Note in Figure 9-7 that reviewers can click and drag a Pin to your prototype, and then attach a comment to it, making it easy for you to locate where changes might need to be made.

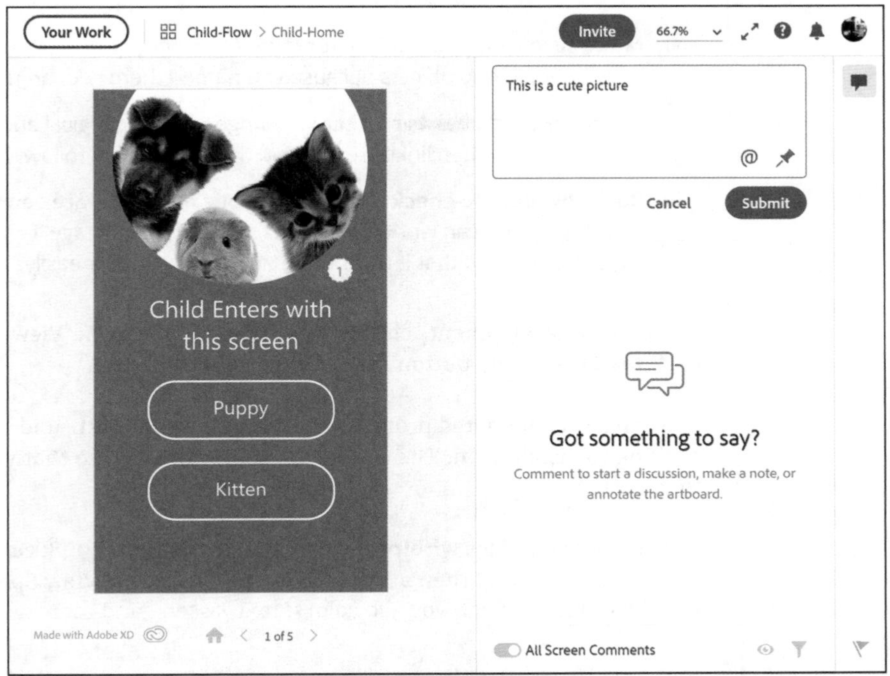

FIGURE 9-7: Reviewers can drag a Pin from the commenting section on the right and attach a comment to it.

Keep in mind that as you update your XD file you will want to enter the Share mode and click the Update Link button to keep your cloud version current.

The Development view

Sharing assets and screen design with your development saves hours of time in documentation. In order to share with your developers, you will want to make sure your file is ready. How will you know it's ready? Check the following:

- **Make sure it is pixel perfect.** This is the time you want to make sure that objects are aligned and not shifted off your pixel grid. You can check by returning to the Design Mode, right-clicking on objects, and selecting Align to Pixel Grid from the contextual menu.
- **Mark any artwork that you want the developers to access for export.** As a default, placed images and art are automatically marked for export. If you want a button created in XD to be exported, select it, right-click, and choose Mark for Export. This feature is discussed in more detail in Chapter 4, "Building Your UI."
- **Verify that your text styles are consistent.** The developers are going to need to create CCS. It is helpful if you are using saved text styles that are named and tagged appropriately.

- **Make sure that you name your layers correctly.** Don't make it difficult for developers to find objects because you named them `rectangle1`.
- **Name your artboards correctly.** Change names to logical and relevant names so that when links are created, they are easy to follow.
- **Basically, double-check and neaten your file up before sending it to your developer.** You can work freely in XD, but at the end stage, invest time cleaning up your file so that it transitions to your developers easily.

To send to development, choose Development from the View setting, and then click the Create Link button in the Properties Inspector.

You can see your shared prototype by clicking on the URL, and you can share it by clicking the chain icon. The chain icon copies the URL so that you can message it to others.

Notice the two modes within the preview URL in Figure 9-8: On the right, you see a comment icon and then a View Specs icon. Click the View Specs icon to see the Developer view, where you see colors, text, assets, and more.

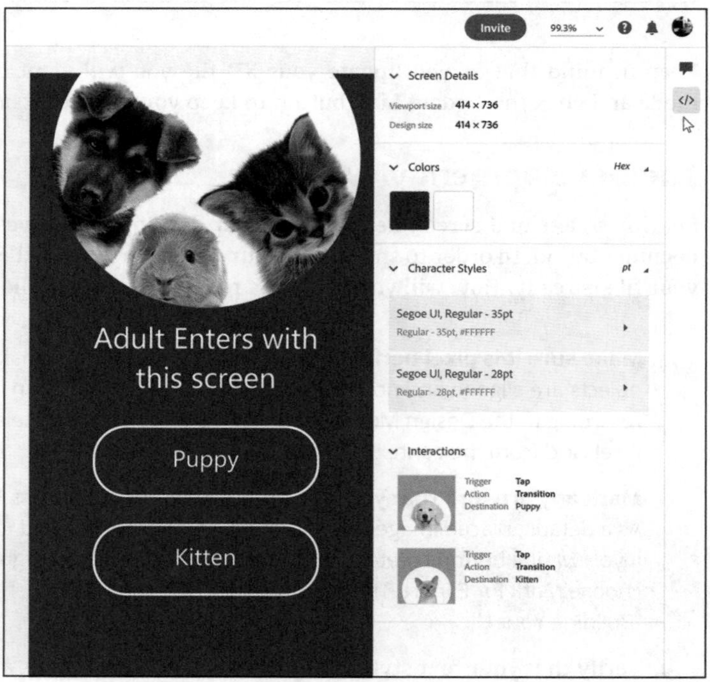

FIGURE 9-8: Click the View Specs in the preview URL to see developer information.

The Presentation view

Perhaps you want to present your design but don't need other functionality. You can use the Presentation mode to create a presentation in full-screen mode.

The User Testing view

The User Testing view is similar to the Presentation mode, where additional functionality is not available. Basically, it's a clean view that offers no distractions so that you can assign tasks and watch users try to reach their goals.

The Custom view

Use the Custom view to have the opportunity to turn off and on features that you think would be helpful in the shared prototype, as shown in Figure 9-9.

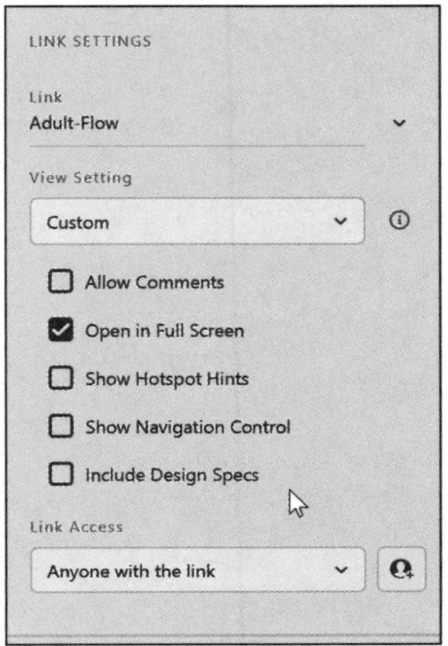

FIGURE 9-9: Set up the Custom view to share your prototype.

Adding collaborators to your XD document

If you are looking for input from others on your team, you can also share your XD document. This option is available in any mode. You can share your document by clicking the Invite to Document icon in the upper right of the XD workspace and invite others to collaborate, as shown in Figure 9-10.

Keep in mind that people you share reviews with do not need Adobe XD, but collaborators will need XD in order to make changes.

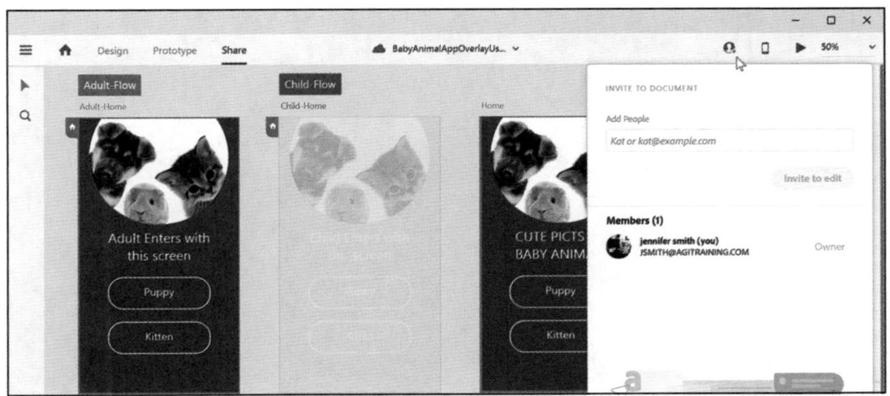

FIGURE 9-10: Invite others to collaborate in your document.

Index

Numbers

A

B

C

E

F

G

H

I

J

K

L

M

N

O

P

Q

R

S

T

U

V

W

X

Z

About the Authors

Jennifer Smith's expertise bridges the gap between design and development of print, web, and interactive design.

Jennifer's career started when she was one of the first creative directors to push the limits of technology and its integration with design. She has since managed and developed projects for companies such as Adobe, Microsoft, Reebok, Nike, and more. Jennifer's expertise lies in interpreting highly technical information and presenting it in a clear and easy-to-understand manner.

In addition to teaching and consulting, Jennifer is also a UX designer who creates applications and websites. She is directly involved in the entire process from research, prototyping, wireframing, testing, and final visual design.

Jennifer has written more than 20 books on technology and design, as well as over 100 publications used for curricula. Some of her other books include: *Digital Classroom: Adobe Illustrator and Photoshop CC*, *Adobe Illustrator Classroom in a Book*, and more. Jennifer is the co-founder of American Graphics Institute, which has classroom training centers in Boston, New York, and Philadelphia.

Christopher Smith is president of American Graphics Institute. He is the author of more than 10 books on digital publishing tools and technologies, including official Adobe training titles and several books on InDesign and Acrobat. At American Graphics Institute, he oversees professional development training programs including live online training courses, certificate programs, and private training workshops. Learn more about American Graphics Institute training programs at `www.agitraining.com`.

Authors' Acknowledgments

Jennifer Smith: Thank you to the highly professional staff at American Graphics Institute (`www.AGItraining.com`). You offered great insight into what people need to know in order to create projects using Adobe tools.

Thanks to all at John Wiley & Sons, Inc., especially Christopher Morris, who applied just enough pressure to help me meet my deadlines, yet was an absolute pleasure to work with.

Also, we would like to thank our technical editor, Patti Scully-Lane, for all the extra insight and effort she put into technically reviewing this book.

Grant, Elizabeth, and Edward — thanks for putting up with our long hours in front of the keyboard night after night.

Thanks to all of Kelly and Alex's friends and pets for permission to use their photos.

Christopher Smith: Thanks to Patrick Murphy for reminding us that it was time to update this book.

My appreciation goes to the team of exceptional instructors at American Graphics Institute for insight and support.

With much gratitude to the John Warnock and Chuck Geschke for their foresight in creating and bringing together great tools for the design community.

Publisher's Acknowledgments

Executive Editor: Steve Hayes
Project Editor: Christopher Morris
Copy Editor: Christopher Morris
Technical Editor: Patti Scully-Lane
Production Editor: Mohammed Zafar Ali
Proofreader: Debbye Butler

Cover Image: © Grandfailure/Getty Images